Produktion und Logistik

Reihe herausgegeben von
B. Fleischmann, Augsburg, Deutschland
M. Grunow, München, Deutschland
S. Helber, Hannover, Deutschland
K. Inderfurth, Magdeburg, Deutschland
H. Kopfer, Bremen, Deutschland
H. Meyr, Stuttgart, Deutschland
T. S. Spengler, Braunschweig, Deutschland
H. Stadtler, Hamburg, Deutschland
H. Tempelmeier, Köln, Deutschland
G. Wäscher, Magdeburg, Deutschland
C. Bierwirth, Halle, Deutschland
K. Schimmelpfeng, Stuttgart, Deutschland
M. Fleischmann, Mannheim, Deutschland
H.-O. Günther, Berlin, Deutschland

Diese Reihe dient der Veröffentlichung neuer Forschungsergebnisse auf den Gebieten der Produktion und Logistik. Aufgenommen werden vor allem herausragende quantitativ orientierte Dissertationen und Habilitationsschriften. Die Publikationen vermitteln innovative Beiträge zur Lösung praktischer Anwendungsprobleme der Produktion und Logistik unter Einsatz quantitativer Methoden und moderner Informationstechnologie.

Kontakt

Professor Dr. Thomas S. Spengler
Technische Universität Braunschweig
Institut für Automobilwirtschaft
und Industrielle Produktion
Mühlenpfordtstraße 23
38106 Braunschweig

Weitere Bände in der Reihe http://www.springer.com/series/12449

Falco Jaekel

Cloud Logistics

Reference Architecture Design

With a foreword by Prof. Dr. Dr. h.c. Werner Delfmann

Springer Gabler

Falco Jaekel
Cologne, Germany

Dissertation University of Cologne, 2017

Produktion und Logistik
ISBN 978-3-658-22836-1 ISBN 978-3-658-22837-8 (eBook)
https://doi.org/10.1007/978-3-658-22837-8

Library of Congress Control Number: 2018948671

Springer Gabler

Printed on acid-free paper

This Springer Gabler imprint is published by the registered company Springer Fachmedien Wiesbaden GmbH part of Springer Nature
The registered company address is: Abraham-Lincoln-Str. 46, 65189 Wiesbaden, Germany

Foreword

The present dissertation pursues the ambitious objective of examining whether, or to what extent and how, the underlying design principles and concepts of cloud computing can be transferred to or adopted for logistics system design. Thereby, the author aims to examine whether physical logistics capabilities can be delivered with the same essential characteristics as cloud computing capabilities. This research question is of high theoretical and practical relevance.

The dissertation at hand focuses on the thus far unclear term "cloud logistics," which has recently emerged in practice and academia, and makes this term the primary object of research. Unlike existing studies, which consider the transfer of design principles and concepts of cloud computing to logistics system design in a vague manner only, this dissertation develops specific propositions for how the cloud principles and concepts can be transferred to logistics system design and which physical logistics capabilities can be delivered with the essential cloud characteristics. This analysis ultimately aims to show how firms can achieve competitive advantage through increased logistical effectiveness and efficiency in an increasingly dynamic environment.

The theoretical research approach adopted allows for integrating existing studies within this still young research field into an overall concept (referred to as a "reference architecture"), anchoring this concept in existing literature, and differentiating from this literature at the same time. The approach is interdisciplinary and multiperspectival and is based on theories and concepts in the fields of logistics, organizational sciences, and computer sciences. The use of the international standard for "architecture descriptions" is of particular importance for the approach because this standard includes an ontology that provides the fundamental structure to transfer the design principles and concepts of cloud computing to logistics system design. In particular, the use of multiple perspectives ("viewpoints") is decisive, where each perspective adopts a separate design approach and, therefore, can be anchored in a different academic discipline. This also allows for breaking up the design principles and concepts of cloud computing into components when transferring them to logistics system design, thus making them manageable in a more differentiated way. In addition, the separation between the design approach ("viewpoints") and results ("views") proves most beneficial. Documenting the reference architecture for "cloud logistics systems" by means of a standardized "architecture description" therefore contributes directly to answering the central research question and also extends existing studies from a conceptual perspective.

The reference architecture for cloud logistics systems developed in this dissertation demonstrates in a coherent and detailed manner how to transfer the design principles and concepts of cloud computing to logistics system design and whether, through this transfer, the delivery of physical logistics capabilities with the essential cloud characteristics can be achieved. In doing so, this dissertation considers infrastructural, institutional, and organizational aspects and convincingly shows the extent to which the design principles and concepts can be transferred and what limitations are associated with this transfer. The set of logistical prototype scenarios and the characteristic mix of coordination mechanisms deserve particular emphasis in this context.

All in all, the present dissertation makes a remarkable, foundational, and indeed directional academic contribution to structuring cloud logistics systems, which is also of the highest practical relevance. I wish this dissertation a lively reception in academia and practice.

Werner Delfmann

Preface

The pursuit of a doctoral degree is a significant investment of lifetime and is often a pivotal decision for one's professional career. My desire to pursue a doctoral degree developed during my early studies. I discovered my affinity for absorbing knowledge in different fields, and I was impressed by the university readily providing access to knowledge. Yet, I also realized that my affinity went beyond the depth of knowledge usually conveyed in lectures. The pursuit of a doctoral degree appeared to me as a possible way to satisfy my desire for deeper knowledge, at least within a certain area. During my later studies, I equally discovered the joy in working in a business environment, primarily through several internships. I therefore settled for the compromise to start a business career after completing my studies but to take a "break" a few years later to pursue a doctoral degree.

The innovative concept of "cloud logistics" was first introduced to me by my doctoral supervisor, Prof. Dr. Dr. h.c. Werner Delfmann, who was working on that topic together with an interdisciplinary working group of the German Logistics Association (BVL). Intrigued by this concept, I adopted cloud logistics as the focus area of my dissertation, specifically pursuing the research objective to design a logistics system in accordance with the principles and concepts of cloud computing in order to determine if and how logistics systems can deliver physical logistics capabilities with cloud characteristics. To this end, I developed a reference architecture of "cloud logistics systems" that can be conceived of as a *gedankenexperiment* of an innovative logistics system. As such, this dissertation aims to provide a foundational structure of knowledge to scholars to enable systematic future research in this area as well as a concrete design proposal to practitioners to enable the successful realization of such an innovative logistics system with the ultimate objective to achieve competitive advantage in an increasingly dynamic business environment. In a broader context, this dissertation is concerned with how the logistics industry and operations may be transformed by the introduction of new technologies, especially by cloud computing, virtualization, service-orientation, Internet of Things, and internet-based platforms.

The present dissertation was developed in the Department of Business Policy and Logistics at the University of Cologne under the supervision of Prof. Dr. Dr. h.c. Werner

Delfmann, whom I would like to thank for his constructive feedback, guidance, and freedom in pursuing my research objective. I thank Prof. Dr. Detlef Schoder for acting as second referee. I thank DHL Consulting for supporting my request and for providing the flexibility for me to pursue a doctoral degree.

Writing a dissertation is an endeavor that is characterized by frequent setbacks that, despite causing high levels of frustration at the time, eventually turn out to be both inevitable as well as necessary. Writing a dissertation is therefore unlikely to be accomplished without continuous support and encouragement from family and friends. I would like to express my utmost gratitude to my parents, who inexhaustibly supported and motivated me, not only throughout the time I spent writing this dissertation but also throughout my entire life so far. I would like to equally express my gratitude to my wife who supported and motivated me every single day and who also allowed me to satisfy my desire to absorb and build knowledge in a field of my interest, despite dreadfully missing me in the many hours that I spent at the library or at my desk in deep thought. I devote this dissertation to them.

Falco Jaekel

Overview

Contents

List of Figures

List of Tables

List of Abbreviations

4PL	fourth-party-logistics-provider
BIC	Bayesian-Nash incentive compatible
BVL	Bundesvereinigung Logistik
CAF	Customized Access Framework
CAP	combinatorial allocation problem
CLS	cloud logistics system
CNKI	China National Knowledge Infrastructure
dAGVA	d'Aspremont and Gérard-Varet and Arrow
DSIC	dominant strategy incentive compatible
EAN	European Article Number
EDIFACT	Electronic Data Interchange for Administration, Commerce and Transport
ERP	enterprise resource planning
GPS	Global Positioning System
GS	Gibbard-Satterthwaite
GVA	generalized Vickrey auction
IaaS	Infrastructure-as-a-Service
IATA	International Air Transport Association
IEC	International Electrotechnical Commission
IEEE	Institute of Electrical and Electronics Engineers
IHIP	Intangibility, Heterogeneity, Inseparability, and Perishability
IML	Institut für Materialfluss und Logistik
IoT	Internet of Things
ISO	International Organization for Standardization
ISO/IEC 42010:2007	Systems and software engineering — Recommended practice for architectural description of software-intensive systems
ISO/IEC/IEEE 42010:2011	International Standard for Systems and Software Engineering — Architecture description
ISSL	International Scientific Symposium on Logistics
ISST	Institut für Software- und Systemtechnik
IT	information technology
JIS	just-in-sequence
JIT	just-in-time
KPI	key performance indicator
LaaS	Logistics-as-a-Service

LITPAP	logistics IT platform and application provider
LSP	logistics service provider
LTL	less-than-truckload
M&A	mergers and acquisitions
MEP	message exchange pattern
MMP	Mall Marketplace
MS	Myerson-Satterthwaite
NAO	network administrative organization
NFC	near field communication
NIST	National Institute of Standards and Technology
OWL	Web Ontology Language
OWL-S	Web Ontology Language for Web Services
PaaS	Platform-as-a-Service
QFD	Quality-Function-Deployment
QoS	quality-of-service
RBV	Resource-based view
RFID	radio-frequency identification
RfP	Request for Proposal
ROI	return on investment
RV	relational view
SaaS	Software-as-a-Service
Scenario Ia	Localized Transportation Capabilities as-a-Service
Scenario Ib	Localized Storage Capabilities as-a-Service
Scenario II	Networked Long-distance (end-to-end) Transportation Capabilities as-a-Service
Scenario III	Localized Storage and Distribution Capabilities with or without Standardized Value-added Capabilities as-a-Service
Scenario IV	Networked Spatially Distributed Storage Capabilities with or without Distribution and Standardized Value-added Capabilities as-a-Service
SCM	supply chain management
SCT	strategic choice theory
SKU	stock keeping unit
SLA	service level agreement
SME	small-and-medium-sized enterprises
SOA	service-oriented architecture
SOA-RAF	Reference Architecture Foundation for Service Oriented Architecture
SOC	service-oriented computing
SVAC	standardized value-added capabilitie
SWSF	Semantic Web Services Framework
TaaS	Transport-as-a-Service
TCE	Transaction Cost Economics
TMS	transport management system
ULD	unit load device
URI	universal resource identifier
URL	Uniform Resource Locator
VCG	Vickrey-Clarke-Groves

VMM	virtual machine monitor
WaaS	Warehousing-as-a-Service
WMS	warehouse management system
WSDL	Web Services Description Language
WSML	Web Service Modeling Language
WSMO	Web Service Modeling Ontology
WSN	wireless sensor network
WWW	World Wide Web
XML	Extensible Markup Language

Part A. Introduction

I. Problem Statement

Throughout the 1990s, cost efficiency was the leading principle that guided firms' decisions about how to source, produce, and distribute their goods and services (Christopher & Peck 2004:2; also see Tang & Tomlin 2008:12; Lee 2004:102; CLSCM 2002:3). To take advantage of national and local differences in labor costs, resource costs, tax or regulatory conditions, firms globalized their value creation activities. To benefit from economies of scale and scope, they internalized activities critical to supporting their core competencies and outsourced large parts of the remaining activities (Prahalad & Hamel 1990:83; Choy et al. 2003:225; see also Quinn & Hilmer 1994:55; Arnold 2000:25f.). This often included the outsourcing of logistics activities to dedicated logistics service providers (LSPs) (Herr & Klaus 2012:315ff.; Razzaque & Sheng 1998:91). To reduce inventory carrying costs, firms implemented lean and complex logistics concepts such as just-in-time (JIT) delivery (Tang & Tomlin 2008:12) and integrated and synchronized their sourcing, production, and distribution activities – not only across functions within their organizations, but also, more importantly, in an interorganizational manner, with their suppliers and customers (Christopher & Peck 2004:2; see also Chandrashekar & Schary 1999:27). These integrating processes were enabled and facilitated by quick innovations in information technology (IT) (Christopher & Peck 2004:2; Simatupang et al. 2002:296) which enabled firms to share information (Srinivasan et al. 1994:1292ff.).

The pursuit of cost efficiency has ultimately led to a fragmentation of value creation systems and transformed them into global networks of organizations that have become increasingly complex, interconnected, interdependent and, thus, vulnerable to disturbances (see e.g. CLSCM 2002; Norrman & Jansson 2004:434; Tang & Tomlin 2008:12; Tang 2006:33). In such networks, even small and localized disturbances can quickly escalate and result in considerable disruptions (Bhatia et al. 2013:5; also see Berdica 2002:117). Scholars therefore deem cost efficiency to be effective in stable environments (Tang & Tomlin 2008:12; Tang 2006:33) but not in dynamic environments because it incurs high "hidden costs" in the event of disruptions (Lee 2004 as cited in Tang 2006:34; also see

F. Jaekel, *Cloud Logistics*, Produktion und Logistik,
https://doi.org/10.1007/978-3-658-22837-8_1

Levy 1997:94). These hidden costs are the opportunity costs of cost efficiency. They manifest in lost revenue due to stock outs or missed market opportunities due to the slow adaptation of production and logistics processes (Lee 2004:104).

Yet, our world has recently become increasingly volatile and uncertain. On the one hand, firms have aimed to increase profitability through frequent technical innovations of their products, which has not only shortened product life cycles, but also made the demand for these innovative products unpredictable due to their "very newness" (Fisher 1997:106). On the other hand, economic crises, environmental disasters, social unrest, and political turmoil occur more frequently and cause more damage (MunichRe as cited by Tang 2006:33; see also Bevere et al. 2013:1/35; SwissRe 2016). While disasters and crises often represent structural breaks that cause abrupt changes and dysfunctionalities, megatrends such as globalization and urbanization cause incremental change. Both abrupt and incremental change reduce the predictability of demand for goods and services and thus the predictability of logistics processes, as logistical demand derives from the demand of goods and services.

Given the existence of vulnerable value creation systems embedded in a turbulent world, Lee (2004:105) posits that cost efficiency will remain necessary, but will not be sufficient for firms to be successful; he therefore calls for rebalancing cost efficiency with effectiveness. Literature provides a variety of concepts to characterize effectiveness in the sourcing, production, and distribution of goods and services, including agility (see e.g. Prater et al. 2001; Christopher & Peck 2004; Lee 2004; Agarwal et al. 2007), which refers to "the ability to respond rapidly to unpredictable changes in demand or supply" (Christopher & Peck 2004:10). Agile value creation systems are thus able to better cope with dynamically changing conditions (Agarwal et al. 2007:443).

In this context, a new logistics notion – cloud logistics – has recently emerged in academic literature, practice-oriented industry magazines, and on web sites of logistics IT vendors and LSPs. Cloud logistics refers to a new mode of provisioning physical logistics capabilities, such as the transportation, storage, and handling of physical goods. Similar to cloud computing, this mode takes an overarching perspective on the delivery of logistics capabilities and is thus detached from firms' specific supply chains and logistical setups. Cloud logistics specifically refers to adopting the design principles and concepts of cloud computing to logistics systems (see e.g. Wang et al. 2012b; Kersten et al. 2012; Pieringer 2012; Li et al. 2013; Leukel & Scheuermann 2014; Teichmann 2014; Ludwig 2014; Ehrenberg & Ludwig 2014; Ballardt 2014; Delfmann & Jaekel 2012). Ultimately, this design approach aims to enable logistics systems to deliver physical logistical capabilities with the same characteristics with which cloud computing systems deliver computing capabilities. These are: on-demand availability, self-service, broad network access, resource pooling, and measured service (pay-per-use).

Delivering physical logistics capabilities with cloud characteristics seems conducive to supporting firms in achieving efficient *and* agile logistical responses. On-demand availability means that logistics capabilities can be made available with little lead time at any time. Rapid elasticity means that the capacity of logistics capabilities can be quickly increased or decreased commensurate with current demand. Thus, making available logistical capabilities in an on-demand and rapidly elastic manner can enable firms to rapidly implement effective logistical responses to newly emerging demand or to unpredictable variations in the demand and/or supply of goods and services. Moreover, such responses are also efficient due to resource pooling and pay-per-use. Thus, cloud logistics seems to be able to reconcile the conflicting goals of logistical efficiency (due to resource pooling and pay-per-use) and effectiveness (due to on-demand availability and rapid elasticity). We therefore conclude that the essential characteristics of cloud computing are highly relevant for the delivery of physical logistics capabilities. Hence, transferring the design principles and concepts of cloud computing to logistics systems is a worthwhile approach to investigating and thus contributing to enabling effective and efficient logistical responses in uncertain and volatile environments.

Due to the recent emergence of the field, existing cloud logistics knowledge is highly fragmented. It collectively falls short of providing a clear conception of what areas of logistics system design are affected by adopting the principles and concepts underlying cloud computing, how the design of logistics systems can be carried out as per these principles and concepts, what the design result would "look like," and if (or the extent to which) logistics systems designed in this manner can in fact deliver physical logistics capabilities with cloud characteristics. In other words, a comprehensive design of logistics systems in accordance with the principles and concepts underlying cloud computing has not yet been accomplished. Existing knowledge tends to focus on certain isolated aspects. For example, several scholars and practitioners focus on virtualization, service-orientation, and the Internet of Things (IoT) (Wang et al. 2012b:see e.g.; Li et al. 2013; Hu & Zhang 2015; Wang et al. 2015b). For designing logistics systems in accordance with cloud principles and concepts, Leukel & Scheuermann (2014) propose a "cloud perspective" and Delfmann & Jaekel (2012) a "cloud paradigm." These contributions take a slightly broader perspective by considering a variety of aspects such as the deployment model, the service model, and the reconfiguration of physical resources. However, both contributions only provide initial insights into the process and outcome of designing logistics systems according to cloud principles and concepts. Some scholars conceive of a logistics system designed in that way as a fourth-party-logistics-provider (4PL) model (see e.g. Ludwig 2014; Leukel & Scheuermann 2014), but the majority associate this design approach with network horizontal LSP cooperation (see e.g. Kersten et al. 2012; Wang et al. 2012b; Teichmann

2014). In sum, these examples clearly emphasize the fragmented and piecemeal character of current knowledge about designing logistics systems as per the principles and concepts of cloud computing.

In addition, cloud logistics knowledge seems largely detached from existing literature and lacks a clear theoretical basis. Scholars and practitioners from logistics and logistics-related computer sciences merely allude to a variety of principles, concepts, and characteristics, such as virtualization, service-orientation, interfirm networks, and auctions, without referencing relevant literature comprehensively and without specifying their exact meanings and relevance for cloud logistics. The great variety of mentioned principles, concepts, and characteristics as well as the interdisciplinary nature of cloud logistics may be pivotal reasons for why a clear and comprehensive basis has not been established yet. Scholars and practitioners usually have profound insights into their specific fields, but may lack deep insights into the other fields relevant for cloud logistics. The tendency to only allude to existing principles, concepts, and characteristics rather than reviewing them in detail and explaining their relevance for cloud logistics is a fundamental issue in current cloud logistics research. As a consequence, proposed arguments and results remain vague and cannot be easily verified and reconstructed. We therefore conclude that cloud logistics currently seems to be floating around like a cloud without anchoring.

The design of logistics systems in accordance with cloud principles and concepts is additionally hampered by terminological ambiguity surrounding the term "cloud logistics." This ambiguity originates in the fact that cloud logistics combines the terms "cloud computing" and "logistics," dropping the term "computing" (Kersten et al. 2012:257; Leukel & Scheuermann 2014:39). Different scholars and practitioners blend the original meanings of both terms in different ways. Besides the conception described above, many conceive of cloud logistics as the use of cloud computing technology for logistics IT systems (see e.g. Gong & Yang 2012; Arnold 2014a; Matkovic et al. 2014). Others conceive of cloud logistics as the fulfillment of e-commerce orders through a decentralized interfirm network of local dealers and LSPs that interact through a cloud-based logistics IT system (see e.g. Thomas & Unruh 2010; Thomas 2011). Still others conceive of cloud logistics as the investigation of logistical questions in distributed cloud computing systems (see e.g. Jaeger & Lindenlaub 2013; Pohlmann et al. 2013). Potential commonalities, differences, and interrelations between these conceptions remain vague. In other words, cloud logistics is a conceptual and terminological jungle. This makes it difficult to clearly delineate the approach for designing logistics systems as per the principles and concepts of cloud computing.

II. Objective

Considering the unclarity of how logistics systems can be designed in accordance with the principles and concepts underlying cloud computing, we set out to attain the following objective:

- Design a logistics system in accordance with the principles and concepts of cloud computing in order to determine if and how logistics systems can deliver physical logistics capabilities with cloud characteristics.

This thesis thus aims to become the first comprehensive contribution for cloud logistics system design. It thus unfolds high relevance for both scientists and practitioners. For academics, the transfer of design principles and concepts from cloud computing to logistics has not yet been accomplished systematically, albeit being mentioned by several scholars. Hence, this thesis closes a gap in current research. For practitioners, the resulting logistics system, which has been designed according to the design principles and concepts of cloud computing, represents a concrete solution template of an innovative system that aims to deliver physical logistics capabilities with cloud characteristics. This innovative logistics system may help firms to compete more effectively and efficiently in dynamic environments characterized by limited logistical predictability. Practitioners can use this solution template as an implementation guideline. Academics can use the resulting system as a foundational structure to advance research in the field.

III. Course of Investigation

This thesis is structured into seven parts, as depicted in Figure 1. Following this introductory part, Part B reviews the fundamentals of logistics systems. It provides a formal definition of logistics systems and identifies three levels of aggregation and five perspectives from which logistics systems can be analyzed and designed. This part considers logistics systems in detail from an infrastructural and functional perspective and from an institutional perspective. The former identifies and describes the core logistics processes, their associated transformations, and the service character of logistics processes. The latter specifically focuses on meso-logistics systems and on the central actors in such systems: LSPs. This part concludes with characterizing the competitive and technological environment of meso-logistics systems.

Part C reviews the fundamentals of cooperative interfirm networks and of networked horizontal LSP cooperations. The part commences with a definition of cooperative interfirm networks and a summary of motives and antecedents for the formation of such networks. This part specifically studies cooperative interfirm networks through the lenses

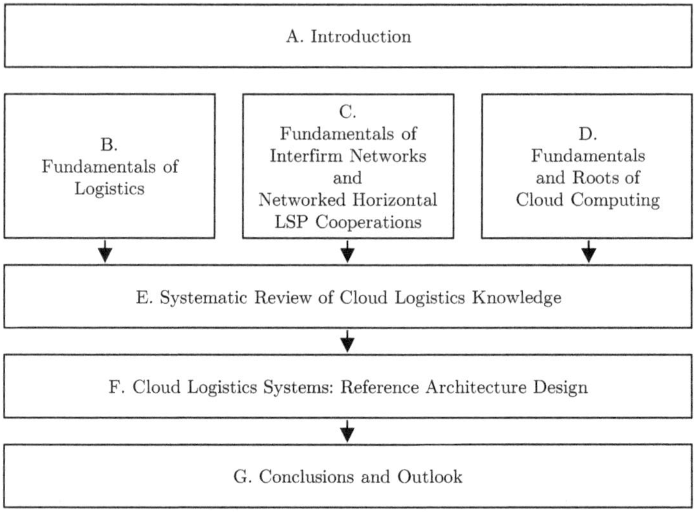

Figure 1.: Structure of Thesis

of Transaction Cost Economics (TCE) with a particular focus on electronic markets and the design of market mechanisms based on game theory, Resource-based view (RBV), and organizational perspective. These concepts provide a theoretical basis for considering meso-logistics systems, especially networked horizontal LSP alliances, from institutional and organizational perspectives in detail.

Part D reviews the fundamentals and roots of cloud computing. It introduces cloud computing by using the NIST definition, including its five essential characteristics, three service models, and four deployment models. Cloud computing is a computing paradigm rooted in several other concepts and technologies, including virtualization and service-oriented computing. These roots are introduced, as they are critical to fully grasping the nature of cloud computing. A particular focus is put on the description of (web) services with well-defined semantics. This part concludes with a review of current research efforts to adopt well-defined semantics for the formal description of logistics services.

Part E systematically identifies and structures current cloud logistics knowledge from both academic and non-academic sources. The review identifies four distinct but interrelated meanings of the term "cloud logistics," which all revolve around a cloud computing technology-based IT system. However, the perspective and role (or function) of this platform differ in each meaning. The knowledge contained in each meaning is consolidated, structured, and (where appropriate) ontologically summarized by specifying the key concepts and their interrelations. Considering the varying conceptions of the term "cloud logistics," the systematic knowledge review provides the basis for designing logistics systems as per the principles and concepts of cloud computing.

Part F designs logistics systems in accordance with the principles and concepts of cloud computing. Specifically, this thesis designs a reference architecture of cloud logistics systems (CLSs). Due to the abstract nature of reference architectures, this architecture focuses on the dimensions of CLSs that should be similar across all potential realizations of such systems. The reference architecture is documented using the International Standard for Architecture Descriptions (ISO/IEC/IEEE 42010 2011b), which provides an ontology for expressing architectures. Key concepts of the standard are stakeholders, concerns, viewpoints, and views. The reference architecture thus specifies the stakeholders of CLSs and their interests relative to this system (concerns). The design process and results are expressed in viewpoints and views, where viewpoints specify the design perspective and required methods, and views document design outcomes by means of propositions. Design decisions are primarily based on contingency theory, but also include strategic choices, made in an economically rational manner. The reference architecture consists of nine viewpoints and associated views. The first conceptualizes CLSs on an aggregate level. The second focuses on CLS infrastructure, which resembles the service model. The third specifies the governance structure. The fourth considers the degree of centralization and the fifth the degree of organizational formalization of CLSs. The sixth expresses the characteristic mix of coordination mechanisms. The seventh, eighth, and ninth focus on specific types of coordination mechanisms, including servitization, standardized semantic service descriptions, and market mechanisms. The third to ninth viewpoints and views collectively specify the deployment model.

Part G concludes this thesis and provides an outlook for further research.

IV. Guidance to the Reader

This thesis aims to provide the first comprehensive account on the design of CLSs. Accordingly, this thesis reviews theoretical fundamentals and existing cloud logistics knowledge comprehensively. This comprehensive review is necessary to clear the terminological and conceptual jungle surrounding the term "cloud logistics," to provide an adequate design basis for the reference architecture of CLSs, and to anchor the reference architecture in existing literature. Moreover, this review is pivotal to cater for the interdisciplinary nature of cloud logistics. The field of cloud logistics thus becomes readily accessible to scholars and practitioners with heterogeneous backgrounds as they are provided with a basis that covers the primary fields of relevance. Thus, unlike existing cloud logistics knowledge, this thesis alludes not only to a variety of principles, concepts, and characteristics but reviews them in detail and describes their meaning and relevance for cloud logistics.

Establishing a comprehensive design basis and adopting a comprehensive approach adds to the overall length of this thesis. In order to cater for different readers' preferences, available time, degrees of familiarity with the topic as well as to enable readers to efficiently navigate, select, and absorb this thesis' content, this thesis comprises several abstracts which depict the key contents and arguments. Parts B – E are each summarized in an abstract at the beginning of the respective part. Given the total length of part F, each chapter within Part F begins with an abstract. Thus, when reading this thesis, each reader can decide at the beginning of the respective part or chapter on whether to focus on that aspect in detail or to move on more quickly. References to the summarized content are provided in each abstract.

Part B. Fundamentals of Logistics Systems

I. Abstract of Part

An Introduction to Analyzing and Designing Logistics Systems

The term "logistics" refers to the flow of objects in time and space, predominantly to the flow of goods and information in a business context (see Section B.II.1). *Logistics systems* can be defined as the structures and processes that serve the purpose of transforming objects in time and space (Delfmann 2012:268f.; see Section B.II.2). To formally define logistics systems in this thesis, we adopt the quadruple of (1) composition, (2) environment, (3) structure, and (4) mechanisms (Bunge 1997:416, 414ff.; see also Bunge 1996:270). The *composition* refers to the collection of components or parts that belong to the system at a given point in time. The *environment* includes all things that do not belong to the system but are nevertheless connected to parts of it. The *structure* comprises the collection of relations among the system's components as well as their relationships to things in the environment. *Mechanisms* are processes (or activities) that are able to cause or prevent change to individual system components or the system as a whole (Bunge 1997:414). Accordingly, we formally define:

Definition 1. *Logistics systems are concrete systems that are composed of artificial and natural material components* (composition)*, which are connected through spatio-temporal object flows* (structure) *and embedded in a logistics-relevant environment* (environment) *where change to the composition and structure are governed by primarily economic, but also social, political, and natural processes* (mechanisms).

Three levels of aggregation are commonly used to analyze and design logistics systems (see Subsection B.II.3.2). *Macro-logistics systems* refer to the logistics systems of national economies and the flows between economies (see e.g. Delfmann et al. 2010:60). *Micro-logistics systems* focus on the logistics systems of single organizations (see e.g. Delfmann et al. 2010:59f.). *Meso-logistics systems* lie between macro- and micro-systems; they refer

F. Jaekel, *Cloud Logistics*, Produktion und Logistik,
https://doi.org/10.1007/978-3-658-22837-8_2

to the combined logistics systems of a group of organizations that collectively represent only a fraction of an economy (see e.g. Pfohl 2010:14f.).

Five perspectives are commonly applied to analyze and design logistics systems (see Subsection B.II.3.3). The *infrastructural perspective* focuses on the physical logistics resources and capabilities necessary to realize or support the flow of objects. The *functional perspective* focuses on the types of logistics processes and on the associated logistical transformations that can be accomplished (Pfohl 2010:14). The *informational perspective* centers on the information (flow) required to plan, realize, control, and monitor the physical flow of objects, especially on the information that precedes, accompanies, and succeeds the physical object flow (Isermann 1998:25; also see Krupp & Klaus 2012:72f.). The *organizational perspective* focuses on the administrative arrangements that determine the division and coordination of labor related to the planning, realization, control, and monitoring of the object flow. The *institutional perspective* focuses on the legal relationships within logistics systems between different organizations.

This thesis focuses on *physical goods* flowing in *meso-logistics systems* and on *horizontal cooperation between LSPs* (see Section B.II.4). Furthermore, this thesis adopts the *five perspectives* presented above in order to analyze and design meso-logistics systems, primarily focusing on the infrastructural, functional, organizational, and institutional perspectives. This focus appears most conducive basis for designing a reference architecture of CLSs – our research objective.

The infrastructural and functional perspectives are closely intertwined, and hence considered together in the following (see Chapter B.III). The composition of the physical infrastructure determines the type and variety of processes that can take place in a logistics system and thus the transformations that this system can bring about. The most important processes are the five *core logistics processes*: (1) storage, (2) handling, (3) transportation, (4) packaging, load unitization, and labeling, and (5) order processing. Fundamental process design can be controlled via few parameters, each of which being influenced by several factors. The (variation in) physical and economic properties of goods to be transformed logistically are pivotal for system design because they influence the design of logistics resources across various processes.

Storage transforms goods in time (see Subsection B.III.2.1). Storage functionality can be designed through the locations of warehouses. Location refers to both the geographical positioning (e.g. in urban areas) and the relative positioning of warehouses in the process of manufacturing and distributing goods. Location choice is critically influenced by consumers' lead time requirements for order fulfillment and the availability of macro-logistics infrastructure. Storage functionality can be furthermore designed through the storage equipment to hold goods in warehouses. Storage equipment design choices are primarily influenced by the (variation in) physical and economic properties of goods to be stored.

Handling is an umbrella term for short-distance movement and short-term storage processes that transform the flow of goods in quantity and composition (see Subsection B.III.2.2). The type of physical handling equipment (e.g. fork lifts) is an important parameter to design such processes. Factors influencing type choice include the (variation in) physical and economic properties of goods, the variation in throughput, and the number of items per customer order.

Transportation transforms goods in space (see Subsection B.III.2.3). The spatial transformations that a logistics system can bring about depend on the type of transportation links (e.g. direct vs. indirect) established between different nodes of the logistics system and on transportation modes (e.g. air, road). Link design is critically influenced by the availability of consolidation opportunities. Transportation mode choice is critically influenced by the modes' costs and performance in interaction with the physical and economic properties of goods.

Packaging refers to the detachable, partial or full wrapping of goods (Pfohl 2010:134; see Subsection B.III.2.4). Logistics-related (industrial) packaging protects goods and facilitates efficient logistical transformations by transforming goods in terms of their storage, transportation, and handling properties. Logistics-related packaging is categorized into master cartons (e.g. boxes, bins) and unit loads (e.g. pallets, containers). Master cartons group several individual products and represent the basic handling unit. Master carton design is primarily influenced by the physical and economic properties of goods as well as by the expected exposure to environmental influences during logistical transformation. Unit loads group several master cartons into a logistical unit for storage, (long haul) transportation, and handling. Unit load design is influenced by the same factors as that of master cartons. However, the properties of logistics resources already deployed in the system are of particular relevance because the reconfiguration of many logistics resources is not economical, especially of long-haul transportation resources, such as vessels and airplanes. The interfaces of unit loads are therefore highly standardized to ensure direct compatibility with different transportation resource types.

Labeling is a design parameter related to packaging. Labels are attached to master cartons and unit loads in order to transfer information about the goods being transformed logistically. Parameters to control label design are location, amount of information, and type (e.g. bar codes). Design choices are critically influenced by requirements in terms of readability, traceability, and governmental regulations.

Order processing aims to enable the firm-internal value creation processes to be executed in a market-conforming manner by transforming customer requirements into firm-internal operative instructions (Straube 2004:154; see Subsection B.III.2.5). Order processing transforms goods in their degree of logistical determinisms. It serves the specific purpose of enabling the information flow that precedes, accompanies, and succeeds the physical flow of goods and thus to enable the planning, controlling, and monitoring of logistics processes (Pfohl 2010:73). The structure of the information flow (sequential hand-overs vs. centralized database) represents an important design parameter for logistics systems, especially in case of meso-logistics systems, which necessitate connecting IT systems of different autonomous firms in order to enable the flow of information across firm boundaries. Design choices are critically influenced by the lead time of information availability and the variation in information requirements of different actors along the physical flow of goods.

In addition to these core logistics processes, *non-logistical production processes* also frequently take place in logistics systems (see Subsection B.III.2.6). These processes differ from logistics processes in that they transform goods in their "form," rather than time and space. In order to quickly fulfill customized orders as well as to increase overall efficiency, non-logistical production processes have become increasingly integrated into logistics systems, especially in warehouses. Due to the variety of these production processes, a comprehensive review and specification of associated design parameters and influencing factors is beyond the scope of this thesis. However, it is important to recognize that the resources and capabilities required for these processes are very often tailored to the unique requirements of a single manufacturing firm or a single type of activity within an industry segment.

The "production process" of logistics processes integrates internal and external production factors, and it consists of a pre-combination and a delivery phase (Freichel 1992:10ff.; see Section B.III.3). Internal factors are under control of the entity which enacts the logistics process. External factors, usually the object to be transformed, are initially not under control of this entity, but are provided by the consumer at a later point in time. In the pre-combination phase, a "capability potential" is established by combining internal production factors. In the delivery phase, the capability potential is accessed by integrating the external production factor with the pre-combined internal factors and additional internal factors, thereby transforming the logistics object.

Logistics processes are conceived of as "services" and, hence, characterized using the same concepts as services (see Section B.III.3). Specifically, they are characterized as (1) intangible and (2) perishable. Logistics processes are *intangible* (Andersson & Norrman 2002), which means that the performance of services cannot be assessed prior to delivery (Lovelock & Gummesson 2004:26). Logistics processes are *perishable* (see e.g. Pfohl 2010:24), which means that they cannot be inventoried after production (Lovelock & Gummesson

2004:31), and, therefore, "unused service capacity of one time period cannot be stored for future use" (Pride & Ferrell 2003:325 as cited in Lovelock & Gummesson 2004:29). Specifically, this means that the capability potential established in the pre-combination phase will be lost, if consumers fail to provide the external production factors at the agreed time.

Institutional Perspective on Meso-Logistics Systems

Considering meso-logistics systems from an institutional perspective shifts the attention to the actors in such systems and the (business) relationships between them (see Chapter B.IV). LSPs and logistics consumers represent the most important actors in meso-logistics systems (see Section B.IV.1). LSPs can generally be defined as "companies which perform logistics activities on behalf of others" (Delfmann et al. 2002:203f.). Logistics consumers are entities that task LSPs with performing logistics activities on their behalf in order to achieve their (business) goals. This thesis will focus on buyer-supplier relationship between LSPs and consumers, and on horizontal cooperative relationships between LSPs in meso-logistics systems. We will later focus on the structure of these cooperative relationships in depth (see Chapter C.VI), after some fundamental theoretical background on such relationships has been provided.

Several dimensions are suggested by logisticians to categorize LSPs, including service type, service variety, geographic coverage, strategy, general problem-solving ability, customer adaptation ability, and modular service design and delivery (see Section B.IV.2). Logistics modularity is a newly emerging research field and will assume an important role in this thesis (see Section B.IV.3). Following the general definition of modularity, logistics modularity can be understood as the structuring of logistics activities into distinct modules, which can be sold separately, or be mixed and matched by consumers in order to create customized solutions, where logical and temporal interdependencies have to be taken into account (process modularity). Delivering logistics processes in a modular manner is associated with several benefits for LSPs, such as competitive advantage in fast-changing customization environments. Still, it may also entail certain disadvantages, such as increasing complexity for customers if they have to choose from various modules. Current research shows that few LSPs have recently started to adopt modularity, which is why literature on concrete methods for creating logistical modules is scarce.

Environment of Meso-Logistics Systems

The environment of a meso-logistics system consists of the organizational environments of all logistics consumers and LSPs in the respective system (see Section B.V.1). The logistics consumers' environments indirectly impact the environments of LSPs via their buyer-supplier relations: Changes in consumers' environments manifest in changed requirements, thus changing the environments of LSPs. An organization's environment consists of physical and social factors outside the organization's boundaries (Duncan 1972:314) that are relevant to its operations (Child 1972:3). The competitive (or economic) and technological environment will be of particular relevance for this thesis. Environmental conditions are typically characterized by the degree of stability (i.e. predictability of changes) and complexity (i.e. the heterogeneity and number of relevant factors) (see e.g. Mintzberg 1979). Dynamic environments often entail a high degree of hostility (i.e. degree of competition) (Mintzberg 1979:269).

The competitive environments of logistics consumers and LSPs have recently become increasingly dynamic (e.g. due to crises, shorter product life cycles), hostile (e.g. due to commoditization and price transparency via the internet), and complex (e.g. due to intensive outsourcing, globalized operations, lean logistics concepts) (see Section B.V.2). In other words, today's economic environment can be characterized by high complexity and limited logistical predictability (see Delfmann & Jaekel 2012:12).

The technological environment of LSPs encompasses all logistics-relevant technologies available for planning, realizing, controlling, and monitoring the flow of goods and information (see Section B.V.3). While technologies that enable the flow of goods have not been subject to major innovations in recent decades (the container was probably the last in this respect), technologies to enable the flow of information have had several innovations in the same period of time, with the internet likely being the most important. The IoT likely represents the most recent innovation that impacts the informational flow in logistics systems. Applying the IoT to logistics systems enables faster and better logistical decisions through automated data capture, real-time information, and real-time location systems (Jones & Chung 2008:119f.).

II. An Introduction to Analyzing and Designing Logistics Systems

1. Logistics: Term Origin and Usage

The term "logistics" is rooted in a military context, where it referred to questions of supplying military operations with armed forces and materials (Krulis-Randa 1977:39ff.; Klaus & Müller 2012:6f.). Today, the term is primarily used to refer to questions of transportation, storage, and handling of goods and information in a business context (Isermann 1998:22; Pfohl 2010:11). This understanding emerged during the 1960s, when such logistical issues came into focus of managers (Schulte 2005:1; Klaus & Müller 2012:8). These issues received increasing managerial attention as many firms grew and expanded into new geographic markets, a situation that required the systematic coordination and monitoring of material flows (Schulte 2005:1), and as computers and quantitative techniques became more available, thus enabling the systematic flow analysis and coordination (Bowersox et al. 1986:6). The overall economic climate that encouraged squeezing profits further shifted managers' interest to logistical issues (Bowersox et al. 1986:6).

The increasing importance of logistical issues for businesses has also sparked scientific interest therein, thus triggering the emergence of a separate, interdisciplinary research field over the last decades, which is still evolving. Delfmann et al. (2010) aim to conceptualize the identity of this field by means of five cornerstones. Mentzer et al. (2004) have tried to establish a unified theory of logistics by developing several propositions on its role in a firm. The field's rather young and still emerging character may explain the plethora of definitions in literature of the term "logistics" (for overviews of these definitions see e.g. Coyle et al. 1988:5ff.; Pfohl 2010:12ff.). Analyzing contemporary definitions shows that the term is not used to refer to a single phenomenon, but to multiple intertwined phenomena. In order to systematize these definitions and systematically capture all of the phenomena, Delfmann (2012:267ff.) distinguishes three "layers" of how the term "logistics" is commonly used: (1) logistics systems, (2) logistics management, and (3) logistics philosophy. These layers build upon one another and are interlinked.

The first and lowest layer of the term's usage – *logistics systems* – refers to a certain real-world subject area that comprises the structures and processes that serve the purpose of transferring objects, mainly goods and information, in time and space (Delfmann 2012:268f.). Structural components of logistics systems include the physical resources necessary to execute logistics processes, such as personnel, warehouses, transportation devices, and IT systems (see e.g. Klaas 2002:21). Basic logistics processes include transportation, storage, handling, packaging, and order processing (Delfmann 2012:268f.; Pfohl 2010:7ff.). The structural components represent input factors to a logistics system, and its output represents the logistical transformations of objects that are accomplished by these processes, as they can satisfy the logistical demand by making available the right object(s) in the right time, location, quantity, and condition and at the right costs (Pfohl 2010:20; Coyle et al. 1988:5).

The second layer of the term's usage – *logistics management* – refers to the entirety of the design and administrative activities related to the planning, realization, control, and monitoring of logistics systems (Delfmann 2012:269; also see Pfohl 2016:21ff.; Bowersox et al. 1986:5). Logistics management pertains to both intra- and interorganizational object flows (Baumgarten et al. 2004:2), and it involves all levels of organizational planning and execution, including strategic, operational, and tactical.

The third layer – *logistics philosophy* – refers to a specific way of thinking, or mind-set, that guides the conceptualization and realization of logistics management (Delfmann 2012:269). Principles included in this philosophy are: (a) *systems thinking*, (b) *total-cost perspective*, (c) *flow- or process-orientation*, and (d) *customer- or service-orientation* (see e.g. Delfmann 2012:267ff.; Pfohl 2010:20ff.; Schulte 2005:3). These principles, however, do not represent a "timeless dogmatic view of the world," but rather a contemporary conceptualization of the awareness and understanding of logistical issues, which is continuously adapted according to changes in the economic, technological, and social environments and based on new scientific insights (Klaas 2002:45). In the context of an increasingly dynamic environment that can be characterized by limited logistical predictability (see below Chapter B.V), it seems that contemporary logistics research and practice increasingly focus on principles related to *logistical effectiveness* to adequately satisfy logistical needs. Research and practice related to the design of logistics systems increasingly focus on, among other principles, (a) *robustness* (see e.g. Tang 2006; Wieland 2012), which refers to the ability of a system "to resist change without adapting its initial stable configuration" (Wieland 2012:890); (b) *resilience* (see e.g. Christopher & Peck 2004; Pettit et al. 2010; Jüttner & Maklan 2011; Ponomarov & Holcomb 2009), which refers to "the ability of a system to return to its original state or to move to a new, more desirable state after being disturbed" (Christopher & Peck 2004:2); (c) *flexibility and agility* (see e.g. Tang & Tomlin 2008); and (d) *modularity* (see e.g. Pekkarinen & Ulkuniemi 2008; Lin et al. 2010; Lin & Pekkarinen 2011; Rajahonka 2013). In addition to focusing on agile logistical responses, this thesis also focuses on modular logistics system design, which is discussed in more detail in Section B.IV.3 below.

2. A Formal Definition of Logistics Systems

Logistics research adopts systemism as a worldview and underlying research approach (Delfmann 2012:267f.; Pfohl 2010:25ff.). Systemism conceives of everything in the world as "either a system or a component of a system, and every system has peculiar (emergent) properties that its components lack" (Bunge 2000:147). In the most abstract terms, a *system* can be defined as a

> "complex object, every part or component of which is related to some other component(s) of the same object" (Bunge 1996:270).

A *concrete system* can be represented through the quadruple of its (1) composition, (2) environment, (3) structure, and (4) mechanisms (Bunge 1997:416, 414ff.; see also Bunge 1996:270), with "concrete" meaning that all components of the system are atoms, organisms, or social groups such as firms (Bunge 1996:270; Bunge 1997:415). The *composition* refers to the collection of components or parts that belong to the system at a given point in time. The *environment* includes all things that do not belong to the system but are nevertheless connected to parts of it. The *structure* comprises the collection of relations among the system's components as well as their relationships to things in the environment. *Mechanisms* are processes (or activities) that are able to cause or prevent change to individual system components or the system as a whole (Bunge 1997:414). Systems are frequently depicted as networks consisting of nodes and edges. Nodes represent system components, and edges visualize the relationships among components and with the environment (Bunge 1996:271).

Logistics research concretizes the (abstract) systemic worldview by focusing on economic systems and by modeling and analyzing these economic systems as spatio-temporal flows of objects in networks, with objects including goods, information, money, and people (Delfmann et al. 2010:58; also see Delfmann 2012:267f.). Above, logistics systems have been generally defined as structures and processes that serve the purpose of transferring objects in time and space. Following this general understanding, the formal definition of systems, and the domain of logistics research, logistics systems can be formally defined as follows:

Definition 1. *Logistics systems are concrete systems that are composed of artificial and natural material components* (composition), *which are connected through spatio-temporal object flows* (structure) *and embedded in a logistics-relevant environment* (environment) *where change to the composition and structure are governed by primarily economic, but also social, political, and natural processes* (mechanisms).

This definition is based on that developed by Klaas (2002:35f.). Logistics systems are *socio-technical systems*, as they comprise both artificial components (e.g. physical resources) and natural components (e.g. people) that interact with each other within spatio-temporal object flows (Isermann 1998:46). Logistics systems are furthermore *open systems*, as they influence and are influenced by the environment in which they are embedded (Isermann 1998:47).

3. Analyzing and Designing Logistics Systems

3.1. Preliminary Considerations

Logistics systems can be analyzed and designed at different levels of aggregation and from different perspectives. Depending on the level and perspective taken, different components, relationships, mechanisms, and aspects of the environment come into the focus of analysis and design. Hence, different analysis and design questions arise and different methods (of potentially different academic disciplines) may be applied (Delfmann et al. 2010:58; Pfohl 2010:14). The following two subsections introduce three common levels of aggregation and five common perspectives.

3.2. Levels of Aggregation: Micro, Meso, Macro

Following the tripartite categorization commonly used in economic research, logistics scholars distinguish three levels of aggregation to consider logistics systems (Delfmann et al. 2010:59f.; Pfohl 2010:14ff.): (1) micro-, (2) meso-, and (3) macro-logistics systems. The main difference between these levels is the number and type of actors (either individuals or organizations) that are the focus of consideration, which is why Pfohl (2010:14ff.) refers to this categorization as "institutional demarcation" of logistics systems. The *micro-level* focuses on individuals or very small business entities, the *meso-level* on constellations of interacting business entities, and the *macro-level* on the interaction between economic sectors, regions, and national economies (Delfmann et al. 2010:59f.). This categorization aims to achieve analytical clarity and demarcation of research areas. However, in reality, these levels are interconnected and each component of a logistics system on one level can itself be conceived of as a network and as part of a superordinate network on another level (Delfmann et al. 2010:60).

Macro-logistics systems refer to the logistics systems of national economies or of a community of states (such as the European Union) (Delfmann et al. 2010:60; Pfohl 2010:14; Klaus et al. 2012:417). System components include organizations and the physical infrastructures (e.g. harbors, airports) that are used to accomplish spatio-temporal object flows (Klaas 2002:37). The structure specifies the spatio-temporal relationships between these organizations. Mechanisms include economic processes that trigger spatio-temporal object flows and political processes of the country's administration, which define the legal framework for spatio-temporal transfer processes (e.g. international trade agreements and Incoterms). The environment includes the macro-logistics systems of other national economies with which spatio-temporal relationships are established (Klaas 2002:37).

Micro-logistics systems, by contrast, refer to the logistics systems of single organizations (Delfmann et al. 2010:59f.; Pfohl 2010:14; Klaus et al. 2012:436). Micro-logistics systems are therefore always intra-organizational systems, limited by the legal boundary of the organization to which they belong (Pfohl 2010:15). System components include the organization's personnel and physical resources that serve the purpose of accomplishing spatio-temporal transformations of objects (Klaas 2002:37f.) within the organization's value creation system. The system structure depends on the specific configuration of the value creation system. Mechanisms primarily include economic processes (within the organization's environment) that trigger the object flow within the organization's micro-logistics system (e.g. customer demand for the organization's products). Mechanisms also include political processes among the organization' decision makers (see Child 1972:1) that define the configuration of the overall value creation system, including its micro-logistical subsystem. The environment comprises the infrastructure of the macro-logistics system, which supports spatio-temporal flows, and the organizations in the macro-logistics system (Klaas 2002:38) that trigger flows in the micro-logistics system or are connected to the organization through flows.

Meso-logistics systems refer to the combined micro-logistics systems of a group of organizations that collectively represents only a fraction of an economy, such as the set of organizations involved in the same distribution channel (see Pfohl 2010:14f.). Meso-logistics systems are interorganizational systems that span organizational boundaries and feature cooperative relationships between multiple autonomous organizations (Pfohl 2010:15f.; Klaus et al. 2012:436). As meso-logistics systems combine multiple individual micro-logistics systems, the components, structure, and environment of a meso-logistics system equal the union of components, structures, and environments of the individual micro-logistics systems. The structure of a meso-logistics system additionally includes spatio-temporal relationships between the combined micro-logistics systems. Mechanisms additionally include interorganizational political processes (see Child 1972:10; Child 1997:58) among the member organizations of a meso-logistics system that structure their spatio-temporal relationships and flows. These relationships may be formalized through contracts.

3.3. Perspectives: Infrastructural, Functional, Informational, Organizational, and Institutional

Although they use partially varying terminologies, scholars adopt largely similar perspectives for analyzing and designing logistics systems, which can be (1) infrastructural (see "technical perspective" in Delfmann et al. 2010:58; see "infrastructure" in Klaas 2002:134ff.; see "freight network" in Krupp & Klaus 2012:72), (2) functional (see "functional sub-systems" in Pfohl 2010:67ff., 167ff.; Delfmann 2012:268f.), (3) informational (see "information network" in Krupp & Klaus 2012:72; Isermann 1998:25), (4) organizational (see "organisational" in Delfmann et al. 2010:58; see "process design" and "formal organizational structure" in Klaas 2002:137ff.; see "institutional network" in Krupp & Klaus 2012: 72f.), or (5) institutional (see "institutional network" in Krupp & Klaus 2012:72f.; see "institutional" in Pfohl 2010:14ff.).

The *infrastructural perspective* focuses on the analysis and design of physical logistics resources and capabilities necessary to realize or support the object flow in networks. This perspective specifically focuses on "the interplay of infrastructures (e.g. roads, railways, and warehouses), machines (e.g. trucks, forklifts, and industrial vehicles), receptacles (e.g. containers and pallets) and people (e.g. picker, forklift operator, and truck driver) in flow systems" (see "technical perspective" in Delfmann et al. 2010:58). This perspective also includes the IT infrastructure and systems that enable the flow of information connected to the flow of physical objects. The basic design problem of logistics infrastructure relates to designing the "logistics network" in terms of number, type, and locations of technical facilities and resources and to sizing technical and personnel capacities (see e.g. Klaas 2002:134ff. and sources therein).

The *functional perspective* focuses on the design and analysis of the types of logistics processes and on the associated logistics transformations that can be accomplished by a logistics system (Pfohl 2010:14; also see Isermann 1998:25f.). The functions of logistics systems are commonly structured along the flow of objects or based on the type of logistics process and associated transformation. If structured along the flow, five *phases* are commonly distinguished: procurement, production, distribution, spare parts, and disposal (Pfohl 2010:167ff.) If structured based on function, five *core processes or functions* are commonly distinguished: transportation, storage, handling, packaging, and order processing (Delfmann 2012:268f.; Pfohl 2010:7ff.; also see Krupp & Klaus 2012:72). The functional perspective is directly linked to the infrastructural perspective because the composition of physical resources and capabilities determines the functions (transformations) a logistics system can bring about.

The *informational perspective* centers on the analysis and design of the information (flow) required to plan, realize, control, and monitor the physical flow of objects, especially including the information that precedes, accompanies, and succeeds the physical flow of objects (Isermann 1998:25; also see Krupp & Klaus 2012:72f.). This perspective is thus concerned with the semantics and syntax of information exchanged as well as the technical standards and protocols used for exchanging it. The information perspective is linked to the infrastructural perspective because the physical infrastructure includes the IT infrastructure necessary to enable and support the information flow. More specifically, electronic sensors allow logistics information about logistics processes to be captured in real time, close to its source, and predominantly electronically (Isermann 1998:25). In turn, the information gathered unfolds its coordinating effect as it is used to plan and control the flow of physical objects (Isermann 1998:24).

The *organizational perspective* focuses on the analysis and design of the administrative arrangements that determine the division and coordination of labor related to the planning, realization, control, and monitoring of the flow of objects. More specifically, the organizational perspective may be divided into "process organization" – which focuses on administrative processes related to the realization of object flows, such as planning, control, and monitoring (Klaas 2002:137ff.) and on the administrative level of transactions that activate the flow of objects, such as ordering Delfmann et al. 2010:58) – and "formal organizational structure", which provides a stable frame for carrying out administrative and operational logistics processes (Klaas 2002:143ff.) by allocating responsibilities to different functions, business segments, or hierarchical levels (Krupp & Klaus 2012:72f.).

The *institutional perspective* focuses on the analysis and design of legal relationships within logistics systems between different organizations. Thus, it refers to the legal constructs that create the formal frame for transactions within logistics systems, especially through the formulation of contracts and the choice of legal organizational forms (Krupp & Klaus 2012:72f.). The design of these legal constructs determines the nature of relationships between actors in terms of the degree of cooperation (Krupp & Klaus (2012:72f.); also see Pfohl 2010:14f.). The institutional perspective thus assumes an important role in meso-logistics systems. From a theoretical point of view, this perspective is related to the continuum between markets and hierarchies as defined by TCE (see Chapter C.III) and hence may also be referred to as "structural governance perspective." The institutional perspective is related to the other perspectives in the sense that legal constructs provide an overarching frame for design choices regarding the infrastructural composition, functional scope, informational flows, and organizational structure of logistics systems.

4. Specifying the Analysis and Design Focus of this Thesis

The preceding subsections have provided a formal definition of logistics systems and introduced three levels of aggregation and five perspectives for analyzing and designing such systems. Depending on the types of objects flowing and the level of aggregation and perspective(s), different design problems arise and different methods are required. This thesis focuses on the *physical goods* flowing in *meso-logistics systems*. More specifically, it focuses on *horizontal cooperation between LSPs* within such systems. Furthermore, this thesis adopts the *five perspectives* presented above in order to analyze and design meso-logistics systems. However, different perspectives are of differing relevance; we primarily concern ourselves with the infrastructural, functional, organizational, and institutional perspectives. We adopt this focus because it appears to be the most conducive basis for designing a reference architecture of CLSs – our second research objective.

The next chapter presents the fundamentals regarding the infrastructural and functional design of logistics systems. The subsequent chapter considers meso-logistics systems from an institutional perspective, especially focusing on LSPs, as they are central actors in such systems. The final chapter analyzes the environment of meso-logistics systems. Fundamentals regarding the structural (legal) relationships between cooperating LSPs (part of the institutional perspective) and the organization of horizontal LSP cooperation (organizational perspective) are presented in the next part of this thesis (Chapter C.VI), after general theoretical foundations regarding the governance structure and organization of interfirm networks are introduced.

III. Infrastructural and Functional Perspectives on Logistics Systems

1. Processes in Logistics Systems and Associated Transformations

The preceding chapter introduced several perspectives from which logistics systems can be analyzed and designed. This chapter considers logistics systems from the infrastructural and functional perspectives. These perspectives are considered together due to their inherent linkages: The composition of the physical infrastructure (in terms of resources and the capacity to deploy these resources in a purposeful manner) determines the type and variety of processes that can take place in a logistics system and thus the functions, or transformations, that this system can bring about.

Figure 2 depicts six basic categories of processes that typically take place in logistics systems, including associated primary transformations. The most important processes in logistics systems are the *core logistics processes.* Storage transforms goods in time, and transportation in space. Handling is an umbrella term that comprises short-distance movement and short-term storage processes that transform the flows of goods in their quantity and composition. Packaging transforms logistics goods in terms of their storage, transportation, and handling properties, and order processing transform goods in their degree of logistical determinisms. In addition to these "pure" logistics processes, *non-logistical production processes* also frequently take place in logistics systems, which differ from logistics processes in that they transform goods in their "form."

The following section briefly considers the processes that typically take place in logistics systems and specifies key design parameters regarding these processes. It also considers the underlying physical resources that bring about these transformations and identifies influencing factors that determine the manifestation of these parameters.

2. Processes, Resource Design Parameters, and Influencing Factors

2.1. Storage

Storage occurs within the nodes of a logistics system. It transforms goods in time, and thus creates time utility (Coyle et al. 2003:285), which can manifest in different ways. As Ballou (1999:246f.) describes, time utility can (a) *coordinate supply and demand* by being able to satisfy (seasonally) fluctuating demand from a (security) stockpile of goods produced earlier rather than from current production; (b) *reduce transportation costs* by storing and consolidating goods prior to transportation over a certain period of time, thus

	Core Logistics Processes					
Transformations	Storage	Transportation	Handling	Packaging, Load Unitization, Labelling	Order Processing	Non-logistical Production
Time	X					
Space		X				
Quantity, Composition			X			
Storage, Transportation, Handling Properties				X		
Logistics Determination					X	
Form						X
	Flow of Goods				Flow of Information	

Figure 2.: Processes in Logistics Systems and Associated Primary Transformations (based on Pfohl 2010:9)

increasing the utilization of transportation resources; (c) *assist in production processes* by storing goods during aging periods (e.g. cheese); and (d) *assist in marketing processes* by storing goods close to customers, thus enabling quick order fulfillment.

Facilities that provide storage space are typically referred to as *warehouses*. Warehouse design is a highly complex issue that takes various design parameters into account (see e.g. Rouwenhorst et al. 2000; Rowley 2000; Nehm & Veres-Homm 2012), but for the purposes of this thesis, considering three basic design parameters is sufficient: (1) location, (2) storage system, and (3) handling system. The handling system is discussed together with handling activities in the following subsection (see Subsection B.III.2.2).

Location refers to both the geographical positioning as well as the relative positioning of warehouses in the process of manufacturing and distributing goods. Warehouses can be positioned as either (a) *sourcing-*, (b) *production-*, (c) *distribution-*, or (d) *transportation-oriented* (see e.g. Pfohl 2010:112ff.; Bowersox & Closs 1996: 499ff.). Depending on their positioning, the warehouses' primary functions differ as does the importance of storage processes therein (see e.g. Coyle et al. 2003:285ff.; Pfohl 2010:112ff.). Production-oriented warehouses serve the primary purpose of providing long-term storage, either before or after production, and hence usually have a comparatively high storage capacity. Sourcing- and distribution-oriented warehouses primarily consolidate inbound flows from multiple suppliers or break up flows towards customers. Storage capacity and duration are often lower than in production-oriented warehouses. Transportation-oriented warehouses serve the primary purpose of sorting and mixing different flows of goods between multiple origins and destinations and hence offer no storage capacity, or only very little for very short periods of time.

Various factors influence the location of warehouses. For instance, identifying a suitable location, among other things, depends on labor costs and availability, land costs, and local taxes (Ballou 1999:438; Bowersox & Closs 1996:407). A full review of factors is beyond the scope of this thesis. However, for our purposes, two factors are of particular importance: *lead time for order fulfillment* and *access to macro-logistics transportation infrastructure* (Ballou 1999:438; Bowersox & Closs 1996:407). It can be argued that the need to fulfill orders with short lead times typically results in warehouses being positioned closer to customers. It can also be argued that the less important the storage processes in a warehouse, the more important the access to macro-logistics infrastructure.

The *storage system* refers to the physical logistics resources installed in a warehouse to enable short- and long-term storage. The nature of a storage system critically depends on two design parameters: (1) physical storage equipment; and (2) building type. Two fundamental categories of *physical storage equipment* can be distinguished: (a) *floor storage* and (b) *racking storage* (see e.g. Schulte 2005:225). While floor storage is self-explanatory, several types of racking systems are described in the literature, including pallet, high-bay, and gravity flow racking (first-in-first-out), shelving racks for boxes, drive-in racks (last-in-first-out), and special racks tailored to the unique requirements of a single good (see e.g. Schulte 2005:226ff.; Pfohl 2010:127ff.; see rack examples in Coyle et al. 2003:330). The nature of the storage system further depends on the *building type* in which the storage equipment is installed. This is because a building can protect storage equipment from environmental influences and *vice versa*, for example by actively controlling interior temperature or humidity or by providing additional security features against pilferage. The factors that primarily influence storage system design are the *physical properties of the goods* to be stored (Bowersox & Closs 1996:433; Ballou 1999:438) in terms of size, weight, volume, value, fragility, and susceptibility to temperature and humidity changes. These properties determine the requirements for the storage equipment. In addition, the *variation in physical properties* is critical, as a high variation makes it less likely that all goods can be stored efficiently using the same type of equipment.

2.2. Handling

Handling refers to all short-distance movement and short-term storage activities related to the transfer of goods between logistical resources, including transfers from transportation to storage resources, from storage to transportation resources, and between transportation resources (Stein 2012:600). These processes transform the flow of goods in terms of quantity and composition and thus provide a "sorting function" by separating, mixing, selecting, and grouping goods (Stein 2012:600; Ihde 2001:2). Handling thus creates value by structuring the flow of goods according to customer requirements (Stein 2012:601). Handling processes can be broken down into four generic steps: (a) unloading received goods from incoming transportation resources, (b) deconsolidating goods (break-bulk), (c) sorting goods to create a new structure in terms of quantity and composition (consolidation), and (d) loading goods onto outgoing transportation resources (Stein 2012:601).

Handling processes occur within the nodes of a logistics network and are therefore intertwined with storage processes (see "warehouse operations" in Coyle et al. 2003:299ff.). The relative importance of handling processes increases with a decreasing importance of storage processes and *vice versa*. Therefore, handling processes are most important in transportation-oriented warehouses and least important in production-oriented warehouses. Nevertheless, despite these differences, both handling and storage processes always take place in every logistics node.

Literature distinguishes between two sub-types of handling: (1) transshipment and (2) order picking (commissioning). The distinction hinges on whether received and deconsolidated goods are stored for a longer period of time before being sorted and shipped again (Stein 2012:601). *Transshipment* refers to the case in which goods are sorted and grouped according to subsequent transportation destinations directly after their reception, without long-term storage. This primarily takes place in transportation-oriented warehouses, which is why they are often referred to as *transshipment terminals* or *hubs*. *Order picking* refers to the case in which goods are stored after reception and picked later from the storage system in order to fulfill a specific (customer) order. Hence, order picking especially occurs in distribution-oriented warehouses.

The *handling system* refers to the physical resources necessary to facilitate short-distance movement and short-term storage. As handling processes take place in logistical nodes, the handling system becomes a design parameter of warehouses and transshipment terminals. The nature of such a system critically depends on two design parameters: (1) short-distance movement equipment and (2) short-term storage equipment. *Short-distance movement equipment* can be defined according to the characteristics of the goods flow: (a) a *discontinuous-flow flexible path* uses hand trucks and forklifts; a *continuous-flow fixed path* uses equipment such as conveyers and draglines; and a *discontinuous-flow fixed path* uses equipment such as monorails, live-roller conveyer, and stacker cranes (see Coyle et al. 2003:333; Pfohl 2010:128f.). Factors that influence the design of the short-distance movement system include the *physical properties of goods* (Coyle et al. 2003:333) and *variation in physical properties and throughput of goods* (Ballou 1999:457). The physical properties determine the requirements for designing the handling equipment in terms of size, volume, and, most importantly, weight. Regarding variation in physical properties and throughput of goods, one can argue that high variation leads to discontinuous-flow flexible path systems, as they are most flexible and can be economically adjusted to varying goods requirements and to capacity fluctuations. More generally, it can be argued that "[m]aterials-handling equipment should be as standard as possible and as flexible as possible to lower costs" (Coyle et al. 2003:313).

Short-term storage equipment can be defined according to whether goods are stored during handling (a) on the *floor* or (b) in *special order picking racks*. Floor storage is most common in transshipment terminals, and special order picking racks are most frequently used in distribution-oriented warehouses. Such racks enable an efficient picking process by holding items in smaller quantities and packing units in a very confined space compared to the main storage equipment of a warehouse (which reduces the need for short-distance movement, thus speeding up order fulfillment and reducing costs) (Coyle et al. 1988:259f.). Factors that influence the setup of dedicated picking racks include *(variation in) physical properties of goods*, *the number of orders*, and *the number of items per order* (Schulte 2005:246ff.). A high variation in physical properties of goods makes the use of order picking racks less likely, as different goods may not be fitted in racks economically. Furthermore, it is reasonable to say that the use of an order picking rack is more likely, the smaller the individual goods, the larger the number of orders, and the more items per order, as these factors all increase the economic benefits gained from special racks.

2.3. Transportation

Transportation provides place utility (Bowersox et al. 1986:22) by moving goods from one place to another (Gudehus 2012:819) to connect dispersed production and consumption processes in the economy (Ihde 2001:1). Transportation refers to the movement of goods *between* the nodes of a logistics network and thus differs from short-distance movements associated with handling processes, which occur within the nodes. Two basic design parameters for transportation in logistics networks are commonly identified: (1) transportation links, and (2) transportation modes (see e.g. Pfohl 2010:150ff.).

Transportation links between a pair of logistics nodes can be established either (a) *directly* or (b) *indirectly* (Pfohl 2010:5f.). Direct links do not include a transshipment point, while indirect links do. The *physical and economic properties of goods* in interaction with *consolidation opportunities* are important factors that determine the design of transportation links. Regarding economic properties, one can argue that the higher the economic value of goods and opportunity costs in the case of delayed delivery, the more likely the use of direct links, as transshipment consumes additional time, thus increasing transit time (Coyle et al. 2003:352ff.). As for physical properties, Ballou (1999:45, italics removed) argues that "the smaller the shipment size [relative to the overall capacity of a single transportation resource], the disproportionately greater will be the benefits of consolidation." The link structure furthermore depends on the availability of opportunities for consolidating multiple flows of goods, which is primarily a function of the geographic demand structure and the overall economic intensity between connected nodes. The more opportunities available, the higher the likelihood for indirect links. However, indirect links can only be established if the efficiency gains from consolidation exceed the costs and opportunity costs incurred by transshipment (Pfohl 2010:7).

Transportation modes are commonly divided into: air, water (maritime, inland waterway), and land (road, rail, pipeline) (see e.g. Pfohl 2010:154ff.; Coyle et al. 1988:330ff.). For the remainder of this thesis, we distinguish between (a) *air*, (b) *water*, (c) *road*, and (d) *rail*, thus consolidating "maritime" and "inland waterway" transport into the aggregate category "water" and neglecting pipeline transport. Important factors that influence mode choice include *cost and performance properties* in interaction with *physical and economic properties of goods*. Transportation modes can be evaluated with several cost and performance criteria (Schulte 2005:170 and sources therein), as depicted in Table 1. The interactions between these modes and the properties of goods to be transported can be illustrated by few examples: oversized goods may require water transport as they cannot fit into airplanes; high value goods warrant fast transportation due to high capital costs; fast moving consumer goods require fast transportation due to high opportunity costs resulting from lost revenue in the event of stock outs; and air transport cannot enable door-to-door links and thus needs to be combined with road transportation (see e.g. Coyle et al. 2003:350ff.).

Criteria	**Definition**	**Assessment (least is best)**			
		Rail	Road	Water	Air
Costs	line-haul rate, accessorial charges	3	4	2	5
Speed	elapsed time of transportation link	3	2	4	1
Availability	ability to directly connect any pair of nodes	2	1	4	3
Capability	ability to transport any type of good	2	3	1	4
Frequency	quantity of scheduled links	4	2	5	3

Table 1.: Comparative Costs and Performance Assessment of Transportation Modes (based on Bowersox & Closs 1996:325f., except for "costs" which is based on Coyle et al. 2003:342, 356)

2.4. Packaging, Load Unitization, and Labeling

Packaging refers to the detachable, partial or full wrapping of goods (Pfohl 2010:134). Hellström & Saghir (2007:198f.) distinguish between three nested, hierarchical levels of packaging: primary, secondary, and tertiary. The first refers to *consumer packaging* and the second and third to *industrial packaging* (see e.g. Coyle et al. 2003:316f.; Bowersox & Closs 1996:436f.). Consumer packaging comes in direct contact with the goods and serves the primary purposes of informing and appealing to customers (see e.g. Coyle et al. 1988:53f.). Industrial packaging, which we focus on, emphasizes logistical aspects.

Industrial packaging interacts with all other logistics processes and thus impacts the overall efficiency and effectiveness of logistics systems (Hellström & Saghir 2007:198f.; Bowersox & Closs 1996:435). It specifically creates (or diminishes) value by transforming the transportation, storage, and handling properties of goods (Pfohl 2010:8f.), thereby serving the following purposes (see e.g. Coyle et al. 2003:315f.; Bowersox & Closs 1996:436ff.; also see Ballou 1999:66f.; Pfohl 2010:134ff.): (a) to protect goods from environmental influences (e.g. punctures, heat, humidity, pilferage) and protect the environment from the influences of goods (e.g. leakage) during storage, handling, and transportation; (b) to facilitate efficient storage, handling, and transportation by reducing or automating handling processes and by increasing utilization of storage and transportation capacities; and (c) to transfer information about goods through labels. These logistical goals can be achieved using two levels of industrial packaging: (1) master cartons and (2) (standardized) unit loads (see e.g. Bowersox & Closs 1996:436f.) in combination with (3) labels. Master cartons, unit loads, and labels thus become critical design parameters of logistics system design.

Master cartons represent the basic handling unit in a logistics system and refer to any type of box, bin, bag, or barrel used to group several individual products (Bowersox & Closs 1996:436ff.). The specific design parameters of master cartons differ across types; for example, boxes need to be designed in terms of physical dimensions, material strength, and shape (Coyle et al. 2003:316). Considering these parameters in more detail is not required for the objectives pursued in this thesis. The design of master cartons depends primarily on the *physical and economic properties of goods*; the *expected exposure to environmental influences*; and the *properties of the storage, handling, and transportation equipment already deployed in the logistics system* (Pfohl 2010:137ff.). For example, physical properties of goods determine the size and shape of cartons; the economic value, fragility, and intensity of exposure to natural forces determine the degree of required protection; and the properties of the existing resources determine the need and economic justifiability for reconfiguring those resources to make them compatible with cartons. Carton design needs to balance these factors to achieve an optimal tradeoff between, *inter alia*, the utilization rate of storage and transportation capacity, packaging costs, the anticipated share of damaged goods, and ease of handling (Pfohl 2010:137f.).

Standardizing the design of master cartons (within a logistics system) is important to enable efficient storage, handling, and transportation (Bowersox & Closs 1996:437). Standardization reduces the need for repacking and restacking goods during handling (see e.g. Bowersox & Closs 1996:445) and for retrofitting a system's storage, transportation, handling, and order picking resources as long as the same or compatible cartons are used for packaging different types of goods. However, standardized master cartons may not align with the physical and/or economic properties of all goods that are stored, handled, or transported by the logistics system. For example, cartons may be too large for certain goods, thus wasting logistical capacity. Thus, the *variation in physical and economic properties of goods* is a critical factor that influences the (economic) ability to standardize. Yet, as homogeneity of goods can often not be achieved, logisticians frequently suggest using modular master cartons of different sizes (see e.g. Bowersox & Closs 1996:438 USB; Pfohl 2010:147ff.). Each size can be used to efficiently package a specific type or number of goods, and all boxes can be grouped efficiently to fit with larger standardized load units (Specter 2012:14).

Unit loads represent the second level of packaging as they combine several master cartons into a single logistical unit for storage, (long haul) transportation, and handling. This grouping process is called *unitization* (Bowersox & Closs 1996:442) and primarily aims to increase operative logistical efficiency by increasing the number of goods that can be transformed at once (which reduces handling steps and time) (Ballou 1999:261), by enabling handling through mechanized or automated equipment (e.g. fork lifts) (Ballou 1999:261; Pfohl 2010:141f.), and by increasing the utilization of storage and transportation capacity. Unitization is commonly achieved by means of (a) *containers* and (b) *pallets* (Ballou 1999:261; Pfohl 2010:142ff.). Containers are large rigid metal boxes typically of cuboid shape that are stackable and can be loaded from one of their front sides. Pallets are portable platform devices on which goods can be loaded (Bowersox & Closs 1996) and secured using stretch/shrink wrapping, rope ties, and strapping. The design of unit loads essentially depends on the same factors as for master cartons, as containers and pallets simply represent another layer of industrial packaging. Hence, containers and pallets are available in different sizes and with different loading properties in order to match various physical and economic properties of goods and different types of transportation resources. For example, air transport uses special pallets and containers (referred to as unit load devices (ULDs)) tailored to different aircraft types. Hapag-Lloyd (2010), an ocean carrier, offers six basic container types (e.g. general purpose, ventilated, and temperature-controlled), thus providing suitable choices for goods with different physical and economic properties.

The specifications of containers and pallets have become highly standardized over the last decades, which is why unit loads are often referred to as *standardized unit loads*. For example, the ocean freight container was introduced in 1956 and standardized by the International Organization for Standardization (ISO) in 1968, the Euro-pallet was introduced in 1961. Standards regulate the specifications that determine the storage, transportation, handling, and stacking properties (such as length, width, height, maximum weights, and stacking interfaces), thus enabling wide operative logistical efficiency by ensuring that logistics resources deployed at different locations are compatible with unit loads. In particular, standardization focuses on the potential to automate and use machines to handle unit loads (Schulte 2005:150) by reducing the variation in physical properties of grouped master cartons. Unit load standards are most often governed by international bodies, such as the ISO and the International Air Transport Association (IATA) (see "Unit-Load-Device Panel" IATA 2014), thus increasing efficiency in international trade, especially in air and ocean transport. Hence, the latitude for designing unique unit loads for international trade is rather limited, as retrofitting transportation resources would be too costly. Therefore, design choices in logistics systems are usually limited to selecting the most suitable standardized unit load. Nevertheless, logistics managers occasionally develop *special standardized unit loads*, which are standardized in terms of their external loading and stacking interfaces, but specialized internally to align with unique goods properties (e.g. special Euro-pallets for car parts). Such unit loads may only become a "standard" in a single logistics channel but, in some cases, can encompass an entire industry.

Labels are attached to master cartons and unit loads. Labeling focuses on the transfer of information about goods (e.g. providing handling instructions, enabling tracking). Design parameters regarding labels include the location of the label, the amount of information it carries, and the label type (Hellström & Saghir 2007:203). Typical label types include European Article Number (EAN) codes, RFID tags, bar codes, and color codes. Factors that influence label design include *governmental regulation* regarding the identification of certain types of goods (Pfohl 2010:139) as well as *readability and traceability requirements* of the logistics system (Hellström & Saghir 2007:203).

2.5. Order Processing

Order processing aims to enable firm-internal value creation processes in a market-conforming manner by transforming market-induced customer orders into firm-internal operative instructions (Straube 2004:154). Customer orders express a customer's requirements and thus represent the input for order processing. In meso-logistics systems, orders usually manifest in contracts that legally bind logistical actors. Logistical order processing serves the specific purpose of enabling the information flow that precedes, accompanies, and succeeds the physical flow of goods and thus to enable the planning, controlling, and monitoring of logistics processes (Pfohl 2010:73): The preceding flow provides information to plan the realization of logistics processes. These planning activities transform goods in their logistical determination, for example, by allocating them to a specific order (Pfohl 2010:7ff.). The accompanying flow provides information required by operative personnel to realize, control, and monitor logistics processes. The succeeding flow comprises information that becomes available only after processes are realized, such as invoicing or any type of feedback.

The structure of the information flow represents an important design parameter for logistics systems, especially in case of meso-logistics systems, which necessitate connecting IT systems of different autonomous firms in order to enable the flow of information across firm boundaries. There are two basic options for realizing the information flow (1) sequential, which means that the IT systems of consecutive firms along a physical flow of goods are interfaced and information is exchanged between these firms like a "baton" in a relay race, and (2) centralized, which means that the IT systems of all firms are interfaced with a single central database through which information is exchanged and which may be operated by a third-party firm (Pfohl 2010:85).

Factors that influence the structure of the information flow are diverse, as they require the adaptation, interfacing, and potential standardization of IT systems of multiple autonomous firms along the physical flow of goods (Schulte 2005:477). Two relevant factors in this thesis are the *lead time of information availability* and the *variation in information requirements* of different firms and IT systems along the physical flow (Pfohl 2010:86). Temporal urgency is especially high in dynamic environments in order to enable rapid responses. Within logistics systems, many information requirements are homogeneous, which is why several international standards for data exchange have been introduced, such as EDIFACT (Pfohl 2010:86). One can argue that increasing temporal requirements and homogeneity results in a more centralized structure.

2.6. Non-Logistical Production Processes

Non-logistical production processes provide utility by transforming goods in their "form" instead of in time and space. In recent years, the boundary between logistics and production processes has become increasingly blurred as more and more production processes have become integrated into logistics systems, especially into warehouses. This development has been facilitated by, among other things, customers' increasing requirements for the rapid fulfillment of customized orders (see Chapter B.V). Hence, manufacturing firms position production activities closer to their customers (manufacturing postponement). In addition, manufacturing firms have started using logistics infrastructures to efficiently handle reverse flows in terms of product repairs and recycling.

Examples of production processes carried out in logistics systems include light manufacturing, final assembly, quality control, product customization, and promotional co-packaging. Due to the variety of these production processes, a comprehensive review and specification of associated design parameters and influencing factors is beyond the scope of this thesis. However, it is important to recognize that the resources and capabilities required for these processes are very often tailored to the unique requirements of a single manufacturing firm or a single type of activity within an industry segment. Two examples support this claim. DHL, a logistics service provider, assembles customized car doors for Audi and delivers them in a just-in-sequence (JIS) manner directly to the production line (Sheffi 2012:131f.). This assembly process is tailored to the unique requirements of Audi. Rhenus, another LSP, has "high-tech personnel" specially qualified for carrying out pre-assembly and configuration work of high-tech products with touchscreens, such as automatic bank and ticket machines (Rhenus 2017). The personnel is trained to carry out a specific process, which is very similar across firms within the same industry.

2.7. Interim Summary

The preceding subsections have introduced processes that take place in logistics systems, key parameters for designing these processes and underlying logistics resources, and factors that influence the manifestation of these parameters. Table 2 summarizes the preceding discussion and shows that many influencing factors are specific to the respective design parameters. Yet, the *physical and economic properties of goods* influence the design of logistics resources in various logistics processes and thus assume a pivotal role in logistics system design.

3. The "Service Character" of Logistics Processes

The previous section introduced core logistics processes. However, so far, we have not focused on the particularities of how these processes are "produced." The logistical production process is depicted in Figure 3. The process integrates (1) internal and (2) external production factors (Freichel 1992:11ff.) to bring about logistical transformations. Internal and external production factors differ in the degree to which the entity, which brings about logistical transformations, has control over these factors. *Internal factors* include physical and immaterial logistics resources necessary for logistical transformations, and they are under the control of the entity that realizes logistics processes. *External production* factors represent the goods or information that are logistically transformed; hence, they are often referred to as *logistics objects.* External factors are initially not under the full control of the entity that enacts the logistics processes, but are contributed by an

Logistics Process and Design Parameter	Typical Design Options	Influencing Factors
Storage		
Location	sourcing-, production-, distribution-, transportation-oriented	lead time for order fulfillment; access to macro-logistics infrastructure
Storage equipment (incl. building type)	floor storage, racking storage	(variation in) physical properties of goods
Handling		
Short-distance movement equipment	discontinuous-flow flexible path, continuous-flow fixed path, discontinuous flow fixed path	(variation in) physical properties of goods; variation in throughput
Short-term storage equipment	floor, special picking racks	(variation in) physical properties of goods; number of orders and number of items per order
Transport		
Link	direct, indirect	physical and economic properties of goods; consolidation opportunities
Mode	air, water, road, rail	physical and economic properties of goods; costs and performance properties of modes
Packaging, Unit Loads, and Labeling		
Master carton	boxes, bins, barrels, modular boxes	(variation in) physical and economic properties of goods; expected exposure to environmental influences; properties of storage, handling, and transportation equipment already deployed in the logistics system
Unit loads	containers, pallets	*same as for master cartons*
Labels	EAN codes, RFID, bar codes	readability; traceability; governmental regulation
Order Processing		
Structure of information flow	sequential, centralized database	lead time of information availability; variation in information requirements

Table 2.: Summary – Logistics System Design Parameters, Design Options, and Influencing Factors

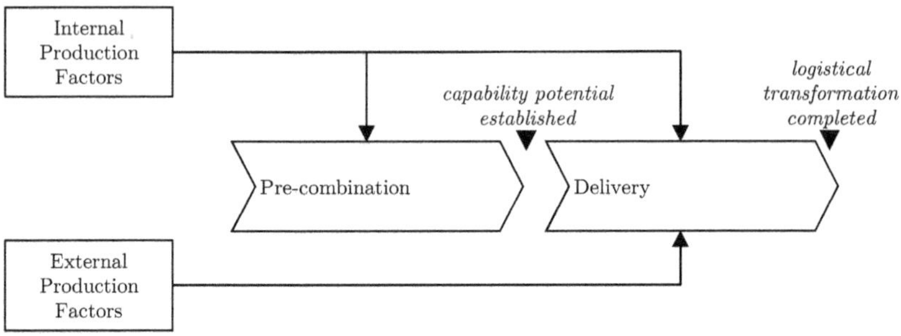

Figure 3.: Logistical Production Process (based on Freichel 1992:13)

external entity at a later point in time. Uncertainty about the properties of the external factors and the time when they become available impairs the ability to adequately plan logistics processes and consistently deliver high quality (Rümenapp 2002:20). The production process can be divided into two phases: pre-combination and delivery (Freichel 1992:10ff.; Pfohl 2010:24). The pre-combination phase aims to establish a "capability potential" by combining internal production factors before the external production factor becomes available, at which point the delivery phase begins. The capability potential is accessed by integrating the external production factor with the pre-combined internal factors and additional internal production factors, thereby transforming the logistics object. The integration of the external production factor is considered characteristic of the logistical production processes.

To investigate the nature of the production process, we focus on the "service character" of logistics processes, which several logistics scholars emphasize (see e.g. Freichel 1992:11ff.; Andersson & Norrman 2002:4; Pfohl 2010:24f.; Bretzke 2012:372f.). Four properties are typically used to specify the character of services in general: (1) intangibility, (2) heterogeneity, (3) inseparability, and (4) perishability (IHIP) (Lovelock & Gummesson 2004; Zeithaml et al. 1985). We exclude *heterogeneity* from the following discussion for two reasons. First, logistics scholars have made no reference to this concept so far, and second, Lovelock & Gummesson (2004:28) conclude, based on a literature review, that heterogeneity is not a property unique to services but equally applies to goods.

Intangibility means that the performance of services cannot be assessed prior to delivery (Lovelock & Gummesson 2004:26). This property is frequently discussed by logisticians with regard to the fact that, prior to realization, the performance of logistics processes cannot be determined, and only an "immaterial promise of performance" can be made by the entity that will enact the process (Pfohl 2010:24; Freichel 1992:12; Rümenapp 2002:19; also see Andersson & Norrman 2002:8). However, regarding the outcomes of core logistics processes (except order processing), Lovelock & Gummesson (2004:27, 31) argue that intangibility is a misleading concept because such processes represent "physical acts to owned objects" and, therefore, include tangible processes (i.e. specifiable) with tangible (either ephemeral or permanent) outcomes. Performance can be determined by means of an immediate "before-after" comparison (Fitzsimmons et al. 1998:374).

Perishability means that services cannot be inventoried after production (Lovelock & Gummesson 2004:31), and, therefore, "unused service capacity of one time period cannot be stored for future use" (Pride & Ferrell 2003:325 as cited in Lovelock & Gummesson 2004:29). Hence, business may be lost if demand exceeds capacity, and capacity will be wasted if demand is lower than expected. In particular, logistics processes are perishable, as the capability potential established during the pre-combination phase cannot be stored (Pfohl 2010:24; Freichel 1992:12). This makes the entity that performs logistics activities particularly vulnerable to demand variations, which is why perishability represents an important influencing factor for organizing logistics systems (Freichel 1992:12).

Inseparability of production and consumption means that customers need to be present during service production (Lovelock & Gummesson 2004:28f.). This property is also associated with logistics processes (Andersson & Norrman 2002:8). However, this property seems not to be suitable for characterizing logistics processes; the external factor must be provided by the owner, but the owner does not need to be present during logistical transformations. More generally, Lovelock & Gummesson (2004:31) therefore argue that for services that represent "physical acts to owned objects" (which include freight transportation according to them), customers can usually be absent during production. In conclusion, logistics processes are not generally characterized by concurrent production and consumption.

In addition to their service character, logistics processes are frequently referred to as "logistics services" because they are carried out by specialized providers on behalf of their customers; thus, logistics processes are "rendered as a service." These specialized providers are referred to as LSPs and are discussed in detail in the following chapter.

IV. Institutional Perspective on Meso-Logistics Systems

1. Types of Actors and Relationships

Considering meso-logistics systems from an institutional perspective shifts the attention to the actors in such systems and the (business) relationships between them (see Section B.II.3). Although many different types of actors may be considered to be part of meso-logistics systems (Hertz & Alfredsson 2003:140), logistics research and practice most commonly structures them into (1) providers and (2) consumers of logistics processes.

Providers are commonly referred to as *logistics service provider (LSP)*, which are defined in various ways in literature and practice. The common denominator of many definitions is the understanding that LSPs are

> "companies which perform logistics activities on behalf of others" (Delfmann et al. 2002:203f.).

LSPs are referred to as "service" providers because performing logistics activities on behalf of others can be conceived of as "rendering a service" to another entity, as services are defined as "the application of specialized competences (operant resources – knowledge and skills), through deeds, processes, and performances for the benefit of another entity or the entity itself" (Vargo & Lusch 2008:26). The (market-conform) delivery of logistics services represents LSPs' primary business purpose (Pfohl 2010:16). Services are usually delivered against financial remuneration with the objective of making a profit. The next section provides a deeper understanding of LSPs by describing several dimensions with which to categorize them.

Logistics consumers are entities that task LSPs with performing logistics activities on their behalf in order to achieve their (business) goals. They usually remunerate LSPs for carrying out logistics activities and thus are LSP customers. A wide range of entities can act as logistics consumers, such as private households and (non-)governmental agencies. However, for-profit manufacturing, commercial, and service firms are the most common logistics consumers.

Meso-logistics systems can be categorized by the types of actors that cooperate during the management or realization of logistics processes. Three typical categories identify cooperation between (1) LSPs and logistics consumers, (2) multiple logistics consumers, and (3) multiple LSPs (Pfohl 2010:15f.).

Relationships between LSPs and logistics consumers are "buyer-supplier" relationships and have received significant attention in academia (see e.g. Selviaridis & Spring 2007; Tate 1996). They can range from discrete, transaction-oriented exchanges to strategic alliances (Rogers & Daugherty 1995:251) and can thus take a variety of structural governance forms along the continuum between markets and hierarchies (see e.g. Daugherty 2011:18f.; Lambert et al. 1999:169). These relationships can be based on informal or formalized agreements (i.e. written contracts) (see e.g. Hertz & Alfredsson 2003:140). For the objectives of this thesis, it is sufficient to recognize that these agreements typically specify customers' logistical requirements as well as the (financial) terms and conditions under which LSPs perform the requested activities.

Cooperative relationships between multiple logistics consumers can take a variety of forms, including logistics purchasing cooperatives (see e.g. Darestani et al. 2012) and types of cooperation where multiple consumers collectively operate a logistics system (see e.g. Leitner et al. 2011). However, cooperation between multiple logistics consumers is not the focus of this thesis and is not considered in more detail.

Cooperative relationships between multiple LSPs represent horizontal logistics alliances because LSPs usually operate on the same level of the logistics market or "logistical supply chain" are thus proximate or distant competitors. This thesis focuses on the structure of these cooperative relationships, and they are discussed in depth (see Chapter C.VI) after some fundamental theoretical background on such relationships is provided. For now, it is sufficient to be aware of their existence.

2. Dimensions for Categorizing LSPs

In practice, a highly heterogeneous set of companies, which perform logistics activities on behalf of others, is subsumed under the term "LSP." This heterogeneity primarily results from the increasingly idiosyncratic requirements placed upon the LSPs by their customers, essentially making each LSP unique, in terms of services provided and geographic coverage. However, this makes it hard to classify providers into a few distinct "pure" types using only a few criteria, although logisticians have presented several plausible classifications (see e.g. Hertz & Alfredsson 2003; Persson & Virum 2001). Still, any classification of types can be considered only an attempt to identify more or less useful generic types of providers, while being well aware that, in reality, LSPs may often not be clearly allocable to these types. Moreover, research and practice partially lack common terminology to denote different provider types. For the purposes of this thesis, it is not crucial to be aware of the different types and their associated names, but rather to know the dimensions commonly used to categorize them, as the concept of cloud logistics impacts some of these dimensions.

Based on existing research, there are seven categories commonly used for classifying LSPs: (1) service type, (2) service variety, (3) geographic coverage, (4) strategy, (5) general problem-solving ability, (6) customer adaptation ability, and (7) modular service design and delivery. The first, second, and fourth recur in several classifications of LSPs (Berglund et al. 1999; also see Rümenapp 2002; Lai 2004; Persson & Virum 2001; Wang et al. 2006; Sum & Teo 1999). The third dimension is common in characterizing LSPs or alliances among them (see e.g. Niebuer 1996; Jharkharia & Shankar 2007; Schmoltzi & Wallenburg 2011). The fifth and sixth dimensions are adopted from another classification (Hertz 1993; Hertz & Alfredsson 2003). Finally, the last dimension derives from a newly emerging research field that focuses on innovative approaches for logistics service design and delivery by LSPs, namely logistics modularity (see e.g. Pekkarinen & Ulkuniemi 2008;

Lin et al. 2010; Lin & Pekkarinen 2011; Rajahonka 2013).

Service type specifies which (non-)logistics processes are performed by an LSP. In principle, LSPs can perform any activity related to the planning, realization, control, and monitoring of logistics processes. In fact, LSPs have started to perform non-logistical activities as well, most often those that directly precede and succeed logistics activities. Logistical activities beyond the core processes and non-logistical (production) activities are commonly referred to as *value-added services*. Examples include inventory management, kitting, network design (Berglund et al. 1999:63), and technology-enabled services, such as track-and-trace and sending shipment notifications (Lai 2004:392). By means of a comprehensive literature review, Liu & Lyons (2011) identify nine categories of activities contemporarily performed by LSPs, as depicted in Table 3. This compilation clearly illustrates LSPs' heterogeneity.

Service variety refers to the range of different logistics activities an LSP performs. In this respect, component services and solutions are typically differentiated, with the former referring only to a single logistics activity and the latter to a composition of at least two activities (Berglund et al. 1999:62f.; also see Rümenapp 2002:51ff.).

Geographical coverage refers to the spatial footprint within which an LSP offers to perform logistics activities. Geographic coverage interacts with service type and variety by specifying the area within which a service can be made available. For example, Cui & Hertz (2011:1005) distinguish between local, international, and global providers; Schmoltzi & Wallenburg (2011:563) differentiate between regional, nationwide, continental, and intercontinental providers.

Strategy refers to the competitive strategies an LSP adopts with regard to the geography, the type and range of services provided. To distinguish between strategies, scholars most often adopt the generic competitive strategies defined by Porter (1980). Conceptual and empirical research suggests that LSPs use different strategies depending on type and variety of services offered (see e.g. Persson & Virum 2001; Rümenapp 2002:299), which is why strategy can be considered a dependent variable. Berglund et al. (1999:63) argue, for example, that providers of single service components often adopt a cost-leader strategy, while those offering a wide service range pursue a differentiation strategy. The chosen strategies also impact providers' financial performance. Wang et al. (2006:806) find in an empirical study that LSPs, which offer a wide range of services and pursue a combined cost and differentiation strategy, financially outperform competitors that pursue either a cost or a differentiation strategy (see similar results in Sum & Teo 1999:594).

General problem-solving ability refers to CLSs' ability to perform difficult, complex, and advanced logistics activities (Hertz 1993:30f.; Hertz & Alfredsson 2003:140f.; also see

Activity Category	Examples
Transportation	inbound transportation, outbound distribution, expedited delivery
Transportation planning and management	route and network optimization, carrier selection, freight forwarding
Warehousing/Inventory	storage with goods reception, spare parts, pick and pack, cross, docking, inventory management
Value-added services	labeling, packaging, kitting, (re-)assembly/installation
Information technology	track and trace, order entry/management system, electronic commerce
Product design and marketing support	packaging design, promotional support
Finance	invoicing, freight bill audit, insurance
Consulting services	logistics planning, supply chain design
Other customer services	customs brokerage, call center operation, procurement of materials

Table 3.: Activities Performed by Logistics Service Providers (based on Liu & Lyons 2011:550ff.)

Håkansson 1982:382). Problem solving ability interacts with service type and variety because different types and compositions of services require different degrees of problem solving. A low problem solving ability is associated with providers of component services, including general transportation, warehousing (Hertz & Alfredsson 2003:140f.), and special transportation for certain types of goods (Hertz 1993:30f.). Conversely, providers that offer time-sensitive, global (express) transportation services or complex bundles of logistics services, including warehousing and value-added services, have a high problem-solving ability (Hertz 1993:30f.; Hertz & Alfredsson 2003:140f.).

The *customer adaptation ability* refers to an LSP's ability to perform logistics activities tailored to customers' idiosyncratic requirements (Hertz 1993:30f.; Hertz & Alfredsson 2003:140f.); also see Håkansson 1982:382). Similar to the ability of problem solving, the ability to adapt to customer requirements interacts with service type, variety, and geography because different types and solutions across varying geographies require different degrees of adaptation ability. Providers of general transportation services and highly time-sensitive express services exhibit low customer adaptability (Hertz & Alfredsson 2003:140f.). As argued above, these services commonly use standardized unit loads. By contrast, providers that offer warehousing services, bundles of warehousing and (non-logistical) value-added services, or that manage complex logistics operations on behalf of their customers, exhibit a high customer adaptation ability (Hertz & Alfredsson

2003:140f.). High customer adaptation is associated with warehousing services because many tailored, value-added (non-logistical) services can be performed in warehouses.

Modular service design and delivery refers to the property by which LSPs structure their service portfolios into distinct modules, which can be combined by their customers to create customized solutions. Modular service design and delivery interacts with all other dimensions and requires at least one service type in at least two different geographies or a at least two services within the same geography, thus enabling service compositions. Modular service design and delivery also interacts with strategy because modular design can be considered a strategy. Moreover, it also interacts with the customer adaptation ability because logistics consumers can combine multiple standardized services to create complex tailored solutions. Finally, it interacts with problem solving ability because it depends on the types of services offered in a modular manner. The next section considers modular service design and delivery in more detail due to its high relevance for this thesis.

3. Modular Service Design and Delivery by LSPs

In general systems theory, modularity can be defined as

> "a continuum describing the degree to which a system's components can be separated and recombined, and it refers both to the tightness of coupling between components and the degree to which the 'rules' of the system architecture enable (or prohibit) the mixing and matching of components" (Schilling 2000:312).

Due to the process character of logistics services, modular design and delivery of logistics services is specifically related to a *process (or service) modularity* (Rajahonka 2013:48; Bask et al. 2010:368), which can be defined as

> "the usage of reusable process steps that can be combined ('mixed and matched') to accomplish flexibility and customization for different customers or situations in service implementation" (Bask et al. 2010:368).

Process modularity exhibits two important characteristics: *Logical interdependencies* and *temporal interdependencies* between service modules, which need to be taken into account when combining them (Böttcher & Klingner 2011:327). Logical interdependencies arise from non-functional service characteristics. They refer to constraints that restrict the choice of modules that can be combined in the sense that module $x$and module $y$cannot be part of the same solution. Temporal interdependencies arise from the procedural nature of services and refer to constraints that restrict the sequence in which modules need to be combined and executed: Module y must be executed after module x. Process modularity is also characterized by services' non-technical aspects, which manifest in the human or "soft" nature of interfaces between services (Bask et al. 2010:366). As interfaces involve people, they may be particularly vulnerable to unforeseen errors, such as information loss due to miscommunication (Voss & Hsuan 2009:545). These general characteristics of process modularity seem applicable for characterizing logistics services. For example, transportation services can resemble logical interdependencies because different modes cannot be combined without an intermediate handling process. An example of temporal interdependencies might be customers requiring that goods are stored prior to final delivery. Finally, operative process realization involves human interfaces as goods are handed over by humans.

Logistics modularity and modular logistics service design and delivery have recently attracted attention from logistics scholars (see e.g. Corsten & Gössinger 2007; Mayer 2007; Pekkarinen & Ulkuniemi 2008; Bask et al. 2010; Lin et al. 2010; Bask et al. 2011; Lin & Pekkarinen 2011; Rajahonka 2013; Rajahonka et al. 2013). Despite its newness, the understanding of logistics modularity aligns well with the general definitions of (process) modularity. Empirical evidence and scholars suggest that logistics modularity manifests in the fact that services can be sold separately or combined into unique service solutions (Rajahonka 2013:42f.; Pekkarinen & Ulkuniemi 2008; Corsten & Gössinger 2007). LSPs may offer highly standardized core logistics services, which they supplement with customized value-adding service modules to fulfill heterogeneous demand (Pekkarinen & Ulkuniemi 2008:92). Scholars essentially conceive of any process related to the planning, realization, control, and monitoring of logistics processes as a separate service module (see e.g. Pekkarinen & Ulkuniemi 2008:96 Lin et al. 2010:202; Rajahonka 2013:41; Bask et al. 2010:365). Furthermore, Bask et al. (2011:392) find evidence for LSPs offering a wide variety of (non-logistics) "modular customised value-adding services", including "assembly of sales displays, pre-delivery inspections or final assembly of products, ticketing and labeling of products, and country-specific packing (adding manuals, etc.)." These are added to

Introducing modularity to logistics service design and delivery is associated with several benefits for LSPs. For example, Rajahonka (2013:41) reports that interviewed LSPs consider modularity a useful approach that may lead to competitive advantage. She also finds evidence that logistics modularity can, among other things, enable outsourcing modules, mass-customization, and a reduction in complexity as services are clearly defined. Pekkarinen & Ulkuniemi (2008:93) collect empirical evidence from LSPs suggesting that modularity leads to faster sales, reduced time-to-market, reduced errors, and increased efficiency and quality. Lin & Pekkarinen (2011:350) find evidence that modular service design and delivery allows LSPs to maintain competitive advantage in fast-changing customization environments without compromising on cost efficiency and flexibility. More specifically, regarding service design, modularity enables LSPs to offer customers more options and to rapidly develop customized solutions by combining service modules, thus being able to respond to heterogeneous and quickly changing customer demands (Rajahonka 2013:42; Lin et al. 2010:200ff.). Such an efficient and fast service design process is important for LSP success (Lin & Pekkarinen 2011:352). Regarding service delivery, modularity enables LSPs to deliver high quality services and better deploy internal resources, to easily measure service quality and internal operations (Lin & Pekkarinen 2011:352) and to deliver customized services in a cost-efficient and flexible manner (Lin & Pekkarinen 2011:353). Rajahonka (2013:43) provides evidence that using repetitive standardized processes as modules leads to cost efficiency and economies of scale in delivery.

Despite various benefits, logistics modularity also entails disadvantages. Some LSPs interviewed by Rajahonka (2013:42) report that modularity can reduce the flexibility to adapt to customer requirements as modules are standardized (in contrast to the ability to fulfill customized orders), increase complexity for customers if there are too many modules to choose from, and lead to an excessive division of labor, which, combined with too little coordination, may lead to overall inefficiencies due to silos.

Several approaches have been proposed to create logistics modules. In principle, these approaches aim to break down complex customer requirements into smaller modules. Three approaches are briefly introduced to provide some understanding of how logistics modularization can be accomplished. Based on an empirically grounded study of LSPs, Lin & Pekkarinen (2011) use the Quality-Function-Deployment (QFD) method to modularize logistics requirements. This method ensures that designed services actually satisfy customer requirements. They specifically develop a modular logistics service platform that consists of four module types: service, process, activity, and organizational modules (for a very similar platform see Pekkarinen & Ulkuniemi 2008; Lin et al. 2010). The first three module types refer to logistics activities at different levels of granularity, ranging from a category of logistics processes (e.g. transportation) to a specific type of logistics process (e.g. air transport) to a specific logistics task within that process (e.g. loading the aircraft). Each of these is also associated with an organizational module, ranging from the LSP that offers the service to a department of that LSP to a specific team within that department, respectively.

Corsten & Gössinger (2007) develop an approach to modularizing logistics processes by taking into account process interdependencies, which consist of logical and behavioral interdependencies (which result from the behavioral uncertainty of actors involved in different logistics processes). The approach defines modules such that interdependencies between them are minimized. This approach has some similarities to the concept of *logistical segmentation* (see e.g. Delfmann 1995) if each module is conceived of as a separate logistical order cycle.

Mayer (2007) develops a (1) concept and (2) instrument to modularize logistics systems. Although he focuses on the modularization of logistics systems from the perspective of a production or manufacturing firm, rather than LSPs, we will nevertheless briefly present his contribution as it seems to be largely applicable to LSPs as well. The proposed *concept* includes seven different types of modules structured hierarchically into strategic, tactical, and operative. Examples of modules include design module, planning module, customer integration module, supplier integration module, operative capability module. The concept also provides a generic template to describe modules based on their inputs, outputs, transformation processes, required resources, and module responsible. The concept additionally stipulates 14 specific design principles for the modularization of logistics systems. These principles are derived from systems theory, new institutional economics, strategic management theory as well as other scientific publications and case studies (Mayer 2007:155). The proposed *instrument* structures the process of modularizing logistics systems into five fundamental steps. Two steps are of particular relevance for the modularization: capability and design analysis as well as structuration. The former includes the analysis of the current and target state as well as the documentation of key requirements of the target state. The latter includes the creation of modules, the definition of module interfaces, and the assignment of modules to organizational entities or external suppliers. Mayer (2007:200ff.) proposes to create modules based on agglomerative cluster analysis, which takes into account the homogeneity of resources, of objectives, and of control instruments.

The previous paragraphs reviewed key contributions on logistics modularity. Current research in this field provides an initial understanding of the modularity phenomenon within the logistics industry. Hence, further research seems highly warranted to further investigate how logistics systems can be modularized as well as to understand the outcomes for both production firms as well as LSPs.

V. Environment of Meso-Logistics Systems

1. Conceptualization of Environment

This chapter focuses on the environment of meso-logistics systems. As meso-logistics systems combine several micro-logistics systems, their environment equals the union of all micro-logistics systems' environments. Hence, a meso-logistics system's environment consists of the environments of participating logistics consumers and LSPs. In this respect, it is important to recognize that logistics consumers' environments also impact LSPs' environments due to their buyer-supplier relation. Changes in consumers' environments manifest in changed logistical requirements, thus changing the environments of LSPs.

The environment of a system is conceptualized as all things that are connected to the components of a system, but not part of the system (see Section B.II.2). As we focus on the environment of logistics consumers and LSPs, we conceptualize the environments of meso-logistics systems as *organizational environments*. An organization's external environment consists of physical and social factors outside the organization's boundaries (Duncan 1972:314) that are relevant to its operations (Child 1972:3). Specific physical and social factors that shape the external environment include competitive, political, legal, technological (Tushman & Anderson 1986:439), and ecological factors (Klaas 2002:37). These factors impact logistics activities and the entities that perform them. For example, legislation provides the formalized institutional frame within which logistics activities can be carried out; societal trends such as urbanization change demand patterns for logistics activities and create pressure on LSPs to reduce carbon emissions; and natural disasters, such as earthquakes and floods, can destroy pivotal logistics infrastructure and thus adversely impact the flow of goods and information. For the purposes of this thesis, competitive and technological factors are of particular relevance, and hence are considered in more detail in the following sections.

Researchers propose several dimensions with which to characterize environmental conditions, including homogeneous/heterogeneous, stable/shifting (Thompson 1967), low/high diversity, and not/highly dynamic (Lawrence & Lorsch 1967) (also see Jurkovich 1974; Child 1972). In this thesis, we characterize environmental conditions by means of two dimensions that are frequently used in organization theory and logistics research: (1) environmental stability and (2) environmental complexity (see e.g. Mintzberg 1979; Duncan 1972; Pfohl & Zöllner 1997). *Environmental stability*, which ranges from stable to dynamic, refers to the degree to which change in environmental factors relevant to an organization's operations can be predicted (Mintzberg 1979:268; Child 1972:3). Thus, change alone does not make an environment dynamic, but only if it occurs in an unpredictable fashion. Such change is particularly problematic because no patterns can be discerned in advance in order to adequately respond to changed conditions (Mintzberg 1979:268). Furthermore, Mintzberg (1979:269) argues that dynamic environments are characterized by a higher degree of *environmental hostility* than stable environments, where "hostility" refers to the degree of competition. *Environmental complexity*, which ranges from simple to complex, refers to the heterogeneity and number of environmental factors relevant to an organization's operations (Child 1972:3; also see Duncan 1972:315). In complex environments, an organization requires sophisticated knowledge about various factors to carry out its work, including for example products, customers, and technology (Mintzberg 1979:268). In simple environments, the relevant factors are few in number and related to few system components (Duncan 1972:315). Given the definitions of stability and complexity, four types of organizational environments can be distinguished: stable-simple, stable-dynamic, complex-simple, and complex-dynamic.

2. Competitive Environment

The competitive (or economic) environment of an organization

> "encompasses the demands made by the market, including the price, characteristics, and features of the product; the location of customers; the time requirements of customers; and the variability in demand. It also refers to the relative importance of each of these attributes and the extent to which these attributes are changing or stable over time [...] [and] might also include those economic [...] trends which shape the global marketplace" (Stock et al. 1999:39).

The competitive environment of LSPs is shaped by two fundamental factors: (1) market conditions experienced by their customers, and (2) rivalry among LSPs.

The economic environment of LSPs is shaped by the *market conditions experienced* by the firms on whose behalf they carry out logistics activities. This is because the demand for such logistics activities ultimately derives from the demand for products and services. Thus, changes in the economic environments of the firms requesting logistics services, also changes the economic environments of firms providing logistics services. Wallenburg & Raue 2011:385 remark in this respect: "As the complexity and dynamics of supply chains have increased, so have the challenges for logistics service providers (LSPs) as key players in the management and execution of logistics" (see similar Hertz & Alfredsson 2003:139).

Margin pressures can be observed in many industries. As a response, logistics consumers have implemented various measures to adjust their cost bases, frequently related to (1) the exploitation of geographical or national costs or policy advantages, (2) lean logistics concepts, and (3) outsourcing.

First, in order to *exploit geographical or national differences* in labor costs, material costs, tax policies, or regulatory conditions, firms procure, produce, and distribute goods on a global scale (see e.g. Midoro & Pitto 2000:32). This transformational process, commonly referred to as *globalization*, has been further enabled and flanked by the availability of new information and communications technology, such as the internet (see e.g. Midoro & Pitto 2000:32). It has led to value creation processes that are geographically (and often also organizationally) fragmented and involve many actors. Hence, "as the number of supply chain partners increases, these global supply chains can become 'longer' and 'more complex'" (Tang & Tomlin 2008:12). Therefore, logistics activities, especially transportation, become increasingly important to connect these distributed actors and ensure a smooth flow of goods. As a results, LSPs increasingly face a requirement to perform logistics activities on a global scale (Midoro & Pitto 2000:32; also see Raue & Wallenburg 2013:21 and sources therein), which has led them to pursue internationalization and globalization strategies through mergers, strategic alliances, joint ventures, and acquisitions (Lemoine & Dagnæs 2003).

Second, to reduce inventory carrying costs, logistics consumers have implemented *lean logistics concepts* such as JIT delivery (Tang & Tomlin 2008:12) and have integrated and synchronized their sourcing, production, and distribution activities – not only across functions within their organization, but also, and more importantly, in an interorganizational manner with their suppliers and customers (Christopher & Peck 2004:2; see also Chandrashekar & Schary 1999:27). Thus, LSPs not only need to be able to perform logistics activities on a global scale, but also with high temporal precision in an increasingly integrated network of economic actors, each with potentially unique requirements.

Third, in order to exploit benefits from economies of scale and scope, firms have retained all activities critical to supporting their core competencies and have increasingly outsourced large parts of the remaining activities, often including logistics (Prahalad & Hamel 1990; Choy et al. 2003:225; see also Quinn & Hilmer 1994:55; Arnold 2000). *Outsourcing* logistics activities to specialized LSPs aims to simultaneously reduce costs and improve service quality (see e.g. Stank & Daugherty 1997:53). While logistics consumers initially outsourced only basic logistics activities, LSPs today perform comprehensive and often customized bundles of logistics activities, frequently also including non-logistics value-added services. Thus, LSPs have become exposed to an increasingly complex bouquet of requirements (see e.g. Midoro & Pitto 2000:32).

Beyond responses to margin pressures, logistics consumers also aim to increase profits by raising revenue through frequent technical innovations of their products, which not only shortens products' life cycles, but also makes the demand for these innovative products unpredictable due to their "very newness" (Fisher 1997:106). As a consequence, LSPs face frequent and potentially unpredictable changes in the logistics requirements placed upon them.

The competitive environment of logistics consumers has furthermore been shaped by (end-)consumers' rising demand for larger product variety and customization (Da Silveira et al. 2001:2). This adds not only to the complexity of production processes, but also influences how logistics activities need to be performed. For example, mass customization is usually achieved through manufacturing postponement, which means that final production steps – e.g. customization – are postponed in the distribution channel and performed close to customers to rapidly deliver customized products (van Hoek 2000:37). This gives LSPs a chance to "penetrate segments of the supply chain with higher value-added operations, such as final manufacturing" (van Hoek 2000:37), while, at the same time, exposing them to non-logistical, potentially more complex requirements.

In addition to gradual changes in the economic environment, crises expose logistics consumers and LSPs to largely unpredictable and often drastic changes. To this day, the most severe economic crises have started in the financial sector and subsequently impacted the real economy (e.g. Mississippi Bubble of 1720, Great Depression of 1929, Global Financial Crisis of 2008). As the demand for logistics activities derives from overall economic activity, crises quickly and often significantly reduce the need for carrying out logistics activities, as demand for products and services slows. As a consequence, prices for logistics activities decline, and LSPs suffer from costs incurred by idle logistics capacity as they are usually remunerated partially or entirely on a pay-per-use basis. For example, this led to bizarre practices by ocean carriers during the 2008 Global Financial Crisis: Carriers implemented "slow steaming" which means that they deliberately reduced the sailing speed of ocean vessels to save fuel and to artificially reduce transportation capacity and thus stabilize transportation rates (Meyer et al. 2012).

The previous paragraphs have primarily focused on the competitive environments of logistics consumers and their impact on LSPs. In addition to these factors, LSPs' environments are also shaped by the *rivalry* among LSPs, which refers to the degree of hostility. Two factors have especially shaped and still determine their competitive environment. First, basic logistics services have increasingly become commodities, thus squeezing providers' profitability and making providers increasingly interchangeable. Schmoltzi & Wallenburg (2011:552), for example, speak of "growing intensity" in competition. In response to this development, LSPs have started to offer value-added services to integrate with their customers and thus become more difficult to interchange. Second, these competitive pressures are further aggravated by the availability of modern communication and information technology such as the internet, which drastically increases transparency of available services and prices. Arnold (2014b:14) argues that "this development cannot be reversed and things will self-evidently become more comparable," for example by means of interlinked online marketplace for logistics services. We thus conclude that rivalry among LSPs has not only become more intense, but the intensity is likely to increase further as more and more transactions are conducted electronically.

Considering the economic environments of logistics consumers and LSPs, one can conclude that the environment has become increasingly dynamic, hostile, and complex. It is more dynamic because the rate and magnitude of unpredictable change has increased. It is more hostile due to margin pressures and increased rivalry within the logistics market. And it is more complex as increasingly sophisticated knowledge becomes necessary to operate in this environment. In other words, today's economic environment can be characterized by high complexity and limited logistical predictability (see Delfmann & Jaekel 2012:12).

3. Technological Environment

LSPs' technological environment encompasses the logistics-relevant technologies available for carrying out logistics activities, with technology understood to be

> "those tools, devices, and knowledge that mediate between inputs and outputs (process technology) and/or that create new products or services (product technology)" (Rosenberg 1972 as cited in Tushman & Anderson 1986:440).

The technological environment encompasses the logistics resources necessary to plan, realize, control, and monitor the flow of goods and information. While technologies that enable the flow of goods have not been subject to major innovations in recent decades (the container was probably the last in this respect), technologies to enable the flow of information have had several innovations in the same period of time, with the internet likely being the most important. Hence, when describing the technological environment in this section, we focus on technologies related to the informational flow associated with logistics activities. More specifically, we focus on one technological innovation of particular relevance to this thesis: the IoT and its relation to logistics.

The IoT is a composition of the terms "internet" and "things" and introduces a "disruptive level of innovation" to today's information and communication technology (Atzori et al. 2010:2788). Semantically and intuitively, the IoT can be understood as "a world-wide network of interconnected objects uniquely addressable, based on standard communication protocols" (Santucci & Lange 2008:6). More specifically, the IoT is a new paradigm whose idea focuses on

> "the pervasive presence around us of a variety of *things* or *objects* – such as Radio-Frequency IDentification (RFID) tags, sensors, actuators, mobile phones, etc. – which, through unique addressing schemes, are able to interact with each other and cooperate with their neighbors to reach common goals" (Atzori et al. 2010:2787, italics original).

Thus, the IoT is not only about equipping objects with miniature computers and connecting them through a common network, but is about making objects interact with each other to perform tasks, ultimately on behalf of humans.

The IoT concept results from the convergence of three visions (see Atzori et al. 2010:2788ff. and sources therein): (a) *"things"-oriented visions*, which emphasize the pervasive use of RFID and wireless sensor network (WSN) technologies to link the physical world to the virtual world for the purpose of sensing, unambiguously identifying, and controlling objects. These sensors thus bridge the *last mile media break* between the physical and virtual world in an automated manner (Fleisch 2010:142); (b) *"internet"-oriented visions*, which emphasize that communication between things is achieved by means of a common standard, the internet protocol; and (c) *"semantic"-oriented visions*, which emphasize the use of semantic technologies to represent, store, search, and organize information generated by objects, to model objects, and to reason over the generated data. The usefulness of the IoT can only be "unleashed" in application domains where these three visions intersect (Gubbi et al. 2013:1646). This is because equipping objects with mini computers (object-oriented) and connecting the computers through a network (internet-oriented) is insufficient to achieve wide interoperation and cooperation among masses of heterogeneous objects. In fact, to achieve "unarchitected" interoperation and autonomous and proactive behaviors of legions of heterogeneous objects, well-defined semantics become necessary – as in the case of semantic web services (see Section D.III.4).

The application of the IoT in logistics, especially the use of RFID, is a topic that has already received significant attention from scholars and practitioners (see e.g. Jones & Chung 2008; Poon et al. 2009; Lee & Chan 2009). The combination promises to manage logistics activities more efficiently due to better monitoring and thus *better visibility* over logistics activities (see e.g. Dittmann 2006). More specifically, RFID's ability to provide timely information and visibility in logistics activities is based on three components: automatic data capture, real-time information, and real-time location systems (Jones & Chung 2008:119f.). As depicted in Figure 4, the ability to automatically collect and disseminate logistics-relevant information in real time leads to faster, more effective decisions, thus driving operational optimization, which can manifest in reduced supply chain inventory and better utilization of logistics resources (Jones & Chung 2008:119f.). Various technical innovations have introduced the IoT to logistics activities, such as smart transportation boxes that monitor their internal temperature and humidity, track their current location via the Global Positioning System (GPS), and transmit this data via the mobile phone network to involved LSPs and shipping firms. The diffusion of RFID technology has been rather slow due to margin pressures in the logistics industry (Dittmann 2006:71). Still, LSPs may be forced to increasingly use IoT, even if they have to absorb large parts of the associated costs if their customers ask for more accurate and timely visibility into their logistics processes.

In addition to increasing visibility, the IoT can also enable autonomous control in logistics systems (see e.g. Scholz-Reiter et al. 2004), which can be defined as "the ability of logistic objects [such as containers, pallets, or even single shipments] to process information, to render and to execute decisions on their own" with the goal of increasing robustness against disturbances and flexibility in dynamic and complex environments (Windt et al. 2008:573). Thus, adopting the IoT in logistics can enable a paradigm shift from centralized hierarchical control of "non-intelligent" logistics objects toward decentralized heterarchical control of "intelligent" objects (Windt & Hülsmann 2007:4f.).

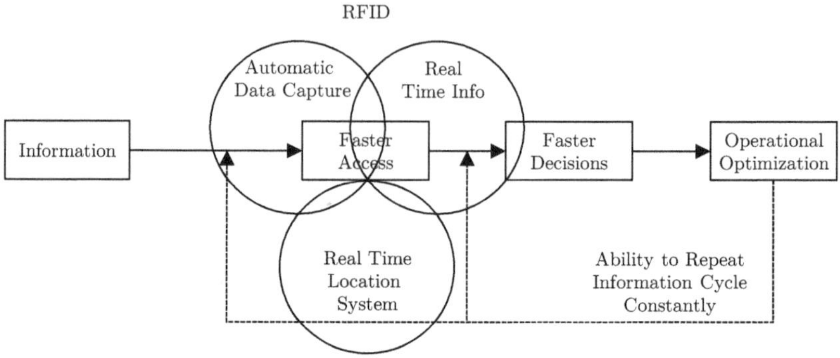

Figure 4.: RFID in Logistics (adopted with minor changes Jones & Chung 2008:120)

The IoT represents a technological innovation that can change and likely improve logistics management and enable autonomous control within logistics processes. Its diffusion and adoption by LSPs, however, depends on LSPs' financial ability and on the willingness of manufacturing, commercial, and service firms to absorb costs incurred by the introduction of this new visibility improving technology.

VI. Summary

Part B has reviewed the fundamentals of logistics. The meaning of "logistics" has been examined via three layers of the term's usage: logistics systems, management, and philosophy. Rooted in systemism, "logistics systems" have also been formally defined as a foundation for the analysis and design of such systems in this thesis. To this end, three levels of aggregation and five perspectives have been introduced; this thesis focuses on meso-logistics systems primarily from infrastructural, functional, and organizational perspectives.

Considering (meso-)logistics systems from the infrastructural and functional perspectives provide a comprehensive understanding of the (non-)logistical processes that can take place in such systems. Our focus especially emphasizes important design parameters of logistics resources and identified crucial influencing factors. The physical and economic properties of goods are identified as the most important factor because it influences resource design across all logistics processes.

Considering meso-logistics systems from an institutional perspective identifies logistics consumers and LSPs to be fundamental actors in such systems. Three types of cooperative relationships between these actors are identified, and this thesis focuses on horizontal cooperation between LSPs. However, the above considerations do not center on the organization of such horizontal alliances. These aspects are considered after the fundamental theoretical aspects of cooperative interfirm networks are established in the next part. In addition, the institutional perspective brings to light several dimensions with which LSPs can be classified; this thesis is particularly concerned with the modular design and delivery of logistics services.

Finally, considering the environment of meso-logistics systems has guided our focus to the competitive and technological environments of logistics consumers and LSPs. Reviewing environmental conditions reveals that today's meso-logistics system environments are dynamic, hostile, and complex.

Part C. Fundamentals of Interfirm Networks and Networked Horizontal LSP Cooperation

I. Abstract of Part

Introduction

The understanding of *interfirm networks* in this thesis is based on Provan & Kenis's (2008:231) definition of *interorganizational networks*, which they define as "groups of three or more legally autonomous organizations that work together to achieve not only their own goals but also a collective goal," where we assume that "organizations" are "for-profit firms" (see Section C.II.1). We adopt this definition as it explicitly emphasizes two features relevant to this thesis: (1) the legal autonomy of member firms, which means that firms have authority over the decision of whether to join an interfirm network; and (2) the concurrent pursuit of own and collective goals, which provides a motive for why networks are created as well as a source of conflict among network members, especially if they are competitors as in the case of horizontal networks.

Firms' motives for erecting or joining interfirm networks can be quite diverse; however, they can ultimately be divided into increasing profitability by increasing revenue and/or reducing costs (Ebers 1999:6f.; see Section C.II.2).

To provide an adequate theoretical basis for the design of logistics systems in accordance with the principles and concepts underlying cloud computing, we adopt four complementary theoretical perspectives (a) *TCE*, (b) *market mechanisms*, (c) *RBV*, and (d) *organizational theory* (see Section C.II.3).

F. Jaekel, *Cloud Logistics*, Produktion und Logistik,
https://doi.org/10.1007/978-3-658-22837-8_3

The Problem of Economic Organization: Transaction Costs and Structural Choice

The fundamental problem of organization is the problem of achieving cooperation among a collection of economic actors whose goals are only partially congruent (Ouchi 1979:833; see Section C.III.1). In a capitalistic society, resources are owned by economic actors (e.g. firms) and they are allocated through (capitalistic) economic institutions rather than the government (Alchian & Demsetz 1972:777). Economic institutions enable cooperation because they facilitate contract-based transactions among actors. These institutions are commonly referred to as *governance structures*, and can be categorized into markets, hybrids, and hierarchies (Williamson 1988:73). The problem of "economic organization" therefore resembles the problem of explaining what type of transaction is organized through what type of governance structure. Transactions are thus the unit of analysis (Williamson 1981:548).

TCE contends that economizing on transaction costs is central to the study of governance structure choice (Williamson 1988:548). More specifically, TCE posits that economizing on transaction costs can be achieved by matching transactions with governance structures in a discriminating way based on the attributes of both (Williamson 1988:548; Williamson 1991). TCE thus treats governance structure choice as a dependent variable. Interfirm networks are subsumed under the "hybrid" governance structure (Leiblein 2003:941), which is why TCE essentially explains the formation of interfirm networks as a function of transaction attributes.

Transaction costs represent the "costs of running the economic system" (Arrow 1969:48), and they can be structured into "search and information costs, bargaining and decision costs, policing and enforcement costs" (Dahlman 1979:148; see Subsection C.III.2.1). TCE suggests that transaction costs are critically determined by the characteristics of economic actors involved in the exchange and by the attributes of the exchange. The behavior of economic actors is characterized by (a) *bounded rationality*, which means that actors are assumed to behave in a way that is "*intendedly* rational, but only *limitedly* so" (Simon 1957:xxiv, italics original) because of cognitive limitations; and (b) *opportunism*, which denotes the strongest form of self-interested behavior, and can be defined as "seeking self-interest with guile" (Williamson 1985:47; Williamson 1979:234; see Subsection C.III.2.2). These assumptions have critical implications for contract-based exchanges because, without bounded rationality, all transactions could be efficiently handled through complete contingent contracts (see e.g. Williamson 1981:553). Without opportunism, all contract adaptations could be efficiently handled using the "general clause," in which transacting parties promise to behave responsibly and without self-interest whenever adaptations are required to respond to disturbances (see e.g. Williamson 1979:241; Williamson 2002:188;

Williamson 1985:48).

Several attributes of transactions are considered determinants for transaction costs; the most important is asset specificity (see Subsection C.III.2.3). *Asset specificity* refers to the extent to which idiosyncratic (non-marketable) investments are required in support of a specific transaction (Williamson 1979:239f.) to achieve "least cost supply" (Williamson 1981:555). Problems of non-marketability arise when the "*specific identity*" of transacting parties has "cost-bearing consequences" (Williamson 1979:239f., italics original), which means that asset specificity has "reference to the degree to which an asset can be redeployed to alternative uses and by alternative users without sacrifice of productive value" (Williamson 1988:70).

A governance structure can be defined as "the institutional framework in which contracts are initiated, negotiated, monitored, adapted, enforced, and terminated" (Palay 1984:265; see Subsection C.III.3.1). A governance structure can therefore be considered a "mode of organizing transactions" (Williamson & Ouchi 1981). Three governance modes are distinguished by TCE: (1) market, (2) hybrid, and (3) hierarchy, where markets and hierarchies represent the poles of a continuum and hybrid is an intermediate mode between these extremes (Williamson 1991:280f.). These modes differ along three dimensions: (a) *contract law*, (b) *adaptation* (autonomous vs. cooperative), and (c) *instruments* (incentive intensity vs. administrative controls) (Williamson 1991:277ff.).

Markets, one extreme of the continuum, are based on *classical contract law* which is characterized by discreteness and complete "presentiation" (Williamson 1979:236). Thus, classical contract law matches the "ideal" market transaction in economics and law (Williamson 1979:236). Markets facilitate *autonomous adaptations* through changes in *prices* (Williamson 1991:278). Markets feature *high-powered incentives*; hence, actions have immediate monetary consequences (output pricing) (Williamson 1991:279). Hence, *administrative controls* are not necessary to control the behaviors of parties.

Hierarchies, the other extreme of the continuum, are supported by the *contract law of forbearance* (Williamson 1991:274ff.), which aims to facilitate long-term economic exchanges. Under forbearance bilateral adaptations are effected through *fiat*. Among the three governance structures, hierarchies are considered the most elastic and adaptive (Williamson 1991:274) due to the combination of fiat and *low-powered incentives*. Thus, greater cooperation can be elicited for rapid *cooperative adaptations*, as managers and employees receive the same compensation "whether they 'do this' or 'do that'" (input pricing) (Williamson 1991:275). To control unwanted side effects of low-powered incentives (e.g. shirking), hierarchies rely extensively on *administrative controls*.

Hybrids, positioned in between markets and hierarchies, are based on *neoclassical contract law* (Williamson 1991:271), which aims to facilitate long-term transactions under the conditions of uncertainty, for which complete presentiation is not possible or economically prohibitive (Williamson 1979:237). For reasons of bounded rationality, these long-term contracts are inevitably incomplete, but they include an additional administrative apparatus to facilitate sequential adaptations to disturbances and changing conditions (MacNeil 1977:865 as cited in Williamson 1979:237). In case of conflicts, third party arbitration is initially preferred over litigation as transacting parties aim for contract continuation and accommodation rather than termination (Williamson 1979:237f.). Positioned between markets and hierarchies, hybrids are characterized by incentives and administrative controls of medium intensity.

TCE maintains that different types of transactions can be coordinated cost-efficiently by using different types of governance structures, as stated in the "discriminating alignment hypothesis" (see Subsection C.III.3.2). Transactions involving assets of low specificity can be cost-efficiently coordinated through markets. Assets of medium specificity can be exchanged at lowest transaction costs via hybrids. Assets of high specificity can be cost-efficiently exchanged via hierarchies.

Role of Information Technology in Economic Organization

The role of IT in economic organization has been investigated widely (see Section C.III.4). The introduction of IT has led to the emergence of two new "electronic" governance structures: (1) electronic markets and (2) electronic hierarchies (Malone et al. 1987; see Subsection C.III.4.1). We focus on electronic markets.

Electronic markets (or marketplaces) are markets where interactions between buyers and sellers are enabled and supported by IT. With the recent proliferation of the internet, electronic markets, often referred to as "platforms," have become a ubiquitous phenomenon. "Platforms fundamentally create value by acting as *conduits* between two (or more) categories of consumers who would not have been able to connect or transact without the platform. Platforms create value by coordinating these groups of consumers and in the economic view this coordination is effected through pricing" (Gawer 2014:1241, italics original). Electronic markets or platforms exploit the *electronic brokerage effect*, which suggests that a central database that connects many buyers and suppliers is equivalent to a broker who is in contact with a large number of buyers and suppliers (Malone et al. 1987:488).

Malone et al. (1987) focus on how the use of IT impacts "coordination costs" and thus governance choice (see Subsection C.III.4.2). Coordination costs are comparatively high in markets and low in hierarchies. The introduction of IT reduces coordination costs by reducing the time and cost of processing and transmitting information (Malone et al. 1987:484; also see Groth 1999:144; Argyres 1999). While this makes both markets and hierarchies more efficient, reducing coordinating costs "lead[s] to an overall shift toward proportionately more market coordination" (Malone et al. 1987:484) as coordination costs represent a relatively larger share of total costs in markets than in hierarchies.

Fundamentals of Mechanism Design

Market mechanisms, such as auctions, achieve coordination through *prices*, and they assume an increasingly important role in structuring economic decisions and exchanges that involve many self-interested actors (see Subsection C.III.5.1). The design of such mechanisms is a subfield of game theory (see Subsection C.III.5.2). Game theory is concerned with the mathematical analysis of settings of *strategic interdependence* ("games") in which two or more decision makers interact and (1) each agent's welfare is a function of his or her own actions and of other agents' actions, and (2) the actions that may be best for an agent may depend on what each agent expects the other agents to do (Mas-Colell et al. 1995:219). Agents are assumed to be *rational*, which means that they maximize their expected payoff (in a self-interested manner), and *intelligent*, which means that they have the same knowledge and information and can make the same inferences as a game theorist can make (Myerson 1991:2ff.).

Game theory distinguishes between *cooperative* and *non-cooperative* games and between games of *complete* and *incomplete* information. In non-cooperative games, agents cannot form coalitions (to achieve outcomes otherwise unattainable) and share the benefits from collusion through side payments afterwards. In games of incomplete information, some relevant information about the game is *private information*, and thus only known to a specific agent but not the others (see e.g. Narahari et al. 2009:40). Each agent's preferences for different outcomes are private information (referred to as agent *type*).

Game theorists are primarily concerned with predicting the *equilibrium* of games, which equals the set of best *strategies* (or actions) available to agents, and with calculating the outcomes given all agents' actions. Game theorists use *solution concepts* to identify a game's equilibrium. Two central solution concepts for solving non-cooperative games with imperfect information are the (1) dominant strategy equilibrium and (2) Bayesian-Nash equilibrium (see e.g. Rasmusen 2007:20; Parkes 2001:27ff.; Dash et al. 2003:41f.; Tadelis 2013:251) In a *dominant strategy equilibrium*, there exists a strategy for each agent that (weakly) maximizes the respective agent's payoff, regardless of the strategies the other agents choose to play in the game. In a *Bayesian-Nash equilibrium*, there exists a strategy for each agent that (weakly) maximizes the respective agent's expected payoff given the (agent's beliefs about the) common prior (statistical distribution of agent types) and given that all other agents also play the strategies that weakly maximize their expected payoffs.

Mechanism design is a subfield of game theory that focuses on non-cooperative games of incomplete information (see Subsection C.III.5.3). A mechanism is "a specification of how economic decisions are determined as a function of the information that is known by the individuals in the economy," which is why almost any kind of market institution can be viewed as a mechanism (Myerson 1988:1; also see Myerson 1983:1770). The theory of mechanism design is concerned with "how to implement good system-wide solutions to problems that involve multiple self-interested agents, each with private information about their preferences" for different outcomes (Parkes 2001:23). The principle insight of mechanism design theory is that not only *resource constraints* but also *incentive constraints* need to be considered when formulating and analyzing economic decisions because, just like the scarcity of raw materials, the need to offer agents incentives to reveal their private information equally imposes constraints on economic decisions (Myerson 1988:1). A mechanism designer basically faces a similar problem as a policy maker (or central authority or social planner) who wishes to implement a certain collective decision that relies on private information held by agents that are affected by the policy decision (see e.g. Tadelis 2013:288).

A *social choice function* refers to the mapping between agent preferences and the collective outcome. However, as agent types are private information and therefore unknown to the policy maker, it is impossible for the policy maker to make a collective decision directly. He or she therefore needs to approach this problem in an indirect manner by inducing a game through a mechanism. This game is designed by the policy maker with the objective to reveal agents' private information and subsequently determine the single collective decision based on their preferences reported. Narahari et al. (2009:7, italics original) therefore conceive of mechanism design as a problem "*reverse engineering* of games or equivalently as the *art of designing the rules of a game to achieve a specific desired outcome.*"

A mechanism designer aims to achieve a specific social outcome. Yet, one may reasonably ask whether or the extent to which this outcome is in fact a "good" social choice (see Subsection C.III.5.4). Of course, whether or not a social choice is economically good has to be measured on a system-wide basis in relation to agent preferences. Agent preferences are commonly modeled as the sum of utility derived from an allocation of goods and monetary transfers (quasilinear preferences). Given quasilinear preference, desirable properties can be stated separately: (1) allocative efficiency, which requires incentive compatibility, (2) budget-balance, and (3) individual rationality.

A social choice function is *allocatively efficient* if the total value over all agents is maximized (Parkes 2001:33); items are allocated to the agents that value them most (Narahari et al. 2009:94). Allocative efficiency requires incentive compatibility, which is why incentive compatibility may be considered a fourth desirable property.

Literature distinguishes between two types of budget balance: strong and weak. A social choice function is *strongly budget balanced* if there are no net transfers into or out of the system (Parkes 2001:33); the system is closed without running a surplus or a deficit (Narahari et al. 2009:95). A social choice function is *weakly budget balanced* if agents can make a net payment to the mechanism but no net payment can be made from the mechanism to agents (Parkes 2001:33). In other words the mediator of such a mechanism (e.g. a platform operator or the policy maker himself) can appropriate some fraction of the payments made by agents, since the sum of agents' payments may exceed the sum of receipts (Tadelis 2013:289f.).

Interim individual rationality requires that the expected utility from participating in the mechanism is equal to or greater than the utility the agent could earn outside the mechanism, given his type and prior beliefs about the distribution of agent types (Parkes 2001:35). In this context, *interim* means that each agent calculates the expected utility from participation *after* being told his type and *prior to* deciding whether to participate (Mas-Colell et al. 1995:893). Individual rationality is also referred to as *voluntary participation constraint* because rational autonomous agents only participate on a voluntary basis if they can expect to be no worse off after participating.

Of course, all of these properties seem desirable, and a policy maker would like to achieve them all simultaneously. However, there is bad news and good news about the theoretical feasibility of certain combinations of these properties. This news manifests in mathematical *impossibility theorems* and *possibility theorems*, respectively (see Subsection C.III.5.5). The feasibility of achieving certain combinations of desirable properties depends on assumptions about (a) *the exchange environment* (e.g. unilateral trading, bilateral trading, combinatorial auctions), (b) *agent preferences* (e.g. strictly hierarchical, quasilinear), and (3) *the choice of the solution concept* (e.g. dominant strategy incentive compatible (DSIC) and Bayesian-Nash incentive compatible (BIC)) (see e.g. Parkes 2001:49ff.; Narahari et al. 2009:87, 93). Hence, a policy maker may face though tradeoffs.

The Resource-based View

The RBV makes the central proposition that firm resources and the specific characteristics of these resources are critical determinants for firms to achieve (sustained) competitive advantage (Barney 1991; Grant 1991; Peteraf 1993; see Section C.IV.1). A firm is said to have competitive advantage if it is (or has the potential to) persistently earn a higher rate of profit than its rivals (Grant 2010:211). In order to investigate the potential of firms to achieve competitive advantage when participating in interfirm networks ("interconnected firms"), we start from investigating the potential of firms to achieve competitive advantage if that are on their own ("sole firms").

From a very general perspective, resources "include all assets, capabilities, organizational processes, firm attributes, information, knowledge, etc. controlled by a firm that enable the firm to conceive of and implement strategies that improve its efficiency and effectiveness (Daft 1983)" (Barney 1991:101). The RBV rests upon the assumption that resources are distributed among firms heterogeneously (*resource heterogeneity*) and that resources cannot be efficiently traded among firms, which is why heterogeneity will be persistent (*resource immobility*) (see Subsection C.IV.2.1).

The potential of a specific resource to enable a firm to achieve (sustained) competitive advantage depends on four characteristics: (a) *value*, (b) *rareness*, (c) *imperfect imitability*, and (d) *non-substitutability* (Barney 1991:105ff.). If resources possess the first two characteristics, firms may achieve competitive advantage over its competitors (see Subsection C.IV.2.2). If they possess all four characteristics, firms have the potential to achieve sustained competitive advantage: a firm's competitive advantage still exists after the competitors' efforts to duplicate its value creating strategy have ceased.

The initial conceptualization of the RBV focuses on firm internal sources of competitive advantage. However, this constrained perspective is not suitable to explain the sources of competitive advantage of an interconnected firm (see Section C.IV.3). Such a firm's competitive advantage depends not only on rents generated from internal resources but also on *relational rents* generated from resources on the network-level through *idiosyncratic interfirm linkages* (Dyer & Singh 1998:660ff.). Dyer & Singh (1998) identify four potential sources of relational rents: (a) *relation-specific assets*, (b) *knowledge-sharing routines*, (c) *complementary resources and capabilities*, and (d) *effective governance*. Gulati (1999:399) refers to those resources generated and embedded on the network level as *network resources*.

Lavie (2006) proposes a conceptual model that aims to explain – on the firm level – the factors that influence competitive advantage of an interconnected firm (see Subsection C.IV.3.3). His model suggests that the resource-based competitive advantage of a focal firm, which participates in an alliance, is composed of four rent types extracted from both internal and network resources: (a) *internal rent*, (b) *appropriated relational rent*, (c) *inbound spill-over rent*, and (d) *outbound spill-over rent* (Lavie 2006:644ff.). The ability of the focal firm to appropriate the relational rent depends on the firm's (a) *relative absorptive capacity*, (b) *relative scale and scope of resources*, (c) *contractual agreement*, (d) *relative opportunistic behavior*, and (e) *relative bargaining power* (Lavie 2006:645ff.). The focal firm's ability to benefit from inbound spill-overs and to minimize outbound spillovers depends on its absorptive capacity, bargaining power, and opportunistic behavior as well as on its partners isolating mechanisms (Lavie 2006:647f.).

Organization of Interfirm Networks

As already described, the fundamental problem of organization can be conceptualized as the problem of obtaining cooperation among a collection of economic actors whose goals are only partially congruent (Ouchi 1979:833; Section C.III.1; see Section C.V.1). While TCE is concerned with *economic* organization and thus with the factors that determine which institutional arrangement (governance structure) is used to achieve cooperation in the sense of carrying out transactions, TCE provides little insight into the nature of administrative arrangements deployed *within* markets, hybrids, and hierarchies. After all, administrative controls and incentives are characterized solely by their degree of intensity. The lack of precise knowledge about the internal administrative arrangements is problematic because organizational behaviors and outcomes cannot be explained and predicted in a satisfactory manner based solely on the legal structure of interfirm networks, but a more fine-grained approach is required – as offered by organization theory – that penetrates interfirm networks beyond their legal structure and thus reveals internal administrative arrangements (Grandori 1997a:898f.; Albers et al. 2016:582ff.).

A firm's internal administrative arrangements are commonly subsumed under the construct of *organizational structure*. Following Mintzberg's (1979:2) definition, we define the organizational structure of interfirm networks as *the total sum of ways in which three or more partnering firms divide the activities that are to be performed cooperatively into distinct tasks and then achieve coordination among them.* Specifically, it is conceptualized by (1) centralization, (2) formalization, and (3) a characteristic mix of coordination mechanisms (Grandori & Soda 1995:199).

In an interfirm network, *centralization* or (decentralization) refers to the locus of authority to make decisions affecting the network, where "locus of authority" specifies the distribution of authority *between* network members (see Section C.V.2).

Provan & Kenis (2008) propose three distinct alternatives for authority to be distributed in interorganizational networks. These alternatives are distinguished along two dimensions: (1) whether authority is "shared" or "brokered" and, if brokered, (2) whether authority resides with network members or an external entity.

Shared governance (or self-governance) is characterized by a low degree of centralization because authority, at least regarding network-level decisions, is more or less symmetrically distributed among network members (Provan & Kenis 2008:234f.).

Brokered network governance is characterized by a high degree of centralization because a single entity has substantial, yet not necessarily unrestricted, authority over pivotal decisions about the whole network (Provan & Kenis 2008:234f.). Provan & Kenis (2008:235f.) further distinguish between authority being brokered by a (2a) lead organization (member-governed) or (2b) network administrative organization (either member or externally-governed). *Lead organization governance* refers to a case in which a single organization – that still provides its own services within the network – assumes the role of a network leader. *Network administrative organization (NAO) governance* refers to a case in which a separate administrative entity is established that, unlike the lead organization, does not provide its own services in the network but exclusively focuses on governing (Provan & Kenis 2008:236). The network is member-governed if the NAO is voluntarily erected by and monitored by network members; it is externally-governed if it is externally mandated (e.g. by the legislator) and network members are forced to join to the network.

Formalization pertains to two related aspects: (1) *the degree of specification* (or predefinition/prescription), which refers to extent to which organizational rules, procedures, and roles that prescribe behaviors and/or outputs are *specified*, regardless of whether they are codified in writing; and (2) *the degree of codification*, which refers to the extent to which organizational rules, procedures, and roles that prescribe behaviors and/or outputs are *codified* in writing (see Section C.V.3).

With regard to interorganizational relationships, *contracts* are the main means of formalization (Grandori & Soda 1995:197f.; van Laarhoven et al. 2000:432). A cooperation contract encodes rules and procedures, often as part of administrative and technical clauses, that regulate contingencies and involved parties' behaviors and/or outputs (Vanneste & Puranam 2010:188; Albers et al. 2016:598). The degree of interorganizational formalization is the degree to which relationships are "explicitly regulated and safeguarded by contractual provisions" (Grandori & Soda 1995:198; also see Gulati & Singh 1998:781; Gulati 1998:294ff.).

To synthesize, we understand formalization in interfirm networks as the degree to which rules and procedures that prescribe contingencies and network members' behaviors and/or outputs are specified and/or adhered to and the degree to which these rules, procedures, and contingencies are codified in a cooperation contract or IT system used in the cooperation.

Coordination mechanisms can be conceived of as processes that cause or prevent change among actors in a way that facilitates the integration of their tasks in order to achieve a specified goal (see Subsection C.V.4.1). Coordination among actors implies communication: "Without information flowing about what everybody should do and how they progress, coordination is impossible" (Groth 1999:46).

Various mechanisms for achieving coordination among members of interfirm networks are discussed in literature. These specific interfirm mechanisms largely derive from a set of few fundamental coordination mechanisms, which may be represented by the well-known taxonomy initially proposed by Mintzberg (1979) and later extended by Groth (1999) by adding a set of technology- and computer-dependent coordination mechanisms. These added mechanisms specifically leverage the capability of IT to create communication and coordination linkages between economic actors. Three characteristics are critical in this respect: (a) *immense and loss-less storage capacity*, (b) *quick and less costly communication*, and (c) *immense and less costly information processing capacity* (Groth 1999:144; also see Argyres 1999). In the following, we will briefly describe the five initial mechanisms and their computer-dependent correspondents.

In *mutual adjustment*, individuals communicate informally with each other in order to coordinate their work (Mintzberg 1980:324; see Subsection C.V.4.2). *Implicit coordination (by database)* is the corresponding computer-dependent variant. Coordination is achieved by recording relevant information in a repository and making this information accessible to the actors involved in an activity (Groth 1999:303). The repository coordinates the actors' behaviors, as any (new) piece of information has the potential to modify the behaviors of other actors who access the repository (Groth 1999:317).

In *direct supervision*, a separate individual has authority and is exclusively occupied with giving orders to other individuals to coordinate their work (see Subsection C.V.4.3). *System-supported supervision* represents the IT enabled extension of the direct supervision coordination mechanism (Groth 1999:330), and can be defined as "conscious direction of work to great depth and/or great breadth in an organization, based on information gathered and presented through computer-based systems" (Groth 1999:331). The IoT is a crucial technology to achieve system-supported supervision.

The mechanism *standardization of work processes* coordinates interdependent tasks by imposing standards on how these tasks are to be carried out (Mintzberg 1980:324; see Subsection C.V.4.4). *Programmed routines* represent a computer-dependent extension of standardized work processes. They are consciously modeled and designed routines that are "programmed *into* computer-based systems" (Groth 1999:278, italics added).

In the *standardization of skills* mechanism, work is coordinated using standard skills and knowledge that have been internalized by individuals in standardized educational or training programs, typically before the work is carried out (see Subsection C.V.4.5). *System-supported skills* is an IT induced extension of the mechanism standardization of explicit skills, and can be considered a "much improved version" of standardization of skills (Groth 1999:260). The mechanism's primary improvements focus on how skills and knowledge are "stored," how complex stored skills and knowledge can be, and how many different skills and how much knowledge can be made available simultaneously to coordinate work. Using computers to support individuals' work increases their effective *span of competence* by exploiting computers' artificial memory, artificial intelligence, embedded knowledge (Groth 1999:257ff.).

The *standardization of outputs* mechanism achieves coordination by imposing standard performance measures or specifications on the outputs of work to be done (Mintzberg 1980:324; see Subsection C.V.4.6). To our knowledge, no computer-dependent extension of this mechanism has been proposed so far. However, we believe that such an extension already exists but has not yet been related to Mintzberg's (1979). We propose *service-orientation* as a computer-dependent extension of standardization of outputs because both achieve coordination in a conceptually similar manner.

Service-orientation is a software design paradigm that comprises several design principles, which collectively govern the design of services (Erl 2008:37f.), where a service can be intuitively defined as some well-defined piece of functionality (Haas 2004). This functionality, among other things, is encoded in a *service description* (Krafzig et al. 2004). Thus, from a coordination perspective, this description specifies the output of a service. The design principles associated with service-orientation collectively aim to achieve *service composability*, which means that multiple basic services can be combined with ease of use to solve larger problems. Specifically, service-orientation includes the principle to reduce dependencies between the service description and the service's implementation. Hence, similar to standardization of outputs, service-orientation aims to achieve efficient coordination of services via explicit service descriptions without specifying how these outputs must be achieved.

Networked Horizontal LSP Cooperation

Until now, we focused on interfirm networks without any reference to logistics. We thus established the basis to investigate the phenomenon of networked horizontal LSP cooperation. Following logistics literature and the general definition of interfirm networks adopted above, we define horizontal LSP cooperation as *a group of three or more legally autonomous LSPs that operate on the same level of the logistics supply chain and work together to achieve not only their own goals but also a collective goal* (see Section C.VI.1). Scholars use rather similar dimensions to categorize alliances among LSPs than for any other interfirm network. Schmoltzi & Wallenburg's (2011) classification is instructive for the purposes of this thesis and includes six dimensions: (a) *contractual scope*, (b) *organizational scope*, (c) *functional scope*, (d) *service scope*, (e) *geographic scope*, and (f) *resource scope*.

The motives of LSPs to engage in horizontal alliances align well with the general motives identified for cooperation, such as reducing costs and increasing productivity (see Section C.VI.2). Antecedents for such alliances are rooted in certain external conditions, such as stiff global competition, increased customer expectations, and rising complexity and dynamics in today's supply chains. These factors especially pressure small and medium-sized LSPs to form alliances.

Opportunistic behavior by LSPs is pervasive in horizontal LSP alliances, even in trustworthy relationships, due to the direct competition among partnering firms (see Subsection C.VI.3.1). Horizontal LSP alliances can be governed by a variety of arrangements positioned between markets and hierarchies (see Subsection C.VI.3.2). However, due to the amplified risk of opportunistic behavior, governance structures of horizontal LSP alliances need to deter opportunism effectively and, thus, minimize transaction costs in order to achieve cooperation success. Formalized contracts, including contractual safeguards, are therefore suggested by several logisticians (see e.g. Cruijssen et al. 2007c:34; Raue & Wieland 2015; Raue & Wallenburg 2013).

Electronic logistics markets do not represent a hybrid governance structure *per se*; after all, they are markets (see Subsection C.VI.3.3). Still, electronic logistics markets are frequently established *within* alliances, for example among LSPs, among shippers, or between LSPs and shippers (see e.g. Sänger 2004:190; Wang et al. 2011:612). Hence, market-based coordination penetrates hybrid governance structures in this case (Hennart 1993). Key dimensions frequently used to classify electronic logistics markets include (a) *value creation logic*, which refers to the relative positioning of market participants in the logistics supply chain; it can be either horizontal or vertical; (b) *openness*, refers to the existence of barriers to controlling the access of buyers and/or sellers to the market; it is generally used to distinguish open and closed markets (Wang et al. 2011:612; Nandiraju & Regan 2003:4; Pankratz 2003:3f.); (c) *operator structure*, refers to the type of logistical actors that establish the market. Markets can be established by buyers and/or sellers of logistics services, or by neutral actors (Bierwirth et al. 2002:335); (d) *functionality* to support and carry out transactions during the information phase, the negotiation and contracting phase (e.g. auctions), and during the delivery phase (e.g. telematics); (e) *services traded*, which refers to the type capabilities available on the market; the vast majority of electronic logistics markets focuses on transportation; and (f) *geography covered*. Critical issues that impede the diffusion of electronic logistics markets include liquidity issues, quality issues, and functional issues, and acceptance issues (see e.g. Sänger 2004).

Following the dimensions of organizational structure for interfirm networks, horizontal LSP networks can be studied along (1) centralization, (2) characteristic mix of coordination mechanisms, and (3) formalization. Regarding *centralization*, empirical evidence shows that the distribution of authority in networked LSP alliances aligns well with the conceptual findings regarding the distribution of authority in interfirm networks (see Subsection C.VI.4.1). Thus, horizontal alliances can be governed in a shared manner by its members or autocratically through a central unit.

Regarding the *characteristics mix of coordination mechanisms* literature identifies a variety of task domains that are structured by the mechanism standardization of work processes, more specifically, by programmed routines as many of these mechanisms are implemented on IT systems used by the alliance (see Subsection C.VI.4.2). Task domains governed include (a) *partner analysis and selection*, (b) *allocation of resources and tasks*, (c) *determination and allocation of cooperation benefits*, and (d) *physical operative logistics activities*.

Formalization is a critical factor in organizing horizontal LSP networks through standardized work processes, programmed routines, and written alliance contracts (see Subsection C.VI.4.3). Although formalization can deter opportunism, it itself can become a source of conflict among partnering LSPs, thus reducing overall cooperation performance (Wallenburg & Raue 2011). Hence, formalization should only be used if formalizing organizational structure also results in other positive effects.

II. Introduction

1. Definition and Types

Interfirm networks have been a focus for management scholars since the 1980s (see e.g. Miles & Snow 1986; Thorelli 1986; Jarillo 1988). To conceptualize this phenomenon, a plethora of definitions has been proposed over the last decades (see e.g. Sydow 1992:79; Ebers 1999:4; Gulati et al. 2000:203; Provan & Kenis 2008:231). The common denominator of many of these definitions seems to be that interfirm networks consist of three or more firms that work together to achieve some purpose. Beyond this basic understanding, however, scholars have varying conceptions regarding the constituent features of such networks. Table 4 depicts frequently mentioned features and associated network types. A more comprehensive overview is provided by Sydow (2010:380). The variety of features clearly emphasizes the heterogeneity of the phenomenon subsumed under the term "interorganizational network."

Our understanding of *interfirm networks* is based on Provan & Kenis's (2008:231) definition of *interorganizational networks*, which they define as

> "groups of three or more legally autonomous organizations that work together to achieve not only their own goals but also a collective goal,"

Feature	Network Types	Sources (Examples)
Temporal stability of relationships	dynamic vs. stable	Miles & Snow (1992);
	strategic (long-term)	Jarillo (1988); Gulati et al. (2000); Tsang (1998)
Strength of relationships	loosely coupled	Lusch et al. (2010)
Distribution of authority	hierarchical vs. heterarchical (egalitarian)	Gulati et al. (2012); Alter & Hage (1993)
Relative positioning of members in value creation processes	horizontal vs. vertical	Gulati (1998); Park (1996)
Formation mode	voluntary vs. mandated	Provan & Kenis (2008); Gulati (1999)
Member autonomy	legally autonomous	Provan & Kenis (2008); Alter & Hage (1993); Jones et al. (1997)
Goals	competitive advantage	Jarillo (1988)
Formalization	formal vs. informal	Granovetter (1985:502)

Table 4.: Features Used to Distinguish Network Types (Selection)

where we assume that "organizations" are "for-profit firms." We adopt this definition, as it explicitly emphasizes two features relevant to this thesis: (1) the legal autonomy of member firms and (2) the concurrent pursuit of own and collective goals. Emphasizing member firms' *legal autonomy* is important, because this feature excludes the situation in which firms can be forced by another entity (other than a legislator) to join an interfirm network. Thus, interfirm networks are assumed to be self-initiated, and member firms are assumed to have authority over the decision of whether to join. Emphasizing member firms' *concurrent pursuit of own and collective goals* is important, because it points toward the fundamental motive for cooperation (a collective goal) and to an important source of conflict and competition among network member firms. Incongruence between individual and collective goals may eventually cause interfirm networks to disintegrate. Conflict may be particularly strong if member firms are proximate or distant competitors, as is the case in horizontal networks.

Interfirm networks are commonly illustrated using the network metaphor, where the nodes of the network represent member firms (composition), and the edges symbolize the relationships established between these firms (structure). These relationships can refer to various types of links and flows, such as information, materials, financial resources, and services (Provan et al. 2007:482; van de Ven 1976:25).

2. Motives and Antecedents for the Formation of Interfirm Networks

Although firms' motives for erecting or joining interfirm networks can be quite diverse, they can ultimately be divided into increasing profitability by (1) increasing revenue and/or (2) reducing costs (Ebers 1999:6f.). Specific motives to *increase revenue* include (a) gaining access to complementary resources or capabilities (e.g. physical assets, information, knowledge, technology) in order to improve market access (see e.g. Hennart 1988:363; Gulati et al. 2000:203) or time-to-market (see e.g. Powell 1987:71), and (b) increasing market power or blocking competition through collusion or depriving competitors of possibly valuable partners (see e.g. Kogut 1988:322; Porter & Fuller 1986:322ff.). Specific motives to *reduce costs* include (a) achieving economies of scale and scope by joining similar resources (see e.g. Gulati et al. 2000:203; Hennart 1988:363 Contractor et al. 1988:9ff.; Porter & Fuller 1986:322); (b) economizing on governance and coordination costs, as under certain conditions, networks may mediate exchanges at lower costs than markets or hierarchies (see e.g. Oliver 1990:45; Hennart 1991:484; Jarillo 1988:39); and (c) reducing financial, political, or other risks by spreading them across multiple organizations (see e.g. Contractor et al. 1988:9ff.; Hennart 1988:364; Porter & Fuller 1986:325).

Although firm motives can contribute to understanding the formation of interfirm networks, motives, considered in isolation, provide only weak guidance for explaining their formation or specific forms thereof because the motives can equally encourage firms to adjust their net value added by either integrating vertically or outsourcing certain activities (Ebers 1999:7). By contrast, *antecedents*, or *contingencies*, are more important for explaining the formation of interfirm networks; they refer to conditions that facilitate or constrain the emergence of interorganizational forms. Through a comprehensive literature review, Oliver & Ebers (1998) identify 17 theories used by organizational scholars to investigate the phenomenon of interorganizational networks. Consolidating antecedents suggested by different theories, these authors compile a list of 13 antecedents including (im-)material resources, dependence, network position, resource munificence, goal congruence, market constraints, stability, conflict, organizational density, asset specificity, trust, and opportunism (Oliver & Ebers 1998:556; also see Grandori & Soda 1995:185ff.). Comparing motives with antecedents reveals significant differences, but organizations' resource endowments rank high in both.

3. Selecting Theoretical Approaches to Study Interfirm Networks

Scholars investigate interfirm networks via various theories. As mentioned, Oliver & Ebers (1998:556) identify 17 theories, including resource dependence, network, institutional, strategy, industrial organization economics, agency, transaction cost, and contingency theory (also see Das & Teng 2000:34; Grandori & Soda 1995:185ff.; Sydow 1992:168ff.).

Given this variety of theoretical approaches, Borys & Jemison (1989:235ff.) conclude that hybrid organizational forms, such as interfirm networks are "theoretical orphans" because there is no single theory that is able to address and integrate the multiplicity of issues they raise. Therefore, research on interorganizational networks may at first appear highly fragmented. As Nohria (1992:3) remarks, the network literature may be perceived as a "terminological jungle in which any newcomer may plant a tree." In fact, this thesis also intends to plant a tree in this already dense jungle; more specifically, in the particular area where IT savvy logisticians dwell. Yet, Oliver & Ebers (1998:565) surprisingly find, that (a) there is a far greater convergence in this apparently highly fragmented field of research as a limited number of theories are constantly reappearing and (b) research on interorganizational networks can be "segmented into a fairly limited number of distinct and theoretically meaningful configurations of perspectives." They argue that each configuration concentrates on a distinct level of analysis and certain types of interorganizational ties and is based on a particular set of theories, antecedents, and outcomes, as shown in Table 5.

For the purposes of this thesis, reviewing the contribution of each theory or of each configuration of research perspectives is of little help and beyond its scope. Instead, we select one configuration, and thus a distinct set of theories, that promises to provide the most suitable theoretical basis for explaining the formation of and studying interfirm networks in relation to our research objective – designing a logistics system in accordance with principles and concepts of cloud computing. To this end, we select *institutional economics and strategy* because this configuration

> "focuses on how to forge networking relations, and how to organize them, so that actors gain access to, and best utilize, such resources which will reduce their dependence or will otherwise improve their competitive position. This research is concerned with resources and their characteristics and it often conceptualizes network ties in organizational, contractual or ownership terms" (Oliver & Ebers 1998:569).

Our theoretical basis consists of four theoretical approaches: (1) *TCE* which is associated with institutional economics; (2) market mechanisms, which are rooted in the field of game theory, but essentially provide deeper insights into market coordination and are directly related to TCE; (3) *RBV*, which is associated with strategic management; and (4) organizational theory.

This selection of theories is useful, as these theories are conceptually complementary and, therefore, can collectively provide comprehensive insight into the phenomenon of interfirm networks. As Barney et al. (2001:626) remarks in this respect: "RBV and TCE are viewed as complementary because the former is a theory of firm rents whereas the latter is a theory of the existence of the firm." More specifically, RBV takes the existence of firms as a given

Configuration Facet	(1) Social Network	(2) Power and Control	(3) Institutionalism	(4) Institutional Economics and Strategy
Theories	network	political power, resource dependence, exchange	institutional	transaction cost, strategy
Ties	political, horizontal	political	social	dyadic, vertical, ownership, contractual
Levels of Analysis	individual	region/industry	societal, groups of individuals	organizational
Antecedents	network position	goal congruence, dependence, conflict	organizational density, trust	market constraints, material resources, stability, resource munificence, asset specificity
Outcomes	–	power/control, centrality, stability, political participation	density, conflict, legitimacy, extinction, persistence, commitment, trust, size	success, take-over, cost/price, make-or-buy, opportunism

Table 5.: Four Substantive Configurations of Research Perspectives (adopted from Oliver & Ebers 1998:564)

and is concerned with explaining differential firm performance and how firms can achieve superior performance *vis-à-vis* their competitors, while TCE aims to find the most cost-efficient institutional arrangement for a given economic transaction and thus treats the existence of firms as a dependent variable. Organizational theory, in an interfirm context, is concerned with the division and coordination of labor among cooperating firms, such that they can attain collective goals effectively. It thus penetrates deeper into interfirm networks than RBV and TCE by unveiling their internal administrative apparatus. The following three chapters introduce TCE, RBV, and key dimensions of the organizational structure of interfirm networks.

III. Transaction Cost Economics and Market Mechanisms

1. The Problem of Economic Organization and the Formation of Alliances

The fundamental problem of organization is the problem of achieving cooperation among a collection of economic actors whose goals are only partially congruent (Ouchi 1979:833). In a capitalistic society, resources are owned by economic actors, such as individuals, households, and firms, and they are allocated through (capitalistic) economic institutions rather than the government (Alchian & Demsetz 1972:777). Economic institutions enable cooperation because they facilitate contract-based transactions among actors. These institutions are commonly referred to as *governance structures*, and can be categorized into markets, hybrids, or hierarchies (Williamson 1988:73). The problem of "economic organization" is therefore fundamentally concerned with the problem of explaining what type of transaction is organized through what type of governance structure. Transactions thus become the unit of analysis (Williamson 1981:548).

TCE is rooted in Coase's (1937) seminal contribution, and contends that economizing on transaction costs is central to the study of governance structure choice (Williamson 1988:548). More specifically, TCE posits that economizing on transaction costs can be achieved by matching transactions with governance structures in a discriminating way based on the attributes of both (Williamson 1988:548; Williamson 1991). TCE thus treats governance structure choice as a dependent variable. Interfirm networks are subsumed under the "hybrid" governance structure (Leiblein 2003:941). Hence, TCE essentially explains the formation of interfirm networks as a function of transaction attributes.

2. Types and Determinants of Transaction Costs

2.1. Types of Transaction Costs

Transaction costs represent the "costs of running the economic system" (Arrow 1969:48), and they can be structured into "search and information costs, bargaining and decision costs, policing and enforcement costs" (Dahlman 1979:148). As TCE interprets the problem of economic organization as a problem of contract-based transactions, transaction costs specifically include costs of "drafting, negotiating, and safeguarding" contracts (*ex ante* transaction costs) as well as costs of maladaptations, haggling incurred if misalignments need to be resolved bilaterally, set up and running of structures for dispute resolution (including courts or other forums), and bonding to secure commitments (*ex post* costs) (Williamson 1985:20f.).

TCE suggests that transaction costs are critically determined by the characteristics of economic actors involved in the exchange and by the attributes of the exchange. Both are investigated in the two following subsections.

2.2. Characterizing Economic Actors: Bounded Rationality and Opportunism

TCE assumes the behavior of economic actors to be characterized by (1) bounded rationality, and (2) opportunism. These assumptions have critical implications for contract-based exchanges because, without bounded rationality, all transactions could be efficiently handled through complete contingent contracts (see e.g. Williamson 1981:553). In addition, without opportunism, all contract adaptations could be efficiently handled using the "general clause," in which transacting parties promise to behave responsibly and without self-interest whenever adaptations are required to respond to disturbances (see e.g. Williamson 1979:241; Williamson 2002:188; Williamson 1985:48).

Bounded rationality means that actors are assumed to behave in a way that is "*intendedly* rational, but only *limitedly* so" (Simon 1957:xxiv, italics original) because of cognitive limitations regarding their ability to formulate and solve complex problems and to process (receive, store, retrieve, transmit) information (Simon 1957). However, apart from these limitations, actors remain "intendedly rational" (Williamson 1981:553). As a result, complete contingent contracting is not feasible, but, in fact, all complex contracts are unavoidably incomplete, thus requiring sequential adaptations in case of unanticipated disturbances due to "gaps, errors, and omissions in the original contract" (Williamson 2002:174).

Opportunism denotes the strongest form of self-interested behavior and can be defined as "seeking self-interest with guile" (Williamson 1985:47; Williamson 1979:234). Opportunism manifests in actors disclosing only incomplete or distorted information in a calculated manner with the objective of misleading or otherwise confusing, thus causing real or contrived information asymmetries (Williamson 1985:47). As a result of opportunism, transacting parties strategically bargain over any incremental gain whenever one makes a proposal to adapt the initial – incomplete – contract (Williamson 1979:242).

2.3. Characterizing Transactions: Asset Specificity, Uncertainty, Frequency, and Complexity of Product Descriptions

The crucial dimensions for characterizing transactions are (1) asset specificity; (2) uncertainty; (3) frequency (Williamson 1979:239ff.), where asset specificity is considered the most important among these three (Williamson 1981:555); and (4) complexity of product descriptions, which we add as a fourth dimension as it is relevant to the structural choice of electronic markets and hierarchies (Malone et al. 1987).

Asset specificity refers to the extent to which idiosyncratic (non-marketable) investments are required in support of a certain transaction (Williamson 1979:239f.) to achieve "least cost supply" (Williamson 1981:555). Problems of non-marketability arise when the "*specific identity*" of transacting parties has "cost-bearing consequences" (Williamson 1979:239f., italics original), which means that asset specificity has "reference to the degree to which an asset can be redeployed to alternative uses and by alternative users without sacrifice of productive value" (Williamson 1988:70). Transaction specific investments create "lock-in" for suppliers and buyers: Suppliers cannot sell to other buyers, as this would reduce the value of their investments, and buyers cannot change suppliers, as specialized investments would reduce production costs (Williamson 1979:240; Williamson 1981:555). Given bounded rationality and opportunism, the presence of transaction specific investments creates problems if contracts are disturbed and need to be adapted because both suppliers and buyers strategically bargain over (potential) marginal gains, thus incurring haggling costs which may even offset potential gains (Williamson 1979:241f.). Therefore, to enable efficient adaptations in the presence of transaction specific investments, governance structures need to employ appropriate safeguards to attenuate opportunistic behavior (Williamson 1979:242). Furthermore, potential savings in production costs from specialized assets must be traded off with the strategic hazards they bring about when contracts are drafted and negotiated (Williamson 1985:54).

Uncertainty refers to disturbances to which transactions are exposed (Williamson 1991:281). Disturbances can originate in various (environmental) domains in which a transaction is embedded, especially including uncertainty resulting from the strategic behavior of transaction parties attributable to opportunism: *behavioral uncertainty* (Williamson 1985:57f.). Given contracts that are incomplete due to bounded rationality, uncertainty triggers the need for "adaptive, sequential decision-making" (Williamson 1985:56f.) because disturbances cause the misalignment of contract-based transactions.

The *frequency* "with which transactions recur" can be categorized as one-time, occasional, or recurrent (Williamson 1979:239, 246). Transaction recurrence is important because more frequent transactions justify erecting more specialized governance structures, because the fixed costs of establishing these structures can be more easily recovered following the logic of economies of scale (Williamson 1985:60f.). Thus, with regard to frequency, economizing on transaction costs requires a tradeoff between economies of scope by tailoring the structure to the needs of the transactions and economies of scale by tailoring the structure only to the extent that can be economically justified from a scale perspective (Williamson 1985:60f.).

Complexity of product descriptions refers to the amount of information required to describe a product's attributes in enough detail to allow potential buyers to make a selection (Malone et al. 1987:486).

3. Governance Structures and Structural Choice

3.1. Governance Structures: Markets, Hybrids, and Hierarchies

A governance structure can be defined as "the institutional framework in which contracts are initiated, negotiated, monitored, adapted, enforced, and terminated" (Palay 1984:265). A governance structure can therefore be considered a "mode of organizing transactions" (Williamson & Ouchi 1981). Three governance modes are distinguished by TCE: (1) market, (2) hybrid, and (3) hierarchy, where markets and hierarchies represent the poles of a continuum and hybrid is an intermediate mode between these extremes (Williamson 1991:280f.). As depicted in Table 6, the modes differ across five attributes, which can be categorized into three dimensions: (a) contract law, (b) adaptation (autonomous vs. cooperative), and (c) instruments (incentive intensity vs. administrative controls) (Williamson 1991:277ff.).

Markets, one extreme of the continuum, are based on *classical contract law* which is characterized by discreteness and complete "presentiation" (Williamson 1979:236), where "presentiation" means that contractual obligations (e.g. delivery of a product or service) are either fulfilled at the time and place of contract closure or, if future fulfillment is agreed to, that "the future has [...] been brought effectively into the present" (MacNeil 1974:589) by actions carried out at contract closure. Thus, classical contract law matches the "ideal" market transaction in economics and law (Williamson 1979:236). To realize discreteness and presentiation, classical contract law (a) considers the identities of transacting parties to be irrelevant as no further dependency between transacting parties is established beyond the current transaction, and periodical contracting between buyers and sellers only results from coincidentally meeting again in the spot market due to congruent preferences regarding performance and price; (b) includes carefully delimitated agreements interpreted in a very legalistic way, where formal terms dominate less formal terms when transacting parties appeal to courts for dispute resolution; and (c) narrowly prescribes remedies such that consequences due to nonperformance are relatively predictable at contract closure and not open-ended should the initial presentiation fail to materialize (Williamson 1979:236f.; Williamson 1985:69; Williamson 1991:271). Markets facilitate *autonomous adaptations* through changes in *prices* – if disturbances are such that price changes can serve as sufficient statistics to convey all relevant informa-

Attributes	Governance Structure Market	Hybrid	Hierarchy
Contract law	++	+	0
Instruments			
- Incentive intensity	++	+	0
- Administrative controls	0	+	++
Adaptation			
- Autonomous	++	+	0
- Cooperative	0	+	++

Table 6.: Distinguishing Attributes of Market, Hybrid, and Hierarchy
++ = strong; + = semi-strong; 0 = weak
(adopted with minor changes from Williamson 1991:281)

tion about the objects to be exchanged in a single numerical value, buyers and sellers can adapt independently of each other to maximize their respective utility (Williamson 1991:278). Finally, markets feature *high-powered incentives* which means that actions have immediate monetary consequences (Williamson 1991:279); for example, immediate consequences of non-performance are carefully spelled out in a contract from the outset. This property is commonly referred to as *output pricing.* As a result, there is no need to install *administrative controls* to control the behaviors of parties.

Hierarchies, the other extreme of the continuum, are supported by the *contract law of forbearance* (Williamson 1991:274ff.), which aims to facilitate long-term economic exchanges for which "the fiction of discreteness is fully displaced as the relation takes on the properties of 'a minisociety with a vast array of norms beyond those centered on the exchange and its immediate processes'" (MacNeil 1977:901 as cited in Williamson 1979:238). The distinguishing feature of forbearance is that bilateral adaptations are effected through *fiat.* While several scholars argue that fiat derives from employment contracts, Williamson (1991:274) proposes a complementary explanation, namely that fiat originates in the denial of transacting parties' access to courts, requiring parties to instead resolve their conflicts internally, and that "hierarchy is its own court of ultimate appeal." Among the three governance structures, hierarchies are considered the most elastic and adaptive (Williamson 1991:274) due to the combination of fiat and *low-powered incentives.* Thus, it is possible to elicit greater cooperation for rapid *cooperative adaptations*, as managers and employees receive the same compensation "whether they 'do this' or 'do that'" (Williamson 1991:275). This property is commonly referred to as *input pricing.* To control unwanted side effects of low-powered incentives (e.g. shirking) and to promote desired behavior, hierarchies rely extensively on *administrative controls.*

Hybrids, positioned in between markets and hierarchies, are based on *neoclassical con-*

tract law (Williamson 1991:271), which aims to facilitate long-term transactions under the conditions of uncertainty, for which complete presentiation is not possible or economically prohibitive (Williamson 1979:237). Specifically, it refers to long-term contracts in which the transaction parties maintain autonomy but become "bilaterally dependent to a nontrivial degree" and in which the identities of parties only matter in the event of premature contract termination or lasting maladaptation (Williamson 1991:271). For reasons of bounded rationality, these long-term contracts are inevitably incomplete, but they include an additional administrative apparatus to facilitate sequential adaptations to disturbances and changing conditions (MacNeil 1977:865 as cited in Williamson 1979:237). Hence, neoclassical contracts resemble "framework" contracts (Williamson 1991:272). Adaptations can be achieved either in an *autonomous* manner as in markets or, if the nature of the disturbance requires, in a *coordinated* manner through *bargaining*. This works because, on the one hand, transacting parties have an interest in adapting quickly to avoid further losses from the current maladaptations to environmental conditions, and, on the other hand, they have an interest in achieving the most favorable allocation of potential residual gains (Williamson 1991:278f.). In case of conflicts, third party arbitration is initially preferred over litigation as transacting parties aim for contract continuation and accommodation rather than termination (Williamson 1979:237f.). However, if consensus cannot be reached, especially if maladaptations become highly consequential for contracting parties and autonomous ownership creates incentives to defect, neoclassical contracting reverts to a more legalistic regime: Parties appeal to courts (Williamson 1991:273). In sum, neoclassical contracting refers to rather elastic agreements that support continuity and efficient adaptations, albeit only up to a certain degree of disturbance (Williamson 1991:271f.). Positioned between markets and hierarchies, hybrids are characterized by incentives and administrative controls of medium intensity.

Because we are focusing on cooperative interfirm networks, providing some further insights into *hybrid* governance structures is instructive. Literature suggests that hybrids can take on a variety of forms (see e.g. Hennart 1993; Park 1996). For example, Ebers & Oerlemans (2016) identify three configurations of hybrid modes by means of latent class analysis using data from buyer-supplier relationships in the German construction industry. Although not directly rooted in TCE, Das & Teng (2001) propose a typology consisting of types of strategic alliance structures, depicted in Table 7, that can provide deeper insights into the nature of hybrids for the purposes of this thesis. Four types of hybrids are identified: (1) equity joint ventures, (2) minority equity alliances, (3) bilateral contract-based alliances, and (4) unilateral contract-based alliances.

Equity joint ventures involve the creation of a separate legal entity to jointly pursue activities within an alliance, which is why it is considered the most hierarchical alliance type, closest to hierarchy itself (Das & Teng 2001:18f.). Governance costs of joint ventures

Distinguishing characteristics	**Alliances structures**			
	Unilateral contract-based alliances	Bilateral contract-based alliances	Minority equity alliances	Equity joint ventures
Ownership structure	No shared ownership involved	No shared ownership involved	One-way or cross-equity ownership	Joint equity
Degree of interfirm integration	Light: working separately according to contracts	Moderate: working jointly for a common goal	Substantial: equity participation	High: working in one entity
Control mechanism	Contract law	Reciprocity	Interest alignment through equity stake	Hierarchical
Duration of alliances	Short- to medium-term	Short- to medium-term	Medium- to long-term	Medium- to long-term
Unplanned alliance termination	Fairly easy: end the contract	Fairly difficult: organizational rearrangement (e.g., of alliance personnel)	Difficult: selling equity stake to the partner or third parties	Very difficult: joint venture to be taken over by one partner or third parties

Table 7.: Distinguishing Characteristics of Strategic Alliance Structures (adopted from Das & Teng 2001:17)

are high (Das & Teng 2001:19). Investments are required for setting up joint ventures in terms of legal entity, management, etc. (Das & Teng 2001:19), and these investments cannot be easily recovered because joint ventures are difficult to terminate and entail high exit costs due to the deep integration of partnering firms (Das & Teng 2001:19). In other words, joint ventures create significant bilateral dependencies. Although joint ownership aligns incentives to achieve overall performance, adaptations (e.g. in terms of strategy) can only be made bilaterally (Das & Teng 2001:19).

Minority equity alliances require partnering firms to mutually take equity positions ("equity hostages") without creating a new legal entity (Das & Teng 2001:19f.). They also tend to incur high governance costs (Das & Teng 2001:20). Exiting such alliances is difficult, as partners have to decide on the procedures to withdraw equity positions (Das & Teng 2001:20), so significant bilateral dependencies are created between partners. Hence, adaptations can only be made by consent, which may be difficult due to shared ownership but split control (Das & Teng 2001:20). However, mutual equity positions partially align partnering firms' interest (they function as a distribution mechanism of alliance gains) and thus reduce strategic bargaining over marginal payoffs (Das & Teng 2001:20).

Bilateral contract-based alliances do not involve any equity or ownership transfer among partnering firms (Das & Teng 2001:20). Such alliances can be terminated more easily than equity-based alliances, which is why partnering firms maintain higher individual strategic flexibility (Das & Teng 2001:21). In other words, less strong bilateral dependencies are created among partners here than in equity-based alliances. However, due to the lack of equity involvement, partnering firms' incentives are less aligned, and bargaining over residual gains is more intense (Das & Teng 2001:21). Adaptations not covered by the provisions made in the cooperation contract "rely heavily on goodwill, reputation, and voluntary cooperation from independent firms" (Das & Teng 2001:21). Moreover, developing and implementing contractual safeguards incurs costs and may discourage the cooperative spirit (Das & Teng 2001:21).

Unilateral contract-based alliances involve a clear exchange of property rights and are characterized by the limited engagement of partnering firms (Das & Teng 2001:21). Such alliances represent the "loosest" alliance form, as they can be terminated with ease (Das & Teng 2001:22). Hence, only limited dependencies are created between partnering firms, which is why they maintain the highest degree of flexibility (Das & Teng 2001:22).

3.2. Structural Choice: Matching Transaction Characteristics to Governance Structures

Focusing on transactions as the basic unit of analysis, TCE maintains that different types of transactions can be coordinated cost-efficiently by using different types of governance structures. This logic is summarized in the "discriminating alignment hypothesis":

> "The discriminating alignment hypothesis to which transaction-cost economics owes much of its predictive content holds that transactions, which differ in their attributes, are aligned with governance structures, which differ in their costs and competencies, in a discriminating (*mainly, transaction-cost-economizing*) way." (Williamson 1991:277, italics original)

Although Williamson (1979:245) initially maintains that transaction organization depends on economizing on the *sum* of production- and transaction costs, his subsequent analysis mainly focuses on economizing on transaction costs and especially on the impact asset specificity has on these costs (Williamson 1991:281ff.). Asset specificity plays a pivotal role in his analysis because any increase in it results in an increasing degree of bilateral dependency among transacting parties, thus posing contracting hazards (Williamson 1991:281f.). Reducing the analysis to transaction cost considerations means that "neither the revenue consequences nor the production-cost" are taken into consideration (Williamson 1991:282). Williamson (1991:282) acknowledges that this simplifies the analysis, but he also argues that asset specificity raises transaction costs for any type of governance structure and that "added specificity is warranted only if these added governance costs are more than offset by production cost savings and/or increased revenues." Hence, asset specificity assumes a central role in explaining and predicting cost-efficient governance structure choice.

Figure 5 depicts the governance costs for markets ($M(k)$), hybrids ($X(k)$), and hierarchies ($H(k)$) as a function of asset specificity (k). Governance costs rise for an increasing degree of asset specificity as bilateral dependency increases in asset specificity (Williamson 1991:282). For a low degree of asset specificity ($k < k_1$) markets result in the lowest cost; for medium specificity ($k_1 < k < k_2$), hybrids do; and for high specificity ($k > k_2$), hierarchies do. Williamson (1991:282ff.) provides the following rationale. Assets of low specificity do not create bilateral dependencies between transacting parties. In this situation, markets have lowest governance costs, and transacting parties can adapt autonomously to changes in market prices. Hybrids and hierarchies are at a comparative disadvantage, as additional administrative controls would incur additional governance costs without any benefit (Williamson 1991:282); opportunism is already curbed by high-powered incentives.

If transactions involve assets of medium specificity, transacting parties become dependent on each other to some nontrivial degree. In such situations, hybrids incur the lowest governance costs. Transacting parties can either adapt autonomously or in a coordinated fashion depending on the nature of the disturbance. Markets are at a comparative disadvantage, as high-powered incentives prevent efficient, coordinated adaptations due to transacting parties' strategic bargaining. Hierarchies are equally at a comparative disadvantage, as the lack of incentives prevents autonomous adaptations, while at the same time imposes unnecessary bureaucratic costs.

Finally, if transactions involve highly specific assets, transacting parties become bilaterally dependent on each other. Hierarchies incur the lowest governance costs because disturbances increasingly become of the kind that require coordinated responses, due to

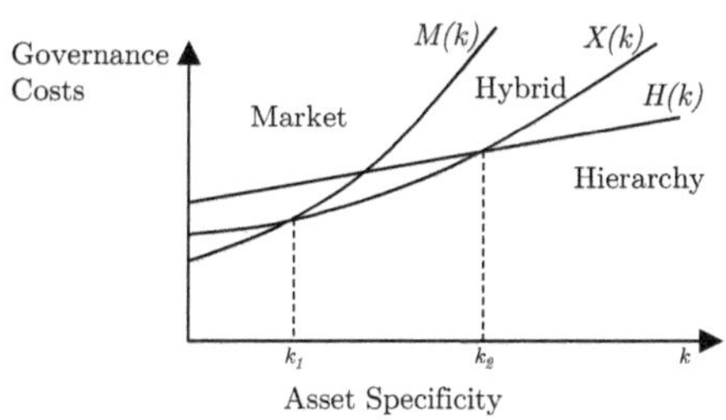

Figure 5.: Governance Costs as a Function of Asset Specificity (Williamson 1991:284)

the high degree of asset specificity (Williamson 1991:282). Efficient adaptations become possible by eliciting cooperation from sellers (employees) by eliminating incentives. However, if actions lack direct monetary consequences, sellers opportunistically reduce their efforts (shirk), which is why buyers must establish comprehensive administrative controls to achieve the desired behavior.

While the previous considerations focus exclusively on the role of asset specificity, Figure 6 depicts the most cost-efficient governance structure as a function of asset specificity and uncertainty, which is measured in the frequency of disturbances. Williamson (1991:291f.) argues that for an increasing disturbance frequency, governance costs for any type of structure increase, but hybrid governance costs are particularly susceptible to an increase in disturbance frequency because adaptations require mutual consent, which takes time. Hence, if frequency increases, hybrids may not be "fast enough" to adapt before the next disturbance occurs. In the event of a high disturbance frequency, either markets or hierarchies offer the lowest costs, as adaptations can be made more quickly, autonomously in markets and through fiat in hierarchies (Williamson 1991:291f.)

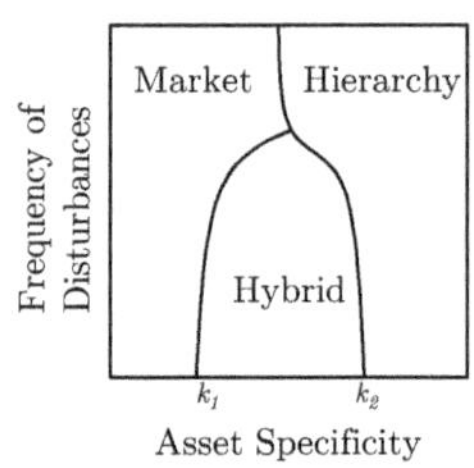

Figure 6.: Changes in Organization Form as a Response to Changes in Frequency (Williamson 1991:292)

4. The Role of Information Technology in Economic Organization

The role of IT in economic organization has been investigated by several scholars (see e.g. Clemons & Row 1993; Bakos 1991; Bakos 1997; Bakos 1998). In this section, however, we focus on the article by Malone et al. (1987), which can be considered the first comprehensive contribution to this area, and their findings have direct relevance for this thesis.

4.1. Impact on Governance Structures: Electronic Markets and Electronic Hierarchies

Malone et al. (1987) distinguish between two types of "electronic" governance structures (1) electronic markets and (2) electronic hierarchies. *Electronic markets* (or marketplaces) are markets where interactions between buyers and sellers are enabled and supported by IT. As Bakos (1991:296, italics original) defines it

> "[a]n *electronic marketplace* (or electronic market system) is an interorganizational information system that allows the participating buyers and sellers to exchange information about prices and product offerings."

Electronic markets exploit the *electronic brokerage effect*, which suggests that a central database that connects many buyers and suppliers is equivalent to a broker who is in contact with a large number of buyers and suppliers (Malone et al. 1987:488).

With the recent proliferation of the internet, electronic markets have become a ubiquitous phenomenon that has attracted the attention of scholars (see e.g. Hagiu 2009; Hagiu & Spulber 2013; Hagiu 2014; Gawer 2014). Examples include Google's Play Store, eBay, and Amazon's Marketplace. However, rather than using the term "electronic markets," scholars more commonly refer to them as "platforms" or as "two-sided" or "multi-sided markets:"

> "Platforms fundamentally create value by acting as *conduits* between two (or more) categories of consumers who would not have been able to connect or transact without the platform. Platforms create value by coordinating these groups of consumers and in the economic view this coordination is effected through pricing" (Gawer 2014:1241, italics original).

Research particularly focuses on the creation of *network effects*, which means that users benefit from an increasing number of other users present on the same platform (see e.g. Gawer 2014:1241).

Malone et al. (1987:490ff.) identify different types of actors' motives for establishing electronic markets. Sellers may do so to tie buyers more tightly to them by trying to reduce the number of alternative providers buyers consider in their procurement process. Conversely, buyers may establish electronic markets to gain access to a larger number of alternative sellers and compare them with ease. Intermediaries such as financial firms may aim to increase revenue by processing a larger volume of transactions. Finally, IT vendors may establish electronic markets to increase their revenue from IT infrastructure and the software they provide. By introducing electronic markets, these actors may become *market makers* (Malone et al. 1987:484).

Electronic hierarchies establish a situation in which either buyers or sellers

> "use information technology to create joint, interpenetrating processes at the interface between value-added stages" (Malone et al. 1987:488).

Electronic hierarchies take advantage of the *electronic integration effect* which means that IT is not only used to increase communication speed, but also to change the underlying processes that create and use this information, leading to tighter coupling between buyers and sellers (Malone et al. 1987:488). Databases are at the core of electronic hierarchies (similar to electronic markets) because they integrate processes across organizational boundaries by enabling continuous and convenient information sharing via the internet (Malone et al. 1987:495).

Malone et al. (1987:493f.) suggest that electronic hierarchies have the potential to particularly improve product development and product distribution, where the former includes a design database to shorten product life cycles, and the latter includes procurement and inventory databases as used in JIT systems to reduce inventory costs. In fact, shortening product life cycles, introducing lean logistics concepts, such as JIT, and integrating and synchronizing value adding processes across organizational boundaries are all actions that have already been observed in the meso-logistics system environment as a means of raising revenue and decreasing costs (see above Section B.V.2).

4.2. Impact on Structural Choice

While the analysis of Williamson (1991) above primarily focuses on transaction costs associated with making adaptations to align with disturbances, Malone et al. (1987) focus on how the use of IT impacts "coordination costs" and thus governance choice. Coordination costs comprise "the transaction (or governance) costs of all the information processing necessary to coordinate the work of people and machines that perform the primary processes" (Malone et al. 1987:485). In markets, they include costs for searching and selecting transacting parties, negotiating contracts, and making financial settlements. In hierarchies, they include costs for managerial decision-making, accounting, planning, and controlling processes. Coordination costs are comparatively high in markets and low in hierarchies: Markets have a costs disadvantage over hierarchies because buyers need to identify and screen multiple alternative suppliers, while a "single" supplier is already predefined in hierarchies, so no search and information costs are incurred (Malone et al. 1987:485). Because of information processing and transmitting costs, products with more complex descriptions are more likely to be obtained through hierarchies than markets (Malone et al. 1987:486f.).

The introduction of IT reduces coordination costs by reducing the time and cost of processing and transmitting information (Malone et al. 1987:484; also see Groth 1999:144; Argyres 1999). Malone et al. (1987:488) refer to the reduction of time and costs as the *communication effect* of IT. While this effect makes both markets and hierarchies more efficient, reducing coordinating costs

> "lead[s] to an overall shift toward proportionately more market coordination" (Malone et al. 1987:484).

They argue that this is because markets benefit comparatively more from lowered (unit) coordination costs than hierarchies do because coordination costs in markets represent a larger share of total costs (sum of coordination and production costs). Specifically, IT reduces the search and information costs in markets by creating transparency into potential buyers and sellers (Malone et al. 1987:489; also see Bakos 1991; Bakos 1997; Bakos 1998).

The role of IT also impacts the types of products and services exchanged through markets. Figure 7 depicts governance choice as a function of the complexity of product descriptions and asset specificity. Regarding the complexity of product descriptions, Malone et al. (1987:489) argue that increasing information processing and communication capability, reduces the costs for handling complex product descriptions, thus shifting the horizontal line up:

> "Databases and high-bandwidth electronic communication can handle and communicate complex, multidimensional product descriptions much more readily than can traditional modes of communication."

As a result, products and services with complex descriptions, previously exchanged via hierarchies, may now be exchanged via markets. Regarding asset specificity, Malone et al. (1987:489f.) argue that IT can support the introduction of new "flexible manufacturing technology," which incurs fewer costs when switching production lines to alternative products, thus making some transaction specific assets, less specific, which moves the vertical

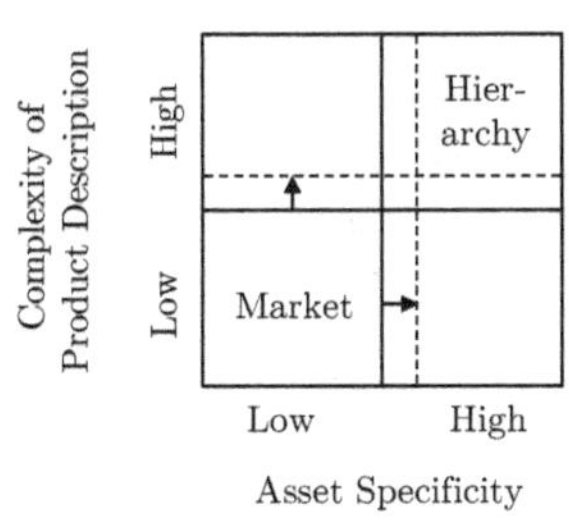

Figure 7.: Product Attributes Affect Forms of Organization (adopted from Malone et al. 1987:487)

line to the right. In sum, reducing coordination costs moves transactions previously organized in hierarchies to markets, especially those that involve products and services with complex descriptions.

5. Market Mechanisms – Game Theory and Mechanism Design

5.1. Preliminary Considerations

Market mechanisms such as auctions achieve coordination through *prices*. Market mechanisms assume an increasingly important role in structuring economic decisions and exchanges that involve many self-interested actors. The rise in importance can be attributed to recent IT innovations that allow connecting many actors and facilitating exchanges among them via new *electronic markets* (Narahari et al. 2009:1). Application areas for market mechanisms are vast and include solving network routing problems with self-interested routers and packets, solving logistics problems with task allocation across multiple self-interested LSPs, supply chain coordination among competing manufacturers (Parkes 2001:1ff.), and allocating resources in grid/cloud computing systems (Narahari et al. 2009:1).

We start this section with a brief introduction to game theory that lays the foundation for mechanism design. Note that game theory and mechanism design usually use rather complex mathematical notations. However, in this thesis, we try to minimize mathematical notations and instead introduce and describe the fundamental concepts of mechanism design verbally. We only revert to mathematical notation if it can provide more concise insight. This largely non-mathematical consideration is sufficient for the aims pursued in this thesis. Mathematical proofs are not provided, but the interested reader is kindly referred to the cited literature.

5.2. A Brief Introduction to Game Theory

A game is a formal mathematical representation of a situation in which two or more decision makers – hereafter referred to as *agents* – interact in a setting of *strategic interdependence*, which means that (1) each agent's welfare is a function of his or her own actions and of other agents' actions, and (2) the actions that may be best for an agent may depend on what each agent expects the other agents to do (Mas-Colell et al. 1995:219). A situation of strategic interdependence can be described through four basic elements (Mas-Colell et al. 1995:219f.):

(i)	*The agents*	Who is involved?
(ii)	*The rules*	Who moves when? What do they know when they move? What can they do?
(iii)	*The outcomes*	For each possible set of actions taken by the agents, what is the outcome of the game?
(iv)	*The payoffs*	What are the agents' preferences (i.e., utility functions) for the possible outcomes?

Game theory is concerned with the mathematical analysis of such situations. Myerson (1991:1) defines it as "the study of mathematical models of conflict and cooperation between intelligent rational" agents. Two pairs of terms in this definition require further attention: (1) *rational* and *intelligent*; and (2) *conflict* and *cooperation.*

Game theory makes two fundamental assumptions about the characteristics of agents: They are *rational* and *intelligent.* An agent is rational if he takes decisions consistently in pursuit of his own goal, which we assume to be the maximization of the expected value of his own payoff measured in some utility scale (Myerson 1991:2). This notion of rationality implies that agents are *self-interested* (Narahari et al. 2009:15). An agent is intelligent if she possesses the same knowledge and information about the game and can make the same inferences about the situation as a game theorist can make (Myerson 1991:4). In this sense, intelligence implies that agents have the capacity to behave *strategically*: Each agent fully takes into account her knowledge or expectations of the other agents' actions when devising her best response (Narahari et al. 2009:16). Comparing these assumptions with TCE shows that self-interested, strategic behavior is "wholly congruent" with opportunism; however, intelligence is only partially congruent with bounded rationality, as it assumes a high computational capacity of actors, thus implying that actors have only very few cognitive limitations (Williamson 1985:51)

Game theory categorizes games according to models of conflict and cooperation (see Rasmusen 2007:21f.). Conflict refers to the question of whether agents' interests can be met simultaneously. In conflict, meeting one agent's interest means that another's can be met to a lower extent only. For example, bargaining over prices is a conflict situation. Cooperation pertains to the question of whether agents are allowed to make credible (binding) commitments. If so, they can coordinate their decisions by forming coalitions to jointly achieve otherwise unattainable outcomes and split the gains from cooperation through side payments. In this respect, it is common to distinguish between *non-cooperative* and *cooperative* games. In market interactions between autonomous firms, games are often non-cooperative with conflicting interests.

Another important dimension for categorizing games pertains to the question of what information is available to agents (see element (ii) above). In this respect, game theory distinguishes between games of *complete* and *incomplete* information. In a game of complete information, the structure of the game (its agents, rules, outcomes, and payoffs) is *common knowledge* (Tadelis 2013:45), which means that every agent knows the structure of the game (or any other piece of information), and every agent knows that every agent knows it, and so on *ad infinitum* (Aumann 1976). In games of incomplete information, some relevant information about the game is *private information* (see e.g. Narahari et al. 2009:40). This means that it is known to a specific agent but not to any other agent in the game.

Until the late 1960s, games of incomplete information were an Achilles' heel of game theory because there was no known way to model "incomplete information" in mathematically reasonable terms. However, a now common and very operational approach to modeling games of incomplete information is to transform them into games of imperfect information, as described in the *Harsanyi doctrine* (Harsanyi 1967; Harsanyi 1968a; Harsanyi 1968b). This doctrine stipulates that the structure of the game is common knowledge except for each agent's preferences for different outcomes of the game. Agent preferences are the private information of each agent and are referred to as each agent's *type*. Moreover, it is assumed that each agent's type is the realization of a random variable whose statistical distribution is common knowledge. This distribution is therefore referred to as *the common prior*. Thus, according to this doctrine, only the actual realization of each agent's type is private information, but due to the common prior, all agents have consistent beliefs about the statistical distribution of agent types (see e.g. Narahari et al. 2009:42). Given the common prior, agents can form and update their beliefs (expectations) about other agents' preferences by means of Bayes' rule, which is why games of imperfect information are denoted as *Bayesian games* (see e.g. Rasmusen 2007:55f.). Given these assumptions, it is possible to model such games in a mathematically convenient manner using statistical instruments. One may be interested in knowing how and when each agent becomes aware of his type. The doctrine suggests that, before the game begins, a pseudo-agent (typically referred to as "nature") draws each agent's type from the commonly known distribution and tells each agent his type in a confidential manner, i.e. unobservable by any of the other agents.

Thus far, we have focused on describing and categorizing situations of strategic interdependence. However, game theorists are primarily intrigued by predicting the best thing for each agent to do in a situation of strategic interdependence and calculating the outcomes of such situations given all agents' actions. In game theory parlance, game theorists try to identify the equilibria of games subject to each agent's set of available strategies. In a game theoretic sense, a *strategy* is defined as a complete contingent plan, or decision rule, that specifies the agent's actions in every possible distinguishable circumstance in which the agent might be called upon to move (Mas-Colell et al. 1995:228) as a function of the information available at that time (Rasmusen 2007:17). An *equilibrium* is defined as a list consisting of a best (i.e. payoff-maximizing) strategy for each agent (Rasmusen 2007:17f.). In order to determine a game's equilibrium, game theorists use *solution concepts* that define an equilibrium based on the strategies available to agents and their payoffs (Rasmusen 2007:18). More intuitively, solution concepts are methods to restrict the *set of all possible outcomes* of a game to a set of outcomes *more reasonable than others* (Tadelis 2013:54). Two central solution concepts for solving non-cooperative games with imperfect information are the *dominant strategy equilibrium* and the *Bayesian-Nash equilibrium*. They can be defined as follows (see e.g. Rasmusen 2007:20; Parkes 2001:27ff.; Dash et al. 2003:41f.; Tadelis 2013:251):

Definition 2 (Dominant Strategy Equilibrium). *In a dominant strategy equilibrium, there exists a strategy for each agent that (weakly) maximizes the respective agent's payoff, regardless of the strategies the other agents choose to play in the game.*

Definition 3 (Bayesian-Nash Equilibrium). *In a Bayesian-Nash equilibrium, there exists a strategy for each agent that (weakly) maximizes the respective agent's expected payoff given the (agent's beliefs about the) common prior and given that all other agents also play the strategies that weakly maximize their expected payoffs.*

Juxtaposing these concepts reveals a key difference in the underlying assumptions about the information required for agents to devise their best strategies. The assumptions underlying dominant strategy equilibria are weaker, which makes this solution concept more robust. In a dominant strategy equilibrium, each agent can devise her best strategy without any information about the other agents or the distribution of agent types. Each agent's best strategy is independent of other agents' choices because it is the best strategy against all strategies that other agents could play (Tadelis 2013:71; Parkes 2001:28). Conversely, a Bayesian-Nash equilibrium demands *strategic reasoning* from agents: Each agent devises his best strategy as a best response to his beliefs about other agents' strategies (see Tadelis 2013:69ff.). More specifically, in a Bayesian-Nash equilibrium, each agent devises her best strategy as a best response to her beliefs about the other agents' best strategies given the distribution of agent types (common prior) and the assumption that all agents play rationally (which means that all other agents will, in fact, play their best response). In other words, an agent needs to justify her chosen strategy based on her beliefs. If some agents have incorrect beliefs about the common prior or behave irrationally, a Bayesian-Nash equilibrium is not reached. This is because, in a Bayesian-Nash equilibrium, each agent's best strategy is only the best response to all other agents' best response strategies, not to all other agents' strategies (see Narahari et al. 2009:26). We refer to these differences and point out their relevance for mechanism design in the following subsection.

5.3. Fundamentals of Mechanism Design

A mechanism is

> "a specification of how economic decisions are determined as a function of the information that is known by the individuals in the economy,"

which is why almost any kind of market institution can be viewed as a mechanism (Myerson 1988:1; also see Myerson 1983:1770).

Mechanism design is a subfield of game theory that focuses on non-cooperative games of incomplete information. The theory of mechanism design is concerned with

> "how to implement good system-wide solutions to problems that involve multiple self-interested agents, each with private information about their preferences" for different outcomes (Parkes 2001:23).

The principle insight of mechanism design theory is that not only *resource constraints* but also *incentive constraints* need to be considered when formulating and analyzing economic decisions because, just like the scarcity of raw materials, the need to offer agents incentives to reveal their private information equally imposes constraints on economic decisions (Myerson 1988:1). A mechanism designer basically faces a similar problem as a policy maker (or central authority or social planner) who wishes to implement a certain collective decision that relies on private information held by agents that are affected by the policy decision (see e.g. Tadelis 2013:288). Specifically, a mechanism designer faces two problems (Narahari et al. 2009:53): (1) eliciting agents' relevant private information about their preferences and (2) aggregating agents' announced preferences into a single collective decision – a *social choice.* While the latter is primarily a mathematical optimization problem, the former represents an economic and organizational problem, and hence is a focus of this thesis.

Figure 8 depicts the formal structure of the mechanism design problem a policy maker faces. Table 8 presents the necessary notation for this figure and the following discussion.

A policy maker initially states a desired mapping from each possible combination of agent preferences (type profile) to a single collective decision. Formally, we can represent this mapping using a social choice function:

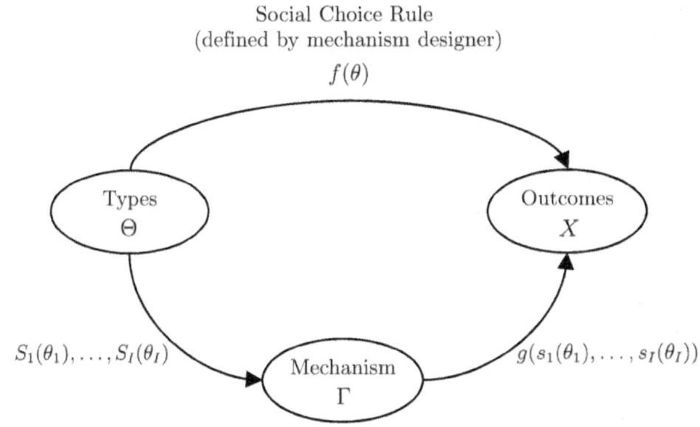

Figure 8.: Stanley Reiter Diagram: Triangular Relationship between Social Choice and Mechanism Implementation (adopted from Reiter 1977:229)

Definition 4 (Social Choice Function). *A social choice function is a function $f : \Theta_1 \times \cdots \times \Theta_I \rightarrow X$ that, for each possible profile of the agents' types $(\theta_1, \ldots, \theta_I)$, assigns a collective choice $f(\theta_1, \ldots, \theta_I) \in X$ (Mas-Colell et al. 1995:859).*

However, as agent types are private information and therefore unknown to the policy maker, it is impossible for the policy maker to make a collective decision directly. He or she therefore needs to approach this problem in an indirect manner by inducing a game through a mechanism. This game is designed by the policy maker with the objective to reveal agents' private information and subsequently determine the single collective decision based on their preferences reported. Specifically, the mechanism represents an institution, protocol, or framework that comprises a set of rules prescribing the strategies available to each agent in this *designed game* and specifies how each possible strategy profile transforms into outcomes (Narahari et al. 2009:68; also see Mas-Colell et al. 1995:866). Formally, we can define a mechanism as:

Definition 5 (Mechanism). *A mechanism $\Gamma = (S_1, \ldots, S_I, g(\cdot))$ is a collection of I strategy sets $S_1, \ldots, S_I$ and an outcome function $g : S_1, \times \cdots \times S_I \rightarrow X$ (Mas-Colell et al. 1995:866).*

The policy maker tries to design the mechanism in such a way that, in equilibrium, the mechanism yields the same outcome as the social choice function, given the same profile of agent types. This is captured by the idea of mechanism implementation. Formally, we can write:

I	set of all agents, numbered $i = 1, \ldots, I$
Θ_i	set of all types of agent i
θ_i	type of agent i, with $\theta_i \in \Theta_i$
Θ	set of type profiles of all agents $\Theta_1 \times \cdots \times \Theta_I$
θ	type profile $\theta = (\theta_1, \ldots, \theta_I)$ with $\theta \in \Theta$
Θ_{-i}	set of type profiles of agents excluding agent i $\Theta_1 \times \cdots \times \Theta_{i-1} \times \Theta_{i+1} \times \cdots \times \Theta_I$
θ_{-i}	type profile excluding agent i, $\theta_{-i} \in \Theta_{-i}$
S_i	set of all strategies available to agent i
$s_i(\theta_i)$	strategy selected by agent i given his type θ_i
X	set of all possible outcomes
$u_i(x, \theta_i)$	utility of agent i for outcome $x \in X$ given his type θ_i
K	finite set of possible allocations
k	an allocation of items to agents
t_i	monetary transfer associated with agent i, if $t_i > 0$ agent i receives money
$E[\cdot]$	expectation operator subject to the common prior

Table 8.: Notations for Mechanism Design

Definition 6 (Mechanism Implementation). *The mechanism* $\Gamma = (S_1, \ldots, S_I, g(\cdot))$ *implements social choice function* $f(\cdot)$ *if there is an equilibrium strategy profile* $s^* = (s_1^*, \ldots, s_I^*)$ *of the game induced by* Γ *such that* $g(s_1^*(\theta_1), \ldots, s_I^*(\theta_I)) = f(\theta_1, \ldots, \theta_I)$ *for all* $\theta_1, \ldots, \theta_I \in \Theta_1 \times \cdots \times \Theta_I$ *(Mas-Colell et al. 1995:867).*

The idea of mechanism implementation brings us to the heart of mechanism design: designing a mechanism that implements a desired social choice function. Narahari et al. (2009:7, italics original) therefore conceive of mechanism design as a problem of

> "*reverse engineering* of games or equivalently as the *art of designing the rules of a game to achieve a specific desired outcome.* The main focus of mechanism design is to design institutions or protocols that satisfy certain desired objectives, assuming that the individual agents, interacting through the institution, will act strategically and may hold private information that is relevant to the decision at hand."

At first glance, the policy maker's design problem may seem hard to solve as there may be many, few, a single, or no mechanism(s) that implement(s) a desired social choice function. However, the *revelation principle* – probably the most positive result in mechanism design – states that we can limit our search to the class of *incentive-compatible direct revelation mechanisms.* Precisely the principle reads:

> "Given a mechanism and an equilibrium for that mechanism, there exists a direct mechanism in which (1) it is an equilibrium for each player to report his or her information truthfully and (2) the outcomes are the same as in the given equilibrium of the original mechanism" (Krishna 2010:62).

The principle holds for dominant strategy equilibria (Gibbard 1973) and Bayesian-Nash equilibria (see e.g. Dasgupta et al. 1979; Myerson 1979). The revelation principle is intriguing because it effectively limits the solution space the policy maker needs to consider in the mechanism design problem: If the desired social choice function cannot be implemented by means of an incentive-compatible direct revelation mechanism, then it cannot be implemented by any mechanism.

A *direct revelation mechanism* limits each agent's strategy set to reporting his or her private information, and the mechanism's outcome function equals the social choice function. Formally, we can define a direct revelation mechanism as a mechanism in which $S_i = \Theta_i$ for all i and $g(\theta) = f(\theta)$ for all $\theta \in \Theta$ (Mas-Colell et al. 1995:868). Myerson (1988:1f.) intuitively explains how direct revelation mechanisms operate:

> "There is assumed to be a mediator who can communicate separately and confidentially with the every individual in the economy. This mediator may be thought of as a trustworthy person, or as a computer tied into a telephone network. At each stage of the economic process, each individual is asked to report all of his private information (that is, everything that he knows that other individuals in the economy might not know) to the mediator. After receiving these reports confidentially from every individual, the mediator may then confidentially recommend some action or move to each individual. A direct-revelation mechanism is any rule for specifying how the mediator's recommendations are determined, as a function of the reports received."

A direct revelation mechanism is *incentive-compatible* if it is each agent's best strategy to truthfully report his private information to the mediator (Mas-Colell et al. 1995:868; Myerson 1988:2). From a different perspective, an incentive-compatible mechanism overcomes incentive constraints by offering the right amount of incentive to agents so that truth telling is in their best interests and consistent with rationality and intelligence assumptions (Narahari et al. 2009:71). It is important to distinguish between two settings of truthful preference revelation (see e.g. Narahari et al. 2009:71): First, reporting his preferences truthfully is each agent's best strategy, regardless of what is reported by other agents in the game. Second, reporting his preferences truthfully is each agent's best strategy whenever all other agents also report their preferences truthfully. These settings directly correspond to the solution concepts of dominant strategy equilibrium and Bayesian-Nash equilibrium respectively.

Given the definition of both solution concepts (see Subsection C.III.5.2), and given the revelation principle, we can define these solution concepts for the implementation of social choice functions via direct revelation mechanisms as follows:

Definition 7 (Dominant Strategy Incentive Compatible (DSIC)). *The social choice function $f(\cdot)$ is truthfully implementable in dominant strategies (or strategy-proof) if* *the direct revelation mechanism* $\Gamma = (\Theta_1, \ldots, \Theta_I, f(\cdot))$ *has a dominant strategy equilibrium* $s_i^*(\theta_i) = \theta_i$ *for all* $\theta_i \in \Theta_i$ *for all* i. *Equivalently, for all* i *and* $\theta_i \in \Theta_i$, $u_i(f(\theta_i, \theta_{-i}), \theta_i) \geq u_i(f(\hat{\theta}_i, \theta_{-i}), \theta_i)$ *for all* $\hat{\theta}_i \in \Theta_i$ *and all* $\theta_{-i} \in \Theta_{-i}$ *(Mas-Colell et al. 1995:871).*

Definition 8 (Bayesian-Nash Incentive Compatible (BIC)). *The social choice function* $f(\cdot)$ *is truthfully implementable in a Bayesian-Nash equilibrium if the direct revelation mechanism* $\Gamma = (\Theta_1, \ldots, \Theta_I, f(\cdot))$ *has a Bayesian-Nash equilibrium* $s_i^*(\theta_i) = \theta_i$ *for all* $\theta_i \in \Theta_i$ *for all* i. *Equivalently, for all* i *and* $\theta_i \in \Theta_i$, $E_{\theta_{-i}}[u_i(f(\theta_i, \theta_{-i}), \theta_i) \mid \theta_i] \geq E_{\theta_{-i}}[u_i(f(\hat{\theta}_i, \theta_{-i}), \theta_i) \mid \theta_i]$ *for all* $\hat{\theta}_i \in \Theta_i$ *(Mas-Colell et al. 1995:883).*

DSIC and BIC implementations inherit their properties from the underlying solution concepts. Comparing DSIC and BIC implementations from the perspective of a policy maker emphasizes that DSIC is more desirable because it yields more robust implementations, which means that the policy maker can be more confident that a socially desired outcome can, in fact, be implemented by that mechanism (Mas-Colell et al. 1995:870). Three specific reasons can be identified for why DSIC is more robust than BIC (Mas-Colell et al. 1995:870; also see Mas-Colell et al. 1995:884f.). First, if agents have a weakly dominant strategy, then the policy maker can be confident that rational and intelligent agents will indeed play this strategy, thus very likely implementing a desired social choice. Unlike in BIC, agents do not need to correctly forecast other agents' best strategies to justify their own best strategy in DSIC. Second, the optimality of a weakly dominant strategy is independent of an agent's beliefs about the distribution of agent types. Thus, in DSIC implementation, agents can have inconsistent and incorrect beliefs about the common prior without influencing the outcome of a social choice function. Third, a mechanism that implements a social choice function in DSIC implements it for any distribution of agent types. In other words, the mechanism is independent of the distribution of agent types. This is very desirable property for a policy maker who does not require any knowledge about agent distribution when designing a mechanism. Conversely, BIC implementation requires the policy maker (and all other agents) to have consistent and correct beliefs about the common prior. In addition, BIC only applies to the case of agent types following a distribution from the exponential family. However, a policy maker can expect to implement a narrower set of social choice functions in DSIC because a social choice function implementable in DSIC is also implementable in BIC but not *vice versa* (Mas-Colell et al. 1995:844). Thus, a policy maker makes a tradeoff to have a narrower scope of implementation with a more robust implementation.

5.4. Properties of Social Choice Functions

Thus far, we have introduced two important solution concepts in which social choice functions can be implemented. However, we have not paid attention to the question of whether the social choice reached in equilibrium is a "good" social choice. In fact, as the policy maker has the authority to choose the social choice function that should be implemented through a mechanism, one may equally ask: Has the policy maker made an economically "good" decision when selecting the social choice function? Of course, whether or not a social choice is economically good has to be measured on a system-wide basis in relation to agent preferences. Thus, to answer this question, we first need to understand the structure of agent preferences.

It is a very common assumption in mechanism design that agents have quasilinear preferences (see e.g. Parkes 2001:31). In an environment of quasilinear preferences, we can write each outcome $x \in X$ as $x = (k, t_1, \ldots, t_I)$, where k refers to an allocation of items to agents from a finite set of possible allocations K, and t_i refers to a monetary transfer to agent i (see e.g. Narahari et al. 2009:93f.; also see Tadelis 2013:289f.). The utility function of agent i can thus be written as $u_i(x, \theta_i) = v_i(k, \theta_i) + (\bar{m}_i + t_i)$, where $v_i(\cdot)$ is agent i's valuation for allocation k and $\bar{m}_i$ is the initial monetary endowment of agent i. Preferences are referred to as quasilinear because they are linear in at least one variable, which, in our case, is money. Quasilinear preferences are convenient because we can easily transfer utility between agents through side payments (Parkes 2001:32). Given quasilinear preferences, we can write a social choice function as $f = (k(\theta), t_1(\theta), \ldots, t_I(\theta))$, where for every $\theta \in \Theta$ we require $k(\theta) \in K$ and $\sum_i^I t_i(\theta) \leq 0$, which means that the mechanism designer cannot externally subsidize the mechanism. $k(\cdot)$ is referred to as an *allocation rule* that specifies the allocation of items to agents as a function of agent types, and $(t_1(\theta), \ldots, t_I(\theta))$ is referred to as a *payment rule* that specifies the financial transfers to be made or received by agents as a function of agent types (see e.g. Narahari et al. 2009:93f.; also see Tadelis 2013:289f.). In other words, the outcome rule $g(\cdot)$ is decomposed into an allocation rule and a payment rule (Parkes 2001:32).

Assuming quasilinear preferences, we can state desirable economic properties of social choice functions separately. Literature identifies the following desirable properties: (1) allocative efficiency, which requires incentive compatibility, (2) budget-balance, and (3) individual rationality.

Definition 9 (Allocative Efficiency). *A social choice function $f(\cdot) = (k(\cdot), t_1(\cdot), \ldots, t_I(\cdot))$ is allocatively efficient, if for all type profiles $\theta \in \Theta$, $k(\theta)$ satisfies $\sum_{i=1}^I v_i(k(\theta), \theta_i) \geq \sum_{i=1}^I v_i(k, \theta_i)$ for all $k \in K$ (Mas-Colell et al. 1995:877).*

Intuitively, a social choice function is allocatively efficient if the total value over all agents is maximized (Parkes 2001:33). In other words, items are allocated to the agents that value them most (Narahari et al. 2009:94). It is important to note that incentive compatibility is a precondition for allocative efficiency. By definition, allocative efficiency depends on true types, not reported types. Thus, a policy maker can calculate the efficiency of an allocation correctly only if the mechanism extracts private information about preferences truthfully. Therefore, *incentive compatibility* may be considered a fourth desirable property with regard to the implementation of social choice functions. One may consider it a property of the mechanism. In the remainder of this thesis, we include incentive compatibility as a fourth economic property.

Literature distinguishes between two types of budget balance: strong and weak.

Definition 10 (Strong Budget Balance). *A social choice function* $f(\cdot) = (k(\cdot), t_1(\cdot), \ldots, t_I(\cdot))$ *is strongly budget balanced if, for all type profiles* $\theta \in \Theta$, $t_1(\cdot), \ldots, t_I(\cdot)$ *satisfies* $\sum_{i=1}^{I} t_i(\theta) = 0$ *(Narahari et al. 2009:95).*

Intuitively, a social choice function is strongly budget balanced if there are no net transfers into or out of the system (Parkes 2001:33); the system is closed without running a surplus or a deficit (Narahari et al. 2009:95). If the implementation of social choice functions is both allocatively efficient *and* strongly budget balanced, then it is Pareto optimal.

Definition 11 (Weak Budget Balance). *A social choice function* $f(\cdot) = (k(\cdot), t_1(\cdot), \ldots, t_I(\cdot))$ *is weakly budget balanced if, for all type profiles* $\theta \in \Theta$, $t_1(\cdot), \ldots, t_I(\cdot)$ *satisfies* $\sum_{i=1}^{I} t_i(\theta) \leq 0$ *(Narahari et al. 2009:95).*

Intuitively, a social choice function is weakly budget balanced if agents can make a net payment to the mechanism but no net payment can be made from the mechanism to agents (Parkes 2001:33). In other words, a weakly budget balanced mechanism allows the mediator of a direct revelation mechanism (e.g. a platform operator or the policy maker himself) to appropriate some fraction of the payments made by agents, since the sum of agents' payments may exceed the sum of receipts (Tadelis 2013:289f.).

Definition 12 (Interim Individual Rationality). *A social choice function* $f(\cdot) = (k(\cdot), t_1(\cdot), \ldots, t_I(\cdot))$ *is interim individually rational if, for each agent* i, $f(\cdot)$ *satisfies* $E_{\theta_{-i}}[u_i(f(\theta_i, \theta_{-i}), \theta_i) \mid \theta_i] \geq \bar{u}_i(\theta_i)$ *for all* θ_i *(Mas-Colell et al. 1995:893).*

Intuitively, interim individual rationality requires that the expected utility from participating in the mechanism is equal to or greater than the utility the agent could earn outside the mechanism, given his type and prior beliefs about the distribution of agent types (Parkes 2001:35). In this context, *interim* means that each agent calculates the expected utility from participation *after* being told his type and *prior to* deciding whether to participate (Mas-Colell et al. 1995:893). Individual rationality is also referred to as *voluntary participation constraint* because rational autonomous agents only participate on a voluntary basis if they can expect to be no worse off after participating.

5.5. Possibility and Impossibility Results

Of course, all of the properties mentioned above seem desirable, and a policy maker would like to achieve them all simultaneously. However, there is bad news and good news about the theoretical feasibility of certain combinations of these properties. This news manifests in *impossibility theorems* and *possibility theorems*, respectively. In this subsection, we outline the most important theorems and create an understanding about the theoretically feasible solution space a policy maker can choose from.

The Gibbard-Satterthwaite (GS) impossibility theorem (Gibbard 1973; Satterthwaite 1975) serves as a starting point for the following discussion, as it applies to a very general situation. Starting from this general situation allows us to emphasize how assumptions about (1) the exchange environment, (2) agent preferences, and (3) the choice of the solution concept impact the feasibility of achieving certain combinations of desirable properties (see e.g. Parkes 2001:49ff.; Narahari et al. 2009:87, 93). The GS theorem can basically be considered an ingenious reinterpretation of Arrow's Impossibility Theorem in the field of mechanism design (see e.g. Reny 2001; Narahari et al. 2009:87). The GS theorem states that given at least two agents with strictly hierarchically ranked preferences (strict total preferences) and at least three alternative outcomes over the set of all agents' preferences, a social choice function can be implemented in DSIC if and only if it is dictatorial (Satterthwaite 1975:193). In other words, no Pareto optimal or efficient allocation can be achieved in this general situation (Parkes 2001:49, 51).

Although it is an impossibility theorem and thus a dismal result for mechanism design, the GS theorem is fortunately less disappointing than one might expect at first glance. This is for two reasons (see e.g. Narahari et al. 2009:87, 93). First, the theorem applies to a very general situation. An impossibility result for a general situation is less severe than for a restricted one because there might be more restricted situations in which this impossibility does not hold (see Parkes 2001:49). Restricted means that additional constraining assumptions regarding the utility environment have to be made, such as quasilinear preferences. Second, the theorem uses a very strong solution concept (DSIC). By settling for a weaker solution concept (e.g. BIC) one can avoid this impossibility as well. Thus, a policy maker can circumvent the impossibility theorem by making additional assumptions or settling for a weaker solution concept (Parkes 2001:51) but must accept other unavoidable and potentially undesirable properties. In other words, a policy maker needs to make a tradeoff between assumptions about the exchange environment, agents' preferences, properties of the solution concept, and desired properties of the social choice function.

For the following discussion, we relax the assumption about unrestricted preferences by assuming quasilinear preferences – a more restrictive preference environment. As pointed out in the last subsection, quasilinear preferences are very common in mechanism design, and applicable to market environments (see Parkes 2001:51), and serve as a good basis for this thesis, which focuses on market mechanisms. Table 9 presents the most important impossibility and possibility theorems for quasilinear preferences. Besides indicating impossibility and possibility, the table also states the solution concept and exchange environment. Some further definitions are needed: In a *simple exchange* environment, buyers and sellers intend to exchange single units of the same good (Parkes 2001:50). An *exchange* environment refers to bilateral trading between buyers and sellers with general valuation functions (including bundle values) (Parkes 2001:54). *Combinatorial auction* refers to a unilateral exchange environment: An auctioneer auctions off a set of items. Bidding agents may bid for bundles of items, which means that agents can express non-linear valuations for bundles of items, in the sense of "I want A only if I also get B" (Parkes 2001:8).

The Hurwicz impossibility theorem derives from several publications (Hurwicz 1972; Green & Laffont 1977; Hurwicz 1975; Hurwicz & Walker 1990). It states that, given quasilinear preferences, it is impossible to implement an allocatively efficient and strongly budget balanced (i.e. Pareto optimal) social choice function in DSIC (for simple exchanges) (see Hurwicz & Walker 1990:694; see Parkes 2001:52). This theorem has a quite negative result, as Pareto optimality cannot be achieved by means of a strategy-proof mechanism, even in the more restrictive environment of quasilinear preferences. However, there may be many situations in market economies in which strong budget balance is not required (Parkes 2001:52) in combination with allocative efficiency.

The Myerson-Satterthwaite (MS) impossibility theorem reinforces the Hurwicz impossibility theorem by expanding the impossibility to include BIC implementation, if interim individual rationality is also required (Parkes 2001:52). The MS theorem states that, given quasilinear preferences, it is impossible to implement an allocatively efficient, strongly budget balanced (i.e. Pareto optimal), and interim individually rational social choice function in BIC (for a simple exchange economy see Myerson 1983, and for a proof of the general impossibility see e.g. Williams 1999). This impossibility theorem is a particularly troubling result because it directly implies that even for a weak solution concept (BIC) we can only achieve *two* out of the three desirable properties simultaneously. Thus, every mechanism designer is confronted with trading off the relative importance of these properties when trying to identify the most appropriate combination of properties for the

	Solution Concept	AE	SBB	WBB	IR	Exchange environment
Impossibility						
Hurwicz	DSIC	×	×			simple exchange
Meyerson-Satterthwaite	BIC	×	×		×	simple exchange
Possibility						
Groves	DSIC	×				
Clarke	DSIC	×		×[1]	×[1]	bilateral trading
GVA	DSIC	×		×	×	comb. auction
dAGVA	BIC	×	×			bilateral trading

Table 9.: Impossibility and Possibility Results for Quasilinear Preferences
AE = *ex post* allocative efficiency; SBB = *ex post* strong budget balance; WBB = *ex post* weak budget balance; IR = *interim* individual rationality;
[1] if additional assumptions hold (based on Parkes 2001:49, 53)

setting under consideration. However, in comparison to the Hurwicz theorem, we are able to achieve allocative efficiency and strong budget balance by settling for a weaker solution concept.

Thus far, we have focused on the bad news in the field of mechanism design. Now we present some good news that states the theoretical possibility of achieving specific combinations of desirable economic properties of social choice functions.

An important positive result in mechanism design is the class, or family, of Vickrey-Clarke-Groves (VCG) mechanisms named after their discoverers, Vickrey (1961), Clarke (1971), and Groves (1973). It is appropriate to speak of a class, or family, of mechanisms because the Vickrey auction is a special case of the Clarke mechanism, which in turn is a special case of the Groves mechanism. Given quasilinear preferences, Groves mechanisms achieve allocative efficiency in DSIC implementation. As a family, they share a common logic of functioning to achieve allocative efficiency in DSIC implementation. Intuitively, Groves mechanisms make truth-telling a dominant strategy by introducing a clever payment rule that imposes a financial burden on each agent, which is equal to the "damage" the agent imposes on all other agents if his preferences change the allocation of goods. In other words, it is a payment rule that causes each agent to internalize the externality he imposes on all other agents (Mas-Colell et al. 1995:878; also see Tadelis 2013:295ff.). Specifically, Groves mechanisms function as follows: They choose an efficient allocation of goods based on agents' *reported* valuations. They also impose payments on agents such that each agent *has to pay* an arbitrary amount and *is paid* the sum of all other agents' reported valuations. As agents can neither influence the arbitrary amount to be paid nor the sum of all other agents' valuations they receive, they can only influence their utility through their announced valuations for the allocation of goods. As each agent derives utility from the allocation of goods and is paid the sum of the total value of all other agents, each agent becomes as concerned with maximizing the other agents' utility as she is concerned with maximizing her own utility. Hence, truth-telling becomes a dominant strategy for all agents (because overstating their own valuation is sanctioned by a lower payment as overstating reduces the sum of other agents' utility). In other words, this payment rule aligns private incentive with social incentives (Tadelis 2013:297). Finally, to emphasize more precisely, we speak of a family of mechanisms because we can vary the arbitrary payment made by agents in different ways: Different choices result in different tradeoffs between budget balance and interim individual rationality (Parkes 2001:42).

The Clarke (Pivotal) mechanism is a special case of the Groves mechanism that, besides achieving allocative efficiency in DSIC, can also achieve interim individual rationality and/or weak budget balance if additional assumptions hold. The Clarke mechanism employs a payment rule that, intuitively speaking, charges each agent the externality that his participation in the efficient allocation of goods imposes on all other agents; otherwise, the agent pays nothing (Mas-Colell et al. 1995:878). This payment is referred to as a Clarke tax. The Clarke mechanism achieves interim individual rationality if we additionally assume *choice set monotonicity* and *no negative externalities.* Choice set monotonicity means that the set of feasible outcomes (weakly) increases if additional agents are introduced to the mechanism (see e.g. Narahari et al. 2009:105). No negative externality means that any outcome that does not involve a certain agent yields a neutral or positive effect on that agent (see e.g. Narahari et al. 2009:105). These assumptions hold in very general environments, which is why Clarke mechanisms usually achieve interim individual rationality (Parkes 2001:44). The Clarke mechanism achieves weak budget balance if we additionally assume that there exists no *single agent effect*, which means that it is possible to remove any one agent from the mechanism without creating a negative effect on the best outcome available to the remaining agents (see e.g. Narahari et al. 2009:104). The Clarke mechanism clearly shows the immediate tradeoff between making additional assumptions and achieving additional desirable properties, given that the assumptions hold.

The generalized Vickrey auction (GVA) results from applying the Clarke mechanism to a combinatorial allocation problem (CAP) (which means that a set of items needs to be allocated among a set of agents, with agents having non-linear valuations for bundles), and within this environment GVAs achieve allocative efficiency, individual rationality, and weak budget balance in DSIC (Parkes 2001:54). It can be shown that the assumptions of choice set monotonicity, no negative externality, and no single agent effect hold in this restricted environment (Parkes 2001:49). It is referred to as a *generalized* Vickrey auction because it remains a second price auction, but applies to a multi-item auction instead of a single item and allows agents to bid for bundles of items. The CAP is a very restricted exchange environment and thus emphasizes how imposing additional restrictions allows for achieving certain combination of desirable properties. It is considered very restricted because it refers to a one sided exchange (Parkes 2001:54): Multiple buyers bid for the allocation of goods, and the auctioneer is obliged to sell a good for the resulting price, regardless of his valuation.

Finally, the d'Aspremont and Gérard-Varet and Arrow (dAGVA) mechanism, named after its discoverers d'Aspremont & Gérard-Varet (1979) and Arrow (1979), achieves allocative efficiency and strong budget balance in BIC. Intuitively, it can be understood as an "expected Groves" mechanism because each agent is charged for the *expected externality* that her reported preferences impose on all other agents, which implies that her payment depends only on her actual preferences and not on the actually reported preferences of other agents (Mas-Colell et al. 1995:886; also see Parkes 2001:55). The dAGVA mechanism directly shows again how a set of assumptions impact economic properties of social choice functions. By weakening the solution concept from DSIC to BIC, we can achieve allocative efficiency and strong budget balance (i.e. Pareto optimality), which cannot be achieved in DSIC, as per the Hurwicz impossibility theorem. However, even with BIC, we cannot achieve allocative efficiency, strong budget balance, *and* individual rationality, as predicted by the Myerson-Satterthwaite (MS) impossibility theorem.

This section has introduced fundamental solution concepts, key assumptions about agent preferences and exchange environments, economic properties of social choice functions, and (im-)possibility results of mechanism design. We emphasize that a policy maker that intends to implement a "good" system-wide solution in a system of private information about preferences needs to account for *resource constraints* and *incentive constraints*. Depending on the desired social choice function to be implemented, a policy maker may face tough tradeoffs between feasible solution concepts, economic properties, and assumptions about preferences and exchange environment, as only certain combinations thereof are feasible given specific (im-)possibility results. In fact, good combinations of desirable economic properties are a critical indicator and prerequisite for the success of a designed mechanism. For example, mechanisms that are not interim individually rational seem unsuitable to facilitating economic decisions among autonomous firms as nobody would participate voluntarily. We refer to the arguments presented above and further investigate the importance of economic properties in Section F.XII.2.

6. Interim Summary

This chapter has studied three fundamental governance structures markets, hybrids, and hierarchies that enable cooperation among economic actors via contract-based transactions. The question of which transaction is processed via which governance structure has essentially been conceptualized as a question about economizing on transaction costs. Transaction costs are rooted in the bounded rationality and opportunistic behavior of economic actors in interaction with transaction characteristics in terms of asset specificity, uncertainty, frequency, and complexity of product descriptions. Economizing on transaction costs can be accomplished by matching transactions with governance structures in a discriminating way depending on their respective attributes. Thus, the formation of interfirm networks, which are considered hybrids, results from economizing on transaction costs.

Furthermore, this chapter has specifically investigated market-based exchanges through formal market mechanisms based on game theory. The feasible set of how resources can be allocated among a set of economic actors and the associated payments that the actors have to make is fundamentally constrained by both resource and incentive constraints, where the former result from a scarcity of resources and the latter from private information held by actors. Two solution concepts, DSIC and BIC, fundamentally differ in the information economic actors require to derive their best strategy. In addition, four fundamental properties have been introduced that can be used to characterize economic decisions made through such formal mechanisms: allocative efficiency, incentive compatibility, budget balance, and individual rationality. Although achieving all properties simultaneously would be desirable, not all combinations of solution concepts and properties are mathematically feasible, hence tradeoffs are inevitable.

IV. Resource-based View

1. Preliminary Considerations: Introduction and Evolution

Over the last 25 years, the RBV has received significant academic attention and has become one of the most influential theoretical frameworks in strategic management for investigating firms' internal sources of competitive advantage (see e.g. Lavie 2006:640; Andersén 2011:87). The RBV makes the central proposition that the firms' resources and the specific characteristics of these resources are critical determinants for firms to achieve (sustained) competitive advantage (Barney 1991; Grant 1991; Peteraf 1993). A firm is said to have competitive advantage if it is (or has the potential to) persistently earn a higher rate of profit than its rivals (Grant 2010:211).

The RBV is rooted in a seminal work by Penrose (1959) who conceptualizes firms as bundles of heterogeneous resources. Significantly later, Wernerfelt (1984) and Rumelt (1984) posit that firm profitability is fundamentally linked to firm resources. Recognizing the implications of this link, Barney (1991) is instrumental in providing comprehensive conceptual insight into the specific conditions under which firms' internal resources have the potential to generate enduring, above-normal firm rents, which is a situation referred to as "sustained competitive advantage."

Although the RBV was initially introduced to investigate sources of competitive advantage of "sole" firms (i.e. firms that are not participating in any interfirm network or alliance), the recent proliferation of alliances has motivated scholars to also use the RBV to investigate sources of competitive advantage of "interconnected" firms (i.e. firms that are participating in an interfirm alliance). Although our focus is on interfirm networks, we start by introducing the RBV for sole firms, as this theoretical approach establishes the basis for the case of interconnected firms. We present the RBV largely following the structure of assumptions and argumentative logic proposed by Barney (1991), as his contribution appears to be the most influential among a variety of notable scholarship (see e.g. Wernerfelt 1984; Rumelt 1984; Dierickx & Cool 1989 Grant 1991; Peteraf 1993). Thereafter, we focus on the RBV for interconnected firms.

2. Competitive Advantage of "Sole" Firms

2.1. Assumptions

Before reviewing the assumptions underlying the RBV, introducing the most fundamental concept of the RBV – resources – is helpful. According to Barney (1991:101), resources "include all assets, capabilities, organizational processes, firm attributes, information, knowledge, etc. controlled by a firm that enable the firm to conceive of and implement strategies that improve its efficiency and effectiveness (Daft 1983)." This definition adopts a very broad understanding of resources as "capabilities" are also conceived of as "resources." For now, we use this broad understanding but later distinguish between resources and capabilities when designing a reference architecture for CLSs.

The RBV rests upon several assumptions, which may be structured into three categories: (1) assumptions regarding the nature of the firm, its managers, and access to new resources; (2) assumptions regarding the endowment and mobility of resources among firms; and (3) assumptions regarding specific resource characteristics.

In the category related to the *nature of the firm, its managers, and access to new resources*, Leiblein (2003:944) formulates two underlying assumptions: (a) firms are *profit maximizing entities* steered by *boundedly rational* managers, which means that management lacks accurate foresight about future profitable opportunities, and (b) firms need to make *up-front investments* to acquire or develop new resources, whose eventual value is uncertain at the time of acquisition or development. Leiblein (2003:944) argues that these assumptions lead to the two fundamental assumptions that underlie the RBV, as formulated by Barney (1991): resource heterogeneity and resource immobility.

Regarding the *endowment and mobility of resources among firms*, Barney (1991:101) assumes that (a) firms are heterogeneous in terms of the set of strategically relevant resources they command (*resource heterogeneity*), and (b) these resources may not be easily transferrable across firm boundaries (*resource immobility*), which is why resource heterogeneity can be enduring. More specifically, resources may be immobile, as they cannot be traded due to ill-defined property rights, high transaction costs (Peteraf 1993:183f.), a lack of or only imperfect factor markets, or tight integration with other firm resources (Das & Teng 2003:288), or as they may be less valuable to alternative users because, they are specialized for the firm's needs and co-specialized with other resources of the firm (Peteraf 1993:183f.).

And in the category regarding *specific resource characteristics*, Barney (1991:105ff.) argues that a resource's potential to enable a firm to achieve (sustained) competitive advantage depends on four characteristics: value, rareness, imperfect imitability, and non-substitutability. Resources are (a) *valuable* if they enable a firm to devise and realize value creating strategies that enhance its effectiveness or efficiency by allowing the firm to seize opportunities or eliminate threats in its environment (Barney 1991:106); (b) *rare* if no, or only few, current or potential competitors control resources of a similar type (Barney 1991:106f.) or if resource supply falls short of demand (Leiblein 2003:944f.); (c) *imperfectly imitable* if they cannot be obtained by firms currently not in possession thereof (Barney 1991:107) due to unique historical conditions connected with obtaining them, causal ambiguity in means-ends relations between resources and firm performance, or social complexity in deploying such resources (Dierickx & Cool 1989 as cited in Barney 1991:107); and (d) *non-substitutable* if there are no other resources – themselves valuable, but neither rare nor imperfectly imitable – that can function as equivalents, where equivalence means that resources can be exploited separately to implement the same value creating strategy (Barney 1991:111f.). Resources that possess all of these properties have the potential to create sustained competitive advantage and are hence referred to as "strategic resources" (Chi 1994:271).

2.2. Argumentative Logic

The argumentative logic that links these assumptions to the attainment of (sustained) competitive advantage is depicted in Figure 9. If a firm possesses valuable and rare resources it can achieve *competitive advantage* over its competitors which means that a firm is

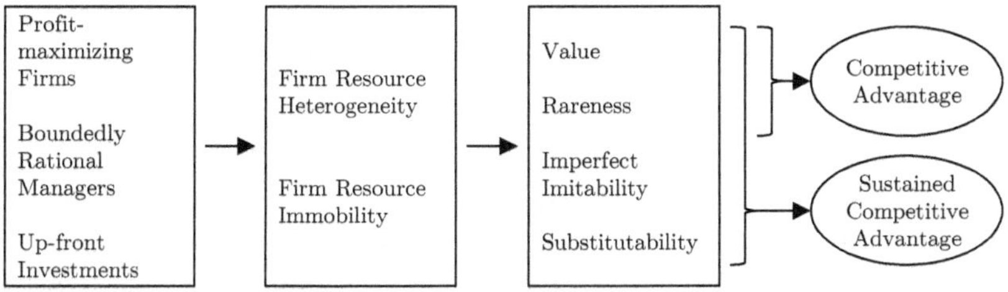

Figure 9.: Resource-based Framework of Competitive Advantage (adopted with changes from Barney 1991:112)

> "implementing a value creating strategy not simultaneously being implemented by any current or potential competitors" (Barney 1991:102).

The possession of valuable resources enables the implementation of a value creation strategy, and rareness of resources hinders competitors' ability to implement the same strategy. However, competitive advantage may erode as competitors imitate valuable and rare resources or find valuable substitutes that are not rare.

If a firm possesses valuable, rare, imperfectly imitable, and non-substitutable resources, it can achieve *sustained competitive advantage* over its competitors, which means that the firm is

> "implementing a value creating strategy not simultaneously being implemented by any current or potential competitors *and* when these other firms are unable to duplicate the benefits of this strategy" (Barney 1991:102, italics original).

Due to imperfect imitability and the lack of substitutes of valuable and rare resources, competitors are not able to duplicate the firm's strategy, and competitive advantage can be sustained. Barney (1991:102) remarks that, in this context, "sustained" does not refer to a period of time, but rather to the ability of competitive duplication: Competitive advantage is sustained if it still exists after the efforts of competitors to duplicate the strategy have ceased; it thus represents an "equilibrium concept."

3. Competitive Advantage of "Interconnected" Firms

3.1. Motives and Antecedents for Alliance Formation

More recently, the RBV has been applied to strategic alliances, and thus to an interfirm context (Das & Teng 2000; Tsang 1998; Eisenhardt & Schoonhoven 1996). Although this research field is still emerging (Das & Teng 2000), existing literature already provides good insight into the motives and antecedents for the formation of interfirm alliances from the perspective of the RBV. Unsurprisingly, both motives and antecedents are related to resources, the focal concept in the RBV.

Motives associated with the formation of strategic alliances pertain to resource flows between firms that are mobilized with the objective of gaining otherwise unachievable competitive advantage by sharing, aggregating, or exchanging resources (Das & Teng 2000). Specific motives frequently identified fall into one of three categories: (1) obtaining resources, (2) retaining resources, or (3) disposing of resources. First, few firms possess all resources necessary to compete effectively in today's dynamic market environment (Ireland et al. 2002:413). Hence, they engage in alliances to *obtain resources* owned by other firms in their environment to create otherwise unachievable rents by combining complementary or similar resources (see e.g. Das & Teng 2000:36f.; Gulati & Gargiulo 1999:1443; Tsang 1998:211ff.; Chen & Chen 2003b:1,4). The motive of obtaining resources refers to resource flows from partners to the focal firm. Second, firms may form alliances to *retain control* over their own resources by combining them with other firms' resources (Das & Teng 2000:37f.). More specifically, firms may form an alliance to retain control over resources that are currently idle by temporarily deploying them within an alliance instead of divesting from them (Das & Teng 2000:37f.). The motive of retaining resources involves temporary flows of resources toward the alliance or partners. A comparison of motives for obtaining and retaining resources reveals that

> "[t]he difference between the two motives, gaining access to additional resources possessed by others, and retaining ones own resources, is that, while obtaining resources is more about creating competitive advantage in the immediate present, retaining resources is concerned more with securing competitive advantage later on." (Das & Teng 2000:38)

Finally, firms may form alliances in order to *dispose of resources*. In this case, alliances are used as a vehicle to gradually transition resources toward partners that may otherwise be difficult to transfer (Tsang 1998:216f.; also see Bleeke & Ernst 1995; Das & Teng 2001:19). The motive of disposing of resources involves permanent flows of resources toward partners.

Antecedents for alliance formation are related to the underlying assumptions regarding resource heterogeneity, immobility, inimitability, and non-substitutability. Tsang (1998:210) argues that

> "[t]he greater the degree of heterogeneity among firms in the market, the higher is the chance of forming alliances which would create rents."

High heterogeneity may facilitate alliance formation, as firms with highly heterogeneous resource endowments likely lack critical resources required to implement value creating strategies. More specifically, Tsang (1998:210f.) argues that alliance formation is more likely between highly heterogeneous firms, as they are more likely to create scare resource combinations from which they can extract (Ricardian) rents. This may especially facilitate the formation of international alliances as firms in different countries tend to be more heterogeneous than those within countries (Tsang 1998:210f.; Shan & Hamilton 1991). Generally, Das & Teng (2000:41) propose that firms with certain resource profiles are more likely to engage in alliancing:

> "The more a firm's resources are characterized by imperfect mobility, imperfect imitability, and imperfect substitutability, the more likely the firm will get involved in strategic alliances."

However, alliances between heterogeneous firms only form, if resource transfers through factor markets and mergers and acquisitions (M&A) are inefficient (Das & Teng 2000:37; also see Chi 1994). Scholars provide several specific conditions for this. Factor markets may be inefficient for obtaining resources if their prices cannot be determined due to information asymmetries (Tsang 1998:216). Conversely, obtaining resources through M&A may be inefficient if the "target" firm owns additional resources that are of only little value to the "acquiring" firm (and thus would become idle after acquisition) and if these resources are difficult to dispose of after acquisition for asset specificity reasons (Das & Teng 2000:37). Similarly, Chi (1994:284ff.) argues that the mode of resource transfer (alliance vs. acquisition) depends on the interactions of the properties of the resources that the buying firm intends to acquire and those the target firm additionally owns. More specifically, the degree of specialization and the severity of measurement difficulties of desired and additional resources determine the transaction mode. Chi (1994) argues that if severe measurement difficulties are associated with either the resources the "acquiring" firm desires or with the "target" firm's additional resources, alliances may be better than outright acquisitions.

Furthermore, Tsang (1998:214f.) argues that inimitability of resources (e.g. tacit knowledge) to be transferred between firms facilitates alliance formation, as alliances allow for gradual transfers of such resources through imitation via close interactions and learning, either openly (with partner consent) or secretly (also see Kogut 1988:323). In addition, if firms aim to retain resources, alliances may be formed if the discounted future value of deploying (currently idle) resources internally will be higher than the current selling price (Das & Teng 2000:38). Finally, Eisenhardt & Schoonhoven (1996) find empirical evidence that firms with vulnerable strategic positions such as being active in highly competitive markets characterized by low margins and a large number of competitors are more likely to cooperate with competitors, for example by sharing costs and risks and thus improving profitability.

The preceding paragraphs have reviewed the motives and antecedents for alliance formation frequently associated with the RBV. Because resources are central to the RBV, both motives and antecedents are linked to them. Motives associated with the RBV align well with the general motives listed above (see Section C.II.2), especially for increasing revenue by gaining access to resources and reducing costs through economies of scale and scope. Antecedents for alliance formation revolve around assumptions about resources, and they complement those antecedents associated with TCE, thus improving our understanding of alliance formation.

3.2. Assumptions

Although the RBV is frequently used to specify motives and conditions for alliance formation, the RBV in Barney's (1991) initial conceptualization, as presented above, struggles to accurately evaluate the competitive advantage of interconnected firms (Lavie 2006). This conceptual shortcoming is rooted in the "proprietary assumption" that *implicitly* underlies the RBV and states that resources need to be either owned or at least fully controlled by a firm to have the potential to create competitive advantage for the firm (Lavie 2006; Foss 1999; Duschek 2004).

However, empirical evidence (see e.g. sources in Lavie 2006:641) and conceptual research (see e.g. Dyer & Singh 1998; Gulati 1999; Foss 1999; also see Duschek 2004) suggest, in principle,

> "that the (dis)advantages of an individual firm are often linked to the (dis)-advantages of the network of relationships in which the firm is embedded [...and] a firm's critical resources may extend beyond firm boundaries" (Dyer & Singh 1998:660).

Therefore, Dyer & Singh (1998:661) aim to complement the traditional RBV by means of a *relational-view* (RV) which focuses on the dyad/network level and on how *idiosyncratic interfirm linkages* may become sources of relational rents and competitive advantage, where a relational rent is defined as

> "a supernormal profit jointly generated in an exchange relationship that cannot be generated by either firm in isolation and can only be created through the joint idiosyncratic contributions of the specific alliance partners" (Dyer & Singh 1998:662).

The importance of rents generated on the network level is also stressed by Gulati (1999:399), who introduces the notion of *network resources*: Rather than being housed within the boundaries of the firms, such resources are inherent in the interfirm networks in which firms are embedded. Network resources can affect interconnected firms' strategic behaviors by changing the opportunity set available to them and they can allow firms to devise and implement value creating strategies. Although Gulati (1999:400) recognizes that network resources can influence the competitive advantage of individual interconnected firms, these influences are not a focus of his research.

Similarly, Foss (1999:2, 8f.) introduces the notion of *network capabilities*, which, just like firm capabilities, represent the collective knowledge of an interfirm network about the production of goods and services and the organization of production processes. Such capabilities are intangible assets, accumulated over time, and are inherently tied to interactions between cooperating firms. Network capabilities reside on the network level and can be a source of (sustained) competitive advantage *vis-à-vis* other networks, depending on their characteristics in terms of value, rareness, imperfect imitability, and non-substitutability, similar to the traditional RBV (Foss 1999:10).

In this context, Lavie (2006) revisits the initial conceptualization of the RBV. He specifically revisits the assumptions underlying the RBV, namely the (1) implicit proprietary assumption and the explicit assumptions regarding (2) resource heterogeneity and (3) resource immobility.

Lavie (2006:641) argues that the *proprietary assumption* no longer holds for interfirm networks and therefore must be relaxed:

> "Ownership or control of resources is not a necessary condition for competitive advantage. A weaker condition of resource accessibility, which establishes the right to utilize and employ resources or enjoy their associated benefits, may suffice."

In fact, the notion of *resource accessibility* seems to have already been adopted by scholars, even without being stipulated formally. For example, Chen & Chen (2003b:4f., italics added) argue that "firms essentially use alliances to *access* valuable resources that they do not own." Regarding *resource heterogeneity*, Lavie (2006:643) argues that, although alliances reduce heterogeneity through resource flows among partnering firms, firms' persistent resource heterogeneity remains a necessary precondition for competitive advantage of interconnected firms because, in the event of resource homogeneity, firms could not gain competitive advantage by joining resources with other firms as alliances would be formed for collusive purposes only. *Imperfect resource mobility* remains an equally critical precondition for competitive advantage of interconnected firms because, in the event of perfect mobility, any resources could be exchanged and accessed through markets, thus making the formation of alliances unnecessary. However, in imperfect mobility, alliances are vehicles for mobilizing resource flows otherwise not feasible, providing access to resources otherwise inaccessible, and enabling the sharing of resource benefits that would otherwise not be shared, thus creating the potential for competitive advantage (Lavie 2006:643). As such, resource heterogeneity and imperfect mobility remain necessary preconditions for competitive advantage of interconnected firms, albeit the formation of alliances reduces heterogeneity and increases mobility.

3.3. Rents of Interconnected Firms

In addition to revisiting the assumptions underlying the initial conceptualization of the RBV, Lavie (2006) proposes a conceptual model that aims to explain – on the firm level – the factors that influence competitive advantage of an interconnected firm. His model, depicted in Figure 10, suggests that the resource-based competitive advantage of a focal firm, which participates in an alliance, is composed of four rent types extracted from internal and network resources: (1) internal rent, (2) appropriated relational rent, (3) inbound spill-over rent, and (4) outbound spill-over rent (Lavie 2006:644, 648). The model suggests that the competitive advantage of an interconnected firm can be enhanced or reduced compared with the competitive advantage that the same firm could achieve if analyzed in terms of internal resources only (Lavie 2006:648). This is because different types of rents can make positive and negative contributions.

A focal firm can extract *internal rents* from its own shared and non-shared resources (Lavie 2006:644f.). These include Ricardian rents resulting from scarcity and quasi-rents resulting from resource specialization. Internal rents arise from positive and negative complementariness of own resources with both partners' shared and non-shared resources. Note that a focal firm's internal rent also depends on the complementariness of non-shared partner resources, as these may positively or negatively influence the value of the firm's own resources (e.g. spillovers of bad reputation).

Appropriated relational rent refers to the share of relational rent that can be appropriated by a focal firm (Lavie 2006:645). Relational rents can be extracted from resources shared intentionally among partnering firms. The RV developed by Dyer & Singh (1998) identifies four potential sources of relational rents: (a) relation-specific assets, (b) knowledge-sharing routines, (c) complementary resources and capabilities, and (d) effective governance. As relational rents are generated on the "network level" in a collective effort and are thus "owned by partnering firms" (Dyer & Singh 1998:675), the competitive advantage of a focal firm depends on the amount it can appropriate. Lavie (2006:645ff.) proposes five factors that influence a focal firm's ability to appropriate relational rents, which are the firm's (a) relative absorptive capacity, (b) relative scale and scope of resources, (c) contractual agreement, (d) relative opportunistic behavior, and (e) relative bargaining power. Among other things, Lavie (2006:646f.) proposes that the stronger a focal firm's bargaining power, the greater its appropriated share of relational rent.

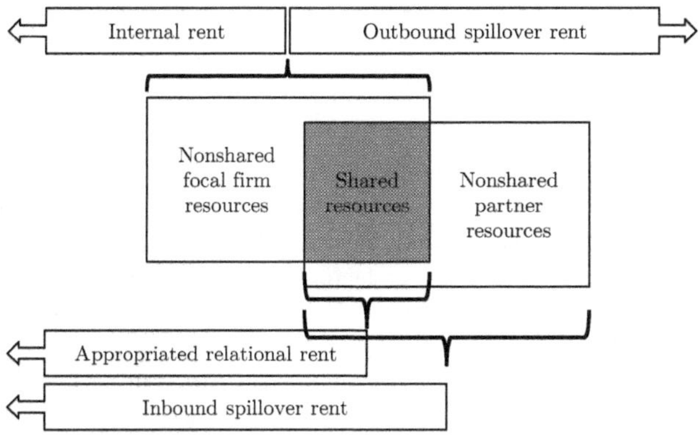

Figure 10.: Composition of Rents Extracted by the Focal Firm in an Alliance (Lavie 2006:644)

Inbound spillover rent refers to unintended gains (e.g. knowledge leakage) exclusively enjoyed by a focal firm that originate in both shared and non-shared resources of alliance partners (Lavie 2006:647). The amount of inbound spillover rent appropriated by a focal firm is positively related to the focal firm's absorptive capacity, bargaining power, and opportunistic behavior (Lavie 2006:647) and negatively related to the strength of isolating mechanisms deployed by the alliance partners to protect their resources (Lavie 2006:647f.).

Conversely, *outbound spillover rent* refers to unintended gains that alliance partners can extract from a focal firm's shared and non-shared resources, thus diminishing the firm's competitive advantage (Lavie 2006:648).

4. Interim Summary

This chapter has investigated potential sources of competitive advantage of "sole" and of "interconnected" firms. A sole firm's potential to attain (sustained) competitive advantage depends on the characteristics of its internal resources in terms of value, rareness, non-substitutability, and non-imitability. The formation of interfirm networks results from firms' heterogeneous resource endowments and imperfect resource mobility. Whether a specific firm establishes or joins an interfirm network depends on the characteristics of its internal resources. A firm's potential to attain (sustained) competitive advantage by participating in an interfirm network depends on the extent to which it can appropriate rents resulting from its own resources, partner resources, and network resources.

V. Organization of Interfirm Networks

1. Conceptualizing the Organizational Structure of Interfirm Networks

As has already been discussed, the fundamental problem of organization can be conceptualized as obtaining cooperation among a collection of economic actors whose goals are only partially congruent (Ouchi 1979:833; Section C.III.1). While TCE is concerned with *economic* organization and thus with the factors that determine which institutional arrangement (governance structure) is used to achieve cooperation in the sense of carrying out transactions, TCE provides little insight into the nature of administrative arrangements deployed *within* markets, hybrids, and hierarchies. After all, administrative controls and incentives are characterized solely by their degree of intensity. Grandori (1997a:898) observes that although interfirm networks have been analyzed from various theoretical perspectives, including TCE, "the view of these interfirm organizational arrangements has remained somehow opaque" (also see Albers et al. 2016:582). This is because legal structure turns out to be a weak proxy for predicting either general administrative arrangements or specific degree of hierarchical control in interfirm alliances (Gulati 1998:783; Reuer & Ariño 2007:315 Albers et al. 2016:583). The lack of precise knowledge about the internal administrative arrangements is problematic because organizational behaviors and outcomes cannot be explained and predicted in a satisfactory manner based solely on the legal structure of interfirm networks, but a more fine-grained approach is required – as offered by organization theory – that penetrates interfirm networks beyond their legal structure and thus reveals internal administrative arrangements (Grandori 1997a:898f.; Albers et al. 2016:582ff.).

A firm's internal administrative arrangements are commonly subsumed under the construct of *organizational structure*. The role which organizational structure plays in solving firms' specific organizational problems is concisely captured by Mintzberg (1979:2, italics and bold original):

> "Every organized human activity—from the making of pots to the placing of a man on the moon—gives rise to two fundamental and opposing requirements: the *division of labor* into various tasks to be performed and the *coordination* of these tasks to accomplish the activity. **The structure of an organization can be defined simply as the sum total of the ways in which it divides its labor into distinct tasks and then achieves coordination among them.**"

The construct of organizational structure has long been a focus for scholars and is used to distinguish between different organizational forms; to analyze, explain, and predict organizational behavior and outcomes; and to make comparative analyses between different forms (see e.g. Pugh et al. 1968).

However, because we are concerned with interfirm networks, we are essentially concerned with the administrative arrangements deployed within interfirm networks rather than in firms. In general, following Mintzberg (1979), the *organizational structure of interfirm networks* can be conceived of as *the total sum of ways in which three or more partnering firms divide the activities that are to be performed cooperatively into distinct tasks and then achieve coordination among them.* Grandori & Soda (1995) review several theories used in interfirm network research and thereby establish three concrete dimensions to conceptualize the organizational structure of interfirm networks. Grandori & Soda (1995:199, italics original) specifically propose to distinguish between networks

> "whether they are *formalized* or not (due to the support of exchange or associational formal contracts); whether they are *centralized* (there is a central coordinating firm) or parity-based; their characteristic *mix of coordination mechanisms*."

Thus, the organizational structure of interfirm network consists of (1) formalization, (2) centralization, and (3) a characteristic mix of coordination mechanisms. The following three sections introduce these dimensions in more detail.

2. Centralization

Centralization (or decentralization) is concerned with "the locus of authority to make decisions affecting the organization" (Pugh et al. 1963:304). With regard to a single organization, the degree of centralization is *high* if authority resides with few or one individual(s) (at the extreme) and *low* if authority is widely dispersed among organizational members (Mintzberg 1980:326; also see Hall 2002:37). Accordingly, with regard to interfirm networks, we consider centralization (or decentralization) to be concerned with the locus of authority to make decisions affecting the network, either as a whole or parts thereof and thus all or a subset of its members, where "locus of authority" specifies the distribution of authority *between* network members rather than *within* member firms. Thus, the degree of centralization is measured on the level of member firms. The degree of centralization in interfirm networks is *high* if decision authority resides with few or a single firm(s) (at the extreme) and *low* if authority is widely dispersed among network members (Grandori & Soda 1995:199; Provan & Kenis 2008:233f.).

Authority can be defined as the "influence [over actions] that is accepted as legitimate by those over which it is held" (Child 1972:9; see similar Gulati et al. 2012:573). Scholars typically distinguish between two types of authority, which differ based on the sources from which they are derived (Pugh et al. 1963:304): (1) formal authority, and (2) informal authority. *Formal, or institutional, authority* ultimately derives from ownership. As Baker et al. (1999:56) explain, authority always resides at the top of an organization, as delegated decision-making power may always be retracted by actors higher up in the hierarchy until one reaches a sole proprietor or a group of shareholders, who own all organizational decision rights. *Informal authority* can derive from a central position in a cooperative network; control over the flow of clients, key resources and technology, expertise, reputation, status; and gatekeeping privileges over the system boundary (see e.g. Gulati et al. 2012:573; Provan & Kenis 2008:235). Informal authority can substitute for formal authority and achieve similar outcomes (Gulati et al. 2012:573; Ouchi 1979:834). In an interorganizational setting, informal authority wielded by one firm over another can become formalized by concluding a written cooperation contract that distributes decision power among partnering firms.

Few scholars provide typologies for the distribution of decision-making authority among network members (see e.g. Park 1996; Provan & Kenis 2008). In the following, we present the model proposed by Provan & Kenis (2008), which is directly relevant to this thesis. They propose three distinct alternatives for authority to be distributed in interorganizational networks. These alternatives are referred to as "governance modes," which can be understood as specific forms of hybrid governance structures, as the distribution of authority directly impacts how adaptations can be made, either through bargaining or fiat. These modes are distinguished along two dimensions: (1) whether network governance is "shared" or "brokered" and, if brokered, (2) whether the network is member- or externally governed.

Shared governance (or self-governance) is characterized by a low degree of centralization because authority, at least regarding network-level decisions, is widely dispersed and more or less symmetrically distributed among network members (Provan & Kenis 2008:234f.; also see similar governance modes, "mutual adjustment" and "alliance," as described by Park 1996:809ff.). Accordingly, conflicts are resolved via negotiation and bargaining (Park 1996:810). Furthermore, the collective of members is involved in and responsible for managing internal network activities and external network relations, for example with customers (Provan & Kenis 2008:234f.; also see similar governance modes "mutual adjustment" and "alliance" as described by Park 1996). This results in many direct interactions among network members, which are a constituent feature of this dense and highly decentralized governance form (Provan & Kenis 2008:233f.). The involvement of members in key decisions, conflict resolution, and network management is critical for sustaining their commitment to accomplishing network-level goals (Provan & Kenis 2008:234), especially because there is no central entity to control whether members' activities are in accord with network-level goals (Park 1996:810).

Brokered network governance is characterized by a high degree of centralization because a single entity has substantial, yet not necessarily unrestricted, authority over pivotal decisions about the whole network, its maintenance, and its survival (Provan & Kenis 2008:234f.; also see Park 1996:812). This entity is furthermore responsible for providing administration for the network and managing network-level activities, which may include facilitating the activities of member organizations in their efforts to accomplish the network's goals (Provan & Kenis 2008:234f.; also see "trilateral" Park 1996:812). Thus, brokered governance is characterized by few direct member-to-member interactions – except regarding operational resource and information flows – as governance is brokered by a single entity (Provan & Kenis 2008:234). Provan & Kenis (2008:235f.) further distinguish between network governance being brokered by a (1) lead organization or (2) network administrative organization.

Lead organization governance refers to a case in which a single organization – that still provides its own services within the network – assumes the role of a network leader. As leader, this organization has exclusive authority over all crucial network-level decisions (Provan & Kenis 2008:235f.). Thus, governance is highly centralized; decision power resides in a single organization and, hence, is asymmetrically distributed among its members (Provan & Kenis 2008:235). The lead organization also provides administration for the network, coordinates network members, and facilitates their efforts in accomplishing the network's goals, which may be closely aligned with those of the lead organization (Provan & Kenis 2008:235; also see Gulati et al. 2012:573). It may also defray the network's administrative costs or receive contributions from network members or network-external entities (Provan & Kenis 2008:235f.), such as membership fees. The lead organization may emerge from the network members due to substantial informal authority (Provan & Kenis 2008:235; also see Gulati et al. 2012:573) or be mandated by an external funding source (Provan & Kenis 2008:235f.). Lead organization governance typically occurs in buyer/supplier relations, where one organization is significantly larger than others, and in horizontal networks, where one organization has sufficient resources and legitimacy to take on a leading role (Provan & Kenis 2008:235).

Network administrative organization (NAO) governance refers to a case in which a separate administrative entity is established that, unlike the lead organization, does not provide its own services in the network but exclusively focuses on governing (Provan & Kenis 2008:236). The NAO may be a government entity, a non-profit, or a unique for-profit entity; it may be very modest in scale, comprising only a single person, or be a formal organization with dedicated staff, directors, and board structures (Provan & Kenis 2008:236). Two different types of NAOs are conceivable: (1) voluntarily erected by network members, or (2) externally mandated (Provan & Kenis 2008:236; also see Park 1996:809).

First, an NAO can be *voluntarily erected* by network members and ultimately controlled through board structures which are staffed with representatives from all, or a subset of, network members (Provan & Kenis 2008:236). Board structures enable members to monitor the NAO's activities, set the boundaries of its authority, and become involved in making (strategic) network decisions, while at the same time leaving more operational issues to the discretion of the NAO directors and staff (Provan & Kenis 2008:236). Alternatively, members may staff the network organization with their own personnel to control and influence its activities (Park 1996:809). Thus, in NAO governance, authority over network-level decisions is only partially centralized, while authority over less important decisions is highly centralized.

Second, the creation of an NAO may also be *mandated externally* (Provan & Kenis 2008:234; also see Park 1996:812). In that case, the NAO represents a third party's interests, and organizations are forced to join the network either by law or in order to avoid high costs associated with non-affiliation (Park 1996:812). Network members thus lose their autonomy to the extent that it is exercised by the NAO. Decision authority is highly centralized, and members cannot influence the NAO's policy; in other words, power is asymmetrically distributed. However, due to high costs and legal regulations, members cannot leave the network. An externally managed NAO may also be established temporarily to resolve conflicts or to support critical tasks that are prone to conflict, such as performance evaluation (Park 1996:812).

3. Formalization

Various definitions of formalization can be found in organizational literature (see e.g. Pugh et al. 1963; Ford & Slocum 1977; Hall 2002; Vlaar et al. 2007). Two distinct but related aspects of formalization are recurrently discussed: (1) the degree of specification (or predefinition/prescription) and (2) the degree of codification.

With regard to the *degree of specification*, formalization refers to "the degree to which rules and procedures within a system are specified and/or adhered to" (Ford & Slocum 1977:562; also see e.g. Pugh et al. 1963; Hall et al. 1967) – regardless of whether these rules and procedures are codified in writing (also see Hall 2002:63f.). Ouchi (1979:834f.) indicates that formalization pertains to both the behaviors of organizational actors and the outputs they produce. More specifically, formalization can pertain to (a) work processes standardized through rules, procedures, work instructions, and policy manuals (Mintzberg 1980:325); (b) the operation of procedures related to decision seeking, conveyance of decisions, instructions, and information, which may include feedback on performance (Pugh et al. 1963:303; also see Pugh et al. 1968:75f.); (c) the specification of roles, authority relations, and penalties for violating rules (Hall et al. 1967:907); and (d) responses to contingencies faced by an organization (Hall 2002:63) or by the collective of firms in a cooperative network (Albers 2010:208).

With regard to the *degree of codification*, formalization generally refers to the overall extent of written documentation (Daft 2001:9) and, more specifically, the extent to which organizational rules, procedures, and roles that prescribe behaviors and/or outputs are *codified* in writing (see e.g. Pugh et al. 1963:303f.; also see Pugh et al. 1968:75f.; Tolbert & Hall 2009:33). Written documents typically include operation manuals, policy manuals, and employee handbooks (Pugh et al. 1968:75). Yet, with the diffusion of IT systems, codification may not only manifest in the existence of physical documents, but also in the "workflows" of software applications that structure value creation, interaction, and decision-making processes (Orlikowski & Robey 1991:154f.) and in data models that structure stored information (Dewett 2001:322f.).

With regard to interorganizational relationships, *contracts* are the main means of formalization (Grandori & Soda 1995:197f.; van Laarhoven et al. 2000:432). A cooperation contract encodes rules and procedures, often as part of administrative and technical clauses, that regulate contingencies and involved parties' behaviors and/or outputs (Vanneste & Puranam 2010:188; Albers et al. 2016:598). The degree of interorganizational formalization is the degree to which relationships are "explicitly regulated and safeguarded by contractual provisions" (Grandori & Soda 1995:198; also see Gulati & Singh 1998:781; Gulati 1998:294ff.). Contractual safeguards include periodic written reports of all relevant transactions, designation of certain information as proprietary and subject to confidential provisions of the contract, arbitration clauses, lawsuit provisions (Parkhe 1993:829), and penalty clauses in case of nonperformance (van Laarhoven et al. 2000:432). In this regard, interorganizational formalization directly refers to TCE (see Chapter C.III).

To synthesize, we understand formalization in interfirm networks as the degree to which rules and procedures that prescribe contingencies and network members' behaviors and/or outputs are specified and/or adhered to and the degree to which these rules, procedures, and contingencies are codified in a cooperation contract or IT system used in the cooperation.

Although formalization is a separate structural dimension, it is important to recognize that it is intertwined with the other structural dimensions by *stating their form* (Kieser & Walgenbach 2007:170). For example, understanding formalization as prescribing behaviors and/or outputs through rules and procedures links formalization to authority and thus to centralization. This is because rules and procedures need to be enacted and enforced by an entity that wields sufficient authority (Ouchi 1979:834; Kieser & Walgenbach 2007:169). Otherwise, behaviors and outputs cannot be effectively controlled through formalization. In this context, it is important to terminologically distinguish between "formal" and "informal" in the sense of, on the one hand, relating to formally assigned authority or informal authority, and, on the other hand, being specified and codified in written form. Formalization is also intertwined with coordination mechanisms. As Grandori (1997b:39) remarks, "[a]ll the described coordination mechanisms can be either formal or informal" in the sense of whether they are codified in writing.

In structuring interorganizational relationships, formalization can manifest in a variety of functions and dysfunctions, as depicted in Table 10, with functions defined as consequences conducive to achieving desired ends, and dysfunctions as consequences obstructing the achievement of goals. These functions and dysfunctions are identified by Vlaar et al. (2007) in a review of intra- and interorganizational literature, and they show how formalization is intertwined with other dimensions of interorganizational structure, for example by explicitly expressing coordination and control. Moreover, they show how formalization can specifically support collective action in interorganizational relationships.

4. Coordination Mechanisms

4.1. Taxonomy of Coordination Mechanisms and the Role of IT Systems

Mechanisms are processes or (activities) that are able to cause or prevent change to individual system components or the system as a whole (Bunge 1997:414; see Section B.II.2). The meaning of "coordination" in organizational theory largely resembles the meaning in common language (Grandori 2000:91), and can be defined as "the act of making all the people involved in a plan or activity work together in an organized way" (Cambridge University Press 2016). Thus, coordination mechanisms can be conceived of as processes that cause or prevent change among actors in a way that facilitates the integration of their tasks in order to achieve a specified goal.

Coordination among actors implies communication: "Without information flowing about what everybody should do and how they progress, coordination is impossible" (Groth 1999:46). Depending on the nature of communication needs, mechanisms can be classified into two basic categories: (1) coordination by feedback and (2) coordination by program (Groth 1999:50; Kieser & Walgenbach 2007:105f.). *Coordination by feedback* implies continuous and ad-hoc communication throughout the working process. The need for communication arises as actors observe the effect of their own and others' actions while working (Groth 1999:50). Coordination is thus achieved in real time. *Coordination by program* implies no or only minimal communication during the process of carrying out work. The need for communication while working is minimized by means of instructions, plans, etc. that are developed beforehand and specify exactly what is to be done or how work is to be carried out (Groth 1999:50). Coordination is thus achieved through advance communication.

	Functions	
1.	Coordination: formalization as a means to achieve concerted action	Division of labor Common language/communication Signaling device Fueling interaction processes
2.	Control: formalization as a means to restrain and direct behavior	Control of the partner organization Control of deviation from objectives Control of progress/pace setting Control from a distance Option to forgo control Trusting the partner
3.	Legitimacy: formalization as a means to persuade and convince stakeholders	Internal legitimacy External legitimacy
4.	Cognition and learning: formalization as a means to make sense	Accuracy expectations Degree of ambiguity Focusing attention ...
	Dysfunctions	
1.	Inhibiting	Creativity Innovation Flexibility Mutual accommodation
2.	Reducing	Commitment and aspirations Initiatives that fall beyond specifications
3.	Driving out	Intrinsic motivation
4.	Inducing the risk of	Areas of unilateral dependence Hold-up problems
5.	Imposing	High costs Incompleteness Limited enforceability
6.	Creating the conditions for	Data manipulation, Organizational strife Short-terminism

Table 10.: Functions and Dysfunctions of Formalization
(adopted with minor changes from Vlaar et al. 2007:442f.)

Besides centralization and formalization, interfirm networks can be differentiated by their characteristic mixes of coordination mechanisms. This raises the question of what mechanisms networks employ to achieve coordination among partnering firms. Various coordination mechanisms are described in literature (see e.g. Mintzberg 1979; also see Kieser & Walgenbach 2007; for interorganizational networks see e.g. Grandori & Soda 1995; Grandori 1997b; Grandori 1997a; Nassimbeni 1998). Table 11 gives a selected overview of mechanisms found in existing network literature: it becomes evident that networks draw on a wide variety of mechanisms.

Author(s)	Identified Coordination Mechanisms in Networks
Grandori & Soda (1995)	Communication, decision, and negotiation mechanisms
	Social coordination and control
	Integration and linking-pin roles and units
	Common staff
	Hierarchy and authority relations
	Planning and control systems
	Incentive systems
	Selection systems
	Information systems
	Public support and infrastructure
Grandori (1997a)	Communication and decision procedures
	Mutual monitoring
	Supervisory hierarchy
	Programming
	Hierarchical decision-making for inter-unit adjustment
	Group decision-making
	Property-rights sharing
	Integration and liaison rules
	Authority by exception and residual arbitration
Nassimbeni (1998)	Direct supervision
	Standardization of input/output
	Standardization of process
	Standardization of skills
	Mutual adjustment
Fjeldstad et al. (2012)	Protocols for division of labor
	Protocols to advertise problems or opportunities
	Protocols to search for potential collaborators
	Protocols for mobilizing network members
	Protocols for allocating costs and revenue

Table 11.: Coordination Mechanisms in Interfirm Networks (Selection)

When considering coordination in interfirm networks in this thesis, we refrain from providing a comprehensive review of the mechanisms described in literature, as this is not helpful for our purposes. Instead, we draw on the well-known taxonomy of coordination mechanisms initially proposed by Mintzberg (1979) and later extended by Groth (1999). On the one hand, this taxonomy has already been deployed in a network context (see Nassimbeni 1998). On the other hand, this taxonomy seems instructive because several of the "specific network mechanisms" described above can be derived from mechanisms in this taxonomy. For example, communication, decision, and negotiation mechanisms can be conceived of as standardization of work processes. In other words, instead of reviewing a variety of specific network mechanisms, we focus on few fundamental coordination mechanisms on which they are based.

The basic taxonomy of coordination mechanisms is depicted in Figure 11. Mintzberg (1979) proposes the initial set of five mechanisms: mutual adjustment, direct supervision, standardization of work, standardization of skills, and standardization of output. Groth (1999) extends this taxonomy by adding a set of technology- and computer-dependent coordination mechanisms, which include implicit coordination, system-supported supervision, programmed routines and hyperautomation, and system-supported skills. However, Groth (1999) does not propose an IT enabled mechanism for standardization of output. We believe that service-orientation, a paradigm for software design (see below Section D.III.3), represents a computer-dependent extension of this mechanism. We therefore add this mechanism to the taxonomy.

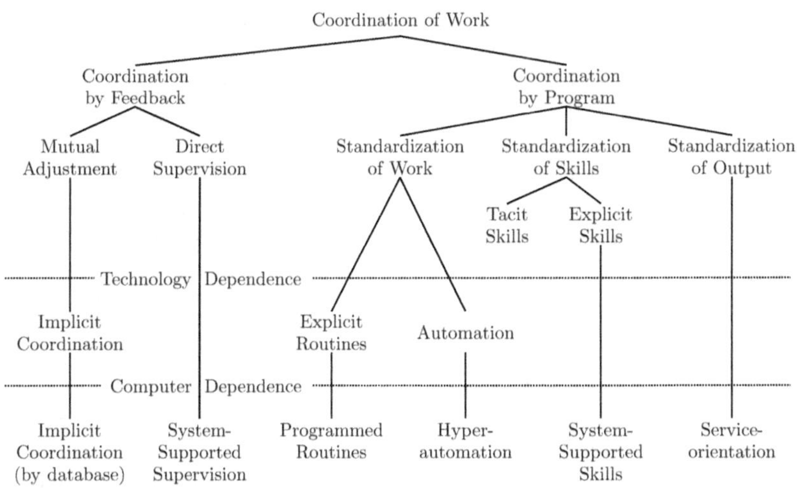

Figure 11.: Taxonomy of Coordinating Mechanisms Extended by the Use of IT (adopted with minor changes from Groth 1999:330)

IT systems can facilitate coordination by creating horizontal communication linkages between various functions or departments in an organization, thus achieving coordination (Daft 2001:37). Similarly, IT systems can be deployed to establish horizontal communication and coordination linkages between firms in networks. In fact, we have already discussed the role of IT systems with regard to electronic markets and electronic hierarchies (see Section C.III.4). Due to the importance of IT systems for coordination, Grandori & Soda (1995:197) even conceive of IT systems as a separate coordination mechanism. Specifically, three characteristics of IT systems are critical regarding communication and coordination: (a) immense and loss-less storage capacity, (b) quick and less costly communication, and (c) immense and less costly information processing capacity (Groth 1999:144; also see Argyres 1999). The computer-dependent mechanisms exploit these characteristics to achieve coordination.

The following sections elaborate on the mechanisms initially proposed by Mintzberg (1979) and their computer-dependent correspondents proposed by Groth (1999).

4.2. Mutual Adjustment and Implicit Coordination (by Database)

In *mutual adjustment*, individuals communicate informally with each other in order to coordinate their work (Mintzberg 1980:324). As there is no individual or organization that continually directs or supervises the collaborative work, mutual adjustment refers to coordination among equals (Groth 1999:51) and therefore relies, in principle, upon negotiation and bargaining (Albers 2005:74, 118). Coordination among equals requires everybody involved to be sufficiently motivated to contribute voluntarily and have a sufficient understanding of their overall goal, the overall structure of the work, and how their individual efforts fit together to accomplish their goal (Groth 1999:51). Although mutual adjustment is based on rather continuous informal communication processes, it can also be applied in more institutionalized forums, such as boards and committees that are summoned on a regular basis, with decisions either made unanimously or by majority rule (Kieser & Walgenbach 2007:113ff.).

Implicit coordination (by database) is the computer-dependent variant of mutual adjustment. Coordination is achieved by recording relevant information in an information repository and making this information accessible to the actors involved in an activity (Groth 1999:303). Implicit coordination is essentially related to mutual adjustment because capturing information in a common information repository allows actors to communicate indirectly with each other, keeping everybody up to date about the activities pursued. In other words, the repository coordinates the actors' behaviors, as any (new) piece of information has the potential to modify the behaviors of other actors who access the repository (Groth 1999:317).

While implicit coordination can basically be achieved by recording information on paper stored in file cabinets, IT features boost its power: Information can now be captured in computer-based *structured databases* that enable automatic indexing, search and retrieval, and electronic communication of information (Groth 1999:303). Structured databases derive their enormous coordinative power from their wide geographical *reach* due to pervasive internet access; their *capacity* to coordinate a large number of actors simultaneously due to computers' large and quickly growing processing power; and the high *speed* with which (large) pieces of information can be stored and retrieved electronically due to the internet's quickly growing bandwidth (Groth 1999:304ff.). In other words, the use of structured databases enables efficient coordination among an almost unlimited number of geographically dispersed actors in real-time (Groth 1999:316).

In interorganizational settings, structured databases are crucial for coupling independent organizations (Groth 1999:316). In principle, they represent IT systems that create horizontal linkages for communication and coordination purposes (see Subsection C.V.4.1). Thus, the implicit coordination mechanism is directly analogous to the use of databases in electronic hierarchies (see Subsection C.III.4.1). Groth (1999:306ff.) states that cooperating firms can either link their databases using online mechanisms or use a common shared database. However, he argues that implicit coordination reaches its full potential only if a single unified database is deployed, avoiding software incompatibility, minimizing synchronization needs, and ensuring full integrity of information.

Regarding the use of shared databases, Groth (1999:307) remarks in his dissertation published in 1999, that a unified single database may not be feasible in the near future, especially in an interfirm context. However, approximately 18 years later, innovations in the IT industry may have brought us much closer to unified interfirm databases. In fact, cloud computing can become instrumental to achieve this goal because it enables to centralize data storage in the internet. Data stored on legions of personal computers and intra-firm servers can now be relocated to few(er) online databases hosted by rather few specialized IT vendors.

4.3. Direct Supervision and System-supported Supervision

In *direct supervision*, a separate individual (usually a manager) is exclusively occupied with giving orders to other individuals in order to coordinate their work efforts (Mintzberg 1980:324). The manager is granted authority rights to direct and monitor the behaviors and outputs of others (Groth 1999:52), which is why direct supervision is similar to coordination by authority or fiat (Albers 2005:75). In interfirm networks, direct supervision implies that member firms have authorized one member firm or a third party to hierarchically direct and supervise their value adding activities performed to achieve the network's goals (Albers 2005:119f.; also see Grandori & Soda 1995:195). Such authority relationships can be established through contractual agreements between network members (Grandori & Soda 1995:195).

System-supported supervision represents the IT enabled extension of the direct supervision coordination mechanism (Groth 1999:330). System-supported supervision can be defined as "conscious direction of work to great depth and/or great breadth in an organization, based on information gathered and presented through computer-based systems" (Groth 1999:331). This means that managers coordinate work through computer systems instead of being on site in person, as systems enable them to remotely direct and monitor subordinates' work in a sufficient degree of detail in real time or with a negligible time lag (Groth 1999:331). Managers give directions to subordinates directly via personal communication (e.g. phone or email), indirectly in the form of new parameters in software applications or new routines, or as an immediate consequence of concrete actions, such as the delivery of goods that subordinates should sell (Groth 1999:331).

Directing work from a distance, however, requires the ability to make sensible decisions that adequately respond to local conditions and developments. To this end, managers extensively rely on computer systems to provide a sufficient information basis. This basis derives from three central abilities of computer systems. First, they can increase information availability by capturing more information and by making this information more readily accessible, for example through structured databases (Groth 1999:319f.). We believe that the concept of the IoT is important for gathering this information as "the IOT is all about sensing the physical world" through a dense network of sensors (Fleisch 2010:130; see Section B.V.3). The IoT is poised to become even more important, especially when sensing costs decrease (e.g. due to technological innovations), sensing efforts intensify in (geographical and organizational) scope, frequency, and richness of data (Fleisch 2010:144). Second, computer systems can concentrate large amounts of raw information into a form that is easier to grasp, for example by means of programmed rules and functions that transform information into compound measures or graphs (Groth 1999:320f.). Finally, computer systems can expose deeper and more complex causal relationships that were formerly unknown or unfathomable and establish more sophisticated feedback loops with sophisticated, computer-based information analysis tools (Groth 1999:322). In sum, managers harness computer systems' abilities to enable system-supported supervision of work, directing and monitoring subordinates' work from a distance.

With regard to interorganizational networks, system-supported supervision can be defined as the conscious direction and monitoring of network members' behaviors and outputs using information gathered and presented through computer systems operated by either a network member or an authorized third party (also see "planning and control systems" in Grandori & Soda 1995:195f.; also see "direct supervision" in Nassimbeni 1998:547). In comparison to implicit coordination, system-supported supervision requires not only capturing information in a common repository, but also integrating the computer systems that gather and process this information. Thus, system-supported supervision creates very strong coupling between organizations.

4.4. Standardization of Work Processes, Programmed Routines, and Hyperautomation

The mechanism *standardization of work processes* coordinates interdependent tasks by imposing standards on how these tasks are to be carried out (Mintzberg 1980:324). Standardization of work requires extensive planning to exactly predefine how work is to be performed, which then eliminates a significant amount of communication and supervision throughout the process of doing the work because behaviors have been explicitly specified beforehand (Groth 1999:48f.). Standards often differ in the scope of work that is covered, the degree of detail to which behaviors are specified, whether they are entirely fixed or include conditional specifications of behaviors, and whether they are binding (Kieser & Walgenbach 2007:115f.). These standards are often codified in handbooks with standard operating procedures, rules, regulations, and protocols (organizational codes of conduct). In interfirm networks, standardization of work processes pertains to activities performed by partnering firms in a cooperative manner in pursuit of the network's goal(s) (Albers 2005:121). Standards regulating these processes are often codified in technical clauses in the cooperation contract (see Section C.V.3) and are thus binding for network members.

IT allows standardization of work to be extended into two new coordination mechanisms: (1) programmed routines and (2) hyperautomation (Groth 1999:277ff.). *Programmed routines* are the computer-dependent extension of explicit routines, which are written accounts of consciously modeled and designed work processes (Groth 1999:165). By contrast, programmed routines represent consciously modeled and designed routines that are "programmed *into* computer-based systems" (Groth 1999:278, italics added). Groth (1999:278) distinguishes between two types of programmed routines: closed and open. Closed routines represent the internal functions of a computer application and therefore remain hidden from users. Open routines are visible to users and they incorporate user dialogues that structure the users' work. Programmed routines are qualitatively different from explicit routines in three ways and therefore warrant a separate mechanism (Groth 1999:278f.). First, many programmed routines can be *fully automated and executed at great speed*, thus eliminating the need for coordinating human activity. Second, programmed routines can be *more numerous and diverse* than explicit routines as work is structured via a user dialogue. Users only need to remember how to operate the computer system and how to relate to the dialogue, instead of being required to learn every detail of every routine. And third, computer systems can *offer assisting features* that give users a broader span of competence and thus the ability to operate more routines and more diverse routines.

Programmed routines are important coordination mechanisms in new forms of both intra- and in interorganizational networks, as described by Fjeldstad et al. (2012:739). They argue that *protocols* (i.e. organizational codes of conduct) structure the exchange and collaboration activities of network members. Examples of such protocols are depicted in Table 11. These protocols refer to programmed routines that are implemented on the network infrastructure – an IT platform – through which all network members interact and coordinate exchanges and collaboration activities (see Figure C.V.4.1).

Hyperautomation is the second computer-dependent coordination mechanism that is related to standardization of work. Specifically, hyperautomation represents "the computer-dependent variation of automation" (Groth 1999:298). In principle, it is similar to mechanical automation, but IT allows the extent and sophistication of automation to advance far beyond the possibilities that were feasible by mechanical means alone, which is why a separate mechanism is warranted (Groth 1999:298). More specifically, IT allows a significantly greater span of organizational activities to be integrated into a single coordinated process, not least because a significant share of administrative processes can either be automated or eliminated (Groth 1999:298f.). For example, because of IT, entire factories and interorganizational and geographically dispersed production processes, such as those typically found in the automotive industry, can be supervised from a single (or few) computer terminal(s). Hence, programmed routines and hyperautomation are directly relevant to the vision of "Industry 4.0," which aims to automate production and administrative processes as well as those pertaining to self-diagnosis, self-optimization, and self-adaptation (Jasperneite 2012).

4.5. Standardization of Skills and System-supported Skills

In the *standardization of skills* mechanism, work is coordinated using standard skills and knowledge that have been internalized by individuals, typically before the work is carried out (Mintzberg 1980:324). This is usually achieved by standardizing the syllabus of educational programs and trainings (Mintzberg 1980:325). As individuals can draw on a standard set of skills and knowledge, the need for communication and supervision during work is minimized (Kieser & Walgenbach 2007:136f.). Groth (1999:52f.) differentiates between two types of skills: *tacit skills* and *explicit skills*. The former refers to skills that individuals may not be aware of and that may never or rarely become explicit, such as social skills learned during childhood. The latter refers to concrete skills that are taught in school, university, or other educational programs, such as math.

Compared with the standardization of work mechanism, which represents a specific standard or program developed by a specific organization for the purpose of coordinating specific tasks within that organization, standardization of skills, by contrast, represents a general program that enables cooperation of a specific kind, often within a specific task domain, independent of the organizational setting (Groth 1999:52; Kieser & Walgenbach 2007:136). In interfirm networks, standardization of skills can also be used to coordinate the network's tasks, either by different member firms contributing staff with similar skills and knowledge, or by standardizing the skills and knowledge of member firm's staff through network-specific training programs (Albers 2005:121f.).

System-supported skills is an IT induced extension of the mechanism standardization of explicit skills. In general, system-supported skills can be considered to be a "much improved version" of standardization of skills (Groth 1999:260). The mechanism's primary improvements focus on how skills and knowledge are "stored," how complex stored skills and knowledge can be, and how many different skills and how much knowledge can be made available simultaneously to coordinate work. Formerly, skills and knowledge needed to be internalized by individuals (often in time consuming learning processes); the capacity to acquire, maintain, and apply a diverse set of skills and knowledge is basically limited by inherent natural human barriers (Groth 1999:257). Now, skills and knowledge can be programmed into computer systems that individuals then use to carry out their work. Given the indefinite and essentially unlimited amount of storage and processing power of computer systems and their allowance of almost unlimited complexity in computer programs (Groth 1999:257), system-supported skills generally breaks down the natural barriers of human capacity. Groth (1999:257, italics original, 260) argues that system-supported skills essentially expand an individual's effective *"span of competence:"* an individual can now perform a broader range of tasks with an acceptable level of quality, which in turn further eliminates the need for coordination and control activities.

Groth (1999:258ff.) proposes three specific factors of computer systems that contribute to this increased span of competence: (a) *artificial memory*, which provides fast information retrieval (e.g. from databases) and enables us to handle more complex decisions and solve problems in a broader field; (b) *artificial intelligence*, which enables computers to make decisions according to predefined rules, especially in situations that cause cognitive overload due to complexity and in which insufficient time is available to consult humans and await their response; and (c) *embedded knowledge*, which refers to the knowledge that is embedded in computer systems by means of its data structure and functions, such as mathematical formulas.

As programmed routines and system-supported skills are both based on computers that support humans in doing their work, it may be helpful to specify the key difference between them. In principle, they differ in the same manner as standardization of work processes and standardization of skills. Programmed routines refer to software that has been developed to perform a specific pre-defined task and is often tailored to the needs of a specific organization, with the sequence of steps already fixed. Conversely, system-supported skills refer to software developed to provide functions, models, calculation tools, and data processing tools that support the coordination of a specific family of tasks. Such software usually includes no or only a roughly pre-defined sequence of steps. Examples include decision support systems, statistical analysis, computerized equipment for chemical analysis, text processing programs, spread sheets, and computer-aided design systems (Groth 1999:256ff.). Groth (1999:260) concludes that such support systems "will make it possible to improve the quality of professional work overall [...] [and] to help most professionals to adhere more closely to professional standards."

4.6. Standardization of Outputs and Service-orientation

The *standardization of outputs* mechanism achieves work coordination by imposing standard performance measures or specifications on the outputs of the work to be done (Mintzberg 1980:324). The imposition of standards is fundamentally similar to achieving coordination by setting goals or making plans, which "specify a desired output—a standard—at some future time" (Mintzberg 1979:148; also see Grandori & Soda 1995). Performance or output standards enable the coordination of work because they need to be achieved in a specified period or point in time. In interorganizational networks, standards can be imposed on the outputs of activities performed in cooperation by network members and on network members' outputs that represent inputs to the network; these standards may be (partially) encoded in the cooperation contract (Albers 2005:121; also see Nassimbeni 1998:547).

To our knowledge, no computer-dependent extension of the mechanism standardization of outputs has been proposed so far. However, we believe that such an extension already exists but has not yet been related to Mintzberg's (1979) taxonomy of coordination mechanisms. We propose *service-orientation* as a computer-dependent extension of standardization of outputs because both achieve coordination in a conceptually similar manner. A detailed description of service-orientation is provided later (see Section D.III.3). In the following, we solely focus on its coordinative power.

Service-orientation is a software design paradigm that comprises several design principles, which collectively govern the design of services (Erl 2008:37f.), where a service can be intuitively defined as some well-defined piece of functionality (Haas 2004). A service's functionality, conditions, and constraints of use are encoded in a *service description* (Krafzig et al. 2004). Thus, from a coordination perspective, this description specifies the output of a service. Three design principles of service-orientation are particularly relevant to its coordinative power: (1) service composability, (2) standardized service descriptions, and (3) service loose coupling.

Service composability is the intended result of applying the remaining design principles of service-orientation, and it means that services need to be designed in a way that they can become effective members of compositions (Erl 2008:74, 393), where compositions represent "coordinated aggregate[s] of services" (Erl 2008:39). Hence, from a coordination perspective, all design principles in service-orientation aim to achieve coordination among a set of services to accomplish some overall objective. The principle of *standardized service descriptions* aims to impose limits on how the functionality, conditions, and constraints are encoded in service descriptions in order to enable services to function together well (Erl 2008:56, 133f., 137f.). Thus, from a coordination perspective, interfaces between different sub-tasks are standardized. The principle of *service loose coupling* aims to minimize dependencies between the service description and the technological implementation of services (Erl 2008:71, 167ff.). Thus, from a coordination perspective, service-orientation imposes standards on the outputs of services but not on how work is performed. If different services are delivered by different providers, service-oriented service design achieves coordination among these providers.

5. Interim Summary

This chapter has penetrated interfirm networks beyond their legal structure to unveil their administrative arrangements. These arrangements are subsumed under the term "organizational structure," which refers to the entirety of ways in which network members divide activities pursued cooperatively into specific tasks and then coordinate among each other in pursuit of a common goal. The organizational structure of interfirm networks is conceptualized by using three dimensions: centralization, formalization, and a characteristic mix of coordination mechanisms. Centralization refers to the distribution of authority among network members. Formalization refers to the specification and codification of rules and procedures that prescribe behaviors and outputs of network members in the cooperation contract or within IT systems used by the interfirm network. Different network forms use different characteristic mixes of coordination mechanisms. Instead of reviewing different network forms and their specific mechanisms, a basic taxonomy of coordination mechanisms has been introduced, from which many network-specific coordination mechanisms can be derived. This taxonomy includes five fundamental coordination mechanisms and their computer-dependent variants.

VI. Networked Horizontal LSP Cooperation

1. Definitions, Classifications, and Types

While LSPs traditionally focus on cooperation with shippers (vertical cooperation), a growing number of providers has started to shift their focus toward horizontal cooperation with their competitors (Carbone & Stone 2005; Cruijssen et al. 2007b; Schmoltzi & Wallenburg 2011:553). Wallenburg & Raue (2011:390) survey LSPs with a legal entity in Germany and find that roughly 60% of total respondents are involved in at least one horizontal cooperation, emphasizing the empirical relevance of this cooperative form.

Various definitions of horizontal LSP cooperation can be found in literature. A selection is depicted in Table 12. In general, these align with the definitions and features of interfirm networks (see Section C.II.1). Yet, because we are focusing on "horizontal" LSP cooperation, these definitions specify the relative positioning in the value creation process in detail: "operating on the same level of the supply chain," where it has to be noted that this supply chain refers to the supply chain of logistics processes. Following these definitions and the general definition of interfirm networks adopted above, we define horizontal LSP cooperation as *a group of three or more legally autonomous LSPs that operate on the same level of the logistics supply chain and work together to achieve not only their own goals but also a collective goal.*

Scholars propose several frameworks to classify and identify different types of LSP cooperation (see e.g. Cruijssen et al. 2007c:24ff.; Verstrepen et al. 2009:234ff.; Schmoltzi & Wallenburg 2011:554ff.). The classification proposed by Schmoltzi & Wallenburg (2011) is instructive for the purposes of this thesis. These authors use six dimensions to classify horizontal LSP cooperation: (a) *contractual scope*, which refers to the degree of formalization of the cooperation contract; (b) *organizational scope*, which refers to the number of cooperating LSPs; (c) *functional scope*, which refers to the number of business functions

Author(s)	Proposed Definition
Cruijssen et al. (2007c:26)	"cooperation between two or more firms that are active at the same level of the supply chain and perform a comparable logistics function on the landside."
Verstrepen et al. (2009:229)	"partnerships between companies that operate at the same level of the market."
Schmoltzi & Wallenburg (2012:54)	"voluntarily initiated, long-term relationships among autonomous LSPs that operate on the same stage of the supply chain as close or distant competitors and that strive for benefits that could not be achieved by the individual companies alone."

Table 12.: Selected Definitions of Horizontal LSP Cooperations

that partnering LSPs integrate in order to manage and support the management of the logistics activities performed in cooperation; (d) *service scope*, which refers to the type of logistics services offered by a cooperation; (e) *geographic scope*, which refers to the geography within which a cooperation offers its logistics services; and (f) *resource scope*, which refers to the similarity (or complementarity) of partnering LSPs in terms of their core competencies, geographical coverage, and corporate culture and structure. Figure 12 provides an overview of these dimensions and of each of their alternative manifestations. The shading implies the relative frequency with which each of these manifestations occurs in their survey of horizontally cooperating LSPs with a legal entity in Germany. It is important to note that the visualization does not imply any limitation in the combinability of these dimensions; for example, a cooperation offering road services can, of course, be multi-lateral.

By means of cluster analysis, Schmoltzi & Wallenburg (2011:564ff.) identify six distinct cooperation types. For example, type 1 refers to cooperation among a large number of LSPs with highly similar competencies and geographic coverage that collectively offer road-based transportation services in large international networks. The cooperation is formalized in an equity-based agreement. Type 5, by contrast, refers to cooperation among very few LSPs that collectively offer non-transportation related value-added services, such as packaging and warehousing. This type involves only a few LSPs that have integrated several of their business functions into the cooperation, and it is based on an equity agreement.

2. Motives and Antecedents

Logistics literature provides good insight into the motives of LSPs for voluntarily establishing cooperative relationships with proximate or distant competitors. These specific motives largely match the general ones for forming an interfirm network described above (see Section C.II.2). In a comprehensive literature review, Cruijssen et al. (2007c:28ff.) identify four categories of motives: (a) *reduced costs and increased productivity*; (b) *improved customer service*, through specialization, complementary goods and services, and the ability to better comply with sophisticated customer requirements; (c) *improved market position* achieved by penetrating new markets, developing new products, serving larger clients, protecting market share, and reducing time to market; and (d) *other*, such as developing standards, overcoming legal or regulatory barriers, and gaining access to superior technology. The relevance of these motives is corroborated empirically by surveys of Flemish and Dutch LSPs (Verstrepen et al. 2009:231ff.) and LSPs with a legal entity in Germany (Schmoltzi & Wallenburg 2011:559f.). Schmoltzi & Wallenburg (2011:559f.) specifically find that external, market-oriented motives (such as improved service quality, enhanced market share, and specialization while broadening service scope) outweigh

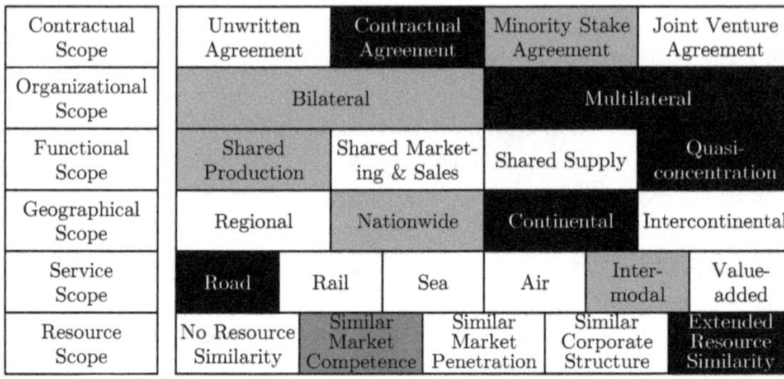

Figure 12.: Overview of Dominating Characteristics of Horizontal LSP Cooperation (adopted from Schmoltzi & Wallenburg 2011:563)

internal, cost-oriented motives (such as increased productivity).

Antecedents for the formation of horizontal LSP cooperation found within logistics literature may be clustered into three categories: (a) *stiff global competition* combined with high fixed costs, rising variable costs (e.g. fuel), and low capacity utilization (Verstrepen et al. 2009:229) leading to low overall industry profitability (Midoro & Pitto 2000:32); (b) *increased customer expectations* (Verstrepen et al. 2009:229), such as customers asking for increasingly complex and customized service bundles with global reach offered by a one-stop shop (Schmoltzi & Wallenburg 2012:53; also see Raue & Wallenburg 2013:217; Midoro & Pitto 2000:32); and (c) *rising complexity and dynamics* in today's supply chains, the resulting challenges LSPs face when managing or carrying out logistics activities in these settings (Wallenburg & Raue 2011:385f.), and the shift in focus toward managing and coordinating the flow of information and goods across multiple supply chain stages (supply chain orchestrators) (Schmoltzi & Wallenburg 2012:53). Implicitly, these antecedents seem to align with the antecedents described by the RBV: resource heterogeneity and imperfect mobility. LSPs lack access to "all" resources necessary to compete effectively, thus alliances become a vehicle to access resources owned by other firms.

Considering these motives and antecedents, horizontal cooperation seems particularly attractive for small and medium-sized LSPs, due to their inherent lack of scale, breadth of capability scope, and geographic reach (Klaas-Wissing & Albers 2010:494). In fact, as a result of the described antecedents, small and medium-sized LSPs have particularly filed for bankruptcy in Western Europe over the last decade (Verstrepen et al. 2009:229). Joining forces with their competitors enables them to compensate for their lack of scale and geographical reach and to successfully compete shoulder to shoulder with larger providers – also in scale driven segments such as less-than-truckload (LTL) (Albers & Klaas-Wissing 2012:181f.; also see Klaas-Wissing & Albers 2010; Krajewska et al. 2008). Horizontal cooperation further allows them to serve larger clients (Cruijssen et al. 2007c:30) and respond to larger tenders (Verstrepen et al. 2009:233). In addition, cooperating with LSPs that have a complementary capability scope enables them to offer sophisticated logistics solutions. In sum, horizontal cooperation offers viable options for small and medium-sized LSPs to compete in an increasingly globalized, dynamic, and complex market environment.

3. Governance

Logistics literature provides some insight into the governance of horizontal LSP cooperation. We specifically characterize the (1) behavior of LSPs in horizontal cooperation, (2) governance structures used in horizontal LSP cooperation, and (3) electronic freight exchanges (electronic logistics markets). Although electronic freight exchanges are not a governance structure of horizontal LSP cooperation, market-based exchanges are still frequently conducted within alliances; for example, Klaas-Wissing & Albers (2010:502) mentions the use of market-based clearing in LTL alliances. Hence, they are included in the following discussion.

3.1. Characterizing LSP Behavior

TCE assumes that economic actors behave opportunistically. Logistics scholars argue that horizontal LSP cooperation is particularly prone to opportunistic behavior of partnering LSPs due to simultaneous cooperation and competition (Schmoltzi & Wallenburg 2012:54; Wallenburg & Raue 2011:386f.; Raue & Wallenburg 2013:218f.; also see Rindfleisch 2000:82; Park & Russo 1996), a phenomenon often referred to as "co-opetition" (Cruijssen et al. 2007c:24; Schmoltzi & Wallenburg 2011:563; also see e.g. Bengtsson & Kock 2000). In fact, the risk of opportunism remains pervasive, even in trustworthy relationships (Verstrepen et al. 2009:239).

Logistics scholars point to several specific sources of opportunistic behavior in horizontal LSP alliances. First, *direct competition* (or rivalry) between partnering firms is considered a crucial source of opportunistic behavior; it is most prevalent if partnering LSPs have similar competences and serve similar geographies (Schmoltzi & Wallenburg 2012:57; Raue & Wallenburg 2013:219). This is because, in these cases, partnering firms compete for the same customers (Wallenburg & Raue 2011:386f.; Midoro & Pitto 2000:37) and may lure a partner's customers away (Schmoltzi & Wallenburg 2012:67). Second, *lower mutual dependence* between LSPs (which results from the fact that partners do not depend on other partners' outputs as inputs for their activities) leads to a quicker escalation of problems and to decreased levels of trust, increasing the risk of opportunism (Rindfleisch 2000:83f. as cited in Schmoltzi & Wallenburg 2012:54; Wallenburg & Raue 2011:387).

Next, *unclarity about the longevity of cooperation* motivates partnering LSPs to opportunistically focus on short-term gains instead of "vague" possible long-term results (Raue & Wieland 2015:409). Finally, an *increasing number of partnering LSPs* increases the likelihood of more diverse interests, goals, and behavioral attitudes, and thus increases the likelihood of opportunistic behavior (Schmoltzi & Wallenburg 2012:56). In conclusion, literature on horizontal LSP cooperation provides ample insight into the varying sources of opportunistic behavior in this kind of cooperation.

3.2. Characterizing Governance Structures

Logistics literature suggests that horizontal LSP cooperation can be governed by a variety of arrangements positioned between markets and hierarchies. As depicted in Figure 12, Schmoltzi & Wallenburg (2011:560f.) find evidence for four different contractual scopes. The vast majority of alliances they observe is based on written contracts, and only 13% use verbal agreements. They also find that equity involvement gains importance with increasing strategic importance of the cooperation. Cruijssen et al. (2007c:25f.) distinguish three types of horizontal LSP cooperation that are positioned between transactions at "arm's length" and "horizontal integration." These three types differ based on the time horizon, the degree of cooperation intensity, and the degree of contract formalization. Both time horizon and cooperation intensity increase when moving from arm's length to horizontal integration. Of these three types, only type 3 is based on a formal contract.

Considering the amplified risk of opportunistic behavior in horizontal LSP cooperation as described above, logistics scholars specifically suggest the use of governance structures that can effectively deter opportunism and, thus, minimize transaction costs and achieve cooperation success. Formalized contracts, including contractual safeguards, are therefore suggested by several logisticians (Cruijssen et al. 2007c:34; Raue & Wieland 2015; Raue & Wallenburg 2013; Albers & Klaas-Wissing 2012:188). Such contractual safeguards can take a variety of forms, including (a) market-oriented penalties and invectives, as used in networked LTL cooperation (Klaas-Wissing & Albers 2010:502); (b) defined processes for carrying out transactions and resolving disputes, procedures for monitoring, and penalties for noncompliance (Raue & Wieland 2015:404 and sources therein); (c) clear rules concerning the protection of intellectual property; and (d) clear policies for financial control and accounting transparency (Verstrepen et al. 2009:243). Furthermore, Albers & Klaas-Wissing (2012:190) suggest combining third party enforceable contractual safeguards with self-enforcing safeguards such as equity participation or financial deposits in the central management unit of an alliance.

Although formalization is considered effective in deterring opportunism, Verstrepen et al. (2009:239) observe that horizontal LSP cooperation is often established in an unstructured process, possibly due to the uncertainty and complexity associated with this type of cooperation. Furthermore, they argue that, in practice, many LSPs often view the writing of contracts as a juridical burden rather than an aid or necessity, primarily because cooperation often starts on a small scale (Verstrepen et al. 2009:242). However, in the long run, they argue, that the lack of a written contract can lead to severe problems in the case of disputes, which is why they recommend the writing of a "cooperation charter," which expresses the most important rules of the cooperation.

3.3. Electronic Logistics Markets

Electronic logistics markets do not represent a hybrid governance structure *per se*; after all, they are markets. Still, we consider electronic logistics markets here because such markets can be established *within* alliances, for example among LSPs, among shippers, or between LSPs and shippers (see e.g. Sänger 2004:190; Wang et al. 2011:612). Hence, market-based coordination penetrates hybrid governance structures in this case (Hennart 1993). Following Bakos's (1991) definition of electronic markets cited above, Bierwirth et al. (2002:335f., translated by the author), define electronic transportation markets as

> "interorganizational transaction systems that allow buyers and sellers of transportation services to specify, negotiate, and exchange these services."[1]

Other scholars provide similar definitions (see e.g. Bloos & Kopfer 2011:192). Although several electronic logistics markets have been established over the last two decades, primarily in the transportation segment and especially for road freight (see lists in e.g. Bierwirth et al. 2002:340; Sänger 2004:248f.V; Nandiraju & Regan 2003:5f.), little research attention has been paid to this phenomenon (Grieger 2003 as cited in Wang et al. 2011:612). This subsection provides a short overview of dimensions for classifying (or designing) electronic logistics markets and points out key issues that impede their diffusion or success. Key dimensions considered are (1) value creation logic, (2) openness, (3) operator structure, (4) functionality including mechanisms for price determination, (5) the services traded, and (6) the geography covered.

Value creation logic refers to the relative positioning of market participants in the logistics supply chain; it can be either horizontal or vertical. A *horizontal* market refers to a market established between LSPs that operate on the same level of the logistics supply chain (Wang et al. 2011:612), say for example forwarders. Horizontal markets serve the purpose of balancing capacity among LSPs (Pankratz 2003:3). A *vertical* market is one established between market participants that operate in different stages of the logistics supply chain, such as forwarders and carriers (Wang et al. 2011:612; Pankratz 2003:3).

[1] "zwischenbetriebliches Transaktionssystem, welches es Anbietern und Nachfragern von transportlogistischer Leistung ermöglicht, diese Leistung zu spezifizieren, zu verhandeln und auszutauschen."

Vertical markets serve the purpose of sourcing or distributing logistics services (Pankratz 2003:3). For example, Sänger (2004:180ff.) distinguishes three types of vertical markets: (a) between shippers and forwards, (b) between forwarders and carriers, and (c) between shippers and carriers (Pankratz 2003:3).

Openness refers to the existence of barriers to controlling the access of buyers and/or sellers to the market; it is generally used to distinguish open and closed markets (Wang et al. 2011:612; Nandiraju & Regan 2003:4; Pankratz 2003:3f.). Open markets are accessible to any buyer or seller as long as certain criteria are satisfied (e.g. financial solvency, service quality). Closed markets involve a fixed group of market participants with high entry barriers and are often established within a long-term cooperation (Pankratz 2003:3f.). Hence, closed markets are often tailored towards the specific needs of the buyers and sellers of logistics services (Wang et al. 2011:612f.). Sänger (2004:189f., 42) differentiates two types of closed markets: private and consortium-led. The former refers to a situation in which the market operators do not compete with each other and do not participate in the market and the latter to one in which individual market operators in the consortium are competitors and participate in the market. Closed markets established by multiple cooperating actors are also termed “collaborative markets” (Wang et al. 2011:612).

Operator structure refers to the type of logistical actors that establish the market. Markets can be established by buyers and/or sellers of logistics services, or by neutral actors (Bierwirth et al. 2002:335), where “neutral” means they are not involved in the actual delivery of logistics services, but only establish the IT system of the electronic market (Nandiraju & Regan 2003:4). Although the market operator structure and market openness are theoretically distinct dimensions, they are nevertheless often intertwined in practice. Neutral markets are generally considered as open (see e.g. Sänger 2004:189ff.), while markets established by buyers and/or sellers of logistics services can be both. However, they seem to be most frequently considered closed, at least for the types of actors who are similar to the market operator(s); for example, a market established by a group of cooperating LSPs may be closed to other LSPs but open to service buyers. The market operator structure also has implications for the existence of transaction-based fees levied by the market operator. While closed markets may be financed by the efficiency gains that participants generate by using the market, neutral markets depend on some kind of transaction-based fee or other income stream in order to pay for running the market at a profit (Sänger 2004:214ff.).

Another dimension for classifying electronic logistics markets refers to the *functionality* offered to support and carry out transactions. Functionality can be structured in three phases of transactions: (a) the *information phase*, which includes engines to find the market, participant registry services, and information services regarding the services offered (e.g. search engine, catalogue of offered services, notification services); (b) the *negotiation and contracting phase*, which includes provider information services (e.g. credit check, quality profile), product information services, price determination services (e.g. auctions), and value-added services (e.g. insurance); and (c) the *delivery phase*, which includes telematics services (e.g. document exchange, track-and-trace) and accounting services (e.g. settlement, provider feedback) (Bierwirth et al. 2002:337ff.; also see Kovács 2009:34f.; Sänger 2004:221f.).

Among the described functionalities, the most important seems to be the mechanism that coordinates demand and supply. Several mechanism types are described in literature, including (a) bulletin boards (Sänger 2004:99; Nandiraju & Regan 2003:2), (b) catalogues (Sänger 2004:199; Kovács 2009:33; Bierwirth et al. 2002:338), (c) (combinatorial) auctions (Sänger 2004:199f.; Bierwirth et al. 2002:338; Pankratz 2003; Nandiraju & Regan 2003:2,7), and (d) spot exchanges (Sänger 2004:202f.; Nandiraju & Regan 2003:2) or broker services (Bierwirth et al. 2002:338).

Electronic logistics markets also differ in the types of *logistics services traded*. It is fair to argue that almost all electronic logistics markets are focused on transportation services (see e.g. Sänger 2004; Bierwirth et al. 2002; Wang et al. 2011; Nandiraju & Regan 2003:3f.). To our knowledge, only marginal research has been published regarding the trading of storage services via electronic logistics markets or the combination of transportation and storage (exceptions are Kovács 2009; Kovács 2010). Regarding the services traded, another distinction can be made in terms of whether buyers publish their service requests or whether sellers publish available capacity (Sänger 2004:213f. Nandiraju & Regan 2003:2). Furthermore, Bierwirth et al. (2002:336) note that transactions can be categorized into one-time transactions and long-term contracts which provide the basis for multiple transactions (Nandiraju & Regan 2003:2).

Finally, electronic logistics markets can be classified regarding the *geographic coverage* within which logistics services are offered (Sänger 2004:218f.; Nandiraju & Regan 2003:4). Transportation markets can focus on selected transportation links, selected areas (of high economic intensity), or whole continents such as Europe (Sänger 2004:218f.).

Based on expert interviews, Sänger (2004:110ff.) identifies several issues that impede the diffusion and success of electronic transportation markets: (a) *liquidity issues*, e.g. in terms of a low likelihood of finding suitable transacting partners; (b) *cost issues*, e.g. in terms of IT system integration or transaction costs; (c) *quality issues*, e.g. in terms of unattractive transportation links, unattractive bulky cargo types, or unreliable LSPs; (d) *functionality issues*, e.g. in terms of inadequate price determination or negotiation mechanisms; and (e) *acceptance issues*, e.g. in terms of increased price and volume transparency for current or potential competitors and customers. In fact, these issues may have been the cause of many electronic markets closing (see list of closed-down markets in Sänger 2004:247; also see Nandiraju & Regan 2003:11).

The most severe of these issues seems to be *market liquidity* ("critical mass problem"). This may result from, among other factors, an overall low number of market participants or a general mismatch between logistical demand and supply, which often results from the fact that only unattractive "overhang" transportation links are handled via electronic markets that could not be handled otherwise. Many scholars also consider the critical mass problem to be pivotal (Wang et al. 2011:612f.; Koch 2010:28ff.; Nandiraju & Regan 2003:3). This problem is particularly severe in neutral markets, as they lack an original cargo volume that can be offered on the electronic market (Sänger 2004:191; Wang et al. 2011:612). A possible solution could be to initially focus on a few selected attractive transportation links and offer the market service without levying transaction fees, even though this runs against market operators' financial concerns, and gradually introduce fees once the market is "liquid" enough (Sänger 2004:231).

Open markets are also particular prone to quality issues because of their inherent openness, which may lead to participation by unreliable transacting parties (Sänger 2004:230f.). As Nandiraju & Regan (2003:3) remark: "In neutral marketplaces the problem of monitoring the execution and performance of business entities is hard." In order to reduce uncertainty about transacting partners, Sänger (2004:230ff.) proposes supplementing price coordination with several attributes to characterize transacting parties (e.g. solvency of shipper, reliability of LSPs) and introducing e-signaling instruments, such as rating systems. Nandiraju & Regan (2003:4) suggest the certification of providers as a prerequisite for participation.

4. Organization

Using the conceptualization of the organizational structure of interfirm networks (see Section C.V.1), we study the organization of networked horizontal LSP cooperation along three dimensions: (1) centralization, (2) formalization, and (3) characteristic mix of coordination mechanisms. Formalization is discussed last for argumentative reasons.

4.1. Centralization

Logistics literature provides only scarce insights into the distribution of decision-making authority within a networked horizontal LSP cooperation. In fact, only two contributions seem to specifically focus on centralization (Midoro & Pitto 2000; Albers & Klaas-Wissing 2012). Both are briefly presented below.

Midoro & Pitto (2000) distinguish between three distinct alternatives for the distribution of decision-making power within strategic alliances among ocean carriers: shared, dominated partner-led, and split. Shared means that authority is widely dispersed among alliance members; dominant partner-led means that a single LSP has authority over key decisions of the alliance; and split means that decision-making power over different domains is allocated to different LSPs.

Following an explorative research design, Albers & Klaas-Wissing (2012) investigate the organizational structure of two networked horizontal LSP alliances that operate in the German LTL segment. They find evidence for two distinct distributions of authority. Each cooperation coordinates and controls day-to-day operational activities of member LSPs through a *central management unit*, which is also responsible for operating the network's central handling hub and IT system. Despite having a central management unit in common, they differ in the degree to which member LSPs can influence operational and strategic decisions made by the central unit. The different authority distributions are referred to as (1) high centralization and (2) low centralization.

In the case of *high centralization*, the central management unit has authority over operative and strategic decisions; it issues and adapts all organizational rules and guidelines (e.g. regarding service quality and operational processes) in an autocratic manner, and LSPs that join the cooperation agree to adhere to these rules (Albers & Klaas-Wissing 2012:191). Beyond that, member LSPs are largely free to decide on their own strategic objectives beyond the cooperation, e.g. in terms of geographic coverage or membership in another cooperation (Albers & Klaas-Wissing 2012:185). The central unit's exclusive decision authority originates in the unit's economic independence and the wide dispersion of voting rights. The central unit is economically independent because it owns and operates the network's central and regional hubs and its IT system and because it levies fees on the operational handling of cargo in these hubs and on administrative services for members (Albers & Klaas-Wissing 2012:184f.). Member LSPs lack control over strategic and operational decisions (Albers & Klaas-Wissing 2012:184) although they voluntarily erected the central management unit and must become a voting-right-holding shareholder of the legal entity that houses this unit. Alternatively, they must pay a non-interest loan of the same value, which is a *de facto* membership fee. This lack of control is due to the large network size and the resulting wide dispersion of voting rights among member LSPs; a single member LSP thus faces an "accept or leave" choice (Albers & Klaas-Wissing 2012:185).

In the case of *low centralization*, the central management unit has no authority over crucial operative and strategic decisions, but authority ultimately resides with the senior managers of the partnering LSPs, who are equity holders of the legal entity that houses the unit (Albers & Klaas-Wissing 2012:186, 191). In this setting,

> "the central management unit takes the role of a *communication and personal networking hub* that facilitates (personal) exchange between partner companies. More specifically, the central management unit's role is one of an enabling moderator rather than being an autocratic decision centre" (Albers & Klaas-Wissing 2012:193, italics original).

In order to enable member LSPs to monitor and become involved in decision-making processes, the central unit comprises several forums, including a management board, committees, and work groups (Albers & Klaas-Wissing 2012:186f.). In fact, authority of the management board is not limited to operative and strategic issues of the cooperation, but additionally constrains the business autonomy of member LSPs by prohibiting them from offering their logistics capabilities in a geography already covered by other members (territory protection) or from simultaneously joining other alliances. Although the central management unit operates the central handling hub and the IT system and levies fees on the operational handling of cargo and delivery of administrative services to members, it is economically dependent on the member LSPs. Any surplus of fees over the central unit's administrative costs is distributed to members and every deficit is covered by them (Albers & Klaas-Wissing 2012:186f.).

Surprisingly, logistics research regarding the distribution of decision-making authority in networked LSP cooperations aligns well with the conceptual findings by Provan & Kenis (2008) regarding the distribution of authority in interfirm networks (see Section C.V.2). The "shared" and "dominant partner-led" modes described by Midoro & Pitto (2000) seem to match Provan & Kenis's (2008) "shared governance" and "lead organization governance," respectively. The low and high degrees of centralization described by Albers & Klaas-Wissing (2012) largely resemble Provan & Kenis's (2008) "NAO governance."

4.2. Mechanisms and the Role of IT Systems

Logistics literature describes a variety of mechanisms that cause or prevent change in horizontal LSP alliances. In the following, we do not discuss any specific mechanism in detail, but provide a basic understanding of, first, the task domains that are governed by mechanisms in horizontal LSP alliances; and, second, the role of IT systems in relation to these mechanisms.

Mechanisms identified in literature pertain to four *task domains*: (1) partner analysis and selection, (2) allocation of resources and tasks, (3) determination and allocation of cooperation benefits, and (4) physical operative logistics activities. *Partner analysis and selection* refers to any mechanism that enables an LSP to identify, evaluate, and ultimately select partners for horizontal cooperation. Selecting the "right" partners is recognized as an important factor for overall cooperation success (see e.g. Raue & Wallenburg

2013:217ff.; Schmoltzi & Wallenburg 2011; Verstrepen et al. 2009:240; Midoro & Pitto 2000). "Right" essentially means that there is fit, or compatibility, between the characteristics of the potential partners. Characteristics frequently used to assess compatibility include geographic similarity (in terms of overlaps in geographies covered), competency similarity (in terms of overlaps in services offered), and cultural similarity (in terms of corporate cultures or management style) (Raue & Wallenburg 2013:218; Schmoltzi & Wallenburg 2011:555f.; also see Cruijssen et al. 2007c; Verstrepen et al. 2009).

Allocation of resources and tasks refers to any mechanism that determines each partnering LSP's (resource) contribution to fulfilling a specific task in the cooperation and/or to assign a specific task to a specific member LSP. Joint route planning and scheduling is a mechanism frequently studied in transportation networks, such as air (see e.g. Yan & Chen 2007; Chen & Chen 2003a; Bilotkach 2007) and road (see e.g. Hernández et al. 2011; Krajewska et al. 2008; Zhou et al. 2011; Cruijssen et al. 2007a). These mechanisms are important for transportation-oriented horizontal LSP cooperation, as they are the crucial means to achieving cost reductions via optimal consolidation of cargo and minimal total distance traveled. For example, Krajewska et al. (2008:1485) find that vehicles can be reduced by 10% and costs by roughly 12% in a bilateral cooperation between LSPs. In a multi-lateral scenario, Cruijssen et al. (2007a:301f.) even report cost reductions of a little more than 30%, and they argue that the potential for costs reduction further increases with the number of cooperating entities.

Determining and allocating cooperation synergies refers to any mechanism that enables partnering LSPs to measure the synergies realized through cooperation and to distribute them among partnering LSPs. These mechanisms are probably the most important ones in organizing alliances in general (Gulati & Singh 1998) and among LSPs (Cruijssen et al. 2007b:135; also see Xu 2012). More specifically, these mechanisms are crucial to "balance cooperation and competition" (see Morris et al. 2007). This is because they determine whether cooperation ultimately yields a net profit or loss for each partner. Therefore, sharing benefits is a common source of conflict in alliances – especially in horizontal alliances – as partnering LSPs usually have precise insights and understanding of the nature of activities performed together (Wallenburg & Raue 2011:387). Moreover, in the event of unequal bargaining positions, smaller partners may feel discriminated against in the allocation of synergies (Wallenburg & Raue 2011:387). To avoid unilateral dominance of a cooperation and unbalanced apportionment of benefits, Verstrepen et al. (2009:240) argue that it may be advisable to choose partners of similar size and market power if fair mechanisms cannot be realized in an unbalanced setting. Achieving a "fair" allocation of synergies is considered crucial for achieving long-term cooperation success, as

> "[m]istrust about the fairness of the applied allocation rule for savings has caused many horizontal logistics cooperation initiatives between shippers

> and/or LSPs to marginalize or disintegrate" (Cruijssen et al. 2007c:31; also see Cruijssen et al. 2007b:135).

However, Verstrepen et al. (2009:241) report that

> "[i]n practice, many partnerships complain about the absence of theoretically sound, and fair allocation methods [... and the valuation and allocation of benefits often represents a...] subtle 'game of give and take', rather than of a hard negotiation process merely based on facts and figures."

In this respect, Cruijssen et al. (2007c:31f.) point out that allocation principles are often simple rules of thumb that are inappropriate for accurately measuring and rewarding marginal contributions of partners and, therefore, inevitably frustrate some partners in the long run, as they may not adequately value the true contributions of each partner. Recent research in cooperative game theory, however, provides promising results for achieving fair allocations. Krajewska et al. (2008), for example, propose a mechanism to allocate synergies using the Shapley value, which remunerates each player according to its average marginal contribution of participation in the cooperation. Xu et al. (2012) propose a mechanism based on the weighted Shapley value that takes into account each partner's bargaining power, contribution, and core stability.

Physical operative logistics activities refer to any mechanism that enables the coordination, control, or monitoring of member LSPs involved in the realization of a physical logistics process, such as transportation, storage, and handling. With regard to LTL networks, Albers & Klaas-Wissing (2012:192) argue that

> "pick-up, sorting, and delivery of LTL shipments are highly programmable in nature and, therefore, tasks that can be standardised [and codified] (apparent in, for example, scheduled line services, process descriptions organisation handbooks, and pre-specified process times)".

Albers & Klaas-Wissing (2012:188), furthermore, point out that formal monitoring mechanisms, in the form of an elaborate set of key performance indicators (KPIs) and frequent reporting, are used because operational efficiency and service quality are of crucial importance in the LTL segment (also see Klaas-Wissing & Albers 2010:502). KPIs can be related to financial aspects, operational performance, customers, and growth (Verstrepen et al. 2009:243).

Considering the mechanisms described in the previous paragraphs, we can conclude that the vast majority represent specified, codified, and thus formalized processes. These formalized mechanisms can be conceived of as variants of the mechanism *standardization of work processes*; in fact, many of these mechanisms are not only standardized work processes but *programmed routines* because they are implemented on IT systems (see

Subsection C.V.4.4). For example, Buijs & Wortmann (2014:206f.) point out that IT systems can be used to facilitate real-time tracking of transportation processes (monitoring mechanism), record transportation related data, and support and provide information for operational planning and decision-making (joint scheduling and planning). This shifts our attention to the *role of IT systems* in horizontal LSP cooperation.

Via a literature review, Cruijssen et al. (2007c:31ff.) identify IT systems as both facilitators and impediments for horizontal cooperation (also see Verstrepen et al. 2009:242). They can be facilitators because they enable information sharing and coordination among member LSPs. Information shared can include transportation orders, inventory levels at production facilities, and performance metrics. Coordination can be achieved through collaborative planning systems, common data standards (e.g. EDIFACT), and through the use of systems that create electronic markets, such as freight exchanges. In this respect, Klaas-Wissing & Albers (2010:502) point out that the IT system of the central management unit in cooperative LTL networks can represent a neutral platform for market-oriented service clearing, including the financial flows between partnering LSPs and between LSPs and the central management unit. Cruijssen et al. (2007c:31ff.) argue that IT can become an impediment, as many LSPs often lag behind in the implementation of IT systems, thus hampering alliances that require significant information sharing. For example, Klaas-Wissing & Albers (2010:502) point out that cooperative LTL networks require strong harmonization of different planning and control systems. Similarly, Buijs & Wortmann (2014) find, in an exploratory study, that technological differences in IT systems used by carriers hinder their integration and thus impede operational planning and control in cooperative road transportation networks. Moreover, Cruijssen et al. (2007c:32) argue that especially alliances that require IT support but may not generate sufficient synergies to warrant investing in new or adjusting existing systems may be particularly hampered. In sum, IT systems are important means to achieve coordination within horizontal networked LSP cooperations, which makes them both a facilitator (if successfully implemented) and an impediment (if not successfully implemented or available) for cooperation success.

In conclusion, logistics literature identifies several task domains that are commonly governed by mechanisms. These include critical administrative processes such as partner analysis and selection, allocation of tasks, and determination and allocation of cooperation gains among partners, and critical physical operative logistics processes. The mechanisms described provide first insight into the set of mechanisms from which different networked LSP alliances can select characteristic mixes.

4.3. Formalization

The previous subsection showed that coordination is, to a large extent, achieved through the use of standardized work processes and programmed routines, thus making formalization a crucial factor of organizing horizontal LSP cooperation. The importance of standardization is emphasized by several scholars. Albers & Klaas-Wissing (2012:188), for example, observe that, in networked LTL cooperations,

> "highly standardised processes [...] are needed in order to provide LTL services; this makes standardisation a major means with which to coordinate work, both within and among these firms."

Although this observation pertains to LTL services, high process standardization seems important for the vast majority of operative physical logistics processes. In fact, Schmoltzi & Wallenburg (2011:63) find evidence that the formalization of day-to-day logistics activities and interactions between partnering LSPs has a substantially positive impact on cooperation commitment and effectiveness. Moreover, the positive impact further increases with an increasing number of LSPs (relational complexity) and business functions involved (functional complexity) and with an increasing overlap between partners' geographic resource setups (geographic complexity) and their logistics competencies (competency complexity).

Wallenburg & Raue (2011) investigate the relationship between formalization, conflict, and cooperation performance, among other things. They find that formalization increases the extent of conflict (defined as the extent to which differences and disagreements about ideas and decisions occur within a cooperation), thus, negatively impacting cooperation performance (Wallenburg & Raue 2011:391). They explain their findings by arguing that

> "[r]ather than clarifying roles and setting guidelines for interaction which target the potential source of conflict, formalization and the corset it provides, is by itself a source of conflict. A potential explanation of this can been seen in the fact that in complex and dynamic environments, as omnipresent today, not all contingencies can be foreseen and be catered for *ex ante*. Yet, when deviations from expectations occur, formalization is likely to involve some parties of the cooperation that will want to stick to rules set out – at least as much as possible – while others may see the necessity to deflect from the guidelines." (Wallenburg & Raue 2011:394, italics original)

They therefore conclude that structure should only be formalized if there are other positive effects of formalization for the relationship.

Raue & Wieland (2015) investigate the interplay between the degree of formalization in organizational structure and the extent of using contractual safeguards in the cooperation contract (see Subsection C.VI.3.2). They find that formalizing day-to-day activities and interactions (process formalization) has a positive impact on cooperation performance if LSPs establish horizontal cooperation to access tangible resources that are directly exploitable, such as transportation networks or warehouses. In these cases, their results specifically suggest that

> "process formalization and contractual safeguarding are substitutes" (Raue & Wieland 2015:414).

Hence, they argue that LSPs have the choice of either relying on elaborate contracts or extensive process formalization in order to reduce opportunistic behavior and improve coordination. In addition, they recommend that LSPs should make this choice subject to their specific situational setting. For example in the case of international cooperation with different LSPs who are subject to different legal systems, elaborate process formalization may be less costly to implement than developing detailed contractual safeguards by involving legal experts form different countries.

To summarize, empirical evidence confirms formalization as an important factor for effectively organizing horizontal LSP alliances in a variety of settings, especially in the case of many partnering LSPs. Nevertheless, formalization cannot be considered the "philosopher's stone" as it can also increase the extent of conflict and adversely impact cooperation performance; formalization should only be used if it brings other positive effects to the relationship. Finally, elaborate process formalization and extensive contractual safeguards are substitutes, thus giving LSPs the opportunity to capitalize on formalization cost-efficiently in different settings.

5. Interim Summary

This chapter has investigated networked horizontal LSP cooperation by using the theoretical approaches introduced in the preceding chapters. Logistics literature suggests that such cooperative networks among LSPs can be defined as a specific type of interfirm network and classified by means of essentially the same features as any other interfirm network. The motives and antecedents for the formation of such networks implicitly align with the motives and antecedents associated with the RBV. Regarding the governance of horizontal LSP networks, logisticians stress the pervasive risk of opportunism, which they aim to "cure" by introducing a high degree of contractual formalization and safeguards. Horizontal LSP cooperation can be based on various governance structures on the continuum between markets and hierarchies. A specific governance structure, electronic logistics markets, has been considered in more detail. Although it is a market in a strict sense, such markets can be established within alliances to coordinate logistical demand and supply. Such markets can be distinguished using a variety of criteria.

Although several markets have been established over the last decades, especially within transportation, large-scale success seems to be absent, most likely due to issues related to market liquidity and quality. Regarding the organization, several distinct distributions of decision-making authority among LSPs have been identified in practice, which largely match those described for interfirm networks in general. Formalization appears to be an important dimension of organizing such cooperative logistics networks, not only in terms of formal contracts, but also with regard to formal mechanisms to coordinate administrative and operative processes. In conclusion, this chapter has illuminated the phenomenon of horizontal LSP cooperation from various perspectives and in a comprehensive manner.

VII. Summary

This part has investigated interfirm networks via three complementary, theoretical approaches, thus providing comprehensive insight into the phenomenon.

The TCE is concerned with the question of what transaction is processed through what governance structure. Structural choice depends on economizing on transaction costs. Transaction costs are rooted in the bounded rationality and opportunistic behavior of economic actors in interaction with the characteristics of transactions in terms of asset specificity, uncertainty, frequency, and complexity of product descriptions. Thus, the formation of interfirm networks, which are considered hybrid governance structures, results from economizing on transaction costs. Certain transactions can be processed more efficiently through hybrids than markets or hierarchies.

The RBV is concerned with the question about of how firms can achieve (sustained) competitive advantage by participating in interfirm networks. Antecedents for the formation of interfirm networks are heterogeneous resource endowments of firms and imperfect resource mobility. Whether a specific firm establishes or joins an interfirm network depends on the characteristics of the firm's internal resources in terms of value, rareness, non-substitutability, and non-imitability.

Organizational theory is not concerned with antecedents for formation of interfirm networks, but rather with the network's internal administrative arrangements that enable the division and coordination of labor among partnering firms in pursuit of a goal. In this respect, centralization, formalization, and the characteristic mix of coordination have been studied in detail.

The last chapter of this part has used the above mentioned theoretical approaches to provide comprehensive insight into the phenomenon of horizontal LSP cooperation.

Part D. Fundamentals and Roots of Cloud Computing

I. Abstract of Part

The NIST Definition of Cloud Computing

Today's understanding of "cloud computing" follows Mell & Grance's (2011:2) definition (see Chapter D.II): "a model for enabling ubiquitous, convenient, on-demand network access to a shared pool of configurable computing resources (e.g., networks, servers, storage, applications, and services) that can be rapidly provisioned and released with minimal management effort or service provider interaction. This cloud model is composed of five essential characteristics, three service models, and four deployment models." Cloud computing has five essential characteristics: (1) on-demand self-service, (2) broad network access, (3) resource pooling, (4) rapid elasticity, and (5) measured service.

On-demand availability means that capabilities become available without delay at any point in time (see Subsection D.II.2.1). *Self-service* refers to a delivery concept in which consumers can access the provider's services via a technological interface without direct involvement of provider staff (Meuter et al. 2000:50). By combining both concepts, cloud computing creates significant consumer flexibility: They neither need to plan ahead nor coordinate with a provider in person to satisfy emerging demands.

Broad network access specifies the delivery channel of computing capabilities (see Subsection D.II.2.2). Cloud computing harnesses the network, primarily the internet, to deliver computing capabilities. This has become possible due to the recent diffusion of ubiquitous, low costs, high bandwidth internet access. Hence, broad network access can be seen as both "a trait of cloud computing and as an enabler" (Williams 2012:6).

Resource pooling characterizes the production process of cloud computing capabilities (see Subsection D.II.2.3). A provider's physical and virtual computing resources are shared

F. Jaekel, *Cloud Logistics*, Produktion und Logistik,
https://doi.org/10.1007/978-3-658-22837-8_4

across different consumers and dynamically assigned and reassigned according to current consumer demand. Pooling ultimately aims to realize economies of scale by increasing the utilization of resources (Zhang et al. 2010:11).

Rapid elasticity means that the capacity of computing capabilities can be rapidly scaled commensurate with consumer demand at any quantity and point in time, thus creating the illusion of an unlimited resource pool (see Subsection D.II.2.4). Elasticity is a dynamic property that refers to both the speed and precision of scaling the capacity of capabilities in order "to adapt to workload changes by provisioning and deprovisioning resources in an autonomic manner, such that at each point in time the available resources match the current demand as closely as possible" (Herbst et al. 2013:24). Elastic resource provisioning would neither be feasible nor economically viable without the pooling of large amounts of resources: The aggregate resource pool needs to be large enough so that individual fluctuations do not exceed its upper capacity limit and providers need to be able to dynamically (re-)assign resources to different consumers in accordance with demand to efficiently use the aggregate capacity (Verma et al. 2010:11).

Measured service means that resources needed for delivering capabilities are measured, and consumers are charged only for those resources actually used (see Subsection D.II.2.5). Usage is monitored, controlled, and reported to provide transparency to providers and consumers.

The NIST definition specifies three service models for cloud computing systems that differ in the type of computing capability provided to consumers (see Section D.II.3): (a) *Software-as-a-Service (SaaS)*, (b) *Platform-as-a-Service (PaaS)*, and (c) *Infrastructure-as-a-Service (IaaS)* (Mell & Grance 2011:2f.).

The NIST definition specifies four deployment models for cloud computing systems: (a) *private*, (b) *community*, (c) *public*, and (d) *hybrid* (see Section D.II.4). They describe alternative arrangement of how cloud infrastructures can be shared among consumers, and they differ in how exclusive the access of a certain group of consumers is (Liu et al. 2011a). Deployment models thus focus on the relationships among actors in cloud computing systems, which is why they represent an institutional perspective.

Hybrid clouds are especially relevant to this thesis as they refer to the cooperation among several providers (see Subsection D.II.4.2). Providers form hybrid clouds to gain access to other provider's resources and thus to and enhance the ability to deliver computing capabilities in an on-demand and rapidly elastic manner. Accessing other providers resources is critical, especially for small providers, to achieve cloud characteristics even if demand fluctuations of consumers outstrip own capacity. Research that focuses on organizational and structural governance aspects of hybrid clouds suggests the importance of three coordination mechanisms: (a) *programmed routines and hyperautomation* to negotiate prices and conditions and constraints of resource exchanges as well as to plan, realize, control, and monitor the allocation of resources to service requests in accordance with agreed service levels; (b) *implicit coordination by database* in order to store and exchange information about currently available resources and capabilities, prices, service-level policies, and demand patterns; and (c) *electronic market-based exchanges* to facilitate resource exchanges among providers and consumers in order to enable on-demand availability and rapid elasticity. Scholars suggest that the exchange needs to be operated by a neutral middleman (Garg et al. 2013:1172) trusted by all market participants (Buyya et al. 2010:25) as the middleman settles all transactions.

Technological and Conceptual Roots

Despite being frequently positioned as something entirely new, cloud computing derives from several other technologies (see Section D.III.1). Two technologies relevant for this thesis are (1) hardware virtualization and (2) service-oriented computing.

Hardware virtualization means the simulation of multiple complete computer systems ("virtual machines") on a single physical computer in order to share physical resources (see Section D.III.2). Each user is assigned a dedicated virtual machine, and physical resources are dynamically (re-)assigned to virtual machines according to current demand, thus increasing the utilization of physical resources. Hardware virtualization creates significant flexibility by removing any dependencies between software, hardware, and geographical proximity(s) of machines and users. Virtualization is a critical technology for cloud computing by enabling resource pooling.

Service-oriented computing (SOC) is a computing paradigm that uses services as a basic element for creating distributed applications (Papazoglou 2003:3; Papazoglou et al. 2008:223; see Subsection D.III.3.1). A *service* "represents a capability of performing tasks that form a coherent functionality" (Haas 2004). A *service composition* represents a "coordinated aggregate of services" (Erl 2008:39) to form a more complex application or realize a complete business process (see e.g. Laskey et al. 2012:37). Cloud computing is based on SOC as it adopts the notion of services: Cloud capabilities are delivered "as-a-service."

SOC revolves around service-orientation as a fundamental design paradigm, with a design paradigm being understood as a set of principles that collectively govern the design of a well-defined functionality (Erl 2008:37f.; see Subsection D.III.3.2). *Service-orientation* consists of various design principles that shape the design of services (Erl 2008:39). Six of these are particularly relevant to the objective of this thesis: (a) *standardized service descriptions*, (b) *service discoverability*, (c) *service reusability*, (d) *service loose coupling*, (e) *service abstraction*, and (f) *service composability* (Erl 2008:70ff.).

A service-oriented architecture (SOA) is basically a means to realize the vision of SOC (Papazoglou et al. 2007:64; see Subsection D.III.3.3). A system architecture can generally be defined as the "fundamental concepts or properties of a system in its environment embodied in its elements, relationships, and in the principles of its design and evolution" (ISO/IEC/IEEE 42010 2011b:2). An SOA is a distinct form of technology architecture geared towards supporting services and service compositions designed according to the design principles associated with service-orientation (Erl 2008:40). Yet, an SOA is not only a technology architecture, but also specifies the constituent entities in an SOA, their roles, their relationships, and interactions (see e.g. Papazoglou 2003:5). An SOA consists of three roles: a service provider, a service consumer, and a service repository (broker) that are arranged in a triangle. They interact by means of three basic operations: (a) *publish*, (b) *find*, and (c) *bind and invoke* (Endrei 2004:27).

The role of the service repository is versatile and can vary significantly in different SOAs. At the one extreme, it may provide only basic functionality limited to storing and publishing service descriptions. At the other extreme, repositories can evolve into *electronic marketplaces* that bring together consumers and providers (Papazoglou 2003:9). The organization that operates the repository acts as a *market maker* that assumes responsibility for maintaining the marketplace (Papazoglou 2003:9).

Web services are an internet technology that has contributed to the emergence of cloud computing. They represent the currently most promising technology based on SOC and SOA (Weerawarana 2005 as cited in Papazoglou et al. 2008:225; Erl 2008:46). Understanding the life cycle of web services is important for the purposes of this thesis. The life cycle consists of nine phases, which further specify the interactions and exchanges among the entities in an SOA (see Subsection D.III.4.1): (1) *development*, (2) *publication*, (3) *discovery*, (4) *composition*, (5) *negotiation and selection*, (6) *invocation*, (7) *execution*, (8) *monitoring*, (9) *unpublication*. Each SOA must include mechanisms to coordinate and interactions and exchanges along the life cycle.

The Semantic Web is a vision that aims to extend today's World Wide Web (WWW) through "the conceptual structuring of the Web in an explicit machine-readable way" (Berners-Lee & Fischetti 1999 as cited in Gomez-Perez & Corcho 2002; see Subsubsection D.III.4.2.1). This enables machines to process and "understand" the data they merely visualize on screens today (Berners-Lee et al. 2001). Of course, computers do not truly "understand" the content they process, but they can manipulate data much more effectively now – especially in ways that are useful and meaningful to human users (Berners-Lee et al. 2001). As a result, the Semantic Web can enable humans and computers to better cooperate and enable computers to better interoperate (i.e. exchange data) and carry out complex tasks on behalf of users (Berners-Lee et al. 2001; Payne & Lassila 2004:14). The crucial feature of interoperation between computers in the Semantic Web is the vision of enabling *serendipitous* interoperation, which means that interoperation is unarchitected and unanticipated (see e.g. Payne & Lassila 2004:14).

For the Semantic Web to function, Berners-Lee et al. (2001) argue that "computers must have access to structured collections of information and sets of inference rules that they can use to conduct automated reasoning" and ways to exchange and share knowledge (see Subsubsection D.III.4.2.2). *Ontologies* evolved as a primary tool to represent, manipulate, and share knowledge (see e.g. Chandrasekaran et al. 1999). "An ontology is a formal, explicit specification of a shared conceptualisation" (Studer et al. 1998:184). Mechanisms can operate on ontologies in order to solve a specified problem or perform a specified task (Chandrasekaran et al. 1999:23f.). Ontologies and their associated mechanisms can be expressed via formal *ontology languages*. Such languages differ in their expressiveness, that is, the scope and complexity of ideas that can be represented and communicated in the language, and in their ability to support complex inferences (Gomez-Perez & Corcho 2002). Therefore, designers need to select ontology languages such that they match with the requirements of the task at hand.

Prior to the advent of the Semantic Web, web services were described by means of *syntactical service description languages*, which only allowed specifying the syntax of services' interfaces (Martin et al. 2005:26; see Subsubsection D.III.4.3.1). However, the use of such languages reduces the potential to automate processes along the web service life cycle. To enable interoperability, consumers and providers need to agree in advance on the syntax and semantics of service interfaces and the messages exchanged between services (Burstein et al. 2005:72). In other words, interoperation between web services needs to be *architected* before interoperation can begin. Semantic web services arise from applying the vision of the Semantic Web to the description of web services. A semantic web service is characterized by a service description that specifies the service's capabilities and conditions and constraints of use in an unambiguous, machine-processable way (McIlraith et al. 2001:46; McIlraith & Martin 2003:90). Semantic web services aim to achieve a higher degree of automation along the entire web service life cycle (Payne & Lassila 2004:14; also see McIlraith & Martin 2003:90; Martin et al. 2005:27).

Two general approaches exist for creating semantic web services (1) semantic annotation and (2) semantic markup (see Patil et al. 2004:553; see Subsubsection D.III.4.3.2). *Semantic annotation* is the creation of semantic web services by adding semantics to an existing syntactical description (Patil et al. 2004:553; Papazoglou et al. 2007:67). *Semantic markup* is the creation of semantic web services by describing services using an *ontology-based description language* that is specifically tailored to the description of web services (see Patil et al. 2004:553; McIlraith et al. 2001). An ontology-based description language is a specific type of ontology language that contains a pre-defined set of domain-independent concepts and relationships that specify how web services and user constraints are to be described (McIlraith et al. 2001:48). These concepts and relations provide a declarative representation of web services (semantic markup).

The SOA-RAF Service Description Model is an abstract model for the description of web services (see Subsubsection D.III.4.3.3). It is abstract as it specifies a generic top-level ontology for the service description, but does not provide a machine-processable description. It thus can be considered a starting point for developing an ontology-based description language (semantic markup). The model structures information to describe services via five basic concepts: (1) *service functionality*, (2) *service policies*, (3) *metrics*, (4) *service interface description*, (5) *service reachability* (see e.g. Laskey et al. 2012:44, 49, 107). This model will be proposed as basis for representing cloud logistics services.

Several scholars, primarily rooted in the realm of computer science, have recently started to use well-defined semantics to describe logistic services (see Subsection D.III.4.4). Adopting semantics to describe physical logistics capabilities is of direct relevance for addressing our research question: designing a logistics system in accordance with the principles and concepts of cloud computing. A brief search in popular scientific databases reveals that current scientific knowledge is not only very scarce but different scholars also adopt different approaches to create semantic logistics services, which emphasizes the newness of this field. Despite following heterogeneous semantic approaches, all contributions use existing web service standards to describe logistics services – even though researchers have already recognized the limited applicability of these standards for describing logistics services (see e.g. Li et al. 2013:1697). The use of existing web service standards therefore appears to be an act of necessity in the absence of logistics specific standards. Considering the present body of knowledge about the semantic descriptions of logistics services, it is fair to conclude that current contributions provide a successful "proof-of-concept" for introducing well-defined semantics to the description of logistics services – nothing more, but also nothing less.

II. The NIST Definition

"The rise of the cloud is more than just another platform shift that gets geeks excited. It will undoubtedly transform the information technology (IT) industry, but it will also profoundly change the way people work and companies operate. It will allow digital technology to penetrate every nook and cranny of the economy and of society, creating some tricky political problems along the way."

— The Economist 2008

1. Origins and Formal Definition

Cloud computing refers to an innovative computing paradigm whose distinctive features are subsumed under the obscure term "cloud." Although the exact origin of the term "cloud computing" appears unclear, the pivotal inception to today's understanding of this paradigm is a speech by Eric Schmidt in 2006, who was then the CEO of Google and proposed a new computing model that would supersede the "old" client/server based computing paradigm, which was largely developed by Oracle:

> "It starts with the premise that the data services and architecture should be on servers. We call it cloud computing – they should be in a 'cloud' somewhere. And that if you have the right kind of browser or the right kind of access, it doesn't matter whether you have a PC or a Mac or a mobile phone or a BlackBerry or what have you – or new devices still to be developed – you can get access to the cloud..." (Eric Schmidt as quoted by Bogatin 2006).

The obscure nature of the term and the lack of a standard definition of cloud computing initially led to significant market hypes, but also to confusion and skepticism (Zhang et al. 2010:8). As Larry Ellison, the CEO of Oracle in 2007, remarked:

> "I have no idea what anyone is talking about [...] It's really just complete gibberish. What is it? [...] When is this idiocy going to stop? [...] We've redefined cloud computing to include everything that we already do [...] I can't think of anything that isn't cloud computing with all of these announcements. The computer industry is the only industry that is more fashion-driven than women's fashion." (Fowler & Worthen 2009:A.1)

The primary reason for the clouds surrounding the "true" meaning of cloud computing might have been its close relationship to other existing computing paradigms, including virtualization, grid computing, and service-oriented computing (Zhang et al. 2010:8f.; Youseff et al. 2008:1 Fowler & Worthen 2009:A.1).

Due to the unclear understanding of cloud computing, developing a concise definition of cloud computing is a question that has received significant attention from computer scientists in academia and practice. As a consequence, the number of definitions proposed grew rapidly after Schmidt's speech. Already in 2008, Foster et al. (2008:1) remark that their definition of cloud computing is "yet another definition to the already saturated list of definitions." In an initial attempt to clear the fog around the term, Vaquero et al. (2009) review 20 definitions to identify all of the concept's features and to develop an all-encompassing definition. Today's understanding of cloud computing is largely based on the definition by Mell & Grance (2011:2) of the NIST, who define cloud computing as

> "a model for enabling ubiquitous, convenient, on-demand network access to a shared pool of configurable computing resources (e.g., networks, servers, storage, applications, and services) that can be rapidly provisioned and released with minimal management effort or service provider interaction. This cloud model is composed of five essential characteristics, three service models, and four deployment models."

The following sections investigate in dental the five essential characteristics, three service models, and three deployment models of cloud computing.

2. Essential Characteristics

The NIST definition of cloud computing specifies five essential characteristics: (1) on-demand self-service, (2) broad network access, (3) resource pooling, (4) rapid elasticity, and (5) measured service (Mell & Grance 2011:2). These characteristics are constituent features of cloud computing systems. Thus, if a computing system delivers computing capabilities with these characteristics, it can be referred to as a cloud computing system. The following subsections discuss these characteristics in detail.

2.1. On-demand Self-service

The *on-demand self-service* characteristic is defined as one where

> "[a] consumer can unilaterally provision computing capabilities, such as server time and network storage, as needed automatically without requiring human interaction with each service provider" (Mell & Grance 2011:2).

This characteristic thus specifies the temporal properties with which computing capabilities can be made available and the mode of interaction between consumers and providers to make them available.

On-demand can generally be understood as making something available "when requested or needed" (Merriam-Webster 2016:"on demand") and "at any time" (Cambridge University Press 2016:"on demand"). It implies a low lead time for computing capabilities to be delivered and the lack of temporal constraints regarding when they can be delivered. As a result, cloud consumers are no longer required to plan far ahead (Armbrust et al. 2010:51; Zhang et al. 2010:7).

Self-service refers to a delivery concept in which consumers can access the provider's services through a technological interface without the direct involvement of the provider's staff (Meuter et al. 2000:50). In cloud computing, consumers can access computing capabilities through *web portals* that allow them to identify, select, order, and consume computing capabilities in a self-service manner (Cisco Systems Inc. 2011:2) and manage and monitor their own resource consumption at any time (Vernier et al. 2011). Cisco, a manufacturer of networking equipment, suggests structuring cloud computing web portals by means of *service catalogs* that contain an overview of the computing capabilities a provider offers (Cisco Systems Inc. 2011:2). Such catalogs describe computing services in a standardized way including, *inter alia*, information about pricing, service level commitments, and terms and conditions of service use (Cisco Systems Inc. 2011:2). To enable self-service, the information included in service catalogs must be sufficient to enable consumers to make decisions unilaterally about which service(s) to purchase to address their needs.

On-demand and self-service are not new concepts, *per se*. However, cloud computing combines these concepts, creating significant flexibility for consumers: They neither need to plan ahead nor coordinate with a provider in person to satisfy newly emerging computing demands. Self-service makes the acclaimed concept of on-demand delivery tangible to consumers (Businesscloud.de 2012). The combination of these characteristics may also be why cloud computing is often described as "user friendly" (Vaquero et al. 2009:51ff.) and "convenient" (Mell & Grance 2011:2).

2.2. Broad Network Access

The characteristic of *broad network access* is defined as:

> "Capabilities [that] are available over the network and accessed through standard mechanisms that promote use by heterogeneous thin or thick client platforms (e.g., mobile phones, tablets, laptops, and workstations)" (Mell & Grance 2011:2).

Cloud computing harnesses the network – especially the internet – as a primary channel to deliver computing capabilities (Zhang et al. 2010:11). This has become possible because network costs have declined while network bandwidth, speed, and geographic coverage have increased significantly over the last decades (Williams 2012:5f.). Today, countless WLAN hot spots and mobile phone networks with high bandwidth provide ubiquitous internet access. Therefore, broad network access can be seen as both "a trait of cloud computing and as an enabler" (Williams 2012:6).

2.3. Resource Pooling

The characteristic of *resource pooling* is explained as follows:

> "The provider's computing resources are pooled to serve multiple consumers using a multi-tenant model, with different physical and virtual resources dynamically assigned and reassigned according to consumer demand. There is a sense of location independence in that the customer generally has no control or knowledge over the exact location of the provided resources but may be able to specify location at a higher level of abstraction (e.g., country, state, or datacenter)" (Mell & Grance 2011:2).

Resource pooling ultimately aims to realize economies of scale by increasing the utilization of resources (Zhang et al. 2010:11). In cloud computing, resource pooling extends to sharing computing infrastructure (e.g. storage capacity, processing power, memory, and network bandwidth) and applications (Mell & Grance 2011:2; Dillon et al. 2010:27ff.). The former is achieved through (1) virtualization, and the latter through (2) multi-tenancy.

Virtualization enables the sharing of computing infrastructure and is considered in more detail below (see Section D.III.2). For now, it is sufficient to understand the basic principle. Virtualization means the simulation of multiple computer systems (referred to as "virtual machines") on a single physical machine (Goldberg 1974:34). Each consumer is then allocated a dedicated virtual machine, thus giving him or her the illusion of actually having a dedicated physical machine at hand (Sugerman et al. 2001:1). Consumers are thus logically isolated by means of virtual machines, but share the same physical hardware resources. Physical resources are dynamically assigned to each virtual machine (and thus to each consumer) according to current demand, thereby increasing physical resource utilization (Armbrust et al. 2010:52f.) and lowering costs (Zhang et al. 2010:7ff.). In fact, cloud computing massively exploits virtualization technology to achieve economies of scale by consolidating infrastructure in very large data centers at low-cost locations (Armbrust et al. 2010:52) while delivering computing capabilities to any location in the world via the internet. To enable consolidation and dynamic (re-)assignment of resources, providers require flexibility in managing resources, which is why consumers are only given the possibility to specify the geographic location of resources at a higher level of abstraction.

Multi-tenancy enables the sharing of applications. It is achieved via a certain design feature in the software architecture that allows multiple users, referred to as "tenants" to simultaneously use the same application and database instance (Bezemer & Zaidman 2010:88), where each tenant is given the illusion of being the only consumer using the software, and other concurrent tenants remain invisible (Wilder 2012). Thus, unlike virtualization, multi-tenancy logically isolates different users on the application level. The larger the number of tenants that share an application, the more likely it is that tenants have slightly varying requirements for the shared application. Multi-tenant applications therefore need to allow tenants to configure and customize at least certain components of the application according to their needs. Generally speaking, multi-tenant applications need to make a tradeoff between standardization and customization. To achieve tenant-specific adjustments while sharing the same application, Mietzner et al. (2008:156ff.) propose that multi-tenant applications should consist of two parts: one that is fixed for all tenants and one that is customizable by each tenant. Customization can be achieved using so-called "application templates," which define how tenants can adjust software components (modules) according to pre-defined alternatives. Service modularity, which is discussed in more detail below (see Section D.III.3), is thus a very important method for enabling application sharing (Azeez et al. 2010).

2.4. Rapid Elasticity

The characteristic of *rapid elasticity* is defined as

> "[c]apabilities [that] can be elastically provisioned and released, in some cases automatically, to scale rapidly outward and inward commensurate with demand. To the consumer, the capabilities available for provisioning often appear to be unlimited and can be appropriated in any quantity at any time" (Mell & Grance 2011:2).

Elasticity seems to be the pivotal characteristic that distinguishes cloud computing from other computing paradigms (Galante & Bona 2012:263). As Owens (2010:46) remarks,

> "[e]lasticity, in my very humble opinion, is the true golden nugget of cloud computing and what makes the entire concept extraordinarily evolutionary, if not revolutionary."

In fact, elasticity is so important that Amazon even integrates the term into the name of its cloud computing services, "Amazon Elastic Compute Cloud."

Elasticity and scalability are often used as synonyms in the context of cloud computing; however, the concepts are different "and should never be used interchangeably" (Galante & Bona 2012:263). In fact, scalability is a prerequisite for elasticity (Herbst et al. 2013:24f.). *Scalability* refers to a system's ability to cope with an increasing workload in a graceful manner or its ability to increase throughput when additional resources are added (see e.g.

Bondi 2000:195; Agrawal et al. 2011:5; Galante & Bona 2012:263; Herbst et al. 2013:25). Some systems have an *upper scalability bound*, which refers to the maximum number of resources that can be added to the system (Herbst et al. 2013:24). Scalability is a static property because it does not specify any temporal aspects regarding how fast, how often, and how many resources can be added or removed from the system (Galante & Bona 2012:263; Agrawal et al. 2011:10).

Elasticity, by contrast, is a dynamic property that particularly focuses on the temporal aspects of the adaptation process when resources are added or removed (Galante & Bona 2012:263; Agrawal et al. 2011:10). It can be defined as

> "the degree to which a system is able to adapt to workload changes by provisioning and deprovisioning resources in an autonomic manner, such that at each point in time the available resources match the current demand as closely as possible" (Herbst et al. 2013:24).

The nature of the adaptation processes can be characterized along two dimensions: (a) *speed of scaling*, which refers to the time necessary to move from an under-provisioned state to an optimal or over-provisioned state (scaling up) or to move from an over-provisioned state to an optimal or under-provisioned state (scaling down); and (b) *precision of scaling*, which refers to the absolute difference between the current resources allocated and those actually being demanded (Herbst et al. 2013:24f.). Cloud computing systems are defined to scale "rapidly," which means that scaling happens with a lead time of minutes (Armbrust et al. 2010:53) or even in real time (Matros 2012:60f.). These prompt resource reconfigurations are technically enabled primarily through virtualization (Verma et al. 2010:11). In terms of precision, Armbrust et al. (2010:53) emphasize that cloud computing systems can add and remove resources at a "fine grain." Thus, elasticity in cloud computing ultimately allows for significantly reducing (and ideally avoiding) under- and over-provisioning of resources by means of rapid and precise capacity adaptation processes.

Rapid elasticity is closely related to resource pooling. Elastic resource provisioning would neither be feasible nor economically viable without significant pooling of resources, as is done, for example, in extremely large data centers with hundreds of thousands of physical machines (Armbrust et al. 2010:52). The illusion of an infinite resource pool can only be achieved by creating resource pools large enough that individual consumers' capacity fluctuations do not reach the upper scalability bound of the system. The illusion of an infinite resource pool can only be achieved in an economically viable manner by allowing providers to dynamically (re-)assign resources to different consumers in accordance with their current capacity demand and thus efficiently use the aggregate capacity of these large resource pools (Verma et al. 2010:11).

2.5. Measured Service

The characteristic of *measured service* is explained as follows:

> "Cloud systems automatically control and optimize resource use by leveraging a metering capability[1] at some level of abstraction appropriate to the type of service (e.g., storage, processing, bandwidth, and active user accounts). Resource usage can be monitored, controlled, and reported, providing transparency for both the provider and consumer of the utilized service" (Mell & Grance 2011:2, footnote original).

Although the idea of pay-per-use appears intuitive, it may be confused with renting resources. However, as Armbrust et al. (2010:53) explain, they are essentially different: Renting resources means paying a negotiated price for the right to use them for a certain period of time, regardless of whether or not the resources are actually used. Conversely, pay-per-use means that actual resource usage is metered, and consumers pay only for the resources that are actually used, regardless of the period of time over which the usage has occurred. The measured service characteristic is anchored in the concept of utility computing, which aims to provide computing capabilities in a pay-per-use manner, similarly to the way utility services are provided for gas, water, and electricity (Buyya et al. 2008:5) without requiring consumers to make any upfront commitments (Armbrust et al. 2010:51f.; Zhang et al. 2010:7).

3. Service Models: SaaS, PaaS, and IaaS

The NIST definition of cloud computing specifies three service models: (1) SaaS, (2) PaaS, and (3) IaaS (Mell & Grance 2011:2f.; also see Youseff et al. 2008:4). Each service model refers to a specific type of computing capability made available to consumers.

SaaS refers to providing consumers with the capability to use a provider's applications deployed on a cloud infrastructure, where a cloud infrastructure can be simply defined as a "collection of hardware and software that enables the five essential characteristics of cloud computing" (Mell & Grance 2011:2). In accordance with the characteristics of broad network access, these capabilities can be accessed by thin and thick client platforms (Mell & Grance 2011:2). *PaaS* refers to providing consumers with the capability to deploy their own or acquired applications on a provider's cloud infrastructure (Mell & Grance 2011:2f.). *IaaS* refers to providing consumers with fundamental computing capabilities such as processing power, storage, and networks, which they can use to run any kind of software, including operating systems and applications (Mell & Grance 2011:3).

[1] Typically this is done on a pay-per-use or charge-per-use basis.

4. Deployment Models: Private, Community, Public, and Hybrid

4.1. Overview

The NIST definition of cloud computing specifies four deployment models: (1) private cloud, (2) community cloud, (3) public cloud, and (4) hybrid cloud (Mell & Grance 2011:3). These models describe alternative arrangements for how cloud computing infrastructures can be shared among consumers (Liu et al. 2011a:1). They primarily differ based on how exclusive the access of a certain group of consumers to a cloud infrastructure is (Liu et al. 2011a:10ff.). Different degrees of exclusivity are manifested in different relationships between consumers. Thus, these models consider cloud computing systems from an institutional perspective by focusing on the relationships between cloud consumers who share a cloud infrastructure and on the relationships between its consumers and providers.

Private clouds make the cloud infrastructure available for the exclusive use by a group of consumers that belong to the same organization. The cloud infrastructure may be owned, managed, operated by the consumers' organization, a third party, or a combination thereof.

Community clouds make a cloud infrastructure available for exclusive use by a group of consumers that belong to different organizations, but which have shared concerns. The cloud infrastructure may be owned, managed, and operated by one or more of the organizations in the community, a third party, or a combination thereof.

Public clouds make a cloud infrastructure available for open use by the general public. The cloud infrastructure may be owned, managed, and operated by a business, academic, or government organization, or a combination of the three.

Hybrid clouds represents a composition of two or more distinct cloud infrastructures that are distinct entities but are connected through some standardized or proprietary technology that facilitates data and application portability. Hybrid clouds thus represent some form of interorganizational cooperation and potentially an interfirm network. As this thesis is concerned with interfirm networks, hybrid clouds are investigated in more detail in the following subsection.

4.2. Hybrid Clouds

While the NIST definition employs the term "hybrid cloud," synonyms frequently used to refer to the composition of multiple distinct cloud infrastructures include "interclouds" (see e.g. Buyya et al. 2010; Bernstein et al. 2009) and "federated cloud" (see e.g. Rochwerger et al. 2011; Moreno-Vozmediano et al. 2012). Although research on hybrid clouds primarily focuses on rather technical issues (see e.g. Tordsson et al. 2012), it nevertheless directly refers to concerns of an organizational and governance nature. In this section, we first consider the motive for the formation of hybrid clouds and then the organizational and governance aspects.

The primary motive for the formation of hybrid clouds is to overcome the inherent capacity and scalability constraints of a single cloud provider or data center (Rochwerger et al. 2011:44 Calheiros et al. 2012:1350) and thus to deliver computing capabilities in an on-demand and rapidly elastic manner (Armbrust et al. 2010:52). In other words, cooperation aims to increase the upper scalability bound of a cloud infrastructure. If demand outstrips a single provider's or data center's available computing resources, service level agreement (SLA) variations or violations may occur (Calheiros et al. 2012:1350). Hybrid clouds can mitigate these effects, at least to some extent. Providers with excess capacity can make their resources available to providers in need of additional resources, thereby enabling those providers in need to achieve consistent service levels in the event of demand peaks and those with excess capacity to increase resource utilization (Rochwerger et al. 2011:44). Small or medium-sized private clouds may especially benefit from hybrid clouds due to their somewhat tighter capacity constraints than those of large public cloud infrastructures (Armbrust et al. 2010:52). The motive for the formation of hybrid clouds aligns well with the motive of gaining access to resources owned by other firms as described with regard to interfirm networks in general.

Load coordination between multiple cloud infrastructures within a hybrid cloud must happen automatically in a JIT manner in order to respond to sudden demand changes and thus offer computing capabilities in an on-demand and rapidly elastic manner (Buyya et al. 2010:13). In order to achieve timely coordination among multiple cloud infrastructures, research on hybrid clouds suggests the importance of three coordination mechanisms: (1) programmed routines and hyperautomation, (2) implicit coordination by database, and (3) electronic market-based exchanges. These are discussed below.

Due to the technical nature of cloud computing systems, *programmed routines and hyperautomation* are pervasive mechanisms in such systems. Specifically in hybrid clouds, programmed routines and hyperautomation are deployed to coordinate the interface between a focal cloud infrastructure and other cloud infrastructures; these mechanisms can be subsumed under the term "cloud coordinator" (Buyya et al. 2010:21ff.). Other terms used to refer to these mechanisms include "SLA resource allocator" (Buyya et al. 2009:602f.), "federation manager" (Moreno-Vozmediano et al. 2012:68), boundary controller" (Badger et al. 2012:4-3), "intercloud resource manager" (Abawajy 2009:786), and "virtual execution environment manager" (Rochwerger et al. 2009:4:8). Scholars describe a variety of detailed mechanisms related to intercloud coordination. However, many of them essentially relate to mechanisms for (a) negotiating prices, conditions and constraints of exchanging resources with other cloud coordinators in the hybrid cloud; and (b) planning, realizing, controlling, and monitoring the allocation of computing resources to service requests in accordance with agreed service levels, where service requests may either come from one's own consumers or other cloud providers.

Besides using programmed routines and hyperautomation, several scholars identify that hybrid clouds also use a central *information registry*, or *market directory* (Buyya et al. 2010:25; Buyya et al. 2009:605), or a *resource catalog system* (Bernstein et al. 2011:294), thus achieving coordination by means of *implicit coordination*. The information stored in the registry includes a holistic and abstract catalog of resources and services available in the hybrid cloud (Bernstein et al. 2011:294). Resources and services may be described by means of semantic technologies (Bernstein et al. 2011:294), which we focus on in detail below (see Section D.III.4). In addition, the registry stores current resource usage and demand patterns as well as resource availability of individual clouds, including the offered prices and service level policies (Buyya et al. 2010:25). Moreover, the registry offers functionalities that allow consumers and providers to search for and locate potential transacting partners with suitable offers and requests, respectively (Buyya et al. 2010:25; Buyya et al. 2009:605; Bernstein et al. 2011:294).

Cloud computing scholars also provide insights into the governance structure of hybrid clouds. Several scholars argue that resource exchanges are facilitated through *electronic marketplaces* (see e.g. Buyya et al. 2009; Buyya et al. 2010; Bernstein et al. 2011; Calheiros et al. 2012). The information registry thus becomes combined with an electronic marketplace, which is referred to as a "cloud exchange" (Calheiros et al. 2012; Buyya et al. 2010:20) or "global cloud market" (Buyya et al. 2009:605; Buyya et al. 2008). These scholars suggest that market coordination is achieved through *fixed prices* or *auctions* (also see Garg et al. 2013; Mihailescu & Teo 2010; Mihailescu & Teo 2012). Market-based exchanges can reduce delays when acquiring resources (Garg et al. 2013:1154), thus facilitating on-demand availability and rapid elasticity (Calheiros et al. 2012:1352). The cloud exchange thus becomes an auctioneer that is in control of the entire trading process. Scholars suggest that the exchange needs to be operated by a neutral middleman (Garg et al. 2013:1172) trusted by all market participants (Buyya et al. 2010:25). Neutrality is also important because the cloud exchange may settle financial transactions (Buyya et al. 2010:25). Figure 13 depicts the structure of a hybrid cloud clustered around a cloud exchange. Observe that cloud coordinators may either negotiate via the cloud exchange (in a 1:m or n:m manner) or directly with coordinators of other clouds (in a 1:1 manner) (Calheiros et al. 2012:1352).

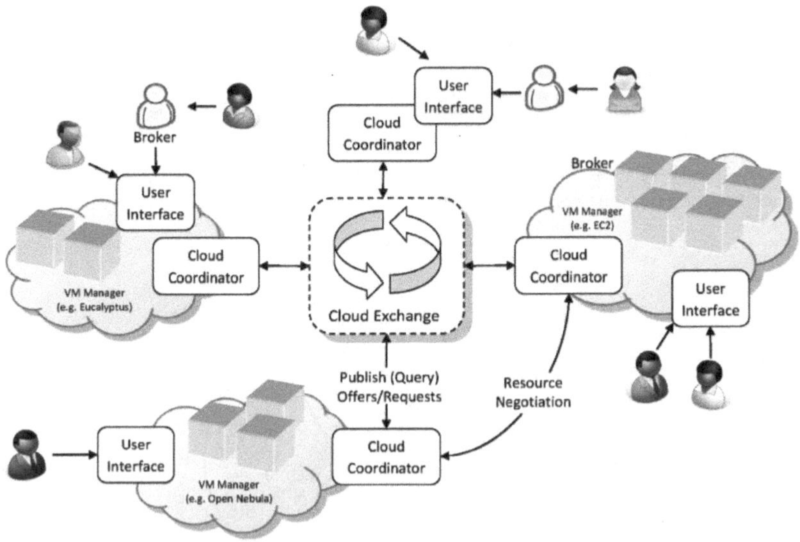

Figure 13.: Hybrid Cloud with Cloud Exchange (adopted from Calheiros et al. 2012:1353)

Exchanging computing capabilities via auctions is not trivial, however, because capabilities may be characterized by multiple attributes. Moreover, the sequential execution requirements of capabilities as well as compatibility problems may constrain the combinability of individual capabilities. Conventional market mechanisms such as spot pricing are only partially suitable for facilitating such complex exchanges. For example, Mihailescu & Teo (2012:626) show that, in a hybrid cloud with spot pricing, users can increase their welfare by behaving strategically. Hence, multidimensional auction mechanisms are necessary to facilitate the exchange of complex services in such complex environments (Blau et al. 2009a:344f.). The development of such mechanisms has recently sparked scientific interest and is ongoing (see e.g. Blau et al. 2009a, Blau 2009, Blau et al. 2010, and van Dinther 2010).

III. Technological and Conceptual Roots

1. Overview

Although cloud computing has often been positioned as "the new thing," it is not an entirely new computing paradigm. In fact, it can be interpreted as an amalgam and descendant of several existing computer technologies and concepts (see e.g. Youseff et al.

2008:1; Buyya et al. 2009:599; Zhang et al. 2010:8f.). Figure 14 shows the most important technological and conceptual roots of cloud computing. The convergence and increasing maturity of these technologies has contributed to the field's emergence (Voorsluys et al. 2011:5f). Hardware virtualization and SOC, the computing paradigm that underlies SOA and web service mashups, are discussed in more detail below because both technologies represent important underpinnings for the concept of cloud logistics, as suggested by cloud logistics knowledge (see Section E.IV.5).

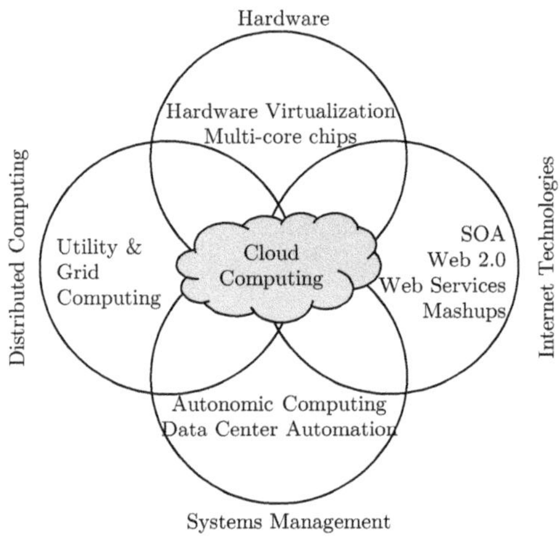

Figure 14.: Technologies that Have Contributed to the Emergence of Cloud Computing (adopted from Voorsluys et al. 2011:6)

2. Hardware Virtualization

Hardware virtualization was originally introduced by IBM in the 1960s to improve resource utilization by enabling concurrent multi-user access to mainframe computers (Sugerman et al. 2001:1). Its fundamental underlying concept is the simulation of a complete computer system on a different computer system (Goldberg 1974:34). The simulated system is referred to as a *virtual machine* and the other as a *real machine*. A virtual machine has the "looks" of a real machine (Rosenblum 2004:34), but not necessarily the same looks as the underlying real machine (Smith & Nair 2005:32). Software developed to run on a real machine can also be run on the corresponding virtual machine (Rosenblum & Garfinkel 2005:34).

To explicate how hardware virtualization works, it is useful to consider the structure of computer systems. Computer systems are composed of hierarchical layers: The lowest layer represents the physical hardware, the following the operating system, and the highest layer software applications (Rosenblum 2004:34). Hardware virtualization interposes an additional software layer – a virtualization layer – between the physical hardware and the operating system, as depicted in Figure 15 (Rosenblum 2004:34ff.). This additional layer is referred to as a virtual machine monitor (VMM). In principle, this is a simulator software that transforms a single real machine into one or more virtual machines, creating the illusion of many existing real machines (Goldberg 1974:36). The VMM controls both real and virtual machines; it starts and stops virtual machines and allocates them physical computing resources. Although they run on the same real machine, virtual machines are logically separated from each other, so the user of each virtual machine is given the illusion of actually having a dedicated physical machine at his or her disposal (Sugerman et al. 2001:1).

Virtualization is cost-efficient (Goldberg 1974:43). This is because multiple users share the same physical computing resources; the VMM balances the load between multiple virtual machines by dynamically allocating physical resources according to demand (Rosenblum & Garfinkel 2005:35).

Moreover, virtualization provides extraordinary flexibility (Goldberg 1974:43) because the VMM decouples software from hardware (Rosenblum & Garfinkel 2005:35). As Rosenblum & Garfinkel (2005:35, italics original) explicate,

> "[a] VMM provides a *uniform view* of underlying hardware, making machines from different vendors with different I/O [input/output] subsystems look the same, which means that virtual machines can run on any available computer. Thus, instead of worrying about individual machines with tightly coupled hardware and software dependencies, administrators can view hardware simply as a pool of resources that can run arbitrary services on demand."

Hardware virtualization enables different users to run different operating systems and applications concurrently while using the same real machine (Goldberg 1974:36). In addition, virtual machines can be migrated from one real machine to another while running and without the user noticing, thus also making them robust against hardware failures (Rosenblum & Garfinkel 2005:35). Furthermore, virtual machines are scalable because real machines can be dynamically added or removed, with the VMM (re-)mapping these resources to virtual machines commensurate with demand. Finally, virtual machines can use the resources of real machines to which they are connected via a network link (Sugerman et al. 2001). Thus, virtualization removes spatial dependencies between the user of a virtual machine and the underlying physical machine (Bernstein & Vij 2010:542).

Hardware virtualization removes any dependencies between software, hardware, and geographical location(s). Any software can run on any hardware, and any hardware can

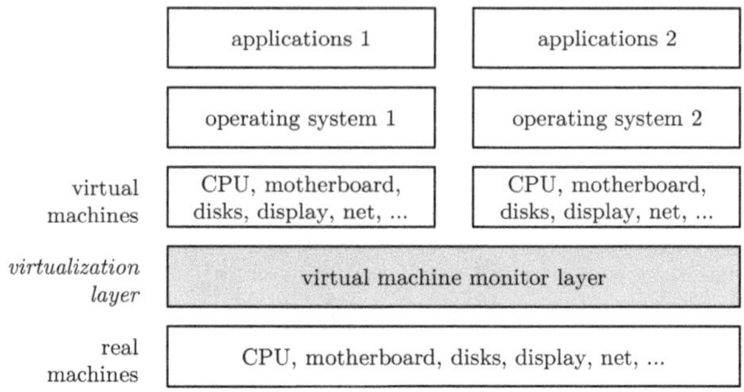

Figure 15.: Concept of Hardware Virtualization (adopted with minor changes from Rosenblum 2004:36)

be used to run any software, regardless of the geographical proximity between users of virtual and real machines. This creates a great amount of flexibility.

Hardware virtualization is a critical technology that underlies cloud computing in two respects. First, hardware virtualization is the key technology for delivering IaaS, which delivers virtual machines to cloud users (Youseff et al. 2008:5). Second, hardware virtualization is critical for enabling *resource pooling* in cloud computing. Cloud computing uses a VMM that dynamically manages the usage and allocation of physical computing resources to the virtual machines of different users (Armbrust et al. 2010:52f.; Zhang et al. 2010:7ff.). We can thus conclude that hardware virtualization has been instrumental in the emergence of cloud computing.

3. Service-oriented Computing

3.1. Definition: Services and Service Compositions

SOC specifies a conceptual model of computing systems, or a paradigm, that uses services as a basic element for creating distributed applications (Papazoglou 2003:3; Papazoglou et al. 2008:223). Cloud computing is based on SOC because it adopts the notion of services: Cloud computing capabilities are delivered "as-a-service."

A *service*

> "represents a capability of performing tasks that form a coherent functionality" (Haas 2004).

This functionality provides recognizable business value by changing the state of the world in a desired way (desired real-world effect) and can be accessed through a well-defined interface (Laskey et al. 2012:29f.). A service can thus be understood as a collection of tasks that lie within a discrete domain of control and are geared towards achieving related (business) goals (Jones 2005:89). Each service is defined within its own distinct functional context and comprises capabilities to achieve goals related to this context; capabilities that can be invoked by an external user are expressed through a *service description* (Erl 2008:39; Subsection D.III.3.2 elaborates on service descriptions).

A *service composition* represents a "coordinated aggregate of services" (Erl 2008:39). Two or more services, each self-contained and delivering some distinct functionality, are combined to form a more complex application or realize a complete business process (see e.g. Laskey et al. 2012:37). Compositions are typically referred to as *composite services*, and, like any other service, they can become members in other compositions (Papazoglou et al. 2007:67). The composition of services usually becomes necessary if requirements cannot be fulfilled by a single existing service. SOC is often associated with the creation of distributed applications because composite services may span multiple domains of control or ownership, as different services that are part of the composition may be distributed over multiple domains of control or ownership (Laskey et al. 2012:21, 70ff.).

Strategic goals associated with this paradigm include increased return on investment (ROI) on IT resources and increased organizational agility (Erl 2008:55ff.; also see Bieberstein et al. 2005). These goals can be attained through the use of distributed applications composed of modular services (Erl 2008:38). SOC promises to make service composition happen in a rapid, ad-hoc, and easy manner, even in heterogeneous environments, enabling the assembly of dynamic and flexible business processes and agile applications that may span several organizations and computing platforms (Papazoglou et al. 2008:223; also see Erl 2008:135).

3.2. Service-Orientation Design Principles

SOC revolves around service-orientation as a fundamental design paradigm, with a design paradigm understood as a set of principles that collectively govern the design of a well-defined functionality (Erl 2008:37f.). *Service-orientation* consists of various design principles that shape the design of services (Erl 2008:39). Six of these are particularly relevant to the objectives of this thesis and are briefly introduced below. These principles are (1) standardized service descriptions, (2) service discoverability, (3) service reusability, (4) service loose coupling, (5) service abstraction, and (6) service composability (Erl 2008:70ff.).

3.2.1. Standardized Service Descriptions

Service-oriented design emphasizes the use of *standardized service descriptions* (Erl 2008:125ff.). In order to explain why service descriptions need to be standardized, we first define service descriptions in general. *Service descriptions*[1] specify or reference the information that is necessary "to use, deploy, manage and otherwise control a service" (Laskey et al. 2012:43). Specifically, a service description typically specifies the "purpose, functionality, constraints, and usage of the service" (Krafzig et al. 2004:81pdf; also see Erl 2008:71) or, using other terminology, the inputs, outputs, preconditions, and effects (IOPE) (Martin et al. 2004). Service descriptions thus provide a comprehensive and important basis for potential consumers to decide whether or not a service is suitable for satisfying their needs (Laskey et al. 2012:43). Still, service descriptions are inherently incomplete (just like any contract), but they may be considered sufficient if they enable stakeholders to access and use the described services solely based on the information provided (Laskey et al. 2012:43).

Service descriptions are based on either a single or inter-linked set of documents (Laskey et al. 2012:43), which may be technical and non-technical (Erl 2008:130). Technical documents provide, for example, information about the service's capabilities, interface, and in- and output data types. Non-technical documents – often referred to as SLAs – specify the service's quality-of-service (QoS) in areas such as performance, accessibility, availability, security attributes, usage statistics, and user ratings (Papazoglou 2003:6; Erl 2008:152f.).

Service descriptions may be standardized along multiple dimensions, including the expression of service functionality, data representation, policies, and the language(s) used to codify them (Erl 2008:133f., 137f.). Standardized descriptions are desirable because they provide a basis for ensuring a meaningful level of interoperability between services by minimizing the need for data transformations (where *interoperability* refers to the sharing of data and the exchange of information between services, see Erl 2008:56). Standardization also increases the potential for service reuse and makes descriptions easily interpretable by humans and processable by machines (Erl 2008:130, 149). Thus, standardized descriptions are crucial to attain the ultimate goal of agile and even ad-hoc service compositions (Erl 2008:135ff.). Nevertheless, standardization may be a goal that is not easily attained, especially if the number of services increases (Erl 2008:132) or consumers put forward different requirements.

[1] "Service descriptions" are sometimes also referred to as "service contracts." However, we prefer the term "service description" because the information it includes basically describes the service. Conversely, the term "contract" implies an agreement between two or more entities. Using "contract" instead of "description" would (a) imply that the conditions of using a service are dictated by the service and do not result from an interaction between consumers and providers and (b) miss the point that a single description can be the basis for multiple contracts (Laskey et al. 2012:44).

3.2.2. Service Discoverability

Service discoverability refers to the ability of humans and machines to easily locate a service that offers a specific desired capability in a specified environment (e.g. a service repository) and, once located, to understand and interpret the service's capabilities, conditions, and constraints of use (Erl 2008:362ff.). Discoverability focuses and depends on the service's "communications quality" (Erl 2008:73). Achieving discoverability requires information about services to be maintained in a consistent format and stored centrally (Erl 2008:362).

Service-oriented design tries to achieve discoverability in order to position services as reusable IT assets (with repeatable ROI) and thus avoid unnecessary creation of services with redundant or overlapping capabilities (Erl 2008:73, 364). It also tries to let users clearly understand a service's capabilities and limitations in order to make sensible decisions about whether or not the service is suitable for satisfying their requirements, thus capturing reuse opportunities (Erl 2008:365). In Section D.III.4, we emphasize how using *well-defined semantics* to create service descriptions can support the (automatic) discovery and interpretation of services.

3.2.3. Service Reusability

Service reusability essentially indicates the potential that capabilities offered by a particular service cannot be used for a solely single purpose, but for multiple purposes in multiple compositions (Erl 2008:254, 258). The principle of reusability goes beyond the question of how and with what information services are described (e.g. standard contract abstraction, discoverability) by focusing on how a service's capabilities need to be shaped to be reusable (Erl 2008:254). This includes considering the question of what capabilities should be encapsulated by (i.e. be enclosed within) each service.

Service-oriented design strongly advocates for service reusability in order to increase the ROI for software and organizational agility (Erl 2008:259). Reusability can increase ROI because services need to be designed only once but can be used multiple times, and it can increase agility because future requirements can be rapidly met by rearranging existing service compositions or creating new compositions from a set of already available services (Krafzig et al. 2004:165pdf; also see Erl 2008:259). Yet, although it is a simple concept, service reusability is challenging to realize and may not necessarily be financially beneficial. This is because designing reusable services requires considering not only a specific set of requirements that should be met immediately, but also potential future requirements that may arise from scenarios in which the service may be used (Krafzig et al. 2004:178pdf; Erl 2008:257). It may also be difficult to determine, find, and understand which capabilities can in fact be reused (Krafzig et al. 2004), which is why service-oriented design often requires subject matter experts who are able to define reasonable service boundaries (Erl 2008:260). Therefore, creating reusable services requires a high(er) initial effort that may not always pay off in the long run (Krafzig et al. 2004:34pdf).

3.2.4. Service Loose Coupling

In general, coupling refers to a relationship or connection among two or more things, where "things" can be organizations, humans, or services (e.g. Beekun & Glick 2001:229; Weick 1976:5; Erl 2008:71). *Service loose coupling* encourages services to depend on each other only to some extent or not at all. In the realm of service design, the degree of coupling is compared with and measured by the degree to which services depend upon each other (Krafzig et al. 2004:70pdf; Erl 2008:71). If services are coupled, modifying one of them requires changing another (Fowler 2001:102). Coupling in SOC can arise from two different types of relationships: (a) those within the boundaries of the service, and (b) those outside of them (Erl 2008:167ff.). The former refers to coupling between a service description and its technological implementation, for example if a service description contains information specifying a particular hardware technology on which the service will be implemented. The latter refers to coupling between a consumer and the service description, for example if a service description contains information or standards specific to a consumer.

Service-oriented design generally aims to reduce or loosen the degree of coupling, but it does not aim to entirely eliminate it (Erl 2008:71). On the one hand, reducing coupling between the service description and the underlying technological implementation enables the independent design of services and creates an environment that promotes the independent and inexpensive evolution of services (Erl 2008:71, 168; Pautasso & Wilde 2009:911; also see Krafzig et al. 2004:34pdf). Loosening the degree of coupling can also reduce overall complexity and risk, enable faster change, increase agility (Krafzig et al. 2004:70pdf), foster resilience (Cerf 2013:96), increase the opportunities for service reuse (Erl 2008:148), and help prevent dependencies that inhibit service composition (Erl 2008:198). On the other hand, coupling cannot be avoided entirely. Some is necessary in order to ensure that services remain interoperable – that is, that they can communicate and exchange information in some way and are thus able to achieve a desired outcome (Fowler 2001:102). In other words, the constraints imposed on interactions between services should be minimal, yet sufficient to enable interoperation (Laskey et al. 2012:85). In this case, services become coupled to a common interaction standard that regulates their interactions and changing this standard impacts all services (Pautasso & Wilde 2009:911). In conclusion, loose coupling is a design principle that must reconcile conflicting aspects in service design: The degree of coupling should be reduced, but not entirely avoided, in order to ensure that service remains interoperable and composable.

3.2.5. Service Abstraction

Service abstraction is concerned with the extent to which information about a service is hidden from or exposed to consumers (Erl 2008:235). It thus influences the design of service descriptions in terms of the scope and granularity of the included information (Erl 2008:72, 215). In principle, abstraction applies to all kinds of information exposed via service descriptions, including information about (a) the functional capabilities of a service, and (b) the QoS in terms of the conditions and constraints of use and interaction requirements (Erl 2008:218). The design principle of service abstraction aims to keep the quantity and detail of information exposed through service descriptions concise and balanced (Erl 2008:215) by hiding all information that is not absolutely necessary for effectively using a service (Erl 2008:212).

Service-oriented design advocates for service abstraction for two reasons. First, hiding information reduces the coupling between a customer and a service (Erl 2008:214). This gives service providers the freedom to independently evolve the (technological and functional) implementation of a service while continuing to provide the specified capabilities to consumers as initially agreed (Erl 2008:212). Second, hiding information is crucial for achieving service reusability. However, hiding too little or too much in the service description may limit the potential of service reuse (Erl 2008:72). Exposing too much information increases the risk of consumers becoming tightly coupled to the service description, minimizing reusability. On the other hand, if too little information is exposed, consumers may not be able to determine whether or not an existing service can meet their goals. We can conclude that the pivotal question about the principle of service abstraction is not whether to apply abstraction, but how to identify the right degree to which abstraction should be applied. In other words, service abstraction balances the needs of other design principles to incorporate information in service descriptions (Erl 2008:214).

3.2.6. Service Composability

Service composability requires services to be capable of becoming effective members in compositions, independent of a composition's size and complexity and regardless of whether compositions are required on short notice (Erl 2008:74, 393). Composability is the intended result of applying other design principles. In fact, all other design principles ultimately contribute to achieving composability, as depicted in Figure 16 (Erl 2008:412). Still, Erl (2008:392) includes service composability as a separate principle to ensure that the other principles are not applied in a way that would prevent composition. In sum, the principle of composability integrates the others to achieve the goals of SOC (Erl

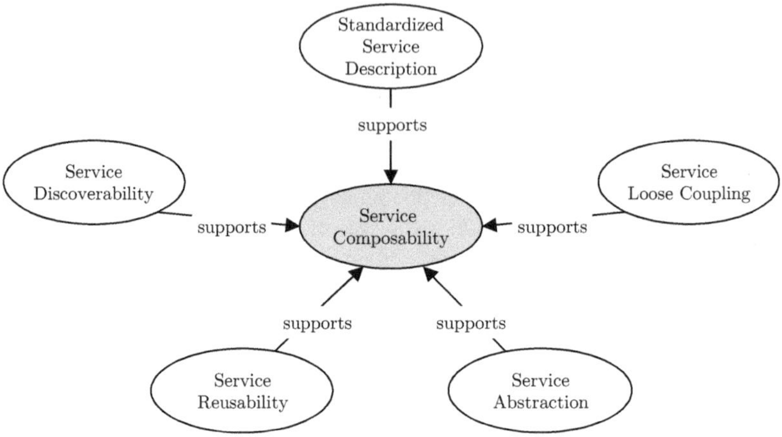

Figure 16.: Relationships between Service Composability and Other Design Principles (adopted with minor changes from Erl 2008:412)

2008:73), including rapid, ad-hoc, and easy composition of services to increase ROI and organizational agility (see Subsection D.III.3.1).

3.3. Service-oriented Architectures

The architecture of a system can generally be defined as the

> "fundamental concepts or properties of a system in its environment embodied in its elements, relationships, and in the principles of its design and evolution" (ISO/IEC/IEEE 42010 2011b:2).

An SOA is basically a means to realize the vision of SOC (Papazoglou et al. 2007:64). More specifically, an SOA is a distinct form of (software) technology architecture geared toward supporting services and service compositions shaped by and designed according to the design principles associated with service-orientation (Erl 2008:40). Yet, an SOA is not only a technology architecture, but it also specifies the constituent entities which participate in a SOA, their roles, their relationships, and interactions (see e.g. Papazoglou 2003:5). They are commonly illustrated with the "SOA triangle," depicted in Figure 17. Accordingly, SOA consists of three types of roles: a service provider, a service customer (requestor), and a service repository and broker. They interact and collaborate by means of three basic operations: *publish*, *find*, and *bind and invoke* (Endrei 2004:27). Service

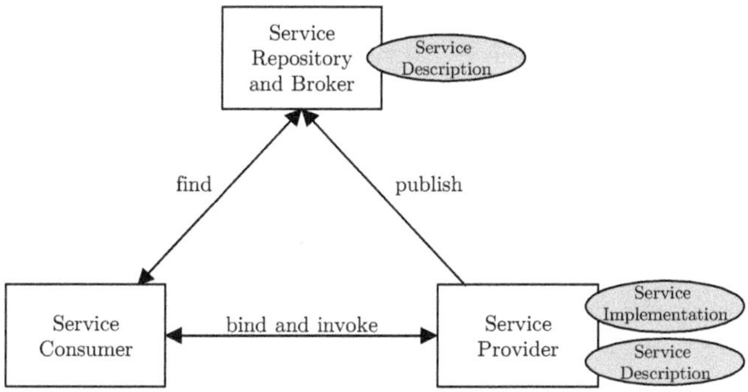

Figure 17.: Basic Roles and Interactions in Service-oriented Architectures (adopted with minor changes from Endrei 2004:26; Papazoglou 2003:5; Erl 2008:368)

providers publish their service descriptions to the service repository (or broker). The repository (or broker) stores these descriptions and makes them discoverable to consumers. Consumers query the service repository (or ask a broker) in order to find a service that fulfills their requirements. Once a suitable service is found, consumers retrieve the service description, bind (i.e. connect) with the service provider, and invoke (i.e. call) the service according to the information provided in the service description. The service provider implements the service using the input information received from the consumer. The distinction between providers and consumers may sometimes be of a logical nature only because any provider can also "act" as a consumer and *vice versa* (Papazoglou et al. (2007:66)) An SOA can therefore be considered an approach to

> "organizing and utilizing distributed capabilities that may be under the control of different ownership domains. It provides a uniform means to offer, discover, interact with and use capabilities to produce desired effects consistent with measurable preconditions and expectations." (MacKenzie et al. 2006:29; also see Laskey et al. 2012:21; Endrei 2004:24)

The *service repository* assumes a central role in organizing and using distributed capabilities. In principle, the repository represents a location where service descriptions are stored and made discoverable to consumers (Krafzig et al. 2004:78pdf; synonyms of repository include "registry" (Endrei 2004:25f.; Martin et al. 2005:32) or "service inventory" (Erl 2008:40)). It represents a pool of existing services that can be used to create composite services (Erl 2008:42). The application of common design principles and standards within a repository is of "paramount importance" in order to ensure a high degree of native service interoperability, thereby enabling repeated, agile, and efficacious service compositions (Erl 2008:40).

The role of the service repository is versatile and can vary significantly in different SOAs and therefore requires further consideration. At the one extreme, it may provide only basic functionality essentially limited to storing and publishing service descriptions. At the other extreme, service repositories can evolve into *electronic marketplaces* that bring together consumers and providers (Papazoglou 2003:9). In this case, service repositories can additionally offer a variety of services to facilitate business transactions (e.g. by offering negotiation protocols and payment services) or to support the community of providers and consumers as a whole (e.g. by offering reliability reporting of services based on customer feedback) (Papazoglou 2003:9; also see Burstein et al. 2005:79). The organization that operates the repository acts as a *market maker* that assumes responsibility for administering and maintaining the marketplace (Papazoglou 2003:9).

3.4. Interim Summary

The preceding subsections have introduced the concept of SOC. They have specifically focused on the fundamental conceptual elements of this computing paradigm: services, service compositions, design principles associated with service-orientation, SOAs, and service repositories. To summarize these subsections, Figure 18 depicts these elements and their interrelations.

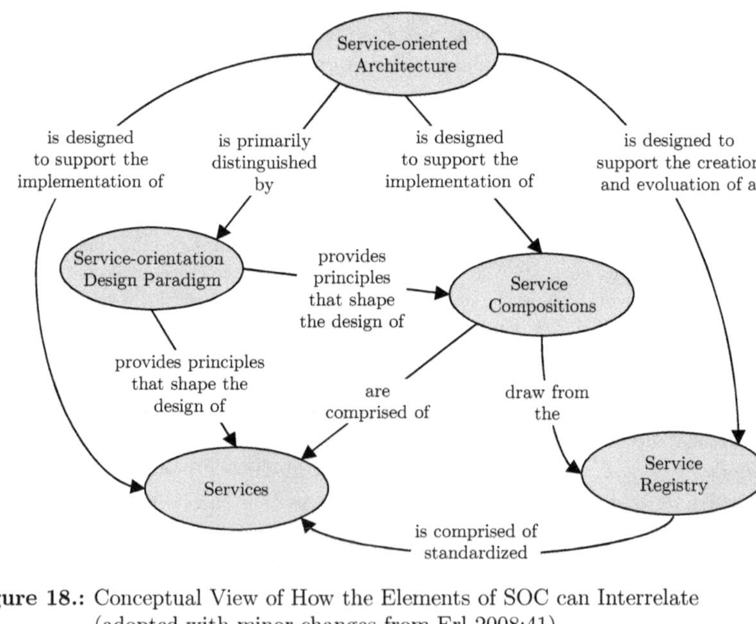

Figure 18.: Conceptual View of How the Elements of SOC can Interrelate (adopted with minor changes from Erl 2008:41)

4. (Semantic) Web and Logistics Services

Web services represent the currently most promising technology based on SOC and SOA (Weerawarana 2005 as cited in Papazoglou et al. 2008:225; Erl 2008:46). In fact, they are another internet technology that has contributed to the emergence of cloud computing (see Section D.III.1). Web services can be defined as

> "self-contained, modular applications that can be described, published, located, and invoked over a network, generally, the World Wide Web"(Gottschalk 2000; McIlraith & Martin 2003:90).

Thus, web services are services, as defined earlier, with the distinct characteristic that they can be published, located, and invoked through the WWW. A *semantic* web service represents an amalgam of a web service and the vision of the Semantic Web (McIlraith & Martin 2003). More specifically, semantic web services arise from the use of well-defined semantics for the creation of service descriptions.

For this thesis', two aspects of web services are important: (1) the life cycle, and (2) semantic service descriptions. The life cycle is discussed in the next subsection. In order to introduce semantic service descriptions, we first introduce the vision of the Semantic Web. Thereafter, we focus on semantic and syntactic web service descriptions as well as on the adoption of these technologies to description of logistics services.

4.1. The Life Cycle of Web Services

The life cycle of web services contains all phases of a web service's existence. Several, albeit very similar, conceptualizations of this life cycle can be found in literature (see e.g. McIlraith & Martin 2003; Tsalgatidou & Pilioura 2002; Patil et al. 2004; Martin et al. 2004; Burstein et al. 2005). In these contributions, the life cycle typically includes all or a subset of the following phases: (1) development, (2) publication, (3) discovery, (4) composition, (5) negotiation and selection, (6) invocation, (7) execution, (8) monitoring, and (9) unpublication. These phases are closely linked to the basic interactions between service providers, consumers, and service brokers as described in the SOA triangle (see Subsection D.III.3.3). In fact, they provide a more detailed understanding of SOA activities and interactions. This link is emphasized in the following by showing that each SOA must include *mechanisms* to coordinate activities and interactions along the life cycle.

Development refers to the process of creating a service description, which is developed by a service provider. Every SOA must include a mechanism that supports the development of service descriptions (Laskey et al. 2012:64). The development phase ends when a provider has created a service description.

Publication refers to the process of announcing the description of a new service to the service repository or updating the description of an already existing service. The publication of service descriptions is initiated by the service provider (Gottschalk et al. 2002:171). Every SOA must include a mechanism that supports the publication of new service descriptions and a mechanism that supports the notification of consumers about changes and updates of existing descriptions (Laskey et al. 2012:64). The publication phase ends once service descriptions are made discoverable to consumers.

Discovery refers to the process of consumers searching for and finding a capability in the service repository (Erl 2008:364). Consumers query the repository with their requirements, and the repository matches the query with registered capabilities and returns the set of candidate services that offer the desired capability (Burstein et al. 2005:75f.). "Candidate" is a service that has not yet been implemented, and only exists as a description (Erl 2008:53). Every SOA requires a mechanism that enables consumers to search for and locate capabilities in the repository (Laskey et al. 2012:64). Consumers interpret the service description and evaluate whether and the extent to which candidate services fulfill their requirements. Every SOA must include a mechanism that allows consumers to access service descriptions in order to inspect service functionality and associated conditions and constraints of service use and to verify the credentials of respective service providers (Laskey et al. 2012:64). The discovery phase ends once the interpretation and evaluation of the service description have been completed.

Composition refers to the dynamic process of combining two or more candidate services into a single *composite service*, with the goal of addressing consumer-specific requirements (Limthanmaphon & Zhang 2003:201; Milanovic & Malek 2004:51). Composition becomes necessary if the consumer's requirements cannot be fulfilled by a single service, but only by the combination of multiple. Service composition is driven by the consumer and may be supported by the repository or a broker. Hence, every SOA requires a mechanism to support the composition of candidate services for both long- and short-lived business transactions (Laskey et al. 2012:72). In order to evaluate the extent to which each composition satisfies their requirements and compare available alternatives, consumers need to aggregate the conditions and constraints of use (e.g. quality of service attributes) of all composition members. Every SOA must include a mechanism to support the aggregation of conditions and constraints in service compositions (Laskey et al. 2012:75). Composition is a *dynamic process* because consumers may (a) combine and re-combine services multiple times until they have identified a candidate composite service that, according to the information contained in the service description, satisfies their requirements and (b) move back and forth between the service discovery phase and composition to identify additional or other services to satisfy their requirements. The composition phase ends with consumers sending a formal request to those providers of the candidate services (Burstein et al. 2005:76f.) part of a candidate composite service.

Negotiation and selection refers to the interaction process between a consumer and the potential service providers he or she engages with to reach a formal agreement – a service contract – on the conditions and constraints of service provisioning (Yao et al. 2006:108; Burstein et al. 2005:74, 76f.), including, for example, performance levels and penalties in case of service failure. Negotiation becomes necessary if (a) constraints and conditions stipulated in the service description or service request are not directly acceptable by the consumer or provider respectively (Paurobally et al. 2007:14:6); (b) service descriptions lack the information necessary for consumers to decide whether a service suitably addresses their concerns, such as costs; or (c) conditions and constraints stipulated by different candidate member services included in the composition are in conflict with each other.

The negotiation process can include multiple interaction rounds (or a sequence of message/proposal exchanges) that, on one hand, enable consumers to determine if and how a potential provider is able to deliver a requested service and, if necessary, continuously refine their requirements and, on the other hand, let providers tailor their offerings in line with consumer needs (Paurobally et al. 2007:14:7; Yao et al. 2006:108; Laskey et al. 2012:29). Hence, every SOA requires a mechanism that enables interactions among consumers and providers through message exchanges (Laskey et al. 2012:72) and a mechanism that supports the resolution of conflicting conditions and constraints (Laskey et al. 2012:41). Ultimately, that may include the escalation of conflicts to human adjudication in order to determine which conditions and constraints are honored (Laskey et al. 2012:75). The service repository supports this process by providing the required negotiation protocols that govern the interaction process and the computing platform on which these protocols are implemented. The selection and negotiation phase ends with consumers and service providers either concluding a formal service contract that reconciles their preferences and constraints or terminating negotiations (Burstein et al. 2005).

Invocation refers to the activities consumers need to perform in order to call a service (Oracle 2002). This includes consumers sending input data to a specified endpoint, such as an Uniform Resource Locator (URL) in the WWW. The invocation phase concludes when consumers have supplied all the information necessary to execute the agreed-upon service.

Execution refers to the process of performing the tasks contained in a service with the objective of producing a desired (real-world) effect as specified in the service contract, thus addressing consumer concerns (see Laskey et al. 2012:49). A service is executed by a provider. The service execution phase ends once the service has met the goals specified in the service contract. However, if the service fails to execute as expected, execution ends once the service is terminated by the provider.

Monitoring runs concurrently with execution. It refers to the process of measuring, reporting, and determining whether service execution complies with the conditions and constraints agreed on in the service contract (see Laskey et al. 2012:102; Burstein et al. 2005). Every SOA must include a mechanism to determine the extent to which the achieved real-world effect is compliant with the conditions and constraints stipulated in the service contract (Laskey et al. 2012:74f.). Monitoring is conducted by the provider, and the results are reported to the consumer. Alternatively, if monitoring mechanisms allow, the state of service execution can also be monitored directly by consumers (Burstein et al. 2005:75). Monitoring becomes especially challenging in the case of composite services, as their properties depend on the individual member services which may be executed under different domains of control and monitored by different organizations, which requires an additional coordination effort (Tsalgatidou & Pilioura 2002:157; also see Burstein et al. 2005:79). If the service is not executed as agreed, that is, the desired goal is not or only partially achieved it may be necessary to compensate the consumer as stated in the service contract (Burstein et al. 2005:75, 77ff.). Hence, every SOA must include a mechanism that, in case of service failure, enables financial transactions to compensate stakeholders involved in a multiple-step process (Laskey et al. 2012:72).

Unpublication refers to the process of removing service descriptions from a repository, either through the service provider or the repository itself. This may be necessary if a service is no longer offered by its provider or needed by consumers (Tsalgatidou & Pilioura 2002:136). This phase ends once the description is no longer discoverable.

The life cycle of web services identifies the entire set of activities and exchanges that must be coordinated between consumers, providers, and the service repository in a service-oriented system.

4.2. The Semantic Web

4.2.1. The Vision of the Semantic Web

The Semantic Web refers to a vision originally formulated by Tim Berners-Lee in the late 1990s and early 2000s. According to proponents of this vision, the Semantic Web does not refer to a "separate" Web, but refers to an extension of today's WWW into which today's WWW will evolve (Berners-Lee et al. 2001). To understand how this vision aims to extend the WWW, we start by considering what the WWW comprises and how it works today.

Most of the content available on the WWW is contained in documents designed for humans to read, not for computers to manipulate in a meaningful manner (Berners-Lee et al. 2001).

This was true in 2001 and is still true to a large extent today. Any piece of information in the WWW is only "one click away" as it can be accessed by typing a URL into a web browser (subject to authorization). Thus, finding relevant information turns out to be the crucial problem. To search the WWW, we type keywords into search engines. Although contemporary search engines, such as Google, deliver useful and relevant content almost instantaneously, web search capabilities are largely constrained by and limited to the mere use of keywords. A keyword search may produce irrelevant search results because some keywords are ambiguous and others lack concise definitions. Therefore, humans ultimately need to interpret search results and judge whether the presented information is, in fact, relevant. If not, they need to refine the search in an iterative manner using other keywords until relevant information is found. The root cause of this limited search capability is the fact that computers cannot "understand" the meaning of what we search and cannot unequivocally judge whether the identified information answers the query. These limitations can be easily illustrated by entering the question "What day is today?" into Google's search engine. Instead of answering this basic question, Google presents multiple links that are likely to contain the answer.

The Semantic Web can be defined as "the conceptual structuring of the Web in an explicit machine-readable way" (Berners-Lee & Fischetti 1999 as cited in Gomez-Perez & Corcho 2002). Thus, the vision of the Semantic Web is giving information on today's WWW a well-defined meaning that is processable by computers (Berners-Lee et al. 2001). Shadbolt et al. (2006:96) describe the Semantic Web as

> "a Web of actionable information — information derived from data through a semantic theory for interpreting the symbols. The semantic theory provides an account of 'meaning' in which the logical connection of terms establishes interoperability between systems."

Establishing a machine-readable conceptual structure of Web content enables machines to process and "understand" the data which they currently merely visualize on screens (Berners-Lee et al. 2001). Of course, computers do not truly "understand" the content they process, but they can manipulate data much more effectively now – especially in ways that are useful and meaningful to human users (Berners-Lee et al. 2001). Berners-Lee (1998) explains:

> "The concept of machine-understandable documents does not imply some magical artificial intelligence which allows machines to comprehend human mumblings. It only indicates a machine's ability to solve a well-defined problem by performing well-defined operations on existing well-defined data."

As a result, the Semantic Web can enable humans and computers to better cooperate and enable computers (software agents and services) to better interoperate (i.e. exchange data) and carry out complex tasks on behalf of users (Berners-Lee et al. 2001; Payne & Lassila 2004:14). The crucial feature of interoperation between computers in the Semantic Web is the vision of enabling *serendipitous* interoperation, which means that interoperation is unarchitected and unanticipated (see e.g. Payne & Lassila 2004:14). Serendipitous interoperation between computers results from a semantic theory that defines logical connections between terms in a machine-processable manner.

Serendipitous interoperation in the Semantic Web leads to a variety of tasks that can be conceived of as being performed in an automated manner by software agents on behalf of humans. Examples in the literature are innumerable and versatile – some intriguing, some odd, some frightening, and some difficult to classify:

> "A revolution is underway in computing, and if you believe pundits such as Vint Cerf, 'father of the Internet,' it won't be long before your bathroom scale surreptitiously transmits your weight to your doctor, who might command a stop to the rocky road ice cream your fridge automatically orders for you from www.groceries.com." (McIlraith & Martin 2003:90)

Although this kind of interoperation between multiple devices (scales, refrigerators, and web pages) and their interactions with humans (the doctor) in this example still seem some time away from becoming a reality (if at all), asking a semantic search engine the simple question of what day it is today results in the engine telling you the actual day. Kngine, a sematic search engine currently in beta status, answers, "01:40 PM Thursday (WEDT) - Time in Bonn, Germany" (Kngine 2015). Despite being a very basic example, it shows how computers can use conceptually structured knowledge that is encoded in a machine-readable way to answer a simple question. More generally, it shows how information processing on the Web can be automated (Payne & Lassila 2004:14).

4.2.2. The Role of Ontologies and Ontology Languages

For the Semantic Web to function, Berners-Lee et al. (2001) argue that "computers must have access to structured collections of information and sets of inference rules that they can use to conduct automated reasoning" and ways to exchange and share knowledge. *Ontologies* evolved as a primary tool to represent knowledge, manipulate knowledge, and share knowledge (see e.g. Chandrasekaran et al. 1999).

> "An ontology is a formal, explicit specification of a shared conceptualisation"[2] (Studer et al. 1998:184; also see Gruber 1993:199).

A "conceptualization" is an abstract model of a phenomenon in the world that specifies

[2]Observe that Berners-Lee & Fischetti's (1999) definition of the Semantic Web ("the conceptual structuring of the Web in an explicit machine-readable way") and the definition of an ontology are very similar. This emphasizes the importance of ontologies for the Semantic Web.

the relevant concepts of that phenomenon (Studer et al. 1998:184). Conceptualizations are typically expressed via taxonomies that use generalization and specialization relationships to organize knowledge about a phenomenon (Gomez-Perez & Corcho 2002:56). Therefore, ontologies often appear as taxonomic trees of concepts, ranging from very general domain-independent concepts at the top to increasingly domain-specific concepts at lower levels of the hierarchy (Chandrasekaran et al. 1999:22). "Explicit" means that the types of concepts used, and the constraints on their use (for example, in terms of attributes, values of attributes, and relationships), are explicitly defined (Studer et al. 1998:184). "Formal" means that the ontology is described in a machine-processable language, which excludes natural language (Studer et al. 1998:184). This is usually achieved through formal *ontology languages*. "Shared" reflects the idea that the knowledge captured in an ontology is consensual, which means it is accepted by a community instead of being private to an individual (Studer et al. 1998:184). In sum, ontologies capture consensual knowledge about the intrinsic structure of a phenomenon within a specific domain in a structured, explicit, and machine-processable way.

Ontologies provide the basis for knowledge manipulation and automated reasoning. Mechanisms or inference processes can operate on ontologies in order to solve a specified problem or perform a specified task (Chandrasekaran et al. 1999:23f.). Mechanisms can come in a variety of forms and with varying capabilities, ranging from basic reasoning strategies and inference rules to complex problem- or design mechanisms that comprise a sequence of complex inferences to achieve desired ends (Chandrasekaran et al. 1999:24). In general, these mechanisms are "proposed as the secret of making intelligent machines" (Chandrasekaran et al. 1999:21), as they enable computers to make meaningful inferences, creating the perception of computers being able to "understand" the processed content. Yet, these inferences are only meaningful to the extent to which mechanisms are programmed in advance.

Ontologies furthermore enable knowledge sharing and knowledge reuse (see e.g. Chandrasekaran et al. 1999:21). Knowledge sharing via publishing or exchanging ontologies is important to enable serendipitous interoperability between machines. Software agents can exchange ontologies in order to establish a shared vocabulary for subsequent interaction. Ideally, agents are equipped with a mechanism that allows them to determine the equivalence between concepts in different ontologies as a starting point for interoperation (Berners-Lee et al. 2001). This equivalence can also be achieved by publishing and storing ontologies in Web-accessible databases that specify the relationships between the different ontologies. For example, a service repository can offer ontology lookup or mapping for the community of providers and consumers (Burstein et al. 2005:79) in order to enable interoperability between services in that repository or facilitate unambiguous communication between customers and providers.

Ontologies and their associated mechanisms can be expressed by means of formal *ontology languages* such as OML, RDF, DAML+OIL, and OWL. Ontology languages differ in their expressiveness, that is, the scope and complexity of ideas that can be represented and communicated in the language, and in their ability to support complex inferences (for an evaluation of ontology languages see Gomez-Perez & Corcho 2002). Given the varying expressiveness and reasoning capabilities of ontology languages, Gomez-Perez & Corcho (2002:54) argue that developers need to clearly understand the representation and inference requirements that an underlying problem domain imposes in order to make a sensible choice regarding a suitable ontology language to support their objectives (also see Berners-Lee et al. 2001).

In conclusion, the Semantic Web represents an extension of today's Web that is based on the conceptual structuring of knowledge contained in the Web in an explicit and machine-readable manner. The use of ontologies and ontology languages is critical for representing, sharing, and processing knowledge; enabling automated reasoning; and achieving interoperability, thus realizing the vision of the Semantic Web.

4.3. Semantic and Syntactic Web Service Descriptions

4.3.1. Semantic vs. Syntactic Service Descriptions

Prior to the advent of the Semantic Web, web services were described by means of *syntactical service description languages*, which only allowed specifying the syntax of services' interfaces, including the syntax of input and output messages, and other data necessary for service invocation (Martin et al. 2005:26). In this context, a widely established language is the Web Services Description Language (WSDL). However, this language allows only for a syntactical service description, which reduces the potential to automate processes along the web service life cycle in many ways. For example, software agents cannot autonomously discover, compose, invoke, and interoperate with web services described merely by syntactical interfaces because such descriptions cannot represent the meaning of the interface's syntax; yet without "understanding" the syntax's meaning, software agents are not able to autonomously determine how to utilize the interface (Martin et al. 2005:28). Hence, human intelligence is necessary to read and understand a syntactical interface and ultimately tell software how to use it. In general, consumers and providers need to agree in advance on the syntax and semantics of service interfaces and the messages exchanged between services for enabling interoperability (Burstein et al. 2005:72). In other words, interoperation between web services needs to be *architected* before interoperation can begin.

Semantic web services arise from applying the vision of the Semantic Web to the description of web services. A semantic web service is characterized by a service description that specifies the service's capabilities and conditions and constraints of use in an unambiguous, machine-processable way (McIlraith et al. 2001:46; McIlraith & Martin 2003:90). Semantic web services aim to achieve a higher degree of automation along the entire web service life cycle, especially in an "open, unregulated, and often chaotic environment (that is, the Web)" (Payne & Lassila 2004:14; also see McIlraith & Martin 2003:90; Martin et al. 2005:27). For example, describing services by means of well-defined semantics can enable software agents to search for and locate a specific capability in a service repository (automated discovery). Of course, this also requires consumers to state their requirements in machine-processable language (Papazoglou et al. 2007:67). Well-defined semantics can also enable software agents to automatically coordinate the inputs and outputs of multiple services and align their respective conditions and constraints of use in order to achieve a desired goal (automated composition) (Martin et al. 2005:36). In fact, semantics can allow agents to break down a high-level customer request and determine the basic capabilities needed to fulfill it (e.g. McIlraith et al. 2001:47). Furthermore, semantic descriptions can let software agents determine what inputs are necessary to invoke a service, what will be returned by the services, and how to execute the service (automated invocation) (e.g. McIlraith et al. 2001:47). To conclude, the use of semantic web service descriptions aims to enable automated, seamless, and unarchitected (i.e. serendipitous) interoperability between web services in heterogeneous distributed environments.

4.3.2. Semantic Annotation and Semantic Markup

In the literature, two general approaches can be identified for creating semantic web services: (1) semantic annotation, and (2) semantic markup (see Patil et al. 2004:553). Both approaches require developing one or more domain-specific ontologies that collectively represent the distinct functional context related to a service's task(s), goals, and conditions and constraints of use (Patil et al. 2004:553). They differ in how these ontologies are associated with or embedded in the service description.

Semantic annotation is the creation of semantic web services by adding semantics to an existing syntactical description (Patil et al. 2004:553; Papazoglou et al. 2007:67). This means that domain-specific ontologies are initially developed and encoded using an ontology language without any relation to the syntactical description of a web service. The concepts included in these domain-specific ontologies are then related to terms in the syntactical service description. Figure 19 shows a conceptual example of how elements of a syntactical service description (in WSDL) are associated with elements of the respective domain model. This process of relating and tagging service descriptions with concepts in ontologies is referred to as semantic annotation (Patil et al. 2004:553).

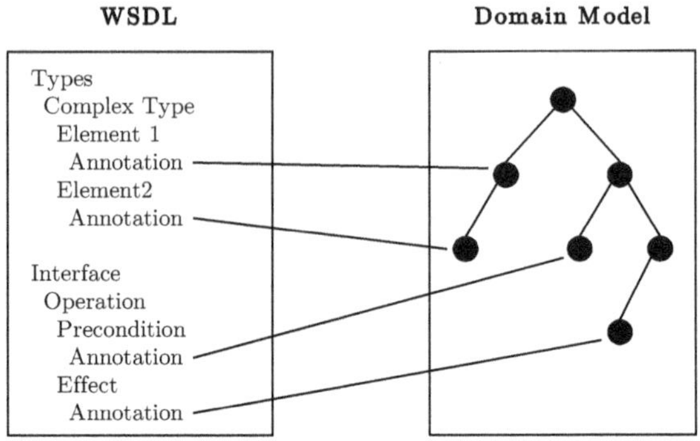

Figure 19.: Semantic Annotation of a Syntactical WSDL Service Interface Description (Akkiraju et al. 2005:5)

Semantic markup is the creation of semantic web services by describing services using an *ontology-based description language* that is specifically tailored to the description of web services (see Patil et al. 2004:553; McIlraith et al. 2001). An ontology-based description language is a specific type of ontology language. "Ontology-based" means that the language contains a pre-defined set of domain-independent concepts and relationships that specify how web services and user constraints are described (McIlraith et al. 2001:48). These concepts and relations provide a declarative representation of web services, which is referred to as semantic markup. In other words, this domain-independent ontology provides the "backbone" (McIlraith et al. 2001:48) or "building blocks" for creating rich semantic descriptions (Martin et al. 2005:27). The distinct functional context of a web service is then represented by one or more domain-specific ontologies that are *subclasses* of the domain-independent ontology (McIlraith et al. 2001:48). Thus, services of different functional contexts inherit the same set of concepts, relationships, and vocabulary, which is important to facilitate automated interoperation (McIlraith et al. 2001:48). In order to account for the particularities of a service's functional context, it is possible to refine and augment the domain-independent ontology as needed. The resulting domain-specific ontologies may themselves be used as a *standardized* semantic service description for all services in the same domain.

4.3.3. The SOA-RAF Service Description Model

Several ontology-based web service description languages have been developed so far. Examples include Web Ontology Language for Web Services (OWL-S), Web Service Modeling Ontology (WSMO), and Semantic Web Services Framework (SWSF). OWL-S is perhaps the most widely adopted of these. It was developed with the particular objective of facilitating the automation of processes along the life cycle (Martin et al. 2004). However, for the purposes of this thesis, it is not helpful to review any of these languages, as they are tailored to the specifics of web services. Instead, for our objectives, understanding the basic idea of ontology-based description languages is more important. To that end, we present an abstract model for the description of web services: the Reference Architecture Foundation for Service Oriented Architecture (SOA-RAF) reference model for service descriptions. The model is abstract in the sense that it specifies a generic top-level ontology, including key sub-classes for the description of services, but does not provide a machine-processable description. It thus differs from the mentioned ontology-based description languages for web services, as it has not yet been implemented. Rather it can be considered a starting point for developing an ontology-based description language (semantic markup).

Figure 20 depicts the basic ontology of the SOA-RAF service description model. The model structures the information to describe services encompassing five basic concepts: (1) service functionality, which specifies what a service can accomplish; (2) service policies, which prescribe conditions and constraints of service use; (3) metrics, which enable determining the degree to which service behavior is compliant with asserted policies and agreed-upon service contracts; (4) service interface description, which defines how to interact with a service through messages; and (5) service reachability, which specifies the technical means for interacting with a service through messages (see e.g. Laskey et al. 2012:44, 49, 107).

Service functionality expresses what stakeholders can expect a service to accomplish, subject to a set of technical assumptions and constraints. It specifies the functions of a service and the real-world effects that can be expected from invoking them (Laskey et al. 2012:49f.; MacKenzie et al. 2006:21). The functions stand for business activities in some domain that bring about real-world effects, that is, changes in the world's state (Laskey et al. 2012:49), with state defined as the condition of an entity at a certain point in time (Laskey et al. 2012:40). The technical assumptions and constraints are domain specific restrictions that determine the limits of the real-world effects that can be achieved by a service; real-world effects must be consistent with these technical assumptions and constraints (Laskey et al. 2012:50). Technical assumptions and constraints include both (immutable) physical constraints imposed by nature and constraints and assumptions (deliberately) imposed by stakeholders.

Service policies prescribe the conditions and constraints under which stakeholders can interact with and thus use a specific service (Laskey et al. 2012:50). Policies are asserted by stakeholders, often by the respective service provider, and they represent stakeholders' (subjective) choices (Laskey et al. 2012:73). The need for policies arises from the fact that the behavior of many (technical) systems is "under-specified" by default: "the scope of

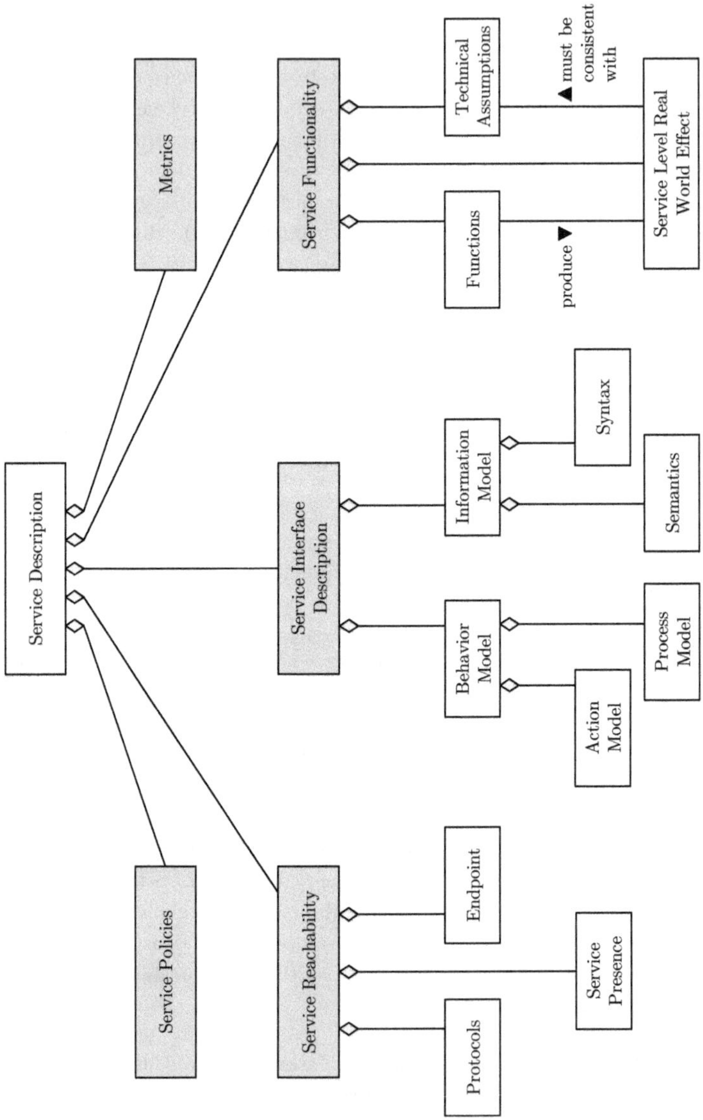

Figure 20.: Basic Ontology of Service Description Model
(adopted with minor adjustments/changes from Laskey et al. 2012:48)

potential behavior is much broader than is actually needed for a particular circumstance" (Laskey et al. 2012:73). Stakeholders therefore assert policies to direct actual system behavior towards desired behavior and performance (Laskey et al. 2012:73). To achieve that, policies may be implemented as technical assumptions that must be satisfied by the real-world effects brought about by the service's functions (Laskey et al. 2012:51).

Metrics are the conditions and quantities that can be measured to characterize a service's functions and associated real-world effects (Laskey et al. 2012:81). Metrics are used to define and measure compliance (Laskey et al. 2012:86) and they provide measurable quantities to determine whether actual service behavior is compliant with established service policies and agreed-upon service contracts (Laskey et al. 2012:52). They thus represent the means to support enforcement of policies and contracts, which, in turn, depends on the potential for measurement (Laskey et al. 2012:51).

Compliance is tracked in a service's compliance record (Laskey et al. 2012:52), which contains operational historical data about the degree of service compliance achieved (independent of a specific service contract) and therefore provides an important information source for consumers to assess whether a candidate service can suitably satisfy their business goals. The record is also used by stakeholders to track the actual performance of services in order to determine whether services were provisioned within the limits as agreed to in the service contract. Some compliance information may be private to the contracting parties, and other information may by publically accessible to all stakeholders in the system (Laskey et al. 2012:52).

The *service interface description* contains all the information necessary to interact with a service in order to achieve the real-world effect described in the service description (Laskey et al. 2012:48; also see MacKenzie et al. 2006:22).). Interaction is achieved through an exchange of messages between a stakeholder and the service. To support these message exchanges, the interface description consists of (1) an information model and (2) a behavior model. The *information model* represents a "characterization of the information that may be exchanged with the service" by defining the syntax (structure) and the semantics (meaning) of messages and data (MacKenzie et al. 2006:16; Laskey et al. 2012:48f.). The *behavior model*, in turn, consists of an action model and a process model (MacKenzie et al. 2006:17f.; Laskey et al. 2012:48f., 54, 64ff., 72). The action model specifies and characterizes the set of permissible actions that can be performed against a service and the real-world effects that result from each of these actions. The process model specifies the process – formally referred to as a message exchange pattern (MEP) – by which messages can be exchanged with the service. This includes the temporal relationships and properties of actions (such as their sequence) (Laskey et al. 2012:67). Moreover, the process model is important for enabling the composition of services. This is because the interface description model specifies the inputs and outputs of services, which need to be aligned in case of service composition (see Martin et al. 2005:36).

Service reachability can be defined as the ability of stakeholders to locate and interact with one another (MacKenzie et al. 2006:29; Laskey et al. 2012:63). If reachability is established, then there exists some kind of (technical) relationship that allows stakeholders to interact, usually by exchanging information via messages (MacKenzie et al. 2006:15). In order to support reachability, service descriptions should specify (a) endpoints, which are reference-able entities, (conceptual or actual) locations, or addresses to which stakeholders can send messages to invoke service actions that have a desired real-world effect; (b) protocols, which define a structured method of communication with the service; and (c) service presence, which indicates whether the service (or parts of its functionality) are available at a certain point in time (Laskey et al. 2012:49, 63).

4.4. Semantic Description of Logistics Services

The previous subsections have focused on the (semantic) description of web services. However, several scholars, primarily rooted in the realm of computer science, have recently started to use well-defined semantics to describe logistic services. This subsection provides a short overview of the current state of this emerging research field. This is of direct relevance for addressing our research question: designing a logistics system in accordance with the principles and concepts of cloud computing.

To gain an overview of existing contributions, we queried popular scientific databases (IEEE, ACM, EBSCO, and Google Scholar) with the terms "semantic logistics services" and "semantic logistics." In sum, the search engines produced several thousand hits in May 2015. For each search engine, we limited our analysis to the first 50 hits. We assessed the relevance of each hit in terms of whether or not scholars have actually used well-defined semantics to describe logistics services, rather than just mentioning semantic logistics services. Specifically, we assessed relevance by analyzing their titles, abstracts, and full content sequentially. If a hit was not considered relevant, it was excluded from the next analysis phase. In addition, to identify further contributions, we searched the bibliographies of the relevant articles. Table 13 lists the resulting six contributions that focus on the semantic description of logistics services. In addition to these contributions, the paper by Leukel & Kirn (2011) is also noteworthy because these scholars suggest to adopt service-orientation, especially OWL-S to intermodal freight routing. However, their contribution is not included in the table because they do not provide semantic descriptions of physical logistics capabilities but develop a prototype system to validate the feasibility of representing such capabilities by means of web services and to enable algorithm-based service composition.

Author(s)	Semantic Approach	Approach Details
Scheuermann & Hoxha (2012)	Mixed - semantic annotation of OWL-S	- OWL-S ServiceProfile used to describe service features - OWL-S ServiceProfile annotated with logistics domain ontology - Logistics domain ontology expressed using OWL 2 DL
Hoxha et al. (2010)	Mixed - semantic annotation of OWL-S	- All OWL-S top-level concepts used to represent logistics services, including ServiceProfile, ServiceModel, and ServiceGrounding - OWL-S top-level concepts annotated with logistics domain ontologies - Logistics domain ontology expressed using OWL DL - OWL-S ServiceModel: Conditions and effects expressed using Semantic Web Rule Language (SWRL) - OWL-S Grounding: expressed using WSDL
Wang et al. (2005)	Semantic annotation	- Syntactic description annotated with the ontology of the OWL-S ServiceModel - OWL-S ServiceModel extended with domain specific ontologies as the expressiveness of OWL-S is deemed insufficient - Logistics domain ontology expressed using OWL
Jung et al. (2008)	Semantic annotation	- Semantic annotation achieved through Semantic Annotations for WSDL - Freight document ontology expressed using Web Service Modeling Language (WSML)
Fagui et al. (2008)	unclear	WSDL and OWL-S and logistics ontology used to describe services, unclear how these are exactly related either as semantic markup or as semantic annotation
Preist et al. (2005)	Semantic markup	- service descriptions based on own ontology that is compliant with an early version of Web Service Modeling Ontology (WSMO) - service descriptions expressed using OWL DL

Table 13.: Approaches to Describing Logistics Services Using Well-defined Semantics

The very limited number of these contributions emphasizes the young character of this research field. Current research on this topic can be characterized as heterogeneous. Nearly every contribution uses a unique approach to generating semantic logistics service descriptions. In fact, researchers combine different approaches typically used for creating semantic web services in order to create semantic descriptions of logistics services: They use an ontology-based description language to annotate a syntactical service description (see Wang et al. 2005; Jung et al. 2008). In other words, they use a semantic markup language for semantic annotations. Other researchers annotate ontology-based description languages with logistics ontologies in order to adapt web service standards to logistics services (see Scheuermann & Hoxha 2012; Hoxha et al. 2010). The lack of a predominant approach emphasizes the newness of the field. Furthermore, although the contributions state their respective approaches to creating semantic logistics services, few of these descriptions and underlying ontologies are shown within the actual publications. It therefore becomes difficult to gain satisfactory insights into the research results in particular and the current state of research on semantic logistics service descriptions in general. Finally, this research field is highly fragmented, as there is very little cross-referencing between the individual contributions; no contribution serves as an "anchor" so far.

Despite following heterogeneous semantic approaches, all contributions use existing web service standards to describe logistics services – even though researchers have already recognized the limited applicability of these standards for describing logistics services. For example, Preist et al. (2005:998) point out that Web Ontology Language (OWL) cannot adequately express and reason with date ranges, which are crucial for scheduling transportation services, for example. To overcome this problem, they propose resorting to other reasoning engines or extensions to OWL. Likewise, Li et al. (2013:1697) remark that existing web service standards (e.g. WSDL and OWL) cannot be applied to logistics directly

> "because the meaning and usage of logistics services are different from those of traditional web services".

Liu & Li (2012:47) put forth a similar argument for cloud manufacturing. The use of existing web service standards therefore appears to be an act of necessity in the absence of logistics specific standards. Considering the present body of knowledge about semantic descriptions of logistics services, it is fair to conclude that current contributions provide a successful "proof-of-concept" for introducing well-defined semantics to the description of logistics services – nothing more, but also nothing less.

IV. Summary

This part has reviewed the fundamentals and roots of cloud computing in two chapters. The first chapter has reviewed the fundamentals of cloud computing and introduced the NIST definition of cloud computing, which consists of five essential characteristics, three service models, and four deployment models. The essential characteristics are the constituent features of cloud computing capabilities. In particular, rapid elasticity has been identified as the pivotal characteristic that distinguishes cloud computing from other computing paradigms. The service models specify what types of computing capabilities can be delivered. The deployment models specify how cloud infrastructures can be shared among a group of consumers and how they can be operated by providers. Particular emphasis has been placed on hybrid clouds because they can be conceived of as interfirm networks if more than two distinct cloud infrastructures are connected. Although cloud computing research tends to focus on rather technical aspects, previous research makes direct reference to the organization and governance of hybrid clouds in terms of the coordination mechanisms deployed (primarily programmed routines, hyperautomation, and implicit coordination) and the institutional framework used to negotiate contracts between participating clouds (including auctions on electronic markets).

The second chapter has reviewed the conceptual and technological roots of cloud computing. Among other roots, virtualization and SOC have been considered in detail because of their relevance to this thesis. Virtualization refers to the simulation of multiple computers systems on a single physical machine, thus achieving resource pooling. SOC is a computing paradigm that revolves around the concept of services, with the objective of creating distributed applications through service compositions. This paradigm consists of six design principles (subsumed under the term "service-orientation") that shape the design of services and that ultimately aim for achieving service composability. SOAs are a specific type of architecture that is based on these design principles and that specifies the fundamental roles and interactions among stakeholder within service-oriented systems. Specifically, roles include consumers, providers, and a service registry that are connected in a "triangular" manner. Two approaches to creating formal description of (web) services have been introduced: syntactic and semantic description. The latter applies the vision of the Semantic Web to the description of services in order to achieve automation along the life cycle of (web) services. A top-level ontology including key sub-classes has been presented in order to emphasize how semantic service descriptions can be created by means of ontology-based description languages. Finally, this chapter has reviewed the initial efforts to describe logistics services by using well-defined semantics and available web service standards.

Part E. Systematic Review of Cloud Logistics Knowledge

I. Abstract of Part

Objective and Method

The term "cloud logistics" has recently emerged in the academic literature of IT and logistics, in more practice-oriented industry magazines, and in practice itself. However, the meaning of the term "cloud logistics" is unclear. Academics and practitioners use the term inconsistently to refer to a variety of phenomena. In principle, this terminological ambiguity results from the fact that the term is a composition of two concepts: cloud (computing) and logistics, and inevitably inherits and blends their meanings (Leukel & Scheuermann 2014:39; Kersten et al. 2012:257).

In order to form a basis for designing a logistics system in accordance with the principles and concepts underlying cloud computing – our research objective – we conducted a broad systematic search and review of existing knowledge associated with the term "cloud logistics." We followed the same method as is typically used for systematic literature reviews (see e.g. Tranfield et al. 2003; see Chapter E.II). Yet, compared to typical reviews, a review regarding cloud logistics faces the opposite problem: We have clarity about the keyword to be searched for, but the phenomenon itself is vague. Therefore, by conducting a systematic review, we aim to structure current knowledge not only to derive research avenues as typically done, but also to determine the underlying phenomenon itself. In other words, we aim to determine the direct meaning of the term "cloud logistics" through a systematic analysis of current knowledge.

Descriptive Analysis and Results: An Emerging Field

Our search in several academic and non-academic search engines identified a total of 274 hits of which 79 were considered intelligible and relevant and, hence, included in our

F. Jaekel, *Cloud Logistics*, Produktion und Logistik,
https://doi.org/10.1007/978-3-658-22837-8_5

review. Each search hit is referred to as "source" hereafter. The term "cloud logistics" initially emerged in 2010 and its use has proliferated and accelerated since then (see Chapter E.III). Due to the emerging character of this research field and our broad search approach, we identified a heterogeneous set of source types, ranging from peer-reviewed journal articles to U.S. patent applications. Because of the broad search approach and heterogeneity of the identified sources, we have exercised special caution in the process of analyzing the sources, because many of them were not subject to peer-review or other neutral validations or quality checks.

Content Analysis and Results: Four Distinct but Related Meanings

By analyzing the content of the sources, we identified four distinct but related meanings of the term "cloud logistics" (see Section E.IV.1). Each meaning uses the term "cloud logistics" to specify a distinct phenomenon, but meanings are related because these phenomena share two elements: All are concerned with a logistical issue in relation to a cloud computing based IT system. The identified meanings and their underlying knowledge have varying relevance for our research objective. Meaning 4 forms a direct basis for our objective to design a logistics system in accordance with the principles and concepts of cloud computing. It is described by 21 sources. Meaning 2 also contributes to this design process and is described by 57 sources. Meanings 1 and 3 have no direct relation to our research objective. They are described by two and three sources respectively.

Meaning 1: A Logistics Perspective on Cloud Computing Systems

The first meaning of "cloud logistics" denotes *a logistics perspective on cloud computing systems* (see Section E.IV.2). It is concerned with logistical questions related to the flow of data and services in networked cloud computing systems. Similar to production processes in the real economy, software consumers increasingly ask for shorter software implementation lead times, and software becomes increasingly fragmented across various computing systems. A single software solution likely consists of various component services, where each service is implemented on a different system. Critical enablers for these developments are SOC (due to the use of modular services) and cloud computing (due to moving storage and processing power to the network).

Meaning 2: Using a Cloud Computing Technology-based Platform for Logistics IT Systems

The second meaning of "cloud logistics" understands the concept as *the use of a cloud computing technology based platform for logistics IT systems* (see Section E.IV.3; Subsection E.IV.3.1). This platform, hereafter referred to as *cloud-based platform*, enables users to deploy logistics-related applications or to use applications already deployed on that platform. This cloud-based platform can be used to support administrative and/or operative logistics processes or other value creation processes in supply chains. Logistics processes or supply chains are particularly suitable contexts for cloud-based platforms because of the enormous division of labor and the large number of actors involved (Hannig 2012:6; Opresnik 2014:73; Stinnes 2013:37f.; also see Colajanni 2012:10; Zhao & Wu 2013:450). The fundamental reason why a cloud-based platform is considered particularly suitable in these contexts is the platform's crucial ability to *connect all economic actors* involved in the same logistics process or supply chain by means of a single IT system (see Urban 2015; Cloud Logistics 2013; Stinnes 2012:8; Schmidt 2013:32). By connecting all economic actors involved in logistics processes or along supply chains, the cloud-based platform becomes the enabler and nexus for new forms of *networked interfirm cooperation* (Stinnes 2012:8; Heaney 2010; Gantzia & Sklatinioti 2014:49f.).

The functionality offered by the cloud-based platform can be categorized into four *generic use cases*: (a) *the outsourcing of (on premise) logistics or supply chain management (SCM) IT systems to a cloud, which is operated by a specialized third-party provider;* (b) *collection, integration, sharing, and synchronization of logistics relevant information generated and utilized by actors in logistics processes or along supply chains;* (c) *management and optimization of collaborative business activities of multiple business partners;* and (d) *support of value exchanges via a virtual marketplace.* These use cases have been empirically derived by the "LOGICAL" research project (see Arnold 2014a:22f.; Arnold 2014b:14; Arnold et al. 2013; Arnold et al. 2012).

The largest benefit of using cloud technology for logistics IT systems may be the ability to enable networked interfirm cooperation and to increase logistics efficiency and effectiveness, especially due to broad network access and on-demand availability (see Subsection E.IV.3.6).

Meaning 3: E-commerce Fulfillment through a Network of Cooperating Local Dealers and LSPs

The third meaning of the term "cloud logistics" understands cloud logistics as *e-commerce fulfillment through a network of cooperating local dealers and LSPs* (see Section E.IV.4). This meaning is related to the second meaning because it relies on a cloud-based platform to support e-commerce fulfillment. However, we introduce a distinct third meaning because, unlike in meaning 2, the platform is not concerned with sharing, managing, or exchanging logistics information and resources, but is primarily concerned with sharing and exchanging *finished goods stocked in the warehouses of local dealers.*

Meaning 4: Logistics-as-a-Service – Interpreting and Designing Logistics Systems through the Lens of the Cloud Paradigm

The fourth meaning of "cloud logistics" refers to *Logistics-as-a-Service – interpreting and designing logistics systems through the lens of the cloud paradigm* (see Section E.IV.5; Subsection E.IV.5.1). Understanding this meaning is directly relevant to addressing our research objective. This meaning follows the principal idea that the term "cloud logistics" refers to logistics systems that belong to a family of systems that all share the same conceptual template: *the cloud paradigm.* The cloud paradigm comprises an abstract interpretation of the design principles and concepts that underlie cloud computing and of the elements part of the NIST cloud computing definition, including essential characteristics, service models, and deployment models (Delfmann & Jaekel 2012:15ff.; also see Leukel & Scheuermann 2014). Those systems that are interpreted and designed according to this paradigm are referred to as "cloud systems" and they provide their respective capabilities in an "as-a-Service" manner. Accordingly, interpreting and designing logistics systems according to the cloud paradigm creates *cloud logistics systems* (CLSs) that provision logistics capabilities as a service, referred to as "Logistics-as-a-Service (LaaS)."

Although several sources offer definitions of the term "cloud logistics," a consensual definition is still missing (Wang et al. 2012b:558; Delfmann & Jaekel 2012:17; Pieringer 2012:45). The majority of sources define cloud logistics by focusing on the *process* of how CLSs can be created (by adopting the design principles and concepts underlying cloud computing for logistics system design) rather than on the *outcome* of this process. Consequently, the understanding of CLSs remains vague. Still, some provide insights into the potential outcome of this design process. These outcomes differ significantly and include horizontal LSP cooperation, on-demand delivery, pay-per-use, virtualization of logistics resources, and accessing logistics resources through standardized software interfaces. In spite of differences, definitions nevertheless agree on that "cloud logistics" is "more than cloud computing in logistics systems" (Leukel & Scheuermann 2014:38), thereby differentiating the fourth meaning from the second.

Cloud logistics knowledge of the fourth meaning can be structured according to elements of the cloud paradigm (Delfmann & Jaekel 2012:15ff.; also see Leukel & Scheuermann 2014): (1) design principles and concepts that underlie cloud computing, (2) essential characteristics of cloud logistics' services, (3) cloud logistics deployment models, and (4) cloud logistics service models.

The *underlying design principles and concepts* are the first element of the cloud paradigm (see Subsection E.IV.5.2). Our review identified three underlying design principles and concepts consensually mentioned: virtualization, service-orientation, and IoT. While virtualization and service-orientation are principles and concepts that underlie cloud computing, IoT is added as a concept for synchronizing the virtual world of computers with the physical world of tangible logistics resources and capabilities. In spite of this consensus, the application and interrelation of these principles and concepts when designing logistics systems partly remains vague. While the IoT is a well investigated topic in logistics research, virtualization and service-orientation are new to logistics system design and require further research attention.

The *essential characteristics of cloud logistics services* are the second element of the cloud paradigm (see Subsection E.IV.5.3). Given that cloud logistics adopts virtualization and service-orientation from cloud computing, one may reasonably ask whether cloud logistics services equally feature cloud computing characteristics. In fact, apart from self-service and broad network access, sources suggest that physical cloud logistics services (e.g. transport) are characterized by on-demand availability, resource pooling, rapid elasticity, and pay-per-us – just like cloud computing services. Although many sources associate physical cloud logistics services with one or more essential cloud characteristics, the majority of sources merely mention the essential cloud characteristics without elaborating on their specific meaning or how they can be achieved – despite being crucial for the cloud paradigm. Furthermore, current knowledge seems to associate physical logistics capabilities with the essential cloud characteristics to motivate the emergence of CLSs as a promising approach and response to better cope with an increasingly complex and dynamic logistical environment. However, it is still unknown whether cloud logistics can, in fact, attain these characteristics and whether attaining them represents a viable solution for overcoming logistics challenges.

Deployment models are the third element of the cloud paradigm (see Subsection E.IV.5.4). They define the institutional composition of cloud systems, and they define the basic nature of interorganizational relationships between involved actors. Cloud logistics knowledge comprises heterogeneous conceptions regarding the deployment models of cloud logistics. Deployment models are either derived from cloud computing or SOA. Both associate cloud logistics with some form of horizontal cooperation among several LSPs, hereafter referred to as *networked horizontal LSP cooperation.*

If derived from cloud computing, deployment models generally follow the common cloud computing logic and terminology in terms of public, private, and hybrid (see Subsubsection E.IV.5.4.1). If derived from SOA, deployment models adopt the constituent roles of SOA: resource providers, consumers, and a service registry operated by a platform provider as well as the triangular role structure and operating logic of SOA (see Subsubsection E.IV.5.4.2). Resource providers and users become interconnected via the platform, which is why the majority of sources conceive of this platform to enable (dynamic) networked horizontal cooperation among LSPs, where dynamic means that the composition of cooperating LSPs changes over time. However, sources provide only few insights into the concrete functionality of the cloud-based platform, the concrete responsibilities, the neutrality, and autonomy of the platform operator.

To explain the formation of horizontal networked cooperation in cloud logistics, the sources provide, amongst others, specific reasons that relate to the goal of achieving on-demand and rapidly elastic provisioning of physical logistics services. Sources (implicitly or explicitly) suggest a link between the size of the logistics resource and capability pool and the ability to achieve on-demand and rapidly elastic service provisioning. Horizontal cooperation between LSPs can increase the size of the disposable resource and capability pool and, thus, support on-demand availability and rapid elasticity. In other words, the goal of delivering logistics services in an on-demand and rapidly elastic manner drives LSPs to cooperate in cloud logistics, despite being distant or proximate competitors.

A critical question in every economic system and form of cooperation pertains to the allocation logic of resources, revenue, and costs (see Subsubsection E.IV.5.4.4). Conceptions regarding cloud logistics are very heterogeneous and provide few substantial insights into these pivotal issues. Ideas primarily diverge regarding the degree of centralization, which ranges from *centralized* to *decentralized* (market-based). If centralized, the cloud-based platform in the CLS uses some kind of algorithm or (intelligent) optimization technique to determine the allocation of resources, revenues, costs, risks etc. among cooperating LSPs. If decentralized, the allocation logic to some extent depends on the choices made or the information provided by the cooperating LSPs, as in the case of market mechanisms such as (combinatorial) auctions.

Service models are the last element of the cloud paradigm (see Subsection E.IV.5.5). Similar to cloud computing, physical cloud logistics capabilities are suggested to be grouped according to their type. These service models adopt the "as-a-service" terminology and can be subsumed under the hypernym LaaS.

The fourth meaning of cloud logistics provides several research opportunities, which can be structured along the elements of the cloud paradigm (see Subsection E.IV.5.7). Virtualization and service-orientation represent *concepts and design principles underlying cloud computing*, which need to be followed in cloud logistics system design. However, current knowledge falls short of actually applying them, thus providing only little insights into the outcome of their application. Physical logistics capabilities are associated with the *essential cloud characteristics*. However, existing knowledge falls short of explaining how these characteristics can be achieved, and how the aim to achieve these characteristics impacts cloud logistics infrastructure as well as governance structure and organizational structure. Networked horizontal cooperation is identified as the predominant *deployment model* of cloud logistics. However, little appears to be known about the governance and organizational structure of this type of cooperation. LaaS is identified as hypernym for cloud logistics *service models*. However, current knowledge does not answer the fundamental question of which physical logistics capabilities can be offered with cloud characteristics. We synthesize these opportunities into the following research objective to be addressed in this thesis: *designing a logistics system in accordance with the principles and concepts of cloud computing in order to determine if and how logistics systems can deliver physical logistics capabilities with cloud characteristics.*

Conclusions: Assessing Commonalities and Appropriateness of Term Usage

Given that the term "cloud logistics" is used to refer to four distinct phenomena, we encounter a terminological question about which of these phenomena represent the direct specific meaning of this term (see Section E.IV.6). We argue that the term "cloud logistics" should be used to refer to either a logistics perspective on cloud computing systems (meaning 1) or a cloud perspective on logistics systems (meaning 4). This is because both meanings blend the terms such that the most important aspects associated with one term is imposed on the direct specific meaning of the other term.

II. Objective and Method

The term "cloud logistics" has recently emerged in the academic literature of IT and logistics, in more practice-oriented industry magazines, and in practice itself. However, the meaning of the term "cloud logistics" is unclear. Academics and practitioners use the term inconsistently to refer to a variety of phenomena. In principle, this terminological ambiguity results from the fact that the term is a composition of two concepts: cloud (computing) and logistics, and inevitably inherits and blends their meanings (Leukel & Scheuermann 2014:39; Kersten et al. 2012:257).

In order to form a basis for designing a logistics system in accordance with the principles and concepts underlying cloud computing – our research objective – we conducted a broad systematic search and review of existing knowledge associated with the term "cloud logistics." We followed the same method as is typically used for systematic literature reviews (see e.g. Tranfield et al. 2003). Systematic literature reviews in management research generally aim to structure and assess the current state of (academic) knowledge about a phenomenon of interest in order to derive future research avenues. They typically start with the selection of an appropriate set of keywords to specify the phenomenon to be examined. Yet, compared to typical reviews, a review regarding cloud logistics faces the opposite problem: We have clarity about the keyword to be searched for, but the phenomenon itself is vague. Therefore, by conducting a systematic knowledge review, we aim to structure current knowledge not only to derive research avenues, but also to determine the underlying phenomenon itself. In other words, we aim to determine the direct meaning of the term "cloud logistics" through a systematic analysis of current knowledge.

Due to the novelty of the term, we did no limit our search to academic peer-reviewed journals and the like, but also incorporated non-academic sources. This approach is supported by Tranfield et al. (2003:215), who argue:

> "Searches should not only be conducted in published journals and listed in bibliographic databases, but also comprise unpublished studies, conference proceedings, industry trials, the Internet and even personal requests to known investigators."

By adopting this broad approach, we aim to capture the entire body of knowledge about cloud logistics and thus establish a solid foundation for the design of logistics systems according to the principles and concepts of cloud computing. Specifically, we searched three categories of sources: (1) academic databases focused on IT (IEEE Xplore and ACM Digital Library), (2) academic databases concerned with management and social sciences (EBSCO, ABI/INFORM, and SSCI), and (3) freely accessible web search engines (Google Scholar and Google Search). In the following, we refer to each search hit produced by any of these sources as a "source."

Because it was our objective to structure knowledge associated with the term "cloud logistics," we confined our search to that exact term. However, we did not limit our search on any other dimension, such as time or search fields. As the term "cloud logistics" emerged only recently and continued to evolve even as this thesis was being prepared, we conducted our search multiple times between February 2012 and August 2015. During this period, we observed a quick growth in the number of search hits. For example, Google Scholar produced 47 hits in December 2013 and produced 164 sources in August 2015. The results presented in the following are based on the latest search conducted on August 12, 2015.

Overall, our search in the specified databases and search engines produced 254 sources. We manually added 20 sources, which we either identified through the analysis of the original sources' bibliographies or were included in previous search results but did not show up in our latest search. We thus arrived at a total of 274 identified sources. We processed the search results in four steps and eventually arrived at 79 sources containing intelligible and relevant information for determining the meaning of the term "cloud logistics." Table 14 summarizes the number of sources identified by search engines/databases as well as the reasons for sources being excluded. In the following, we describe the steps we used to process the identified sources in more detail.

Search Engine/Database	**Sources**
Google Scholar	164
Google (15,900 limited to first 70)	70
ABI/INFORM	7
ACM Digital Library	7
IEEE Xplore	4
EBSCO	2
SSCI	0
Manually added	20
Total Identified	**274**

Reasons for Exclusion	
Duplicate	42
Broken web link	4
Foreign language	3
"Cloud logistics" not used	71
Irrelevant company information	32
Citation w/o contribution	28
Unclear usage	12
"Cloud logistics" used in keywords only	3
Total relevant	**79**

Table 14.: Identified and Relevant Sources

In the first step, we tabulated our results in a Microsoft Excel data sheet. During this initial step, we captured basic information about each source, such as author(s), title, and year of publication. In addition, we removed 42 duplicates and categorized sources according to their type such as journal articles (peer-reviewed and non-peer-reviewed), conference article, and industry magazine article. A full list of source types is provided in the next chapter (see Table 16).

Second, we downloaded each source and saved it as a .pdf file. During this step, we excluded four sources due to a broken web link and three sources that were published in foreign languages of which the author of this thesis has no or only insufficient command (Chinese and French). Of the remaining 225 sources, we could only gain access to the full content of 166 sources of them, despite using the institutional access of three German universities (University of Cologne, University of Bonn, and RWTH Aachen University). For 47 sources, we could only gain access to the abstract, thereof 36 sources stored in a Chinese database (China National Knowledge Infrastructure (CNKI), `http://en.cnki.com.cn/`). For 12 sources, we could only gain access to the sources' preview (i.e. the first and second page made available by the publisher as a picture), 11 of which were published on Scientific.Net (`http://www.scientific.net/`).

Third, we searched the accessible content of each source for the term "cloud logistics" via a keyword search in Adobe Reader. If the term was found, we included the source for further analysis in our knowledge search. If the term was not found, we excluded the source from our review. We chose the existence of the term "cloud logistics" as the primary inclusion/exclusion criterion in order to eliminate any bias regarding what *we* believed the term to mean. We eliminated 71 sources in which the term "cloud logistics" was not used.

Fourth, we extracted and captured the information found in relevant and intelligible sources and added it to our Excel data sheet. Information was recorded in the columns of the data sheet and each column was used to capture a specific aspect of knowledge (e.g. definitions used, figures included, concepts discussed). As our understanding of the meaning of the term evolved throughout the process of reading and extracting knowledge, we (re-)adjusted the labels and content of these columns. Hence, it was necessary to reallocate knowledge and read sources multiple times in order to restructure the knowledge according to the new columns in the data sheet.

During this step, we removed another 75 sources. Thirty-two sources were excluded because they contained irrelevant information about LSPs or vendors of logistics IT. Twenty-nine of the sources provided company-related information about a US-based LSP named Cloud Logistics, including financial results and Twitter, Facebook, and LinkedIn accounts. Three sources referred to websites of another LSP named Cloud Logistics.

However, these websites did not contain any information about why the company was called Cloud Logistics or what the term means. Twenty-eight sources were excluded because they cited another source that contained the term "cloud logistics" in the title, but the source itself did not contain any relevant or intelligible information about this term. Twelve sources were excluded because the term "cloud logistics" was used without any clear meaning or context. Three sources were excluded because they used the term cloud logistics in their keywords but did not provide any relevant or intelligible information about its meaning. We thus arrived at a final set of 79 sources with relevant and intelligible information for determining the meaning of the term "cloud logistics."

III. Descriptive Analysis and Results: An Emerging Field

Before deriving the meaning of the term "cloud logistics," it is appropriate to provide some "descriptive analysis" of when the term is used and by whom and of the (non-)academic types of sources (Tranfield et al. 2003:218).

Figure 21 shows the number of sources per year. The figure emphasizes a growing and accelerating use of the term "cloud logistics" by researchers and practitioners. In 2010, the term initially occurred in an interview with a supply chain engineering expert from a UK-based think tank. In 2011, it appeared in two articles published in two German practice-oriented logistics magazines (*"MM Logistik"* and *"Beschaffung aktuell"*). Moreover, the term was used in a call for papers for the 6th International Scientific Symposium on Logistics (ISSL) conducted by the Bundesvereinigung Logistik (BVL), a Germany-based logistics think tank. It was also used in an article in a German logistics industry magazine (*"LOGISTIK HEUTE"*), which reported about the 28th Deutschen Logistik-Kongress conducted by the BVL in October 2011 in Berlin. Finally, the term was used on the website of a research project named "Cloud Logistic" (IMA RWTH Aachen University 2011). In 2012, the term became more widely used and surfaced in both a practice-oriented magazine and academic articles. In May 2012, another article on cloud logistics was published in the industry magazine (*"LOGISTIK HEUTE"*). However, this article is closely related to the earlier report on the 28. BVL Kongress, as both refer to the same example of cloud logistics. Also in May 2012, cloud logistics was the focus of a paper presented at the IEEE 16th International Conference on Computer Supported Cooperative Work in Design (CSCWD) in Wuhan, China. It was the focus topic of two papers presented at the 6th ISSL, which was held a few weeks later in Hamburg, Germany, and was the focus of a paper presented at the International Symposium on Management of Technology (ISMOT) in late 2012 in Hangzhou, China. Throughout the year, two more articles were published in Chinese journals (non-peer reviewed periodicals, called "journals"). The term was also used that year in two publications of a German IT provider

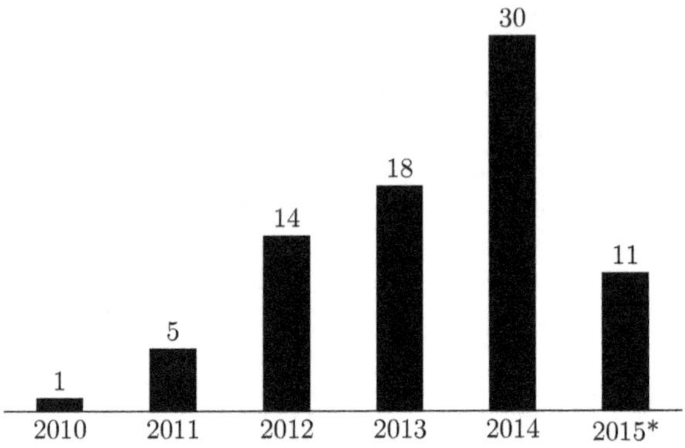

Figure 21.: Number of Sources per Year (*until August 12)

for logistics software (AXIT GmbH) and in another article in a German practice-oriented logistics magazine (*"LOGISTIK HEUTE"*). Finally, it was used in two presentations of research projects: LOGICAL (an international research project including 13 partners from six EU countries) and Logistics Mall (EffizienzCluster LogistikRuhr). As of 2013, the usage of the term has quickly proliferated and is not discussed in detail. Considering the distributed emergence of the term, we can conclude that it emerged independently in the realms of logistics and IT, in the UK, Germany, and China, and almost simultaneously in practice and academia.

Table 15 shows the number of sources produced by authors affiliated with an institution (e.g. university or firm) in their respective country. The table clearly shows that the term "cloud logistics" is primarily used in sources authored by Chinese (35), German (33), and American (3) researchers and practitioners. All other countries provided a single source each. Yet, sources from four of these countries are anchored in the sources of Chinese and German authors. Therefore, one can conclude that, albeit the term emerged in various countries, its "frequent usage" is in fact limited to only a few.

No. of Sources	Country of Author Affiliation
35	China
33	Germany
3	United States of America
1 each	Egypt, Hungary, India, Italy, Poland, Serbia, Sweden, United Kingdom

Table 15.: Number of Sources by Country

Table 16 shows the number of sources by source type and academic or non-academic affiliation. Given our broad search approach, we identified a heterogeneous set of source

Source Type	No. of Sources
Academic Sources	
Conference Papers	18
Journal Articles (peer-reviewed)	9
M.Sc. Thesis	1
Subtotal	**28**
Non-academic Sources, Gray Literature	
Industry Magazine Articles	16
Journal Articles (not-peer reviewed)	14
Book Chapter	4
Company Publication	3
Website - IT Provider	3
Website - Other	3
Website - LSP	2
Presentation	2
US Patent Application	2
Call for Conference Papers	1
Whitepaper	1
Subtotal	**51**
Grand total	**79**

Table 16.: Number of Sources by Type and Affiliation

types, ranging from peer-reviewed journal articles to U.S. patent applications. Although non-academic sources outnumber academic ones, it is fair to conclude that cloud logistics is relevant to both practitioners and researchers. As one would expect from a young research field, conference articles outnumber articles published in peer-reviewed journals. Only one source is published in a ranked peer-reviewed journal (*"Journal of Network and Computer Applications"*). In fact, this journal can be considered a high-quality journal as it has a 5-year impact factor of 2.223.

Because of the broad search approach and heterogeneity of the identified sources, we have exercised special caution throughout the process of analyzing the sources, because many of them were not subject to rigorous peer-review or other neutral validations or quality checks.

IV. Content Analysis and Results: Four Distinct but Related Meanings

1. Overview of Meanings and Relevance for Research Objectives

By analyzing the content of the sources, we identified four distinct but related meanings of the term "cloud logistics." Figure 22 provides an overview of the identified meanings. Each meaning takes "cloud logistics" to specify a different phenomenon, but they are related because these phenomena share two elements: All are concerned with a logistical issue related in some way to a cloud computing based IT system.

The identified meanings and their underlying knowledge have varying relevance for our research objective: designing a logistics system in accordance with the principles and concepts of cloud computing. Meaning 4 forms a direct basis for this objective, and meaning 2 also contributes to this design process. Meanings 1 and 3 have no direct relation to our research objective.

From a quantitative perspective, sources are unevenly distributed across these four meanings. Meaning 1 is described by two sources, Meaning 2 by 53 sources, Meaning 3 by three sources, and Meaning 4 by 17 sources. Four sources described both Meaning 2 and Meaning 4. An overview of these sources is depicted in Table 17. The following sections

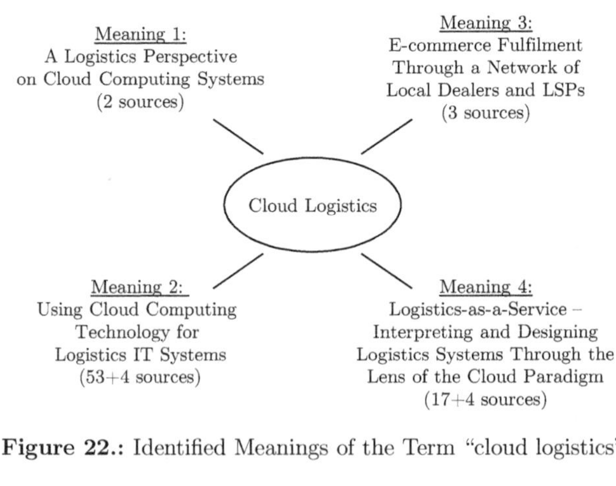

Figure 22.: Identified Meanings of the Term "cloud logistics"

Meaning	Identified per Source	Number
1	Jaeger & Lindenlaub 2013; Pohlmann et al. 2013	2
2	Arnold 2014a; Arnold 2014b; Axit AG 2012; Banyai 2014; Böhmer 2012; Delfmann et al. 2011; Cloud Logistics 2013; V&V Dabelstein Logistik GmbH 2015; Deng & Fang 2014; DHL 2014; Diao 2014; Ding 2014; Diwan 2015; Du 2014; Gantzia & Sklatinioti 2014; Geng & Li 2012; Gong & Yang 2012; Hannig 2012; He et al. 2015; Heaney 2010; IMA RWTH Aachen University 2011; Jiang et al. 2013; Jiao 2014; Jörgl 2012; Kiewitt 2013; Li & Huang 2013; Li & Ding 2013; Li et al. 2014a; Liao & Fu 2014; Long 2014; Matkovic et al. 2014; Nie & Fan 2015; Nowicka 2014; Opresnik 2014; Pagel 2014; Pan & Tong 2014; QuantumID Technologies Inc. 2015; Schmidt 2013; Shen & Qian 2014b; Shen & Qian 2014a; Stinnes 2012; Stinnes 2013; Sun & Tao 2013; Sun & Liu 2013; Sun et al. 2015b; Tummel et al. 2014; Urban 2015; Wang et al. 2012a; Wang 2014; Wang et al. 2015c; Zhang & Yuan 2013; Zhang & Sun 2013; Zhao & Wu 2013	53
3	gaxsys GmbH 2013; Thomas & Unruh 2010; Thomas 2011	3
4	akw/jö 2011; Ballardt 2014; Cloud Pallet Storage 2015; Delfmann & Jaekel 2012; Hu & Zhang 2015; Kersten et al. 2012; Leukel & Scheuermann 2014; Li et al. 2013; Ludwig 2014; Mosavi 2012; Pieringer 2012; Sun et al. 2015a; Teichmann 2014; Wang et al. 2012b; Wang et al. 2015b; Wang et al. 2015a; Zhang et al. 2014	17
2 & 4	Böhmer 2013; Colajanni 2012; Ehrenberg & Ludwig 2014; Li et al. 2014b	4

Table 17.: Overview of Identified Sources by Meaning

describe each meaning using the knowledge contained in the identified sources.

2. Meaning 1: A Logistics Perspective on Cloud Computing Systems

The first meaning of the term "cloud logistics" denotes *a logistics perspective on cloud computing systems.* This meaning is described in two sources, both of which are United States patent application publications created by employees of SAP AG, a German software vendor which received ownership of both of the patents (Jaeger & Lindenlaub 2013; Pohlmann et al. 2013).

The inventors derive the meaning of cloud logistics from the definitions of the words "cloud" and "logistics." They develop their understanding of the term by specifying the process and their intentions in blending these definitions:

> "[C]loud logistics denotes integrated planning, implementation, and control of internal and network-wide flow and storage of material (hardware and virtualized components), and complex software solutions and components thereof including the information flow and required services along the value-chain and product life cycle for the purpose of conforming to customer and business requirements including the mode of delivery." (Jaeger & Lindenlaub 2013:2)

This definition somehow reads like a typical definition of logistics, albeit with a focus on logistics in computing systems. As such, it is concerned with the flow of objects in networks to satisfy needs (Jaeger & Lindenlaub 2013:2; also see Delfmann et al. 2010:58). Objects flowing in cloud computing systems are "deployment templates,"[1] which are services in the sense of SOC (see Subsection D.III.3.1), and "virtual software appliances," which are software solutions composed of two or more templates (Jaeger & Lindenlaub 2013:1f.). These templates are "stored" in a "template warehouse" and assembled into software solutions in a "cloud factory" before they are "transported" (or distributed) to customers (Jaeger & Lindenlaub 2013:1, 3). Services that support or realize the storage and flow of data and services are denoted as "cloud logistics services" (Jaeger & Lindenlaub 2013:4). They may furthermore provide support in assembling, configuring, and deploying virtual appliances (Pohlmann et al. 2013:3). In sum, the definition of cloud logistics and the use of this "classic" logistics terminology clearly points to the consideration of "classic" logistics questions in cloud computing systems.

The implementation of contemporary enterprise software solutions faces the intrinsic conflict between ever growing complexity and customers' demand for shorter implementation cycles (on-demand availability) (Jaeger & Lindenlaub 2013:1). Virtualization and service-oriented design are technologies that overcome these challenges. However, these technologies have important implications for the structure of the computing environment and the implementation of software solutions: Value creation processes become increasingly fragmented and distributed as software solutions divide into many distinct services, each of which can be deployed on spatially dispersed machines connected via the network. In addition, software solutions are increasingly deployed in distributed cloud computing environments (see Subsection D.II.4.2) (Jaeger & Lindenlaub 2013:3). Each environment can be used to deploy distinct parts of larger software appliances. Therefore, within these distributed cloud computing systems, the delivery of software solutions becomes a matter of logistics, as it depends on the flows that connect the fragmented and distributed value creation stages. Jaeger & Lindenlaub (2013:2) remark in this respect:

> "As a number of applications employed in a business and along with it the

[1] "Deployment templates are generally pre-assembled, pre-configured, and pre-tested for their respective purposes [...] [They] can essentially be 'plugged-in' to one another and they function for their intended purpose [...] [They] are generally gathered and assembled into a virtual software appliance" (Jaeger & Lindenlaub 2013:3).

> technical heterogeneity tends to increase causing solutions and computing environments to become quite complex and thereby raising interesting logistical challenges."

By focusing on the management of the flow of software solutions between the point of initial installation/configuration and the point of delivery, it becomes possible to overcome the intrinsic conflict between complexity and the need for on-demand implementation of new software (Jaeger & Lindenlaub 2013:1).

The relevance of logistical issues in distributed computing environments seems natural; after all, the relevance of logistical problems arises from temporally and spatially dispersed production and consumption processes and is therefore independent of whether flows in networks pertain to virtual flows of services in computing systems or to physical flows of goods in the real economy. In fact, logistical questions in distributed computing systems are highly relevant due to the strict temporal requirements imposed on the availability of computing resources and data: Any delay immediately impacts the productivity of personnel using the distributed applications. As virtualization, service-orientation, and cloud computing are likely to become more important in the future, so will answering logistical questions in networked computing environments.

In order to better understand and to tackle such virtual logistical challenges, future research should initially focus on how results from physical logistics systems can be instructive in overcoming logistical issues in distributed computing systems. However, research should also focus on identifying the constituent features of "virtual logistics systems" and on the potential factors that may determine the design of such systems. For example, inventory holding costs critically determine the structure of physical logistics systems. However, such costs are not incurred by data, unless data access is charged on a temporal basis. In addition, costs for data storage are comparatively low and are likely to decline further as technology evolves. Furthermore, modern network technology allows for the transfer of large amounts of data anywhere in milliseconds, thus reducing the importance of spatial dispersion. As a result, logistical problems in distributed computing systems may become primarily concerned with aligning the temporal availability of data with the temporal availability of processing power to avoid any delays in the flow and transformation of information.

3. Meaning 2: Using a Cloud Computing Technology-based Platform for Logistics IT Systems

3.1. Cloud-based Platform as Enabler and Nexus for Networked Interfirm Cooperation

The second meaning of "cloud logistics" understands the concept as *the use of a cloud computing technology based platform for logistics IT systems.* As mentioned in the overview, the majority of sources (53+4) view the term "cloud logistics" in this way. Although many sources use this meaning, few provide a formal definition. Shen & Qian (2014a:1848) define cloud logistics as "an innovative logistics-service cloud mode based on cloud computing." Another definition reads: "Cloud Computing platform is the infrastructure of Cloud Logistics" (Zhao & Wu 2013:450). Gong & Yang (2012) state "'Cloud logistics' is the application of 'cloud computing' to the logistics industry; it is the derivative of 'cloud computing'" (see similar Arnold 2014a:16; Li & Huang 2013:2825). Matkovic et al. (2014:12) speak of "cloud computing logistics."

The existence and use of a cloud computing technology based computing platform – hereafter referred to as *cloud-based platform* – is the common denominator of all sources that employ the second meaning. A computing platform is generally understood as an environment in which users can run software (Section D.II.3). Hence, the term "platform" implies that economic actors can either deploy software of their choice on this platform or use software already deployed in order to support their respective value creation processes. By adding the prefix "cloud" to the word "platform," the sources stress the use of cloud technology, yielding, for example, "cloud logistics platform" (Shen & Qian 2014a:1848; Diwan 2015:8; Wang et al. 2015c:85; Urban 2015; Geng & Li 2012:633), "cloud logistics service platform" (see Sun & Liu 2013; Sun et al. 2015b), and "cloud platform" (Colajanni 2012; Stinnes 2012:8).

The sources consider the use of a cloud-based platform in two different, but related contexts: administrative and/or operative logistics processes (see e.g. Zhao & Wu 2013:449f.; Matkovic et al. 2014:18; Cloud Logistics 2013) and general value creation processes in supply chains/ networks, especially including logistics processes (see e.g. Stinnes 2012:8; Sun & Liu 2013; Ding 2014; Urban 2015; Colajanni 2012; Kiewitt 2013; Schmidt 2013; Jiang et al. 2013). Regarding logistics processes, a cloud-based platform can be described in several specific logistics contexts: e-commerce in China (Shen & Qian 2014b:495; Shen & Qian 2014a; Du 2014; Zhang & Yuan 2013; Wang 2014; Pan & Tong 2014; Sun et al. 2015b), cross-border e-commerce (Wang et al. 2015c), LTL shipping (IMA RWTH Aachen University 2011; Tummel et al. 2014), multi-modal transnational transport (project "LOGICAL" in Arnold 2014a), global transport (project "Cloud Logistic" in Cloud Logistics 2013), and air logistics (Zhao & Wu 2013).

Several sources suggest logistics processes or supply chains as particularly suitable contexts for cloud-based platforms because of the enormous division of labor and the large number of actors involved (Hannig 2012:6; Opresnik 2014:73; Stinnes 2013:37f.; also see Colajanni 2012:10; Zhao & Wu 2013:450). Hannig (2012:6) argues that the logistics industry and LSPs

> "are simply predestinated for the use of cloud. Their business is marked by a high degree of division of labor, many parties involved in processes and growing internationalization and can thus profit from the advantages which cloud computing offers to a very high degree." (see similar Opresnik 2014:73)

The fundamental reason why a cloud-based platform is considered particularly suitable in these contexts is the platform's crucial ability to *connect all economic actors* involved in the same logistics process or supply chain by means of a single IT system (see Urban 2015; Cloud Logistics 2013; Stinnes 2012:8; Schmidt 2013:32). Actors are connected in a "social network" like fashion (Cloud Logistics 2013; Stinnes 2012:8). In support of this, Stinnes (2013:37f.) argues that cloud computing technology offers more than reduced IT costs in logistics and SCM contexts because both focus on interfirm cooperation and process coordination among many (i.e. hundreds) distributed actors on a global level, and, unlike traditional enterprise resource planning (ERP) systems, cloud-based platforms are especially designed for supporting information flows between firms (also see Matkovic et al. 2014:18). Similarly, Urban (2015) argues that cloud-based platforms are best suited to support communities of firms: An LSP can join a platform that is already being used by the largest Fortune 1000 companies and major LSPs. Thus, this new LSP can immediately interact with all connected companies.

By connecting all economic actors involved in logistics processes or along supply chains, the cloud-based platform becomes the enabler and nexus for new forms of *networked interfirm cooperation* (Stinnes 2012:8; Heaney 2010; Gantzia & Sklatinioti 2014:49f.), which has never been seen before (Diwan 2015:1). Colajanni (2012:10) argues that "[c]loud-based platforms are inherently collaborative". Matkovic et al. (2014:19) mention the cloud's "real-time collaboration capabilities." In this context, many sources describe or associate cloud logistics and a cloud-based platform with some form of interfirm cooperation or collaboration (see e.g. Cloud Logistics 2013; Matkovic et al. 2014; Jiang et al. 2013; Banyai 2014; Geng & Li 2012; Zhao & Wu 2013; Shen & Qian 2014b; Li et al. 2014a; Sun et al. 2015b; Nie & Fan 2015; Tummel et al. 2014; IMA RWTH Aachen University 2011; Arnold 2014b; Urban 2015).

Thus far, we have focused on the platform's ability to enable cooperation among many economic actors in logistics processes or along supply chains, without focusing on the platform's functionality to support these alliances. Some sources directly specify the platform's functionality by means of the platform's name. For example, "intelligent logistics integrated management platform" (Jiang et al. 2013), "logistics information platform" (Geng & Li 2012:631), "public logistics information platform" (Wang et al. 2012a), and "public information and management platform" (Ding 2014). According to those sources, a platform's functionality revolves around information and management. However, a platform's name represents only a vague indicator of its specific functionality.

We structure existing cloud logistics knowledge about platform functionality into four *generic use cases*: (1) the outsourcing of (on premise) logistics or SCM IT systems to a cloud, which is operated by a specialized third-party provider; (2) collection, integration, sharing, and synchronization of logistics relevant information generated and utilized by actors in logistics processes or along supply chains; (3) management and optimization of collaborative business activities of multiple business partners; and (4) support of value exchanges via a virtual marketplace. These four generic use cases are derived from a survey (interviews and questionnaire) of approximately 32 LSPs conducted by the "LOGICAL" research project (Arnold 2014a:22f.; Arnold 2014b:14; Arnold et al. 2013; Arnold et al. 2012). They are suitable for structuring cloud logistics knowledge because they can provide a comprehensive picture of how a cloud-based platform can be used in logistics systems.

In the following four subsections, we describe each of these use cases in detail using the information included in the cloud logistics sources. While the sources suggest that cloud-based platforms represent an enabler and nexus for networked interfirm cooperation, they do not provide any indication of the specific scope and intensity of cooperation. Hence, we will complement current knowledge by commenting on the nature and degree of cooperation intensity likely associated with these use cases.

3.2. Generic Use Case 1: The Outsourcing of Logistics IT Systems to a Third Party Operated Cloud

The first generic use case refers to the outsourcing of logistics IT systems to a cloud operated by a specialized third party provider (Arnold 2014a:22; V&V Dabelstein Logistik GmbH 2015). Both LSPs and customers (firms with logistics demands) can outsource their logistics IT systems in this way.

Firms that are considering whether to outsource their logistics IT systems to a third party operated cloud or to use available cloud-based logistics IT are naturally concerned with the type and scope of available application functionality. In principle, the sources suggest that cloud-based platforms can provide the same functionality as any other non-cloud-based logistics IT system. Functionalities mentioned in the sources cover most common functionalities of logistics management software, including transport management system (TMS), warehouse management system (WMS), and SCM software (see e.g. Arnold 2014a:22; Hannig 2012:7, 21; also see Kiewitt 2013:29; QuantumID Technologies Inc. 2015, Cloud Logistics 2013). Sun et al. (2015b:357) and Banyai (2014:14) specifically describe the functionality of a cloud-based (virtual) supply chain management system. Yet despite this broad conceivable usage of cloud-based logistics IT systems, the current availability of cloud-based logistics applications is rather limited (Kiewitt 2013:29) and the market in transparent (Hannig 2012:21).

This first use case of cloud-based platforms involves a limited scope of interfirm cooperation because it is limited to sharing computing resources. The intensity of cooperation specifically depends on the deployment model in which the cloud-based platform operates. In fact, if the platform is operated as a private cloud, no interfirm cooperation occurs. Conversely, public, community, and hybrid clouds involve interfirm cooperation, as computing resources are shared across multiple firms. In a community cloud, firms usually have knowledge about with which other firms they share computing resources. However, firms using a public cloud do not have this information. Two sources describe a specific setting that allows computing resources to be shared between cloud-based logistics platforms. Long (2014) proposes a basic process for sharing computing resources between platforms based on the RESERVOIR model for hybrid clouds (see e.g. Rochwerger et al. 2009; Rochwerger et al. 2011). Zhao & Wu (2013) consider shared computing resources among actors involved in air logistics systems, including airlines, air cargo stations, air cargo agents, and transport companies.

3.3. Generic Use Case 2: Collection, Integration, Sharing, and Synchronization of Logistics-Relevant Information via a Cloud-based Platform

In the second generic use case, a cloud-based platform collects, integrates, shares, and synchronizes logistics-relevant information generated and used by actors involved in the same logistics processes or by actors part of the same supply chain (Arnold 2014a). In contrast to the previous use case, actors are not only connected to the same platform and share its computing resources, but are also enabled by the platform to interact with each other. In the following, we structure the existing cloud logistics knowledge along the platform's key functions in this use case: (1) collection, (2) integration, (3) sharing, and (4) synchronization of information.

Collection means that information is gathered from different sources and by different actors and subsequently sent to the cloud-based platform. Collection can cover all kinds of logistics processes, including transport, handling, packing, storage (Zhao & Wu 2013:449), and it connects the online commercial flow with the offline logistics processes (Shen & Qian 2014a:1848). The collection of information in cloud logistics is based on the IoT. For example, Nie & Fan (2015:367ff.) argue that the IoT enables real-time mapping between the virtual and physical worlds and thus represents a critical technology for automatically feeding the platform with information generated by distributed logistics and production processes. Many other cloud logistics sources similarly associate a cloud-based platform with the IoT (Li et al. 2014a:25; Zhang & Yuan 2013; Li & Huang 2013:2825; He et al. 2015:1721; Ding 2014; Deng & Fang 2014:1639; Colajanni 2012; QuantumID Technologies Inc. 2015; Geng & Li 2012:633; Sun et al. 2015b:262). From the perspective of telecommunication providers, the growing need for communication technology may present a promising business opportunity in selling these technologies to manufacturers and LSPs (see Li & Huang 2013:2825).

Integration means that information generated by different actors or commonly stored and maintained in their IT systems becomes now stored and maintained in a single, common database hosted on a cloud-based platform (Matkovic et al. 2014:18; Li et al. 2014a:23; Li & Huang 2013:2825; He et al. 2015:1723; V&V Dabelstein Logistik GmbH 2015; Banyai 2014:14). The platform can represent a unique integration space for information (Arnold et al. 2013:1058) or a "unified logistics chain data service center" (Jiang et al. 2013). Furthermore, integration includes platform functionality to analyze, evaluate, and process information stored and maintained in the common database (Shen & Qian 2014b:495; Stinnes 2013:39; Li et al. 2014a:23f.; Matkovic et al. 2014:18; Nie & Fan 2015:367f.; Li & Huang 2013:2825; Gong & Yang 2012; Geng & Li 2012:633). Hence, Sun et al. (2015b:364) describe the cloud-based platform as offering "big-data" services as support for supply chains.

Sharing means that actors use the cloud-based platform to make information generated by them accessible to others and to access information made available by others. Information can be shared between actors involved in the same logistics process or supply chain (V&V Dabelstein Logistik GmbH 2015; Cloud Logistics 2013; Urban 2015). Some information may be private and only accessible to selected actors (Stinnes 2012:8; Stinnes 2013:38), and other information may be public and accessible to every actor connected to the platform (Teichmann 2014:62 shared public in figure). The platform's information sharing capability is emphasized in many sources; for example, "[c]loud computing platforms can effectively promote data sharing" (Zhao & Wu 2013:451), "the logistics information platform is a general information exchange platform" (Matkovic et al. 2014:18), "multiple partner information exchange" (Jiang et al. 2013), and "users can share information via the cloud platform" (Li & Huang 2013:2825). He et al. (2015:1723) express the platform's information sharing capability directly through its name: "information sharing platform."

Other sources just use the terms "information sharing" or "information exchange" in some (loose) connection with the term "platform" (Gong & Yang 2012; Sun & Liu 2013; Nie & Fan 2015; Long 2014:149).

Synchronization means that there is no delay in the dissemination of information among actors connected to the platform. It refers to the platform's information transmission capability (Shen & Qian 2014b:495). Synchronization results from storing and maintaining information in a single, common database on the cloud-based platform. As a result, all actors connected to the platform can always access the latest "copy" of every information piece, and changes in the database become visible to all actors instantaneously (Stinnes 2013:38; Stinnes 2012:8; see Cloud Logistics 2013). Urban (2015) describes this property as follows: "Cloud centralizes data, making it easy to have a single, up-to-date reality for all partners." Thus, the cloud-based platform synchronizes not only the virtual and physical worlds via the IoT, but also all economic actors via a common database. Moreover, considering the platform's combined capability to integrate and synchronize information, Stinnes (2013:39) argues that the platform is more than a central information hub that merely "forwards" information between actors.

Although many sources identify these four basic functions of a cloud-based platform, only a few sources provide concrete information about what type of information the platform actually collects, integrates, shares, and synchronizes. Most sources are vague in this respect; for example, Jiang et al. 2013 simply speak of "logistics information," and Shen & Qian (2014b:495) argue that the platform can satisfy all logistics information requirements of all actors involved, without specifying what these requirements are (see similar Zhao & Wu 2013:449). By contrast, Li et al. (2014a) specifically point out that the platform can handle information related to logistics activities, such as logistics service information, freight information, railway station information, and aviation port information. Two other sources mention information such as order sheets, stock keeping units (SKUs), customs documents, packing lists, RFID-data, and payment instructions (Stinnes 2013:38; Stinnes 2012:8). Diwan (2015:8) points out that cooperating logistics companies can use the platform to exchange information about available services and capacities. Thus, one can argue that the platform can handle any type of *logistics-relevant information*, which refers to any piece of information necessary for planning, executing, controlling, and monitoring logistics processes in an interorganizational context and any piece of information that firms exchange with their clients and suppliers to coordinate fragmented and distributed value creation processes.

Considering the platform's ability to connect a large number of economic actors and support the information flows between them, it seems that this platform can tackle a critical and persistent challenge in logistics and supply chain research: achieving end-to-end visibility of logistics and production processes that are fragmented among various firms (Urban 2015; Heaney 2010; Gantzia & Sklatinioti 2014:18; also see Diwan 2015:8). Prior to the emergence of cloud computing, information was typically stored and maintained in many heterogeneous IT systems primarily designed for internal firm use and therefore difficult to connect with each other (Stinnes 2013:36; Matkovic et al. 2014:12f., 17). Arnold (2014b:12) finds evidence, for example, that many small LSPs use various IT systems that lack any automated interfaces, so data can only be exchanged through manual conversion and transportation (Arnold 2014a:19). This leads to the existence of "information islands" (Deng & Fang 2014:1639; Wang et al. 2015b:322; Wang et al. 2015a:68; Long 2014:147; Wang et al. 2012b:558), a lack of database synchronization and system connectivity (Diwan 2015:7), and thus a lack of information sharing with suppliers and clients (Sun & Liu 2013; Li et al. 2014a:22; Matkovic et al. 2014:12f.). These well-known IT problems impede successful SCM because roughly 80% of information necessary for successfully managing supply chains is stored in the IT systems of external business partners (Stinnes 2013:36). Cloud computing can overcome these obstacles of incompatibility (Gantzia & Sklatinioti 2014:49f.). Information is no longer passed like a "baton" between actors involved in logistics and other value creation processes, but, instead, all relevant information is integrated at a single location: a database on the cloud-based platform, creating end-to-end visibility.

The second use case (collection, integration, sharing, and synchronization of logistics relevant information via a cloud-based platform) can involve varying scopes and intensities of interfirm cooperation, depending on the information handled on the platform. Sharing real-time traffic information in an urban area to avoid congestion is certainly cooperation of a weaker scope and intensity than sharing current and projected inventory levels with clients in all warehouses on a global scale.

In conclusion, there is a significant body of cloud logistics knowledge concerned with the collection, integration, sharing, and synchronization of logistics relevant information via a common cloud-based platform. One can argue that cloud logistics sources provide a comprehensive understanding of the platform's functions in these regards. It is particularly interesting that the use of a cloud-based platform is claimed by sources to contribute to creating end-to-end visibility in organizationally fragmented and geographically distributed logistics processes and other value creation processes in supply chains.

3.4. Generic Use Case 3: Management and Optimization of Collaborative Value Creation Processes

The third generic use case of the cloud-based platform refers to using the platform for the management and optimization of collaborative value creation processes (Arnold

2014a:22f.). Value creation processes comprise both production processes and logistics processes. "Collaborative" means that processes are fragmented among at least two organizations and that some coordination between these organizations is required in order to achieve the desired overall outcome. Thus, the management and optimization of collaborative processes inevitably spans organizational boundaries and, therefore, requires some form of interfirm cooperation. The extent and form of cooperation depends on the type and scope of processes managed and optimized via the platform and on the allocation of decision authority between cooperating firms.

This use case can be considered an extension of the previous use case because the platform is used not only for collecting, integrating, sharing, and synchronizing information, but also for supporting or carrying out the management and optimization of fragmented value creation processes by exploiting this information. The information maintained on the platform thus provides the foundation for the platform-based management and optimization of logistics processes and supply chains on a strategic and operational level (Matkovic et al. 2014:12, 16f.; Diwan 2015:8; Banyai 2014:14; Sun et al. 2015a:109; Nie & Fan 2015:369). Integrating information into a single database provides decision makers with immediate and accurate information (Sun et al. 2015a:109; Nie & Fan 2015:369). Using a single database can therefore enable logistics companies to "take more informative decisions" (Diwan 2015:8) and speed up decision-making (Matkovic et al. 2014:17). This use case therefore involves a stronger degree of interfirm cooperation than the previous one.

Several sources indicate the platform's management and optimization capability by incorporating the word "management" directly into the platform's name; for example "public information and management platform" (Ding 2014), "4PL management platform" (Zhang & Sun 2013), "intelligent logistics integrated management platform" (Jiang et al. 2013), "manage platform" (Li & Huang 2013:2825), and "cloud logistics information and management platform" (Diao 2014). Specific management activities mentioned in the cloud logistics sources include "scheduling and directing of logistics resources" (Ding 2014); "scheduling of all regional logistics" (Sun & Tao 2013); "dynamic combination, intelligent scheduling, and monitoring" of logistics resources (Sun et al. 2015b:362); order allocation, order processing, and disposition of shipments (IMA RWTH Aachen University 2011); and the design of complex logistics processes based on pre-defined process modules (Böhmer 2012:11). Considering these activities, one can conclude that the cloud-based platform can cover the entire management cycle, including planning, realization, control, and monitoring of value creation processes – and especially logistics processes.

Several of the sources emphasize the cloud-based platform's interorganizational management and optimization focus by stating that the platform "integrates" logistics resources and capabilities of different economic actors (Wang et al. 2012a; Shen & Qian 2014b:496; Gong & Yang 2012; Sun & Tao 2013; Zhao & Wu 2013:450). In this context, integration specifically means that the platform supports the implementation of interorganizational (logistics) processes both within and between supply chains (Matkovic et al. 2014:17) by using different actor's resources and capabilities. Interorganizational management and optimization via the cloud-based platform is emphasized in Delfmann et al.'s (2011:2) definition of "cloud logistics"

> "The term 'Cloud Logistics' describes an environment of 'virtual' systems that facilitate supply chains' overall coordination and use of distributed resources, capacities, processes, and services from supply chain partners. These systems are based on advanced information and communication technologies that leverage modern Internet services." (see similar Delfmann & Jaekel 2012:17; Kersten et al. 2012:257)

An important organizational question regarding the interorganizational management and optimization of value creation processes pertains to the allocation of authority between cooperating firms regarding management decisions, the choice of decision rules and optimization algorithms, and the setting of optimization goals (see Section C.V.2). Despite the importance of authority distribution, the sources pay a minuscule amount of attention to it. Several of the sources touch this aspect, but do not consider it carefully. Some of the sources suggest a rather "centralized" management and optimization approach, although no specific optimization algorithm is proposed. For example, Li et al. (2014a:25ff.) argue that the platform matches logistics supply and demand and optimizes the allocation of resources (also see Li et al. 2014b:77; Wang et al. 2015c:85). Nie & Fan (2015:369) argue that the cloud warehousing mode can achieve the "global optimum." Jiang et al. (2013) mention a "centralized management mode," and Banyai (2014:13) argues that the cloud-based platform can be used as an "automated multi-objective [...] logistics optimisation system." By managing and optimizing value creation processes through an algorithm, the platform effectively exerts central authority over these processes and the involved actors. In other words, cooperating actors submit to the authority of the cloud-based platform. A more "decentralized" approach is described in the "Cloud Logistic" research project (see IMA RWTH Aachen University 2011; Tummel et al. 2014). Similar to the centralized approach, the disposition of LTL shipments is optimized globally using an algorithm implemented on the cloud-based platform. However, this algorithm is developed in a decentralized manner: The LSPs that are part of the cooperation raise their requirements in several joint workshops. The algorithm has been subsequently validated in a field game. Therefore, it is possible to minimize conflicts of interest and ensure successful and sustainable cooperation.

Considering these sources, we can conclude that there are heterogeneous and, to some extent, vague conceptions about how the cloud-based platform either supports or carries out the management and optimization of collaborative value creation processes. While the use of a cloud-based platform makes a centralized approach feasible from a technical perspective, such an approach may not be acceptable for legally autonomous firms; they may not be willing to submit to the central authority of the platform if centralized management and optimization is in any way disadvantageous for them. In this context, Nie & Fan (2015:369) argue that, in order to encourage cooperation, the optimization algorithm must balance the interests of all parties, especially regarding the distribution of profits (IMA RWTH Aachen University 2011).

Before moving on to the next use case, we provide three concrete examples of how the cloud-based platform is used in practice. They provide a good understanding of the different settings in which cloud-based platforms can be used. First, several sources describe the use of a cloud-based interorganizational logistics management and optimization platform for e-commerce in China (Jiao 2014; Shen & Qian 2014b:495; Shen & Qian 2014a; Du 2014; Zhang & Yuan 2013; Wang 2014; Pan & Tong 2014; Sun et al. 2015b; He et al. 2015; Liao & Fu 2014). Given the number of such sources, we can describe this particular context in more detail. While e-commerce is a broad logistics field, cloud logistics sources are specifically concerned with logistics operations related to business-to-business and consumer-to-consumer *online trading platforms* (see Jiao 2014:222; Geng & Li 2012:631; Sun et al. 2015b:359; He et al. 2015). Sellers on these platforms are usually small and medium-sized merchants and consumers that lack own logistics capabilities for size reasons; they therefore rely on third party LSPs to deliver their goods sold on the platform (see Jiao 2014:222f.). Due to the large number of geographically distributed buyers and sellers, the small sizes of shipments, and the highly fragmented logistics industry in China, pick-up and delivery of shipments is very inefficient due to a lack of economies of scale. Sun et al. (2015a:359) describe this problem as "fragmented demand" and "fragmented supply". In order to improve logistics efficiency, sources concerned with e-commerce in China propose to introduce of a cloud-based logistics management platform – cloud logistics. This platform may be either operated by the online trading platform operator (see Jiao 2014:223) or by an LSP (see Shen & Qian 2014b; Liao & Fu 2014:100). An LSP frequently mentioned as a platform operator is "XingChenJiBian" (also referred to as "Cloud express" or "Stars express" in some sources), which operated such a platform as a partner for Alibaba.com (see e.g. Shen & Qian 2014b; Liao & Fu 2014:100; Shen & Qian 2014a:1848). This cloud-based logistics platform essentially aggregates and optimally matches logistics demand with supply (Shen & Qian 2014b:497). Specifically, it aggregates the logistics demand from all online trading platform transactions on one side and the logistics resources from many small LSPs on the other. The cloud-based management platform then classifies shipping orders by time, type, region, and urgency and allocates them to LSPs with the objective of achieving logistics efficiency (Shen & Qian 2014b:497), for example through systematic collaboration in last-mile distribution

(Sun et al. 2015b:364; Geng & Li 2012:632). By matching the supply with the demand, the cloud-based platform also creates real-time visibility between merchants and LSPs and "by so doing form[s] the cooperation between small and medium-sized merchants and small and medium-sized warehousing and express companies" (Jiao 2014:223). Thus, the cloud-based platform is the enabler and nexus for networked interfirm cooperation in Chinese e-commerce.

Second, "Cloud Logistic," a joint project between the RWTH Aachen University, an LSP, and a logistics IT vendor (Tummel et al. 2014; IMA RWTH Aachen University 2011), describes another interesting context in which a cloud-based management and optimization platform is used. The research project focuses on establishing a horizontal networked cooperation between small and medium-sized LSPs for operating a line-based LTL network between regions in Germany. Small and medium-sized LSPs usually suffer from low utilization in the LTL segment as they lack the customer base to consolidate multiple shipments. The cloud-based platform aggregates customer requests from all participating LSPs and then optimizes the allocation of shipments to LSPs. LSPs have a monopoly on operating a specific freight route between regions in Germany. The research project develops a business model that enables sustainable cooperation between LSPs. Elements of this business model include (a) rules for allocating service requests and costs and for appropriating revenue; (b) different types of contracts, including escalation measures if services deviate from contracts; (c) ways to protect customer data; and (d) controlling measures. Benefits of this networked cooperation include increased use of transportation capacity, fewer empty runs, less traffic, and increased competitiveness of small and medium-sized LSPs in an increasingly competitive market environment.

Third, another type of cloud-based logistics cooperation is described in the "LOGICAL" research project (Arnold 2014b; Colajanni 2012; Arnold 2014a). LOGICAL is an EU funded project involving a total of 13 partners from six EU countries (Arnold 2014a:22). The project aims to establish cooperation between small and medium-sized LSPs and key logistics hubs in Central Europe in order to promote collective multi-model transportation between these hubs (Arnold 2011). Diwan (2015) conceives of this research project as project being concerned with *logistics e-clusters.* Logistics clusters are generally defined as cooperation between multiple LSPs that aim to achieve operational benefits that result from co-location. Such clusters usually form around major transportation hubs. Logistics e-clusters represent cooperation between a community of LSPs that is enabled through a common interorganizational IT system – in this case, the cloud-based platform. However, unlike "normal" logistics clusters, e-clusters can cover a greater geographic scope because the use of a common cloud platform allows geographically dispersed logistical actors to be connected.

3.5. Generic Use Case 4: Virtual Marketplace

The platform's fourth generic use case refers to establishing a virtual marketplace as part of the cloud-based platform. According to Arnold et al. (2013:1058), such virtual marketplaces can provide a forum for trading a great variety of services – essentially any type of service offered by actors in the logistics community connected to the platform. Based on the analysis of the cloud logistics sources, we have identified two different types of marketplaces that differ in the types of services that are traded: (1) markets for logistics-related IT services on which vendors of logistics IT services offer their IT capabilities to LSP and/or consumers (i.e. firms with demand for logistics capabilities); and (2) markets for physical logistics services on which LSPs offer their logistics capabilities to consumers.

A *virtual marketplace for logistics-related IT services* is described as part of the cloud-based platform in a few of the sources (Böhmer 2012; Böhmer 2013; Kiewitt 2013; DHL 2014). However, all of these sources describe the same marketplace: "the Logistics Mall," which was established in a joint research project between the two Fraunhofer Institutes (Institut für Materialfluss und Logistik and Institut für Software- und Systemtechnik) and Logata GmbH, a privately held IT vendor that operates the Logistics Mall (Böhmer 2013:42). Beyond the sources identified in our review, the Logistics Mall is also described in several other contributions, including Holtkamp et al. (2010), Steinbuß & Weißenberg (2013), Daniluk et al. (2014), Logata GmbH (2012), Logata GmbH (2013), and Weißenberg & Springer (2014). These sources are not included in our review because they do not use the term "cloud logistics." Nevertheless, to describe the Logistics Mall concisely, we draw on these additional sources.

The Logistics Mall consists of two components: the Mall Marketplace (MMP) and the Customized Access Framework (CAF). The MMP is a web shop where logistics IT vendors can offer their services. LSPs and logistics consumers can purchase logistics IT services in the MMP and run these services in the CAF (Böhmer 2013:42). As the CAF is based on cloud computing technology, logistics IT services are delivered in a SaaS manner with typical cloud characteristics (e.g. pay-per-use, rapid elasticity). The CAF therefore resembles the first generic use case (outsourcing of logistics IT to the cloud), as the mall provides not only a marketplace, but also a third party operated platform for running logistics IT services. Furthermore, the Logistics Mall offers a graphical modeling tool ("Logistik Prozess Designer") that enables LSPs and customers to create customized logistics IT solutions to support complex logistics processes (Böhmer 2013:42; Steinbuß & Weißenberg 2013; Daniluk et al. 2014:31f.). Interoperability between different vendor's services is currently achieved through standardized business objects, which are described through a domain model (Böhmer 2013:42; also see Böhmer 2012:11; Weißenberg & Springer 2014) represented as an ontology (Steinbuß & Weißenberg 2013:11f.). Future research directions of the Logistics Mall focus on the use of Semantic Web technology for enhancing service composability (Steinbuß & Weißenberg 2013:15; see Subsection D.III.4.3).

A *virtual marketplace for physical logistics services* is identified as part of the cloud-based platform in several of the cloud logistics sources. Arnold (2014a:22) states that such a marketplace, in principle, represents an online sales channel for LSPs. Arnold (2014b:14) also points out that virtual marketplaces for logistics services become increasingly attractive for customers and LSPs if they interconnect a large community of LSPs, which offer not only basic transportation services, but also a variety of logistics services. Hence, he hypothesizes that, in the future, multiple virtual logistics marketplaces may become coupled, just like today's marketplaces for consumer goods and travel booking. Likewise, Teichmann (2014:41) argues that virtual marketplaces in cloud logistics offer not only basic transportation services, but also handling, storage, value-added services. He also argues that, in cloud logistics, multiple virtual marketplaces may be interconnected in order to counteract the current fragmentation observed in freight exchanges. Colajanni (2012:4) only mentions a "cloud marketplace for logistics services" without providing any further details. Li & Ding (2013) mention a "Logistics Supermarket" in relation to (last-mile) distribution services. However, it remains vague what this supermarket actually is, partly because we could only gain access to the abstract of this source. Although the Logistics Mall is currently limited to logistics IT services, it should also include physical logistics services offered by LSPs (Logata GmbH 2012:4; Holtkamp et al. 2010:307). Although several of the sources identify a virtual marketplace as part of a cloud-based platform, it must be stressed that they usually merely mention the existence of a virtual marketplace, without describing its functioning.

A pivotal question regarding the functioning of any marketplace pertains to the underlying allocation or market mechanism, the price mechanism for "matching" supply with demand (see Section C.III.5). With regard to the virtual marketplace for logistics IT services in the Logistics Mall, IT providers, LSPs, and customers have the ability to negotiate prices (Logata GmbH 2012:6). However, a specific mechanism has not been described. With regard to the virtual marketplace for physical logistics services, slightly more knowledge can be found on the type of mechanism. Arnold (2014b) argues that "comfortable-matching instruments" are important for increasing the attractiveness of virtual marketplaces for LSPs and customers. Yet, what these instruments are is not specified. Teichmann (2014:61) argues that the price mechanism in virtual cloud logistics marketplaces needs to incorporate multiple criteria, instead of simply comparing the quantity of demand and supply. However, the only criteria mentioned are environmental regulation standards and carbon emissions, thus providing little information about operative logistics criteria. Wang et al. (2015a:75) generally mention the use of market mechanisms and combinatorial auctions for allocating resources in cloud logistics (also see Wang et al. 2015b:324). Still, no specific mechanism for cloud logistics is proposed. Arnold (2014a:26) provides deeper insights into the matching of supply and demand, arguing that, in the "LOGICAL" research project, logistics demand and supply are centrally matched by the cloud-based platform based on location, time, and other service characteristics. If a match is found, both the requester and provider of the service are notified via email. They can then decide whether or not to engage in pricing negotiations. Thus, pricing or price calculation techniques are not published on the platform. In sum, current cloud logistics knowledge draws a blurry picture of what kind of market mechanism is or should be used for matching logistics (IT) demand with supply.

Finally, just like in the other use cases, the virtual marketplace also connects different actors involved in logistics and other value creation processes and allows them to carry out economic exchanges. Yet, as markets are – by definition – a non-cooperative mode of facilitating economic exchanges (see Subsection C.III.3.1, Section C.III.5), the pure use of a marketplace involves no interfirm cooperation. However, this use case can still involve interfirm cooperation if market participants jointly control market access or agree to common standards of services traded on the market. For example, in the Logistics Mall, logistics IT services and their vendors are certified by the Fraunhofer Institutes (in a neutral manner) to ensure service quality and that all services adhere to common (interface) standards in order to enable interoperability and composability of the services offered by different vendors in the marketplace (Logata GmbH 2012:9). Teichmann (2014:61ff.) conceives of freight exchanges as the first of three phases of a "market-oriented form of collaboration" on the meso-logistics (interfirm) level. Thus, virtual marketplaces can, but do not need to, involve interfirm cooperation.

3.6. Benefits and Concerns of Using Cloud-based Platforms in Logistics Systems: Essential Cloud Characteristics

The largest benefit of using cloud technology for logistics IT systems may be the ability to enable networked interfirm cooperation and to increase logistics efficiency and effectiveness. In addition, the cloud logistics sources identify several other benefits and concerns associated with the use of cloud technology for logistics IT systems. These benefits and concerns apply to any of the previously mentioned use cases and can be structured into two categories: (1) benefits and concerns associated with outsourcing in general and (2) benefits and concerns specifically associated with the use of cloud computing technology

With regard to *outsourcing in general*, the sources discuss benefits and concerns similar to those discussed regarding the outsourcing of any other kind of business process. On the one hand, outsourcing of logistics IT allows LSPs and customers to focus on their respective core competencies (Kiewitt 2013:28; see similar Thomas & Unruh 2010; Schmidt

2013:33; also see Matkovic et al. 2014:19) while satisfying their respective IT demand from specialized providers (Gantzia & Sklatinioti 2014:52), which may include complex applications (V&V Dabelstein Logistik GmbH 2015). On the other hand, there is strong empirical evidence that LSPs and shippers are concerned with their loss of control over the respective cloud-based logistics IT systems (Hannig 2012:19; Schmidt 2013:34; Axit AG 2012:19).

The *benefits* for LSPs and logistics consumers associated with using *cloud computing technology for logistics IT systems* are closely related to the essential characteristics of cloud computing: (1) on-demand self-service, (2) broad network access, (3) resource pooling, (4) rapid elasticity, and (5) measured service.

With regard to *on-demand* availability, several of the sources argue that cloud-based logistics IT systems become quickly usable and require little implementation effort (see e.g. Hannig 2012:12, 22; Axit AG 2012:18; Schmidt 2013:34; Stinnes 2013:36f.; V&V Dabelstein Logistik GmbH 2015). This enables logistics companies to collaborate in changing conditions as software is provided on-demand to a heterogeneous set of devices (Diwan 2015:8). Moreover, because of the on-demand availability of a cloud-based platform, some of the sources argue that using cloud technology enables quick integration of new actors into logistics or other value creation processes with little implementation effort (Hannig 2012; Axit AG 2012; Schmidt 2013; V&V Dabelstein Logistik GmbH 2015; Kiewitt 2013). This quick integration becomes possible because new actors do not need to install hardware or software on their computers, but can immediately access the cloud-based platform over the internet. Moreover, each actor only needs to establish a single IT interface between the internal ERP system and the cloud-based platform in order to interact with any other actor already connected to the platform. As a consequence, the use of a cloud-based platform enables the integration, sharing, and synchronization of information not only among many actors, but also among a quickly changing set of actors. As Arnold (2014b:14) remarks in this respect, cloud computing enables LSPs to quickly establish new collaboration and thus to quickly respond to changing customer requirements.

Broad network access allows for flexible and worldwide access to logistics management software and data available on the cloud-based platform via the internet (see e.g. Kiewitt 2013; Cloud Logistics 2013; also see Gong & Yang 2012; Matkovic et al. 2014:19; Colajanni 2012:10). Empirical evidence suggests that this flexible and mobile access is of particular relevance to LSPs (Hannig 2012:18; Schmidt 2013:34; Axit AG 2012:8; Gantzia & Sklatinioti 2014:52f.). This is reasonable because logistics processes are inherently distributed in space, often spanning the entire globe, and the data necessary for managing such processes are usually generated and used at different locations.

Due to *resource pooling*, the use of cloud-based logistics IT can reduce IT operating costs because virtualization lets computing resources be shared in the cloud (see e.g. Kiewitt 2013:29; also see Stinnes 2013:36). Yet, at the same time, individualized cloud-based logistics solutions can be created through the composition of basic modules (service-orientation) (Kiewitt 2013:29).

Rapid elasticity lets logistics IT be provided in a flexible manner commensurate with demand (gaxsys GmbH 2013; Stinnes 2013:36; Arnold 2014a:17; Matkovic et al. 2014:19; Gantzia & Sklatinioti 2014:50). Users can quickly adjust processing power and storage space in order to respond to changes in business needs. For example, cloud technology enables an LSP to set up an additional customer warehouse in the IT system within a few hours (Kiewitt 2013:29). Colajanni (2012:10) speaks of a "dynamic supply chain" because IT can be scaled with minimal waste of time and capital on cloud-based platforms.

Finally, *pay-per-use* requires LSPs and customers to pay only for the computing resources and applications actually consumed on the cloud-based platform (Stinnes 2013:36). Pay-per-use is particularly emphasized as a key benefit for small-and-medium-sized enterprisess (SMEs) because these firms often lack the financial resources for making high initial investments and for covering the ongoing maintenance costs that are typically required for sophisticated logistics management software solutions (Kiewitt 2013:27f.; Hannig 2012:12; Arnold 2014b:12; Matkovic et al. 2014:19; Gantzia & Sklatinioti 2014:50f.). Therefore, SMEs often default to the use of Excel sheets (Kiewitt 2013:27f.) or forego new software releases (Hannig 2012:12). In addition, cloud-based systems are kept "up-to-date" by their cloud providers, which makes it unnecessary for users to have maintenance personnel (Schmidt 2013:33; Matkovic et al. 2014:19) and also makes upgrading less complex and time consuming (Gantzia & Sklatinioti 2014:53f.). Cloud technology can therefore help protect SMEs from technical obsolescence (Hannig 2012:17). Although cloud technology has been primarily mentioned here as important for SMEs, cloud-based logistics IT is also increasingly important for large LSPs (King & Wenger 2013:44). Reasons include costs pressures, remote access, ease of availability, and possibilities for enabling real-time cooperation (Matkovic et al. 2014:19).

Concerns mentioned in the sources regarding the use of *cloud computing technology for logistics IT systems* pertain to data security and data privacy. Data security refers to protecting data stored in the cloud against unauthorized access, such as hacking attacks. Privacy refers to the amount and type of information collected, processed, and disseminated in the cloud and the physical location of storage (due to legislation differences). Strong empirical evidence suggests that data security is the primary concern of LSPs and customers (see e.g. Hannig 2012:19; Axit AG 2012:19; Schmidt 2013:35; Kiewitt 2013:30; also see Stinnes 2013:39; Arnold 2014a:19f.; also see Zhao & Wu 2013:449; Colajanni 2012:5ff.). Hence, outsourcing companies should include security as a key dimension when selecting a cloud provider (Stinnes 2013:39). In a survey of LSPs, Arnold (2014a:19f.) reports, almost all rated security as "very important." He also argues that instruments ensuring security and privacy are a central and necessary precondition for a broad introduction of cloud computing technology to logistics systems. By contrast, in another empirical study involving several LSPs, Gantzia & Sklatinioti (2014:46f.) find that LSPs perceive neither security nor confidentiality/privacy as an issue (see similar survey results for data privacy in Hannig 2012:19). In this respect, several of the sources argue that concern regarding data security is "rather a diffuse gut feeling and is based mainly on missing information and a lack of knowledge" (Hannig 2012:18; see similar argument in Kiewitt 2013:30). Arnold (2014b:13) argues that using cloud technology "professionalizes" security and privacy and that IT providers must deliver security and privacy as a basic component of their cloud offerings. V&V Dabelstein Logistik GmbH (2015) points out that frequent security updates reduce the risk of data loss.

In conclusion, the cloud logistics sources provide a comprehensive overview of benefits and concerns associated with the use of cloud computing technology for logistics IT systems. Some concerns and benefits are rather general and not specifically related to logistics. Others are specifically described in a logistics context. In particular, this includes the benefits of the essential cloud characteristics. In this respect, broad network access and on-demand availability are cloud characteristics that demonstrate their power in logistics processes due to these processes' distributed natures and the large number of potentially changing actors. This implies that cloud computing technology is particular suitable for logistics IT systems.

3.7. Interim Conclusion and Research Opportunities

Figure 23 summarizes the most important concepts associated with the second meaning of cloud logistics and the interrelations between these concepts. Existing cloud logistics knowledge provides a comprehensive conceptual understanding of how cloud-based platforms can be deployed in logistics systems. We have not identified a source that describes a use case other than the four described above, thus supporting the conclusion that these four use cases are collectively exhaustive.

Despite the already good understanding of these generic use cases, further research is warranted to (1) better understand the diffusion process of cloud-based platforms; and (2) further detail the functionality of the platform, especially regarding the mechanisms that govern the interorganizational management and optimization of logistics processes and the mechanisms that govern value exchanges in a virtual marketplace. Further research regarding the *diffusion process* seems desirable because, despite compelling benefits from a conceptual perspective, the diffusion of cloud-based logistics platform has materialized

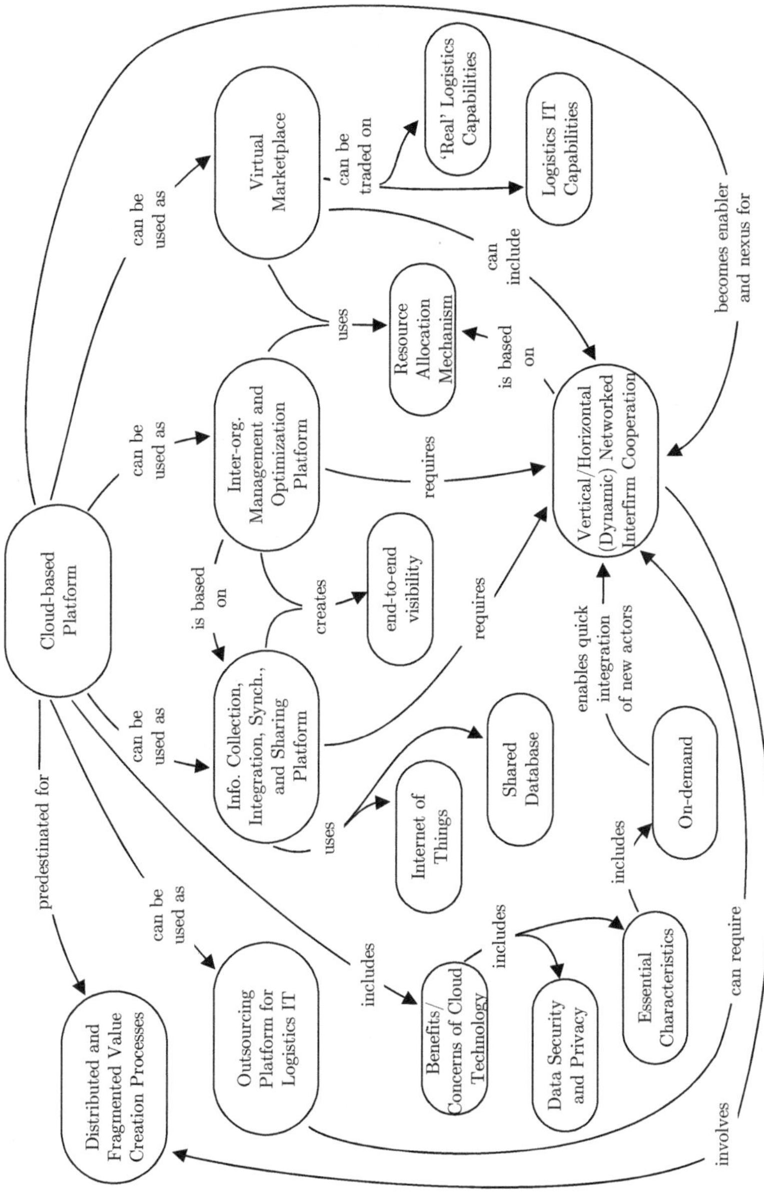

Figure 23.: Summary of Meaning 2: Concepts and their Interrelations

very slowly. Hence, research should specifically focus on validating whether anticipated benefits associated with such platforms can in fact be realized. In addition, research should develop managerial recommendations for overcoming potential implementation barriers, thus enabling practitioners to actually realize benefits.

Further research should also help understand the *mechanisms* that govern the interorganizational management and optimization of logistics processes and the value exchanges carried out on a virtual marketplace. While it seems "easy" to identify mathematically optimal solutions for allocating scarce resources to a set of service requests, such solutions may not be optimal for all participating members in terms of individual profit-maximization. In other words, mathematically optimal solutions may not be feasible in the case of profit-oriented, legally autonomous firms that are free to decide whether or not to participate in a cooperation clustered around a cloud-based platform. Future research should therefore focus on developing specific rules that fairly structure the cooperation, especially in terms of resource allocation and revenue appropriation, where "fairly" means that potentially conflicting interests of partnering firms are balanced in a reasonable and sustainable manner to achieve long-term cooperation. Game theory and mechanism design may provide the tools for developing such mechanisms.

4. Meaning 3: E-commerce Fulfillment through a Network of Cooperating Local Dealers and LSPs

The third meaning of the term "cloud logistics" understands cloud logistics as *e-commerce fulfillment through a network of cooperating local dealers and LSPs.* This meaning can be primarily seen in two articles published in practice-oriented logistics magazines (Thomas & Unruh 2010; Thomas 2011). It is also related to the second meaning because it relies on a cloud-based platform supporting e-commerce fulfillment. However, we introduce a distinct third meaning because in one respect the use of this platform differs from the second meaning. We emphasize this difference after describing how the cloud-based platform is used.

In the third meaning, cloud logistics refers to a novel business and fulfillment model in e-commerce, which was invented and pursued by gaxsys GmbH, a German IT vendor. Specifically, cloud logistics is defined as

> "the IT-based optimization of logistics activities associated with the web shops of brand manufacturers through the real-time integration of local point-of-sales (e.g. retailers, chain stores etc.) into the fulfillment of web shop orders as per a nearest point-of-sale principle."[2] (translated from German, Thomas & Unruh

[2] "Cloud logistics ist die IT-gestützte Optimierung der logistischen Wertschöpfungskette der Online-

2010; Thomas 2011).

The operating logic of cloud logistics can be explained in six steps (Thomas & Unruh 2010; Thomas 2011; also see gaxsys GmbH 2015). First, a customer orders goods from a brand manufacturer's web shop. Second, the customer's order is transferred from the web shop to a cloud-based logistics information and management system. Third, using this system, order fulfillment is offered to a network of local dealers that cooperate with the brand manufacturer. Alternatively, the order is first offered to the dealer located closest to the customer's delivery address. If this dealer is not willing or able to fulfill the order, the radius of dealers offered the order is gradually increased. This operating logic is referred to as "Artikelbörse" and is also incorporated into the name of the company (gaxsys GmbH) that operates the cloud-based logistics information and management system, which is referred to as the "gax-system," with "gax" standing for "global article exchange." Fourth, the gax-system notifies the LSP that the goods are available for pick-up at the local dealer's location and that they should be delivered to the customer. The local dealer is responsible for handling return shipments. Fifth, the brand manufacturer remunerates the local dealer, and, finally, the brand manufacturer invoices the customer. Thus, cloud logistics refers to a "networked fulfillment model" that integrates the e-commerce activities of brand manufacturers with a network of distributed local dealers by means of a cloud-based logistics information and management system. Local dealers play a central role in this fulfillment model because they act as a network of decentralized storage and fulfillment locations for orders received through the manufacturer's web shop (Thomas & Unruh 2010).

The inventors of cloud logistics consider this new fulfillment model a response to current and future logistical challenges in the e-commerce segment, such as the need for increased speed, efficiency, and reliability (gaxsys GmbH 2013). According to them, this e-commerce model has several benefits for the actors involved (see Thomas & Unruh 2010; Thomas 2011; also see gaxsys GmbH 2015). For example, brand manufacturers can transform their websites from mere information sources for potential customers into sales channels. In addition, customer loyalty can increase as customers more often order directly from a manufacturer's web shop instead of Amazon, for example. Local dealers benefit from increased sales as they become integrated into the e-commerce activities of brand manufacturers; this integration also counteracts the online-offline channel conflict. Customers benefit from shorter delivery times as goods are shipped from locations usually closer to their delivery addresses.

As mentioned, the third meaning is related to the second because it relies on the use of a

Aktivitäten von Markenherstellern durch die Einbindung von stationären Fachhändlern über in Echtzeit und nach entfernungsabhängigem Verteilerprinzip über Kundenbestellungen aus Marken-Webshops."

cloud-based platform. Specifically, just like in the second meaning, the platform is used to connect all actors involved in e-commerce fulfillment, enable the sharing of logistics-relevant information, manage logistics processes, and support facilitation of economic exchanges between the actors. However, in one respect, the third meaning's use of the platform is "radically different" (Thomas 2011) from the second meaning's: In the third meaning, the platform is primarily concerned with sharing and exchanging *finished goods stocked in the warehouses of local dealers*. Hence, the primary resource pool that can be accessed through the platform is a pool of finished goods rather than a pool of computing or logistics resources and capabilities.

Figure 24 summarizes the key concepts of this meaning and their relationships. Note that we refer to this fulfillment model as "e-commerce fulfillment through a network of cooperating local dealers and logistics service providers." We deliberately drop the "web shop of the brand manufacturer" aspect because we believe this operating logic can (at least in theory) be implemented by any type of web shop. Given this, future research should evaluate the possibility of expanding the scope of this fulfillment model, for example, in terms of types of products (e.g. food), specific areas of local retailers (e.g. urban areas), and specific groups of local retailers (e.g. franchisees). Specifically, research should focus on how service levels (e.g. same day) and operational efficiency in last-mile delivery perform compared to "traditional" e-commerce setups. Can this decentralized setup compete shoulder to shoulder with the scale of Amazon?

5. Meaning 4: Logistics-as-a-Service – Interpreting and Designing Logistics Systems through the Lens of the Cloud Paradigm

5.1. Derivation and Definitions

The fourth meaning of "cloud logistics" refers to *Logistics-as-a-Service – interpreting and designing logistics systems through the lens of the cloud paradigm*. This meaning follows the principal idea that the term "cloud logistics" refers to logistics systems that belong to a family of systems that all share the same conceptual template: *the cloud paradigm*. Understanding this meaning is directly relevant to addressing our research objective. The cloud paradigm comprises an abstract interpretation of the design principles and concepts that underlie cloud computing and of the elements included in the NIST definition of cloud computing, including essential characteristics, service models, and deployment models (Delfmann & Jaekel 2012:15ff.; also see Leukel & Scheuermann 2014). The systems that are interpreted and designed according to this paradigm are referred to as "cloud systems" and they provide their respective capabilities in an "as-a-Service" manner. Accordingly, interpreting and designing logistics systems according to the cloud paradigm

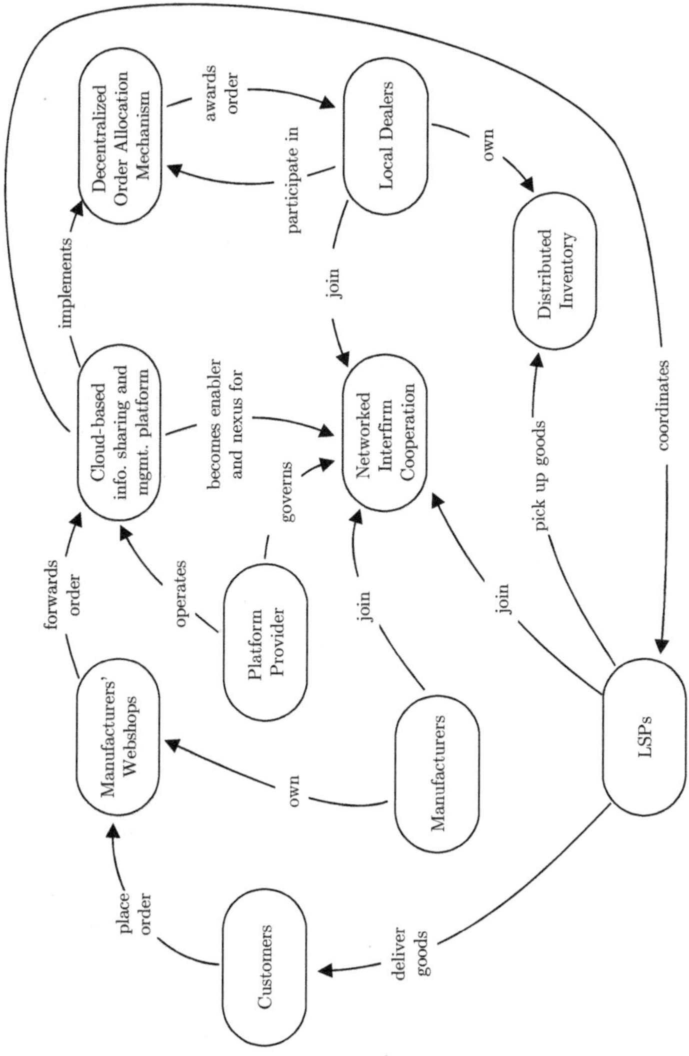

Figure 24.: Summary of Meaning 3: Concepts and their Interrelations

creates *cloud logistics systems* (CLSs) that provision logistics capabilities as a service, referred to as "LaaS." Other systems of this cloud family include cloud manufacturing and cloud computing.

The fourth meaning was first introduced in an industry magazine article that reported on the "28. Deutscher Logistik Kongress" in Berlin in 2011. In the article, the managing director of a German online food retailer, Mohammed Mosavi says,

> "we do not require an own warehouse, we only do cloud logistics"[3] (akw/jö 2011:45).

According to this article, "cloud logistics" refers to a new logistics outsourcing model. In an interview published in May 2012, Mosavi further elaborates by saying that, by using the concept of cloud logistics, their company follows the basic idea of reducing fixed logistics costs and aim to work with transaction-based, variable logistics costs instead (Mosavi 2012:16). Accordingly, LSPs are remunerated for every shipment actually transported. Although no explicit reference is made to applying the principles of cloud computing to the logistics domain in either of the sources, Mosavi actually applies the essential characteristic of pay-per-use to logistics.

The fourth meaning appeared explicitly for the first time in two conference papers by Wang et al. (2012b) and Delfmann & Jaekel (2012) published independently, yet almost simultaneously in May and June 2012, respectively. Although both papers share a very similar conception of cloud logistics, they derive its meaning in different ways. Wang et al. (2012b) derive it from cloud manufacturing which, in turn, derives from cloud computing. Delfmann & Jaekel (2012) derive their understanding of cloud logistics from the cloud paradigm, which also derives from cloud computing. Both derivations are summarized below because they provide the conceptual anchor for subsequent publications that understand cloud logistics in the same way.

Wang et al. (2012b:558) derive the meaning of cloud logistics from cloud manufacturing. To explain this derivation, we briefly consider the concept of cloud manufacturing, which represents "the manufacturing version of cloud computing" (Xu 2012:75). Mirroring the NIST definition of cloud computing, Xu (2012:70) defines cloud manufacturing as

> "a model for enabling ubiquitous, convenient, on-demand network access to a shared pool of configurable manufacturing resources (e.g., manufacturing software tools, manufacturing equipment, and manufacturing capabilities) that can be rapidly provisioned and released with minimal management effort or service provider interaction" (Xu 2012:79).

[3] "wir brauchen kein eigenes Lager, wir machen nur Cloud Logistics"

Using cloud manufacturing as a conceptual anchor, Wang et al. (2012b:558) define and understand cloud logistics as follows:

> "According to the characteristics and objective demonstrated by the cloud manufacturing mode, logistics services should carry out appropriate changes in the resource forms and functions to reflect the characteristics of virtualization, centralized management, intelligent integrated scheduling and distribution on-demand. That forms a new logistics mode – cloud logistics, which matches the cloud manufacturing mode."

Thus, cloud logistics mirrors the conceptual blueprint of cloud manufacturing. Similar derivations of the concept of cloud logistics can be found in other sources (Li et al. 2013:1697; see Wang et al. 2015a:71; Wang et al. 2015b:323).

Delfmann & Jaekel (2012:15ff.) derive their understanding of cloud logistics from the cloud paradigm. Starting from the NIST's definition of cloud computing, they argue that cloud computing describes a specific mode of service provisioning characterized by a set of underlying design principles and concepts, essential characteristics, service models, and deployment models. Yet, these sources can also be interpreted in a more abstract manner – detached from cloud computing. This abstract interpretation is referred to as the "cloud paradigm" (Delfmann & Jaekel 2012:15). Thus, cloud computing can be conceived of as interpreting and designing computing systems through the lens of the cloud paradigm. Of course, the paradigm may also be used to interpret and design systems in other domains, such as logistics and manufacturing systems, thus yielding cloud logistics systems and cloud manufacturing systems, respectively. The idea of interpreting and designing logistics systems through the lens of the cloud paradigm is taken up as a conceptual anchor in several later publications (Pieringer 2012:44f.; Leukel & Scheuermann 2014; Teichmann 2014; Ludwig 2014; Ballardt 2014:16ff.; DHL 2014).

With regard to the cloud paradigm, one more journal article seems worth mentioning, because it develops a perspective that is conceptually similar to the cloud paradigm and also serves as input for Delfmann & Jaekel's (2012) conceptualization of cloud logistics. This article is, however, not part of our knowledge review, because it does not use the term cloud logistics. Leukel et al. (2011a:16) propose the concept SaaS by taking a "Cloud Perspective on Supply Chain Systems." They start with Vaquero et al.'s (2009) definition of cloud computing and consider supply chain systems through the elements of this definition, including resource types, virtualization, service-orientation, resource pool size, resource reconfiguration, pay-per-use, and SLAs. In 2014, they take a similar cloud perspective on logistics systems in order to derive the concept of cloud logistics (Leukel & Scheuermann 2014). Dimensions considered are virtualization, reconfiguration, deployment model, pay-per-use, and SLA. Therefore, one can argue that the cloud paradigm and the cloud perspective are conceptually alike.

Table 18 provides an overview of the "cloud logistics" definitions found in the sources associated with this fourth meaning.

Author(s)	Proposed Definition (sorted chronologically)
Mosavi (2012:16)*	"With cloud logistics, we follow the basic idea of reducing fixed logistics costs and working transaction-based, variable logistics costs instead."[4]
Wang et al. (2012b:558f.)	"cloud logistics (CL) is a new mode of logistics service which is networked, intelligent, low-cost, highly reliable, highly scalable and is new IT technology based. CL makes integration of existing technologies such as cloud computing, Internet of Things, embedded system technology, and high-performance computing technology. After all kinds of logistics resources and capabilities are virtualized and service-oriented, a unified, centralized, intelligent management and operations should be made, in order to achieve effective coordination and a win-win situation. Finally, CL will provide readily available, on-demand, safe, reliable, high-quality and affordable services for the entire supply chain through the network.
Delfmann & Jaekel (2012:17)	"The adoption of the cloud paradigm to the logistics domain will be denoted as 'Cloud Logistics'."
Kersten et al. (2012:259)	"The cloud logistics concept describes a system of autonomous and interconnected logistics service providers focused on the common overall objective of customer satisfaction and service lead time."
Pieringer (2012:44f.)*	"Cloud logistics follows the concept of 'cloud computing' (IT sector) and adopts this concept for the domain of logistics systems"[5]
Li et al. (2013:1696)	"Cloud logistics is a service mode for logistics service provisioning and management based on cloud computing."
Leukel & Scheuermann (2014:39)*	"The cloud perspective is suitable for working out cloud logistics. This perspective provides the basic concepts of cloud computing which are then analyzed in terms of their suitability and relevance for logistics. Due to the greater closeness to logistics as a business area, Logistics-as-a-Service (LaaS) evolves as a new type of cloud service. Such services utilize standardized software interfaces to access logistics activities – namely the spatio-temporal transformations of goods, such as transport, handling, storage, and the associated management activities (planning, controlling)."[6]

[4] "Mit Cloud Logistics verfolgen wir den Grundgedanken, von fixen Logistikkosten wegzukommen und stattdessen mit transaktionsbasierten, variablen Logistikkosten zu arbeiten."

[5] "Cloud Logistics orientiert sich am Konzept des 'Cloud Computing' (IT-Sektor) und überträgt dieses in den Bereich der Logistiksysteme."

[6] "Zur Herausarbeitung von Cloud Logistics eignet sich die Cloud-Perspektive. Diese liefert bereits die Grundkonzepte des Cloud Computing, die sodann auf Eignung und Relevanz für die Logistik untersucht werden. Aufgrund der größeren Nähe zur Logistik als Geschäftsfeld entsteht mit Logistics-as-a-Service (LaaS) ein neuer Cloud-Diensttyp. Solche Dienste greifen über standardisierte Softwareschnittstellen auf Logistiktätigkeiten zu – also die raum-zeitlichen Gütertransformationen, beispielsweise Transport, Umschlag, Lagerung, sowie die damit verbundenen Managementtätigkeiten (Planung, Steuerung)."

Teichmann (2014:59)*	"Cloud logistics abstracts from 'Software as a Service (SaaS)' towards 'Everything as a Service (XaaS) and tries to adopt features of cloud computing for logistics"[7]
Ludwig (2014:49)*	"The adoption of the cloud paradigm in the logistics domain will be denoted as 'Cloud Logistics'. (Delfmann & Jaekel 2012)"[8]
Ehrenberg & Ludwig (2014:7)*	"Cloud logistics virtualizes logistics resources and processes (e.g. cargo, transportation-, handling-, storage-, loading-, and quality control processes) and thereby allows for a flexible, interorganizational design, control, and monitoring of dynamic business processes."[9]
Ballardt (2014:19)*	"Cloud logistics adopts the cloud computing concept from IT to logistics in order to aggregate autonomous logistics units more efficiently into a superordinate system (Delfmann & Jaekel 2012)."[10]
Wang et al. (2015a:71)	"The Internet of things provided logistics information source for cloud computing platforms through the perception of logistics facilities and all kinds of information resources. Cloud logistics service mode virtualized the logistics resources to form cloud services. Finally, it provide modern logistics service mode by logistics resources information sharing of cloud services, cloud service discovery, service resource combination and coordination."

Table 18.: Overview of Cloud Logistics Definitions; * = translated by the author

Analyzing these definitions reveals the shared understanding that cloud logistics refers to logistics systems that are – in some way – conceptually similar to cloud computing. Surprisingly, however, the majority of definitions define cloud logistics by focusing on the *process* of how CLSs can be created rather than on the *outcome* of this process. Consequently, the understanding of CLSs remains obscured and vague. Still, some definitions provide insights into the potential outcome of this design process and thus into the nature of CLSs.

For example, Wang et al. (2012b) use a broad, cluttered set of adjectives to characterize CLSs: such systems provide on-demand and affordable delivery of virtualized and service-oriented logistics resources and capabilities. Kersten et al. (2012) focus on the institutional composition of CLSs and describe them as cooperations among LSPs that collectively

[7] "Cloud Logistics abstrahiert von 'Software as a Service (SaaS)' hin zu 'Everything as a Service (XaaS)' und versucht Merkmale von Cloud Computing in die Logistik zu übertragen."

[8] "Die Übertragung des Cloud-Prinzips auf die Logistik wird als 'Cloud Logistics' bezeichnet. (Delfmann & Jaekel 2012)"

[9] "Cloud Logistics virtualisiert logistische Ressourcen und Abläufe (z. B. Frachtgut, Transport-/Umschlags-/Lagermittel, Beladungs- und Qualitätsprüfungsvorgänge) und erlaubt damit eine flexible, unternehmensübergreifende Gestaltung, Steuerung und Kontrolle von dynamischen Geschäftsprozessen."

[10] "überträgt es [cloud logistics] den 'Cloud-Computing-Gedanken aus der IT auf die Logistik, um autonome logistische Einheiten effizienter in einem übergreifenden System zusammenzufassen' (Delfmann & Jaekel 2012)"

aim for customer satisfaction and on-demand service delivery. Mosavi (2012) argues that such systems denote a new logistics outsourcing model that follows the pay-per-use principle, and Leukel & Scheuermann (2014) mention that physical logistics activities can be accessed through standardized software interfaces. Ehrenberg & Ludwig (2014) argue that, because of cloud logistics, dynamic business processes can be designed, controlled, and monitored in a flexible, interorganizational manner. Ballardt (2014) states that CLSs are a superordinate system to efficiently aggregate autonomous logistics units. Finally, Wang et al. (2015a) argue that CLSs comprise virtualized logistics resources. In sum, considering the variety of definitions and aspects considered therein, one can conclude that a formal definition is still missing (Wang et al. 2012b:558; Delfmann & Jaekel 2012:17; Pieringer 2012:45). In spite of these differences, these definitions nevertheless agree on the fact that "cloud logistics" is "more than cloud computing in logistics systems" (Leukel & Scheuermann 2014:38), which clearly differentiates the fourth meaning from the second.

In the following section, we structure cloud logistics knowledge in its fourth meaning according to elements of the cloud paradigm (Delfmann & Jaekel 2012:15ff.; also see Leukel & Scheuermann 2014): (1) design principles and concepts that underlie cloud computing, (2) essential characteristics of cloud logistics' services, (3) cloud logistics service models, and (4) cloud logistics deployment models. Note that the essential characteristics, service models, and deployment models are adopted from the NIST definition of cloud computing.

5.2. Underlying Design Principles and Concepts: Virtualization, Service-orientation, and Internet of Things

The underlying design principles and concepts are the first element of the cloud paradigm. Our analysis of cloud logistics sources identified three underlying design principles and concepts: virtualization, service-orientation, and IoT. While virtualization and service-orientation are principles and concepts that underlie cloud computing, IoT is added as a concept for synchronizing the virtual world of computers with the physical world of tangible logistics resources and capabilities.

Service-orientation and virtualization are identified in many of the sources. For example, Wang et al. (2012b:558) argue that, in cloud logistics, "all kinds of logistics resources and capabilities are virtualized and service-oriented." Ehrenberg & Ludwig (2014:7) argue that cloud logistics extends the concept of virtualization from computing resources and capabilities to logistics resources and capabilities. Delfmann & Jaekel (2012:17) consider the feasibility of abstracting, virtualizing, and encapsulating logistics resources and operational capabilities into services to be a necessary precondition for cloud logistics. Hu & Zhang (2015:128f.) state that cloud logistics services result from the process of resource virtualization and service-orientation. Similar statements can also be found in several other sources (see e.g. Li et al. 2013:1697; Wang et al. 2015b:322; Leukel & Scheuermann 2014; Sun et al. 2015a:111 (only abstraction)); Ballardt 2014:17; Zhang et al. 2014; Colajanni 2012:4 (service engineering)).

Although many sources identify virtualization and service-orientation as underlying design principles of cloud logistics, surprisingly little is known about virtualized and service-oriented logistics resources and capabilities. Hu & Zhang (2015:128f.) and Wang et al. (2015b:322) specify basic process steps for virtualization and service-orientation without providing any details on the specific activities in these steps. Li et al. (2013:1697, 1702) also describe the process of creating virtualized and service-oriented logistics resources and capabilities, yet they provide a concrete example: They express a logistics resource from a steel logistics center using the Extensible Markup Language (XML) schema and describe the service's interface using a common standard for web service descriptions (WSDL). Despite providing a concrete example, Li et al. (2013:1696) argue that the virtualization of logistics resources and capabilities is more complex than of computing resources because such resources and capabilities exhibit a higher degree of variability, geographical distribution, heterogeneity, and morphological diversity. Delfmann & Jaekel (2012:18) point out that logistics capabilities cannot be delivered over the internet, but are always connected to physical flows in physical networks, which makes "true virtualization" impossible. They furthermore point out that the different types of logistics resources and capabilities differ in their complexity of virtualization and service-orientation (Delfmann & Jaekel 2012:17). We thus conclude that there is consensus about adopting virtualization and service-orientation cloud computing and applying them to logistics, but little practical knowledge of *how* these principles are applied and *what* the outcome might be.

In addition, to virtualization and service-orientation, many of the sources identify the IoT as a design concept that underlies cloud logistics (Wang et al. 2012b:558f.; Delfmann & Jaekel 2012:19; Li et al. 2013:1696ff.; Li et al. 2014b:75f.; Teichmann 2014:57; Hu & Zhang 2015:128f.; Sun et al. 2015a:112; Wang et al. 2015a:69). By means of several technologies (e.g. RFID), the IoT synchronizes the physical and virtual world by bridging the "last-mile" gap between the worlds (see Section B.V.3). The IoT allows physical logistics resources and capabilities to be connected to the cloud logistics system (Li et al. 2013:1698). Furthermore, because of this synchronization capability, some sources associate the IoT with resource virtualization. Wang et al. (2012b:561) and Li et al. (2014b:75) both position virtualization and IoT within the same layer of their cloud logistics architecture – the layer that connects the physical with the virtual world, referred to as the "perception layer" and "virtual sensing layer" respectively. Teichmann (2014:57) argues that the IoT represents the "technical basis" for logistics virtualization.

In conclusion, sources agree on the set of design principles and concepts that underlie cloud logistics: virtualization, service-orientation, and IoT. In spite of this consensus, their application and their interrelation in logistics system design partly remains vague. While the IoT is a well investigated topic in logistics research, virtualization and service-orientation are new to logistics system design and require further research attention. Additional research is also needed to better understand the interplay and relationship between these three design principles and concepts.

5.3. Essential Characteristics of Cloud Logistics Services

The essential characteristics of cloud logistics services are the second element of the cloud paradigm. Cloud computing services feature five essential characteristics (see Section D.II.2). Given that, cloud logistics adopts virtualization and service-orientation from cloud computing, one may reasonably ask whether cloud logistics services also feature the same characteristics. Furthermore, one may ask how these characteristics can be achieved and why they can be desirable in a logistics context.

To investigate these questions, we initially mapped all characteristics used to specify the service provisioning in cloud logistics to the essential characteristics of cloud computing, depicted in Table 19. Note that we separated the "on-demand and self-service" characteristic into two because sources addressed on-demand availability and self-service independent of each other. Table 20 provides an overview of the characteristics identified by which source.

Essential Cloud Characteristics	Identified Characteristics of Cloud Logistics
On-demand	- on-demand (Delfmann & Jaekel 2012) - focus on service lead time (Kersten et al. 2012) - readily available, on-demand (Wang et al. 2012b) - quick realizability (Teichmann 2014) - ad-hoc storage (Cloud Pallet Storage 2015) - establish service in very short period of time (Hu & Zhang 2015)
Self-service	- self-service (not feasible as human interaction between customer and LSPs necessary in handover of goods) (Delfmann & Jaekel 2012) - self-service for online contract closure, not for physical flow of goods (Ballardt 2014)
Broad Network Access	- broad network access (not applicable to physical flows, but only to access to cloud-based platform via internet) (Delfmann & Jaekel 2012) - platform accessible via a variety of user terminals (Wang et al. 2012b)

	- web portal access, intelligent terminal access, and PDA access to cloud-based platform (Li et al. 2014b) - cloud-based platform accessible via various devices, incl. PC, mobile phone, thin client, other terminals (Sun et al. 2015a)
Resource Pooling	- resource pooling (Delfmann & Jaekel 2012) - shared pool of logistics skills and resources (Kersten et al. 2012) - physical resource pool, affordable (Wang et al. 2012b) - resource sharing (Li et al. 2013) - resource pooling (Leukel & Scheuermann 2014) - logistics resource sharing (Li et al. 2014b) - resource pooling, reduced costs (Teichmann 2014) - sharing unused transportation resources (Tummel et al. 2014) - resource sharing (Zhang et al. 2014) - shared warehouse/yard (Cloud Pallet Storage 2015) - resources and capabilities sharing, efficient, low costs (Hu & Zhang 2015) - sharing of basic logistics resources, cloud service resource pool (Sun et al. 2015a)
Rapid Elasticity	- scalability, elasticity (Delfmann & Jaekel 2012) - highly scalable, highly reliable (Wang et al. 2012b) - scalability and flexibility of cooperation between resources (Li et al. 2013) - structural elasticity (Ballardt 2014) - reconfigurability and scalability (Leukel & Scheuermann 2014) - scalability, flexibility (Ludwig 2014) - scalability, elasticity (Teichmann 2014) - scaling resources locally (Tummel et al. 2014) - size matters, therefore large warehouse (Cloud Pallet Storage 2015) - reliable, flexible (Hu & Zhang 2015)
Measured Service	- pay-per-use based on SLA (Delfmann & Jaekel 2012) - "no transactions, no costs" based on SLA (Mosavi 2012; akw/jö 2011) - pay-per-use model (Li et al. 2013) - pay-per-use and SLAs (Leukel & Scheuermann 2014) - pay-per-use (variable instead of fixed costs) (Teichmann 2014) - pay as you store (Cloud Pallet Storage 2015)

Table 19.: Identified Characteristics in Cloud Logistics

A few characteristics used to describe cloud logistics services did not match any of the cloud characteristics and, hence, could not be mapped. Characteristics not mapped include personalized, secure (Zhang et al. 2014), distribute risks, flexible supply chains, ensure logistics quality (Pieringer 2012:44), safe, personalized, high quality (Hu & Zhang 2015:128), networked, intelligent, safe, high quality (Wang et al. 2012b:558f.), and robustness (Delfmann & Jaekel 2012:20; Teichmann 2014:65).

Essential Cloud Characteristics	Delfmann & Jaekel 2012	Kersten et al. 2012	Mosavi 2012; akw/jö 2011	Wang et al. 2012b	Li et al. 2013	Ballardt 2014	Leukel & Scheuermann 2014	Li et al. 2014b	Ludwig 2014	Teichmann 2014	Tummel et al. 2014	Zhang et al. 2014	Cloud Pallet Storage 2015	Hu & Zhang 2015	Sun et al. 2015a
On-demand	x	x		x						x			x	x	
Self-service	(x)					(x)									
Broad Network Access	(x)			(x)				(x)							(x)
Resource Pooling	x	x		x	x		x	x		x	x	x	x	x	x
Rapid Elasticity	x			x	x	x	x		x	x	x		x	x	
Measured Service	x		x		x		x			x			x		

Table 20.: Identified Cloud Characteristics by Source (Legend: x = identified; (x) = identified for cloud-based platform, not for physical flow of goods)

Analyzing both tables and comparing the ratio between the sources mapped and not mapped clearly suggests that cloud logistics services are associated with the same essential characteristics as cloud computing services. Yet, not all characteristics are associated with physical cloud logistics services. Self-service and broad network access pertain to computing services delivered via the cloud-based platform only (see Delfmann & Jaekel 2012; Wang et al. 2012b; Ballardt 2014; Li et al. 2014b; Sun et al. 2015a). Customers can use the platform to initiate business relations, design logistics services, or conclude contracts in a self-service manner (Delfmann & Jaekel 2012:18). Conversely, physical logistics services cannot be provisioned in a self-service manner, as they always require some form of human-to-human interaction when handing over goods from a customer to an LSP or among LSPs (Delfmann & Jaekel 2012:18; Ballardt 2014). As for broad network access, customers and LSPs can access the cloud-based platform using a wide variety of devices from any location connected to the internet (Wang et al. 2012b; Li et al. 2014b; Sun et al. 2015a). Conversely, physical cloud logistics services cannot be provided through broad network access, which would mean that cloud logistics services are available in any location in the world. According to current knowledge, physical cloud logistics services can be provisioned in an on-demand, rapidly elastic, and pay-per-use way from a shared pool of logistics resources and capabilities (resource pooling).

Two examples found in the sources are briefly presented because they vividly show how cloud characteristics are associated with physical cloud logistics services. First, as described above, Mosavi justifies pay-per-use by arguing that his company does not require own warehouses because it utilizes cloud logistics and aims to move toward transaction-based logistics costs by using cloud logistics (akw/jö 2011:45; Mosavi 2012:16f.). Second, Cloud Logistics Limited, a UK-based LSP, provides an intuitive example for on-demand and pay-per-use storage (Cloud Pallet Storage 2015). "Cloud Pallet Storage" is a large, secured warehouse located in the vicinity of London with excellent transportation links to all southern UK ports, major traffic routes, and London. Customers can store palletized goods in an ad-hoc manner for short and long periods of time, and they "only pay when [they] store." More specifically, customers are charged according to a "pallet rate," which includes, *inter alia*, the warehouse lease, lifting equipment, staff overhead costs, racking, and internal infrastructure. Cloud Pallet Storage furthermore offers picking services on the box/bag level and can arrange delivery and collection of goods in and around London.

Although many of the sources associate physical cloud logistics services with one or more essential cloud characteristics, the majority of sources merely mention the essential cloud characteristics, usually without elaborating on their specific meaning or how they can be achieved. Still, a few sources provide deeper insights regarding resource pooling, rapid elasticity, and pay-per-use. Leukel & Scheuermann (2014:42) point out that resource pooling requires the demand of different logistics customers to align in space. Li et al. (2013:1697) argue that virtualization is an important driver for resource sharing (Wang et al. 2012b:562). Furthermore, the type and extent of resource pooling depends on the specific deployment model chosen (see next subsection) (Delfmann & Jaekel 2012; Leukel & Scheuermann 2014). With regard to rapid elasticity, Leukel & Scheuermann (2014:41) argue that logistics' scalability and elasticity depends on the type of logistics resources, and logistics' scalability and elasticity is lower compared to cloud computing because they also depend on temporal, spatial, organizational, and technical logistics conditions (Leukel et al. 2011a:17; also see Delfmann & Jaekel 2012:18; Teichmann 2014:57). With regard to pay-per-use, Delfmann & Jaekel (2012:19) also point out that the precise measurement of physical logistics processes is a prerequisite for enabling pay-per-use, and precise measurement can be achieved through the IoT. In conclusion, existing cloud logistics knowledge provides few insights in the essential characteristics of cloud logistics – even though they are central to the cloud paradigm.

Although it is not entirely clear how essential cloud characteristics can be achieved, they appear to be desirable for better handling of adverse environmental conditions, especially in a competitive environment (see Section B.V.2). In fact, the sources seem to associate cloud logistics with the essential cloud characteristics in order to motivate the emergence of cloud logistics as a promising approach and response to better cope with these conditions. Delfmann & Jaekel (2012:21, 12) conceive of cloud logistics as a departure from contemporary logistics design principles (e.g. efficiency, interfirm supply chain integration) and therefore as a suitable concept for tackling logistics challenges that arise from megatrends, structural breaks, and other crises that manifest in increasingly complex, uncertain, volatile, and, thus, less predictable logistics environments (also cited by Pieringer 2012:44f.; Ballardt 2014:19; Ludwig 2014:49; Teichmann 2014:60). Kersten et al. (2012:256f.) consider cloud logistics a risk management strategy for coping with supply chain demand risks that result from shortening product life cycles, the global dispersion of supply chains, new information technology, and increasingly individualized demand, as typically encountered in today's dynamic and volatile business environments. Wang et al. (2012b:558) see cloud logistics as a response to growing demand for personalized logistics services. Li et al. (2013:1696) argue that the cloud logistics service mode was nurtured by "dynamically arising logistics demands" in conjunction with the emergence of cloud computing, which lets demand be satisfied for individual and complex computing services. Ludwig (2014:47ff.) positions cloud logistics as a basis to better respond to customer- and industry-specific logistics requirements in the 4PL segment. Mosavi (2012:16) motivates the use of cloud logistics from increasing margin pressures in online food retailing. In sum, the common denominator for cloud logistics seems to be its potential to overcome logistics challenges.

In conclusion, cloud logistics not only adopts the design principles that underlie cloud computing (virtualization and service-orientation), but also uses essential cloud characteristics in logistics service provisioning. However, it is still unknown whether and the extent to which cloud logistics can in fact attain these characteristics and whether it represents a viable solution for overcoming logistics challenges.

5.4. Deployment Models: Predominantly Networked Horizontal LSP Cooperation (Hybrid Deployment Model)

Deployment models are the third element of the cloud paradigm. Following the NIST definition of cloud computing, deployment models specify which organizations or users are entitled to access the shared pool of resources and capabilities and which organizations provide and manage this pool (see Section D.II.4). Thus, deployment models define the institutional composition of cloud computing systems, and they define the basic nature of interorganizational relationships between involved actors.

Cloud logistics sources comprise heterogeneous conceptions regarding the deployment models of cloud logistics. Deployment models can be structured into those derived from cloud computing and those derived from SOA. Both associate cloud logistics with some form of horizontal cooperation among several LSPs, hereafter referred to as *networked horizontal LSP cooperation*. In addition to describing these deployment models, the sources also provide reasons for why they conceive of horizontal LSP cooperation to emerge in cloud logistics. Furthermore, they describe several mechanisms that govern the resource allocation in these horizontal cooperations. In other words, existing cloud logistics knowledge provides insights into both the governance structure and the organizational structure of CLSs.

Based on these considerations, we consolidate existing cloud logistics knowledge regarding deployment models into (1) deployment models derived from cloud computing, (2) deployment models derived from SOA, (3) reasons for the formation of networked horizontal LSP cooperations in cloud logistics, and (4) resource allocation mechanisms.

5.4.1. Deployment Models Derived from Cloud Computing

Several cloud logistics sources identify private, public, and hybrid deployment models. However, the specific definitions of these models vary among sources.

Using the NIST definition of cloud computing deployment models, Leukel & Scheuermann (2014:41f.) define a private logistics cloud as the exclusive allocation of a single LSP's logistics resources and capabilities to a single customer, for example a warehouse outsourced to a LSP dedicated to a single customer. A public cloud refers to a single LSP sharing its logistics resources and capabilities across multiple customers. A hybrid logistics cloud is a mixed deployment model that consists of resources and capabilities that are for the exclusive use of a single customer and resources and capabilities that are shared across multiple customers.

Delfmann & Jaekel (2012:18ff.) describe three pooling scenarios that are also related to the NIST cloud computing deployment models (also see Teichmann 2014:58ff.). The first is related to a public cloud, as a single LSP operates a pool of resources that is dynamically shared across multiple customers. The second scenario is related to a hybrid model. Multiple LSPs cooperate horizontally by sharing (pooling) their logistics resources and capabilities to satisfy the demand of a combined customer base or customers of a third party. The third scenario is not related to any cloud computing deployment model, but refers to a situation in which arbitrary companies of arbitrary industries share resources directly connected to logistics resources, for example inventory in warehouses. This model relates to the third meaning of cloud logistics (see Section E.IV.4). Teichmann (2014:58ff.) points out that any of these three scenarios require firms to cooperate, at least indirectly. More generally, Teichmann (2014:59, 61f.) argues that cloud logistics is not "conceivable" without interfirm cooperation, and he believes cloud logistics enables new platform and market-based collaborations between LSPs (see Subsection E.IV.3.5).

Kersten et al. (2012:259, 261) describe a scenario that resembles a hybrid model although they do not use any of the cloud computing terminology: cloud logistics is "a system of autonomous and interconnected logistics service providers focused on the common overall objective of customer satisfaction and service lead time," and LSPs cooperate "even though they may be direct competitors or substitute providers."

Li et al. (2014b) and Li et al. (2014a) provide a very different interpretation of public, private, and hybrid deployment. They argue that these terms basically pertain to the entity that initially pays for the development of a cloud-based (information and management) platform and subsequently operates it. "Public" refers to the government, private to a large LSP with sufficient financial resources, and hybrid to a cooperation between the government and an LSP, with the LSP operating the platform and leading the cooperation while being guided by the government and supported by government funds. Given that these sources are affiliated with Chinese universities, this interpretation of the deployment models seems to derive from the structure of the Chinese economy.

5.4.2. Deployment Models Derived from SOA

Instead of deriving the deployment model of cloud logistics from cloud computing, several other sources start from SOAs. These sources suggest that cloud logistics adopts not only service-oriented design for logistics resources and capabilities (see Subsection E.IV.5.2), but also the constituent roles of an SOA: providers, customers, and a service registry operated by a platform provider (see Subsection D.III.3.1). This triad of actors is identified in several cloud logistics sources (Wang et al. 2012b; Hu & Zhang 2015; Sun et al. 2015a:111; Li et al. 2014b:75). In fact, two of these sources utilize the "SOA triangle" (see Figure 17) to describe the actor structure and the "operating principle of cloud logistics," as depicted in Figure 25. Cloud logistics thus operates in a similar way as an SOA. In other words, a platform provider hosts a service registry on a cloud-based platform. Then, LSPs publish their logistics capabilities as services to the registry, and customers query this registry to discover services to address their logistical needs. They can create customized logistics solutions by composing multiple services, which is why Wang et al. (2012b:558, capitals original) conceive of this platform as a "One-stop Service Platform." Finally, LSPs provide the desired services directly to the customer.

Adopting the SOA triangle as a basis for cloud logistics has two interesting implications. First, interpreting and designing logistics systems through the lens of the cloud paradigm involves a cloud-based platform, which interconnects resource providers and users as described in the second meaning of cloud logistics (see Section E.IV.3). Hence, the fourth and second meanings of cloud logistics are related with regard to the use of a common cloud-based platform. More specifically, Wang et al. (2012b:559) position "knowledge" as a central element in the SOA triangle. According to them, knowledge is used to manage logistics processes efficiently. Hence, one can argue that the cloud-based platform

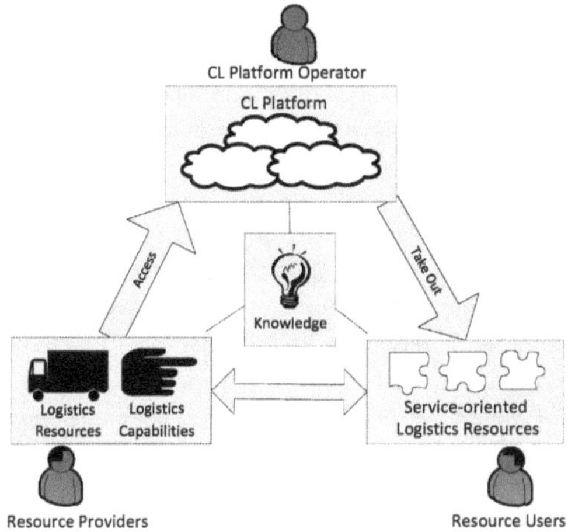

Figure 25.: The Operating Principle of Cloud Logistics
(reprinted from Wang et al. 2012b:559; see similar Hu & Zhang 2015:128)

within this meaning resembles the second generic use case of cloud-based platforms in logistics systems as it enables to collect, share, and synchronize information (see Subsection E.IV.3.3).

Second, CLSs involve an intermediary actor between resource providers and users: the platform provider. However, current cloud logistics sources provide no clear insights about the specific responsibilities of this actor. Sources emphasize the (legal) autonomy of the platform operator. Colajanni (2012:4) considers the platform provider an "independent" facilitator in supply chains. Delfmann & Jaekel (2012:20) call for evaluating the need for a "neutral" platform provider that coordinates different actors in the system or enforces common policies regarding data interfaces, service standards, performance measurement, and data privacy. Li et al. (2014b:76) note that the platform operator can be a professional company.

The sources conceive of the cloud-based platform as an enabler for and nexus of a network horizontal cooperation between LSPs, which is similar to the role of the cloud-based platform in the second meaning of cloud logistics. Wang et al. (2012b:559) associate cloud logistics with "multi-agent collaborative interaction," where cooperation is achieved through a cloud-based platform that enables "different forms of collaborative approaches" and "many-to-many" service provisioning. Hu & Zhang (2015:129f.) argue that LSPs cooperate in order to fulfill "complex logistics tasks," which represent compositions of two or more services in the sense of service-oriented design. Different services in a composition are provided by different LSPs: "It is easy to find that logistics cloud service combination involves many logistics service provider" (Hu & Zhang 2015:130). Figure 26 depicts this SOA-like cooperation in CLSs. Similarly, Wang et al. (2015a:69f.) argue that logistics processes are complex and therefore should be represented through a composition of services, with different service modules performed by different LSPs. Hence, following the SOC idea, complex logistics processes are performed by multiple LSPs collaboratively.

In addition to associating cloud logistics with some form of horizontal networked cooperation, a few sources further characterize this cooperation as dynamic, which means that the composition of cooperating LSPs changes over time. Wang et al. (2012b) conceive of the cloud-based platform as "highly open," with LSPs free to join, regardless of firm size and geographic location. Other sources similarly describe networked LSP cooperation as "strong[ly] dynamic" (Wang et al. 2015a:77) and as a "dynamic network collaboration" (Wang et al. 2015b:325). Kersten et al. (2012:260) argue that "behavior rules determin[e] the emergent network structure of the logistics service providers." In sum, these sources imply that the composition of cooperating LSPs is not stable, but changes over time.

In addition to using SOA as a basis for networked horizontal cooperation between LSPs, SOA can also be a basis for a 4PL model, as shown in two of the sources. Ludwig (2014:47ff.) considers the use of virtualized and service-oriented logistics resources and capabilities which are integrated on a central cloud-based platform, to be conducive to integrating and coordinating a large number of sub-LSPs into customized end-to-end logistics processes, as typically performed by 4PLs. Similarly, Leukel & Scheuermann (2014:42) point out that a 4PL can integrate and coordinate the logistics services of other LSPs into complex logistics services.

In conclusion, several sources conceive of the actor structure of cloud logistics as equal to the actor structure of SOA. This conception introduces not only a cloud-based platform that interconnects resource providers and users, but also a new intermediary actor: the

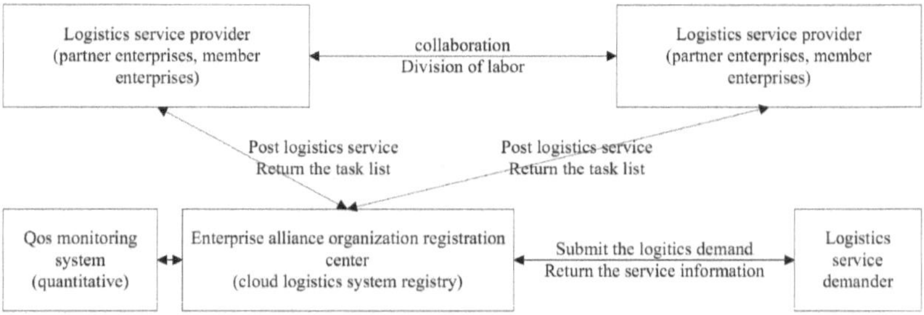

Figure 26.: Logistics Cloud Service Cooperation Architecture (reprinted from Hu & Zhang 2015:129)

platform provider. The majority of sources conceive of this platform as enabling (dynamic) networked horizontal cooperation among LSPs. A few sources conceive of it as enabling a 4PL to integrate services from multiple other LSPs. In either case, the current sources provide very few insights about the concrete functionality of this platform, the concrete responsibilities of the platform operator, and the autonomy of the platform provider. Future research should focus on these aspects as this would help to understand the operating logic of CLSs, given the centrality of the platform and its operator.

5.4.3. Reasons for the Formation of Horizontal Cooperation between LSPs in Cloud Logistics

Given that many of the sources consider cloud logistics to involve some form of networked horizontal cooperation between LSPs, we naturally encounter the question of the reasons that sources provide to explain the formation of this cooperation and whether the reasons differ among sources.

On one hand, the sources list several very general factors that are conducive to the emergence of interfirm cooperation, such as increasing market pressure (Tummel et al. 2014; Hu & Zhang 2015:129; Wang et al. 2015a:76; Wang et al. 2015b:325), rising customer requirements (Ludwig 2014:47; Hu & Zhang 2015:127; Arnold 2014b:14), achieving competitive advantage, or an IT platform's ability to better share information and coordinate with partners (Teichmann 2014:58f.; Arnold 2014b:14; Hu & Zhang 2015:130). These factors generally align with the motives and antecedents described for interfirm networks in general and in logistics (see Section C.II.2; Section C.VI.2)

On the other hand, several sources position on-demand and rapid elasticity in cloud logistics service provisioning as the primary factor driving interfirm cooperation. Teichmann (2014:58) argues that cooperation between LSPs is necessary to achieve scalability and elasticity in the provisioning of physical logistics services because cooperation increases the size of the resource pool, which, in turn, increases scalability and elasticity. Hu & Zhang (2015:127) argue that a single LSP often has insufficient resources to fulfill logistics requirements, which is why LSPs aim to gain access to complementary resources through cooperation. Leukel & Scheuermann (2014:41) point out that cloud systems comprise "large pools" of equal or similar resources and capabilities, and they are managed in order to satisfy any incoming customer request. However, if the resource pool proves to be insufficient to satisfy incoming requests, third parties can be contracted. Cloud Pallet Storage (2015) argue: "Size does matter that's why Cloud has a large secure warehouse" to offer ad-hoc storage. In conclusion, all of these sources (either implicitly or explicitly) suggest a link between the size of the logistics resource and capability pool and the ability to achieve on-demand and rapidly elastic service provisioning. Horizontal cooperation between LSPs can increase the size of the disposable resource and capability pool and, thus, support on-demand availability and rapid elasticity. Finally, computer scientists explain the formation of and need for hybrid computing clouds using the same rationale: Resource pool size is critical for achieving on-demand availability and rapid elasticity (see Subsection D.II.4.2).

In conclusion, to explain the formation of horizontal cooperation in cloud logistics, the sources give some generic reasons but, and more importantly, also provide specific reasons that relate to the goal of achieving on-demand and rapidly elastic service provisioning. In other words, it seems that the goal of delivering logistics services in an on-demand and rapidly elastic manner drives LSPs to cooperate in cloud logistics, despite being distant or proximate competitors. It is interesting to observe that the argument for achieving on-demand and rapidly elastic provision is also posited by cloud computing scholars to explain the need for and emergence of hybrid deployment models. This corroborates the importance of horizontal cooperation in cloud logistics.

5.4.4. Resource Allocation Mechanisms in Cloud Logistics

A critical question in every economic system and form of cooperation pertains to the allocation logic of resources, revenue, and costs. This, of course, applies to cloud logistics as well (see Hu & Zhang 2015:127ff.). Analyzing the sources reveals different ideas about resource allocation logic in cloud logistics; these ideas primarily diverge regarding the degree of centralization, which ranges from *centralized* to *decentralized* (market-based). When centralized, the cloud-based platform in the cloud logistics system uses some kind of algorithm or (intelligent) optimization technique to determine the allocation of resources, revenues, costs, risks etc. among LSPs in the system. When decentralized, the allocation logic to some extent depends on the choices made or the information provided by the cooperating LSPs.

Several sources either explicitly mention "centralized management" or "optimization" or describe a logic that equals a centralized management approach (Wang et al. 2012b; Li et al. 2013 Hu & Zhang 2015; Tummel et al. 2014:672; Sun et al. 2015a:111).

Other sources propose and describe a more decentralized allocation logic. Kersten et al. (2012:262f.) investigate the performance of a cloud logistics system through an agent based simulation approach, which allows the implementation of distributed control in systems. LSPs are modeled as agents, which are autonomous and goal-oriented entities that have some capacity to make decisions according to their logistics resources and capabilities. Delfmann & Jaekel (2012:19) position cloud logistics as based on decentralized self-control mechanisms, including auctions. Teichmann (2014:61) implicitly suggests that cloud logistics uses a decentralized allocation logic by arguing that cloud logistics is based on market-based collaboration. Arnold (2014a:26) proposes that the cloud-based platform should centrally match supply and demand based on the conditions and constraints of services and then notify both consumers and LSPs about the match to let them negotiate prices in a decentralized manner. In a review of resource allocation methods in cloud service, Wang et al. (2015a:75) mention several decentralized allocation mechanisms, including combinatorial auctions and multi-agent negotiation mechanisms (also see Wang et al. 2015b:324f.).

Considering current cloud logistics knowledge, one can conclude that current conceptions are very heterogeneous and provide few substantial insights into the allocation of resources, revenue, and costs in cloud logistics. Given the general importance of this topic, further research is critical for developing the concept of cloud logistics. In fact, Wang et al. (2015b:326) argue that future research on cloud logistics should develop a "cloud logistics service collaboration mechanism" (Wang et al. 2015a:78).

5.5. Service Models: Logistics-as-a-Service (LaaS) as Hypernym

Service models are the last element of the cloud paradigm. In cloud computing, capabilities of a similar category are grouped into the same service model; for example, storage and processing are grouped into IaaS. If we consider logistics systems through the lens of the cloud paradigm, we equally run into the question of what groups of logistics capabilities can be offered by CLSs and how they can be grouped.

Within cloud logistics research, Delfmann & Jaekel (2012:19f.) point out that no comprehensive service model has been proposed for cloud logistics yet. But one needs to developed (Leukel & Scheuermann 2014:40f.). Considering the current sources, it appears that, LaaS has evolved as a hypernym for all logistics services provisioned by a cloud logistics system (see Colajanni 2012:4; Leukel & Scheuermann 2014:39ff.; Ludwig 2014:50). Starting with LaaS, Leukel & Scheuermann (2014:40f.) hypothesize that more specific cloud logistics service models are likely to align with the terminology of typical logistics services. Thus, LaaS may be further categorized into transport, handling, warehousing, and value-added-services (see also Teichmann 2014:61). In fact, Wang et al. (2012b:559) already share this conception, as they introduce the "as-a-Service" terminology to specific logistics processes: "Transport-as-a-Service (TaaS)" and "Warehousing-as-a-Service (WaaS)". Thus, the current sources provide a starting point for systematically categorizing cloud logistics services.

In addition to the question of categorization, we encounter another, more fundamental question: What type of logistics services can actually be provisioned in a cloud-like manner? This question, however, does not seem to be addressed adequately in the cloud logistics sources. Wang et al. (2012b) argue that anything that can be virtualized and packaged into a service can become a cloud logistics service (see similar Hu & Zhang 2015:128). Yet, we believe that simply virtualizing and packaging any type of logistics capability into a cloud service will not necessarily make it more readily available to customers, enable resource pooling or pay-per-use, or allow customers to rapidly adjust the capacity according to their demand.

We thus conclude that cloud logistics service models may be structured according to the type of logistics capability, but only if these capabilities can be provisioned in a cloud-like manner. Considering the current knowledge, further research is warranted to understand how cloud logistics capabilities can be structured into service models and to investigate what logistics capabilities can in fact be provisioned in a cloud-like manner.

5.6. Examples of Cloud Logistics Architectures

The previous paragraphs have described the fourth meaning of the term cloud logistics. Based on current cloud logistics knowledge, we have identified the design principles and concepts that underlie cloud logistics, its essential characteristics of service production and provisioning, its deployment models, and its service models. Taken together, these elements provide an initial understanding of how logistics systems can be interpreted and designed through the lens of the cloud paradigm. While many of the sources focus on a specific element only, two partially integrate these elements into more general models of CLSs, referred to as "architectures." Both architectures are briefly presented, as they are instructive for the objectives of this thesis.

Figure 27 depicts the architecture of a cloud logistics system proposed by Wang et al. (2012b). The architecture resembles that of a computer system, for it is composed of several hierarchical layers (see Section D.III.2). The *physical resource layer* comprises all kinds of physical logistics resources and capabilities required for carrying out physical logistics activities. The *perception layer* identifies and senses physical logistics resources and capabilities by means of IoT technologies, and it supports the collection and classification of information as a basis for intelligent resource management. The *virtual resource layer* virtualizes physical resources and capabilities and aggregates them into a virtual resource pool. The *core service layer* packages virtualized resources and capabilities into services, which are published to a service registry and become available to all actors in the cloud system. The *application interface layer* provides interfaces for connecting the platform to other logistics IT systems. The *application layer* provides different logistics software tailored to the specific needs of different types of actors connected to the platform. Actors connected include manufacturers, distributers, transport companies, warehousing companies, and banks.

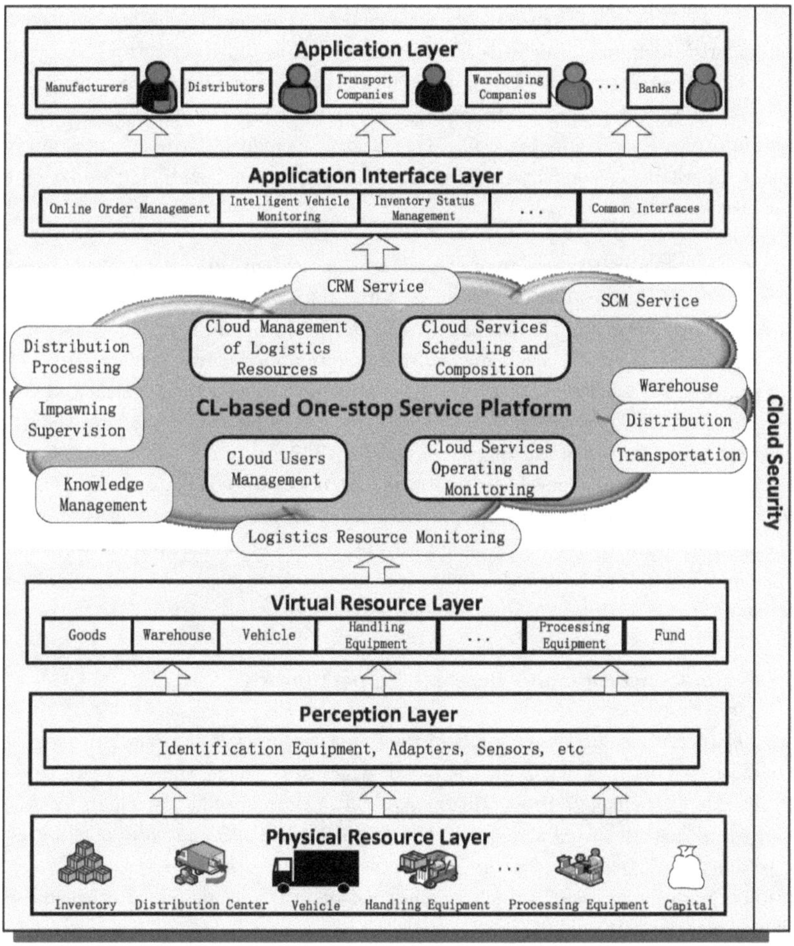

Figure 27.: The Architecture of a CL-based One-stop Service Platform for a Logistics Center (reprinted from Wang et al. 2012b:561)

Figure 28 depicts the architecture of a cloud logistics system proposed by Li et al. (2014b). Again, the cloud logistics architecture is composed of several hierarchical layers, like a computer system. The *virtual sensing layer* virtualizes logistics resources and capacity. It also enables logistics resource perception, recognition, and information collection through the IoT as a basis for intelligent resource management. The *core service layer* includes all services offered, including basic logistics services and value-added services. The *operation mode layer* defines alternative modes in which the cloud-based platform can be run. The *platform service layer* provides interfaces for actors to access and use the platform (e.g. web portal access). Participants using the platform include LSPs (e.g. transportation company, warehousing company), logistics service demanders, and platform operators (e.g. a professionalized company).

Although the two architectures may appear, to some extent, different at first glance, a few important similarities can be identified. First, both architectures extend the typical layered modeling approach for computing systems to non-computing elements, such as the perception layer, virtual sensing layer, and operation mode layer. Second, both architectures introduce a specific architectural layer for synchronizing the physical and virtual worlds via the IoT: "perception layer" and "virtual sensing layer". Hence, they both recognize the need for synchronizing the physical with the virtual world in cloud logistics. Third, both position the cloud-based platform as a nexus for connecting different actors in the cloud logistics system, and this platform hosts a service registry that comprises all services offered in the system. The similarities provide valuable insights for developing a general cloud logistics system model, which is of direct relevance to answering our research objective.

5.7. Interim Conclusion and Research Opportunities

Figure 29 summarizes the key concepts of the fourth meaning of cloud logistics and their interrelations. This meaning positions cloud logistics as "more than logistics software from the cloud" (Leukel & Scheuermann 2014:38). More specifically, cloud logistics is positioned as "logistics from the cloud" (Pieringer 2012:44). Research opportunities related to this meaning can be structured along the elements of the cloud paradigm: (1) the adoption of underlying design principles and concepts, (2) essential cloud characteristics, (3) deployment models, and (4) service models.

Several of the sources identify virtualization, service-orientation, and the IoT as *underlying design principles and concepts* which are to be applied to the logistics domain. Despite consensus on this set of principles and concepts, existing knowledge provides negligible insight into what happens if these are actually applied to logistics systems design. While the IoT is a concept that has received considerable attention over the last years,

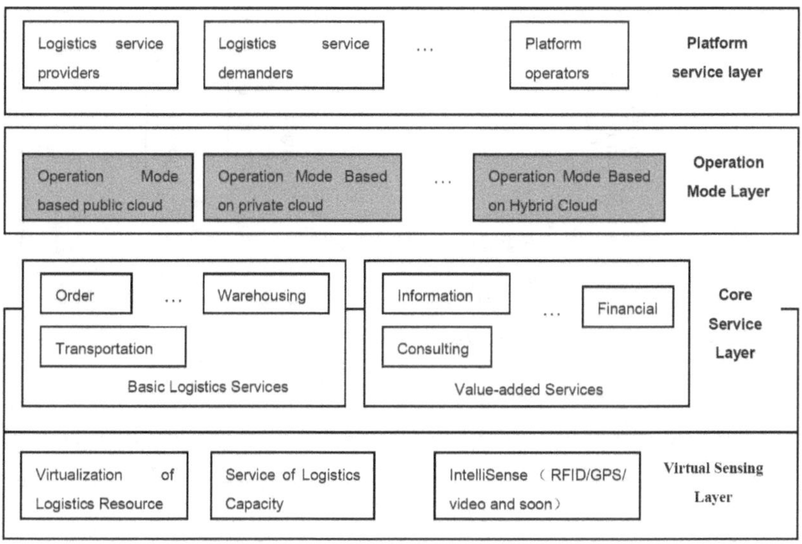

Figure 28.: Regional Logistics Platform Architecture based on Cloud Computing (reprinted from Li et al. 2014b:75)

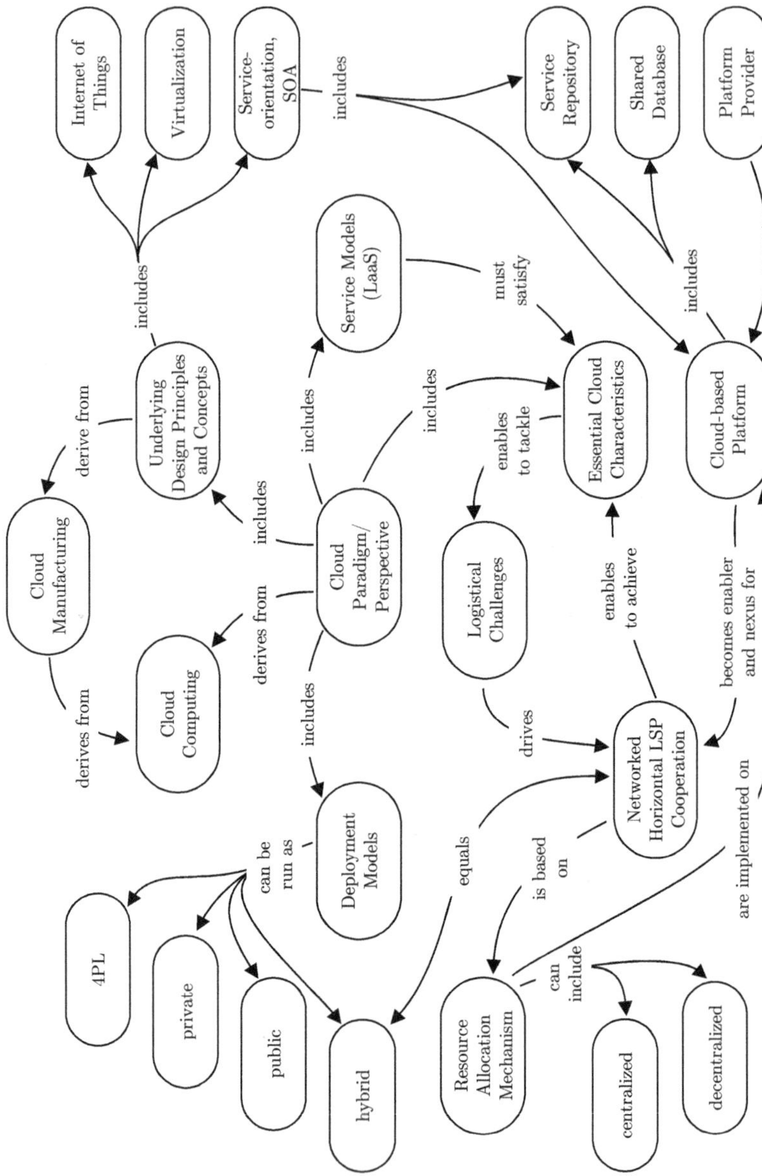

Figure 29.: Summary of Meaning 4: Concepts and their Interrelations

independent of cloud logistics (see Section B.V.3), the application of virtualization and service-orientation to physical logistics capabilities are still in their infancy, also in a more general logistics context (see Subsection D.III.4.4). Moreover, the interaction between virtualization, service-orientation, and the IoT remains opaque. Hence, it seems as if scholars and practitioners agree on what has to be done, but it has not been accomplished so far. The lack of knowledge regarding these principles and concepts is a critical problem, because many scholars define cloud logistics based on the application of these principles and concepts to logistics systems design.

The vast majority of sources associate one or more of the *essential cloud characteristics*, on-demand availability, rapid elasticity, resource pooling, and pay-per-use with physical logistics capabilities. However, existing knowledge falls short of explaining how these characteristics can be achieved. While resource pooling and pay-per-use seem intuitive and may already be largely followed in logistics practice today, on-demand availability and rapid elasticity represent rather "new" logistical requirements and require further attention. Specifically, existing cloud logistics knowledge does not provide any insight into the particularities of how these characteristics may affect the physical infrastructure, the governance and organizational structure of CLSs.

In addition, the majority of sources describe networked horizontal LSP cooperation as the predominant *deployment model* of cloud logistics. In spite of this shared understanding, however, little appears to be known about the governance and organizational structure of this type of cooperation. First, the position on the continuum between markets and hierarchies remains largely unclear. Different institutional arrangements are suggested for structuring exchanges, especially for allocating resources and service requests. Suggestions range from auctions (close to markets) to centralized optimization (close to hierarchy). Furthermore, assuming that the actor structure in cloud logistics follows the "SOA triangle," cloud logistics comprises consumers, LSPs, and a platform provider. While both consumers and LSPs are "well known" logistical actors, little, apart from its "neutrality," is known about the role of the new logistical actor: the platform provider. In particularly, it is unclear how authority is distributed between the platform operator and the network of cooperating LSPs. Besides the governance structure, similarly little is known about any potential administrative arrangements as part of the organizational structure, which are necessary for coordinating the creation and delivery of virtualized, service-oriented physical logistics capabilities with cloud characteristics by a network of cooperating LSPs.

Current cloud logistics knowledge provides some proposals for "labeling" the physical logistics capabilities, which are delivered with cloud characteristics, and for categorizing them into distinct *service models*. The overarching LaaS label is a helpful anchor point. The hypothesis that more specific service models may align with typical logistics processes also appears reasonable. However, the more fundamental problem related to cloud logistics service models is the question of which types of physical logistics capabilities can actually be delivered with cloud characteristics. This is because the goal of achieving resource pooling, on-demand, rapid elasticity, and pay-per-use may likely limit the scope of the capabilities that can be delivered in this manner. Given this, little is known about the service models of cloud logistics. In fact, determining the service models requires first identifying the infrastructural design of cloud logistics and the physical logistics capabilities that can be delivered with cloud characteristics.

Beyond these four rather specific research opportunities, cloud logistics research is furthermore in the need of becoming anchored in existing logistics, organizational, and cloud computing literature. Sources allude to a variety of concepts, principles, and characteristics such as virtualization, service-orientation, the IoT, networked horizontal LSP cooperations, auctions, central optimization, on-demand, resource pooling, rapid elasticity, and pay-per-use without creating a direct reference to existing academic knowledge in these fields and without explaining how they exactly relate to cloud logistics. Thus, cloud logistics is a concept floating like a cloud without being anchored. This hampers research because theories and research fields that can contribute to the design of CLSs remain opaque. Future research therefore needs to anchor the design of CLSs in existing literature.

Considering and synthesizing the research opportunities described above, we propose the following aggregated research opportunity, which we have adopted as research objective of this thesis: *Based on existing cloud logistics knowledge as well as based on existing logistics, organizational, and cloud computing literature, design a logistics system in accordance with the principles and concepts of cloud computing in order to determine if and how logistics systems can deliver physical logistics capabilities with cloud characteristics.* This involves designing both a service model as well as a deployment model for such logistics systems.

6. Conclusions: Assessing Commonalities and Appropriateness of Term Usage

The meaning of the term "cloud logistics" is inevitably ambiguous, as it is the composition of "cloud (computing)" and "logistics" and inevitably inherits and blends their meanings in some way. The preceding content analysis of our review has identified not one, but identified four, distinct, yet related meanings of the term. To conclude the systematic knowledge review, we briefly (1) outline commonalties between these meaning and (2) assess the appropriateness of the term usage.

The meanings are distinct, as each refers to a different phenomenon; they are related because they have some elements in *common.* All revolve around a cloud-based IT system, and they consider this IT system in some logistical context. Meaning 1 takes a logistical perspective on IT, and the other meanings deploy the platform within logistics systems. Furthermore, Meanings 2, 3, and 4 conceive of the cloud-based platform as an enabler and nexus for some form of networked interfirm cooperation within logistics systems. In this regard, Meaning 2 describes the platform in a general way without a specific logistical context, and Meaning 3 describes the cloud-based platform in an e-commerce context. Meaning 4 uses the platform for hosting a service repository and a shared database with logistics-relevant information. The cloud-based platform thus becomes an element for designing logistics systems following the principles and concepts of cloud computing.

Given that the term "cloud logistics" is used to refer to four distinct phenomena, we encounter a terminological question about which of these phenomena represent the direct specific meaning of this term.

Selecting an existing or creating a new term that denotes a specific phenomenon in a specific context is a matter of terminology. Yet, terminology cannot be right or wrong, *per se.* It is rather a question of whether a term is conducive to or appropriate for precisely, and probably also intuitively, representing a specific phenomenon in a specific context. Of course, assessing this is "slippery terrain." Different individuals are likely to have different conceptions and pursue different objectives and, hence, may select different terms to refer to the same phenomenon. Still, in the following paragraphs, we try to build an argument for why *we* believe that using the term "cloud logistics" is only appropriate in two out of the four identified meanings. Specifically, we deem the term "cloud logistics" conducive to referring to the phenomena in Meanings 1 and 4.

Meanings 1 and 4 both blend the meanings of the terms "cloud (computing)" and "logistics" in a reasonable and intuitive manner to describe the respective underlying phenomena. This is because they blend the terms such that the most important aspects associated with one term are imposed on the direct specific meaning of the other term. By taking a logistics perspective on cloud computing systems (Meaning 1), spatio-temporal transformations of data flows are brought to the fore. By taking a cloud perspective on logistics systems (Meaning 4), key underlying design principles, essential characteristics of service delivery, service models, and deployment models of logistics systems are brought to the fore. Hence, Meanings 1 and 4 can be considered terminological twins; they represent two sides of the same terminological coin. In addition, specifically regarding Meaning 4, understanding CLSs as a cloud perspective on logistics systems aligns well with the concept of cloud manufacturing. In fact, combining cloud manufacturing with cloud logistics can lead to the concept of "cloud supply chain" (see "Supply Chain as a Service" Leukel et al.

2011a).

Meaning 2 denotes the use of a cloud-based platform for logistics IT systems as cloud logistics. Using the term "cloud logistics" to refer to this phenomenon is inapt – despite being understandable, as it seems to represent a convenient abbreviation for IT vendors and computer scientists (who have published most of the sources containing this meaning) to promote the diffusion of cloud technology in general and their cloud technology based logistics software specifically. This term usage is inappropriate because, if "cloud logistics" means that cloud computing technology is used in logistics, then a consistent use of the word "cloud" would mean that any context in which cloud computing technology is utilized would be given the prefix "cloud." Consequently, with the diffusion of cloud technology, the word "cloud" would become omnipresent and thus carry little information, thus resulting in many unwieldy new terms such as "cloud smart city logistics" (Nowicka 2014:266). We therefore conclude that – despite its wide usage – "cloud logistics" is unsuitable to denote the use of cloud computing technology for logistics IT systems.

Meaning 3 defines "cloud logistics" as the use of a cloud-based platform to support e-commerce fulfillment through a network of local dealers and LSPs. Using "cloud logistics" to refer to this phenomenon is only partially understandable and also misleading, for two reasons: First, similar as in Meaning 2, the word "cloud" is used as a prefix to imply that cloud computing technology is used in the IT system of a logistics system, specifically in e-commerce fulfillment. Second, inferring the underlying phenomenon from the term "cloud logistics" is largely impossible. This is because the primary resource pool in this meaning neither comprises logistics nor computing resources, but contains finished goods that are used to fulfill orders placed in web shops. We therefore conclude that the term "cloud logistics" is unsuitable for referring to the relevant phenomenon. A more suitable term to describe this innovative logistics model may be "distributed interfirm inventory cloud," as the web shop owner can access a geographically distributed inventory pool owned by a network of local dealers in an on-demand manner via the cloud-based platform.

To summarize, the term "cloud logistics" should be used to refer to either a logistics perspective on cloud computing systems or a cloud perspective on logistics systems.

Part F. Cloud Logistics Systems: Reference Architecture Design

I. Design Objective, Method, and Basis

1. Abstract of Chapter

Fundamental Concepts

Fundamental concepts relevant to our research objective include (1) architectures, (2) reference architectures, (3) (reference) architecture descriptions, (4) stakeholders, and (5) stakeholder concerns.

A *system* can be defined through a quadruple of composition, structure, mechanisms, and environment. A *system architecture* represents a conception of a system's composition, structure, and mechanisms in its environment in our minds (see Subsection F.I.2.1).

A *reference architecture* can be intuitively conceived of as a more abstract or generic architecture: an architecture that does not pertain to single system, but to a system domain (Hassan & Holt 2000:150; see Subsection F.I.2.2). Thus, reference architectures ignore system- and context-specific architectural variations and model only the most significant architectural elements (and their relationships). Reference architectures are commonly derived from a set of existing architectures by extracting and capturing the commonalities and most significant architectural aspects while disregarding system- and context-specific architectural variations (Cloutier et al. 2010:20f.). They also provide guidance for supporting the development of specific system architectures.

Architecture descriptions express the architecture of a system (ISO/IEC/IEEE 42010 2011b:2); reference architecture descriptions express the architecture of a family of systems (see Subsection F.I.2.3). Explicitly expressing architectures particularly aims to document essential system characteristics for potential clients, owners, and operators and to facilitate

F. Jaekel, *Cloud Logistics*, Produktion und Logistik,
https://doi.org/10.1007/978-3-658-22837-8_6

communication among parties involved in development, deployment, and maintenance of a system (ISO/IEC/IEEE 42010 2011b:8f.).

Stakeholders are parties that have an interest in the system under consideration (ISO/IEC/IEEE 42010 2011b:2; see Subsection F.I.2.4). The interests of stakeholders are expressed as *concerns*. A *concern* is defined as "interest in a system relevant to one or more of its stakeholders" (ISO/IEC/IEEE 42010 2011b:2). A concern is connected to any influence on the system that originates in the system's environment, including "developmental, technological, business, operational, organizational, political, economic, legal, regulatory, ecological and social influences" (ISO/IEC/IEEE 42010 2011b:2). Concerns can manifest in many different forms: These include *what* a system does (i.e. functionality) and *how* the system does it (i.e. system qualities) (Lago et al. 2010:22).

Objective: A Reference Architecture Expressed as Architecture Description

This thesis aims to design a logistics system according to the design principles and concepts underlying cloud computing. To this end, this thesis develops a reference architecture, which is formally expressed via an architecture description following the International Standard for Systems and Software Engineering — Architecture description (ISO/IEC/IEEE 42010:2011) (see Section F.I.3). By designing a reference architecture, we specifically aim to synthesize and abstract existing cloud logistics knowledge to derive the elements, their relationships, potential design tradeoffs, and limitations that are common to all CLSs. This approach seems most reasonable to advance the research field, given that knowledge is highly fragmented and spans different academic fields. Hence, by designing a reference architecture, we aim to create an intermediate point of reference for both practitioners and researchers. For the former, it serves as an abstract solution template for implementing a (prototype) CLS. For the latter, it establishes an initial structure of knowledge and system design to guide future research efforts. The reference architecture of CLSs is therefore highly relevant for both scholars and practitioners.

Method: Contingency and Strategic Choice Theory

As this thesis aims to design a reference architecture of CLSs, reviewing the process of how architectures are designed is instructive. To this end, we compare the process models for the design of software architectures and for the structuring of organizations.

The process of designing software architecture is typically referred to as "architecting," and this process is carried out by a "system architect" (see Subsection F.I.4.2). The general design process of software architectures consists of three distinct activities: (a) *analysis*, (b) *synthesis*, and (c) *evaluation* (Hofmeister et al. 2007).

(Structural) contingency theory posits that the structuring of an organization exclusively depends on the manifestation of contextual factors to which the organization is exposed and that there must be some form of "congruence" between the organization's structure and its environment if the organization is to be effective (see e.g. Drazin & van de Ven 1985:514ff.), where "effective" means that an organization "satisfies the interests of one or more constituencies associated with the organization" (Tsui 1990:458). *Strategic choice theory* (SCT) extends contingency theory by introducing *human agency* as another critical factor (besides the environment) that influences organizational structure. More specifically, SCT posits that a "dominant coalition" has authority over decisions important to the organization as a whole, including its structure (Child 1972:1f., 13f.). In fact, choice exercised by the dominant coalition is not limited to the internal strategies adopted by an organization, but also extends to interorganizational relationships (Child 1972). The structure of interorganizational forms depends on the agreed-upon terms of cooperation among the parties involved (Child 1997:58). The crucial question pertains to the degree of (informal) authority that decision makers in one organization can exert over other legally autonomous organizations (Child 1972:9f.).

The process for the dominant coalition to exercise strategic choice consists of three steps: (a) *evaluation of the organization's current situation*, (b) *formulation of the organization's goals*, and (c) *definition of strategies to attain goals*. Depending on choices made, the organization achieves a certain level of effectiveness, which feeds back into the evaluation of the organization's situation, thus establishing a circular structuring process (Child 1997:58). Surprisingly, although software architectures and organizations are very different in nature, their design processes are very similar, and both processes are guided by environmental constraints and stakeholder concerns (see Subsection F.I.4.4). Given these similarities, we conclude that the methods of organizational theory and design can be applied to designing a reference architecture for CLSs.

To make design decisions about the reference architecture of CLSs, we draw on contingency theory and SCT (see Subsection F.I.4.5). Thus, we analyze the environment and, based on the (presumable) manifestation of environmental conditions, develop *propositions* for the design parameters of the reference architecture so that we achieve congruence between presumable conditions and design parameters. Thus, propositions aim to achieve organizational effectiveness by addressing stakeholder interests. If needed, we complement these design decisions by exercising choice in an economically rational manner from the perspective of the presumable dominant coalition in CLSs.

Basis: The International Standard for Systems and Software Engineering — Architecture Description (ISO/IEC/IEEE 42010:2011)

The International Standard for Systems and Software Engineering — Architecture description (ISO/IEC/IEEE 42010:2011) establishes an ontological model for describing architectures (ISO/IEC/IEEE 42010 2011b:v; see Section F.I.5; Subsection F.I.5.1). This model establishes conventions about the content and the content's structure necessary for documenting an architecture in a formal architecture description. This model can be used to express the reference architecture for CLSs because it aims to also be applicable to systems from domains other than computing (ISO/IEC/IEEE 42010 2011b:4).

According to the ontological model, architecture descriptions need to identify the system-of-interest, the system's stakeholders, and their concerns (see Subsection F.I.5.2). Moreover, architecture descriptions need to include (1) architecture views and (2) architecture viewpoints. An *architecture view* expresses "the architecture of a system from the perspective of specific system concerns" (ISO/IEC/IEEE 42010 2011b:2). A view describes the proposed solution of a design process. An *architecture viewpoint* (or perspective) establishes "the conventions for the construction, interpretation and use of architecture views to frame specific system concerns" (ISO/IEC/IEEE 42010 2011b:2). Viewpoints and views are closely related and sometimes even used synonymously, although they are distinct from one another: "A viewpoint is *a way* of looking at systems; a view is *the result of applying* a viewpoint to a particular system-of-interest" (ISO/IEC/IEEE 42010 2011b:20, italics added; see similar Hilliard 1999:4f.).

System architectures are commonly expressed using multiple views (Hilliard 1999:1). Hence, architecture descriptions typically include multiple views. The use of multiple views aims to manage complexity in system design and documentation through the "*separation of concerns*," which means that different sets of concerns are addressed by different views (Hilliard 1999:1, italics original). However, a separation of concerns may not always be possible, especially concerns related to system properties, as "[p]roperties are harder to trace to specific elements of the design or the implementation because they are often cross-cutting concerns, or they affect too many elements" (Kruchten 2004:2). Hence, concerns pertaining to system properties may need to be framed by and addressed in multiple viewpoints and views.

The International Standard establishes a suitable basis for designing and expressing the reference architecture of CLSs for three reasons (see Subsection F.I.5.3). First, expressing the architecture using a standardized ontological model allows to *communicate* the gist of CLSs unambiguously to a heterogeneous set of scholars and practitioners from different domains. Second, the use of multiple viewpoints/ views allows breaking down the overall design problem into manageable sub-problems. It also allows anchoring CLSs in different academic fields because different viewpoints/ views can adopt different methods from different domains depending on the problem at hand. Finally, the separation between viewpoints and views is particularly suitable because it allows for a clear distinction between the principles and concepts of cloud computing applied to logistics system design (viewpoints) and the results of applying them (views).

The process of creating an architecture description can be structured into three steps (see Section F.I.6): (a) *identifying stakeholders and their concerns*, (b) *selecting and defining viewpoints and recording the rationale for their selection*, and (c) *producing one view per viewpoint and recording the rationale for architectural decisions.*

2. Fundamental Concepts

The research objective of this thesis is to design a logistics system in accordance with the principles and concepts underlying cloud computing, which can be subsumed under the cloud paradigm. The following subsections establish a few fundamental concepts relevant to accomplishing this objective.

2.1. Systems and System Architectures

Recall that a *system* can be represented through a quadruple of composition, structure, mechanisms, and environment (see Section B.II.2). A *(system) architecture* provides information about the fundamental manifestation of a system's composition, structure, and mechanisms in its environment. It can be defined as the:

> "fundamental concepts or properties of a system in its environment embodied in its elements, relationships, and in the principles of its design and evolution" (ISO/IEC/IEEE 42010 2011b:2).

Some further explanations of this definition may be helpful. The term "fundamental" emphasizes that a system architecture constitutes only the essence of a system in relation to its environment (ISO/IEC/IEEE 42010 2011b:4). However, what is essential to a system can pertain to any or all of the following aspects: (a) system elements, (b) how these elements are arranged or interrelated, (c) what principles govern the organization or design of a system, and (d) what principles govern the evolution of a system (ISO/IEC/IEEE 42010 2011b:4). An architecture commonly aims to understand and control only those system elements that are essential to a system's behavior, costs, and risks (ISO/IEC/IEEE 42010 2011b:19). Depending on a system's nature, these elements can differ (ISO/IEC/IEEE 42010 2011b:19). Furthermore, the phrase "fundamental concepts or properties" emphasizes two different philosophies of how the term "architecture" can be understood. First, an architecture is an abstract conception of a system in one's mind (fundamental concepts), and, second, an architecture is a property of that system (fundamental properties) (ISO/IEC/IEEE 42010 2011b:19). In this thesis, we understand the term "architecture" as a conception of a system in our minds. Thus, a system architecture represents a conception of a system's composition, structure, and mechanisms in its environment in our minds, which particularly focuses on the elements that determine its behavior, costs, and risks.

Although the term "architecture" was initially introduced in the context of building design and construction, it has been adopted in many other contexts. Today, it is commonly used in relation to IT systems; in fact, the above definition has been obtained from IT literature. The term has also been adopted by organizational theory and logistics research. Despite its use in disparate contexts, its meaning usually aligns with the definition provided above. For example, Schmoltzi & Wallenburg (2011) use the term "cooperation architecture" to categorize horizontal LSP alliances in six fundamental dimensions, including geographies covered and services offered (see Section C.VI.1). Gulati et al. (2012) use the term "organizational architectures" to refer to different designs of interorganizational structures, and they use the term "architect" to refer to the organizations that are pivotal in designing these structures. Gulati & Singh (1998) speak of "architecture of cooperation" to investigate the influence of two fundamental factors (coordination costs and appropriation concerns) on the degree of hierarchical controls in contractual interorganizational relationships. Fjeldstad et al. (2012) use the term "architecture of collaboration" to identify and specify three main (organizational) elements that are critical for multi-actor collaboration. These examples show that, largely independent of the context, the term "architecture" commonly refers to fundamental aspects of a system, which is in line with the understanding described above. In an organizational context, the term architecture refers to the governance structure, interorganizational structure, and other key aspects of interfirm cooperation, such as geographies covered or services offered.

2.2. Reference Architectures

A *reference architecture* can be intuitively conceived of as a more abstract or generic architecture: an architecture that does not pertain to single system, but to a family of systems or an entire system domain (Hassan & Holt 2000:150). Thus, reference architectures ignore system- and context-specific architectural variations and model only the most significant architectural elements (and their relationships) in a specific domain "independent of the technologies, protocols, and products that are used to implement a specific solution for the domain" (Laskey et al. 2012:9; also see Angelov et al. 2012:418; Nakagawa et al. 2012:297; Cloutier et al. 2010:20).

Reference architectures can be designed at many different levels of abstraction (Laskey et al. 2012:9f.). The processes of abstracting and choosing a suitable level of abstraction require great care. Cloutier et al. (2010:21) remark that abstracting does not mean simply leaving out detail, but leaving out system- and context-specific details, while keeping those that are essential for conveying the true meaning. The design of reference architectures therefore needs to balance the conflicting priorities between, on one hand, making reference architectures sufficiently generic so that they are reusable and applicable to multiple system architectures and, on the other, providing sufficiently concrete and specific architectural information that can actually guide architectural design (Cloutier et al. 2010:23).

While the term "reference architecture" is well-established (yet often used inconsistently) in the field of software systems engineering (see e.g. Arsanjani et al. 2007; Hassan & Holt 2000), it is new to the field of general systems design. Cloutier et al. (2010:17) therefore attempt to refine the concept for general systems engineering by deriving the following working definition from the existing literature:

> "Reference Architectures capture the essence of existing architectures, and the vision of future needs and evolution to provide guidance to assist in developing new system architectures."

This definition essentially consists of two elements: The first identifies where the content of reference architectures can be generated; the second specifies the purpose for which reference architectures can be used. Both are briefly considered below.

First, reference architectures are commonly derived from a set of existing architectures by extracting and capturing the commonalities and most significant architectural aspects contained in multiple system architectures while disregarding system- and context-specific architectural variations (Cloutier et al. 2010:20f.). The creation and evolution of reference architectures is an iterative, continuous, and actively managed process that heavily relies on feedback from actually implemented systems (Cloutier et al. 2010:20ff.). For example, the NIST Cloud Computing Reference Architecture was developed in an iterative process that involved organizations from industry, academia, and standards development (Liu et al. 2011a:2).

Second, reference architectures provide guidance for supporting the development of specific system architectures. They are a tool that facilitates system engineering, standardization, and evolution (Nakagawa et al. 2012:297). They provide a common framework for describing, discussing, and developing system-specific architectures within a specific domain (Liu et al. 2011a:2). In other words, reference architectures provide a "solution template" that serves as a starting point for all entities involved in the design, operation, and maintenance of systems within a specific domain. By providing a common solution template, reference architectures specifically aim to achieve reusability, enable interoperability among many existing and evolving systems, reduce risk for developers through pre-defined abstract architectural elements, and provide a knowledge repository that facilitates knowledge transfer and provides guidelines for system design (Cloutier et al. 2010:24f.; also see Angelov et al. 2012; Nakagawa et al. 2012:300).

2.3. Architecture Descriptions

Architecture descriptions express the architecture of a system-of-interest (ISO/IEC/IEEE 42010 2011b:2); reference architecture descriptions express the architecture of a family of systems. An architecture description represents a concrete work product, or an artifact, which is the result of a human work process (ISO/IEC/IEEE 42010 2011b:5). In other words, an architecture description is a work product that documents the outcome(s) of a design process.

Expressing architectures in an explicit manner can serve various purposes. These include documenting essential system characteristics for potential clients, owners, and operators and facilitating communication among parties involved in development, deployment, and maintenance of a system (ISO/IEC/IEEE 42010 2011b:8f.).

2.4. Stakeholders and Stakeholder Concerns

Stakeholders are parties (individuals, teams, organizations, or classes thereof) that have an interest in the system under consideration (ISO/IEC/IEEE 42010 2011b:2). Typical stakeholders include system users, operators, owners, suppliers, and designers (ISO/IEC/IEEE 42010 2011b:12). Each system has at least one but can have multiple stakeholders.

The interests of stakeholders are expressed as *concerns*. Specifically, a *concern* is defined as "interest in a system relevant to one or more of its stakeholders" (ISO/IEC/IEEE 42010 2011b:2). Concerns arise throughout a system's life cycle from, among others, system requirements, design choices, and operating considerations (ISO/IEC/IEEE 42010 2011b:6). A concern is connected to any influence on the system that originates in the system's environment, including "developmental, technological, business, operational, organizational, political, economic, legal, regulatory, ecological and social influences" (ISO/IEC/IEEE 42010 2011b:2). As a consequence, stakeholder concerns can manifest in many different forms: These not only include *what* a system does (i.e. functionality) but also *how* the system does it (i.e. system qualities) (Lago et al. 2010:22). An important group of concerns that stakeholders ascribe to a system derive from the system's purpose(s) (ISO/IEC/IEEE 42010 2011b:4) such as achieving certain (business) goals. Specific examples of concerns include requirements, objectives, constraints, or aspirations that stakeholders have for an architecture (Rozanski & Woods 2011:22; ISO/IEC/IEEE 42010 2011b:6;). Further examples include system properties, performance, resource utilization, reliability, quality of service, autonomy, flexibility, agility, modularity, control, privacy, business goals and strategies, and customer experience (ISO/IEC/IEEE 42010 2011b:6; also see Lago et al. 2010:22; Akerman & Tyree 2006:815). In sum, considering these examples, it is evident that concerns can be very broad or very fine-grained in scope (Emery & Hilliard 2009:34).

2.5. Interim Summary: Linking Fundamental Concepts

The preceding subsections have introduced fundamental concepts relevant to designing logistics systems as per the cloud paradigm. Figure 30 concludes this section by depicting the relationships between these concepts. *Stakeholders* are parties with interests in a system. Their interests are represented by *concerns*, which are connected to environmental influences. A *system* is a complex object that is embedded in an *environment*. The environment is the context that determines the setting and circumstances of all influences upon the system, which also includes the interactions of the system with its environment and with other systems (ISO/IEC/IEEE 42010 2011b:2, 4). A system exhibits an *architecture* which can be expressed by means of an *architecture description*. A system can belong to a *family of systems*. The commonalities of all systems in a family are included in a *reference architecture*. The reference architecture can be expressed by means of a *reference architecture description*.

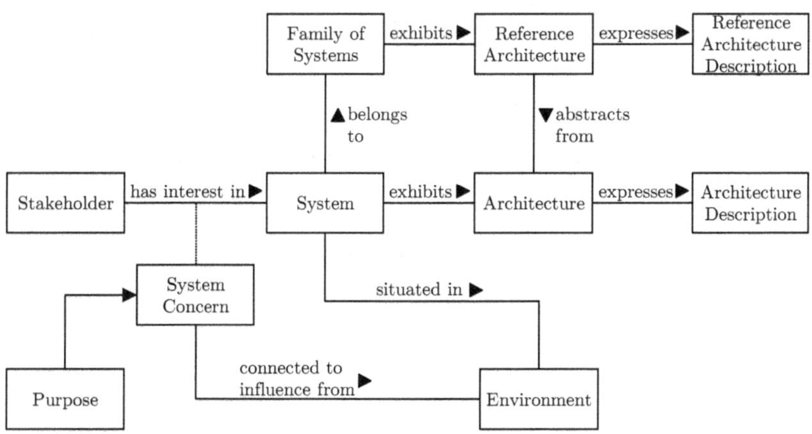

Figure 30.: Context of Reference Architecture Description (based on ISO/IEC/IEEE 42010 2011b:3).

3. Objective: A Reference Architecture Expressed as Architecture Description

The research objective of this thesis aims is to design a logistics system in accordance with the cloud paradigm. As identified in the systematic knowledge review, this requires adopting the principles and concepts of virtualization, service-orientation, and the IoT for the logistics domain as well as developing a deployment model and service model that enables the delivery of physical logistics capabilities with cloud characteristics. Such a comprehensive design has not yet been accomplished. Hence, we encounter a general question about the level of detail, or abstractness, of the proposed design results. This thesis aims to develop a reference architecture for CLS, which equals a higher level of abstraction.

By designing a reference architecture for CLSs, we specifically aim to synthesize and abstract existing cloud logistics knowledge in order to derive the necessary elements, their relationships, potential design tradeoffs, and limitations that are common to all CLSs. Synthesizing and abstracting existing knowledge is the most reasonable approach, given that knowledge is highly fragmented and spans different academic fields. Hence, by designing a reference architecture for CLSs, we aim to create an intermediate point of reference for both practitioners and researchers. For the former, it serves as an abstract solution template for realizing a specific CLS. For the latter, it establishes an initial structure of cloud logistics knowledge and system design to guide future research efforts. This initial structure is particularly useful due to the fragmented nature of this young research field. We thus conclude that designing a reference architecture for CLSs is the most suitable approach to advance the field of cloud logistics research. Furthermore, the proposed reference architecture does not aspire to be the last word on this subject, but among the first, as the evolution of reference architectures is an iterative, continuous, feedback-based, and actively managed process. The reference architecture proposed by this thesis therefore needs to be conceived of as preliminary and subject to change.

The reference architecture of CLSs is expressed in a formal manner through an architecture description. More specifically, we express the reference architecture following the ISO/IEC/IEEE 42010:2011. The International Standard is introduced in a condensed form later in this introductory chapter (see Section F.I.5). Thereafter, we provide three specific reasons for why we follow this standard (see Subsection F.I.5.3).

4. Method: Contingency and Strategic Choice Theory

4.1. Toward an Approach for Architecture Design

As described above, the term "architecture" is rooted in the construction industry but has evolved to be a fundamental construct in computer science and also surfaced in organizational literature. In an organizational context, the term has been used to refer to fundamental aspects of governance structure, interorganizational structure, and other key aspects of interfirm cooperation. As this thesis aims to design a reference architecture of CLSs, reviewing the process of how architectures are designed is instructive. In this respect, we briefly review a general model for the design of software architectures in the next subsection. Thereafter, we review strategic choice theory, which aims to explain the structuring of an organization as a function of both constraints imposed by the environment and choice exercised by the organization's decision makers. Surprisingly, both processes are very alike, thus suggesting that the design methods of organizational theory can be applied to designing a reference architecture of CLSs. In the subsequent subsection we specify how architectural decisions in the process of designing a reference architecture are made.

4.2. A General Model for Software Architecture Design

The process of designing software architecture is typically referred to as "architecting." This process is carried out by a "system architect," who is a person, team, or organization responsible for the "process of conceiving, defining, expressing, documenting, communicating, certifying proper implementation of, maintaining and improving an architecture throughout a systems life cycle" (ISO/IEC/IEEE 42010 2011b:1; also see ISO/IEC 42010 2007:3).

Hofmeister et al. (2007) review existing models for architecture design and integrate them into a "general model for software architecture design," depicted in Figure 31. This model specifies a process consisting of three distinct activities: (1) analysis, (2) synthesis, and (3) evaluation. Although the model indicates a clear sequence of these activities, they are often not carried out sequentially due to the complexity of many design activities, but rather "repeatedly, at multiple levels of granularity, in no predictable sequence, until the architecture is complete and validated" (Hofmeister et al. 2007:112).

Architectural concerns and contexts (or environmental influences) represent the inputs to the design process; both may be subsumed under the term "stakeholder concerns" (see Subsection F.I.2.4; also see Lago et al. 2010:20). As inputs, they critically determine the design process and outcome.

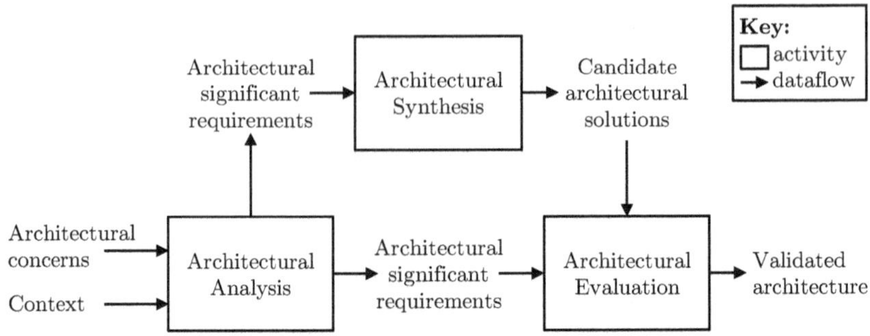

Figure 31.: Architectural Design Activities (adopted from Hofmeister et al. 2007:113)

Architectural analysis is the first activity in the design process; it aims to identify and define the problems the architecture must solve (Hofmeister et al. 2007:113). This identification is driven by the architect, who must uncover the stakeholders' goals, preferences, and needs (Lago et al. 2010:22) and understand environmental influences. Once uncovered, the architect analyzes the stakeholder concerns and environmental influences to identify those that affect the system's architecture, and, hence, must be incorporated into the design process. Such concerns and influences are termed "architectural significant requirements" (Hofmeister et al. 2007:113) and basically define the dimensions the architect must consider during system design, implementation, and operation (Lago et al. 2010:20). In other words, they can be conceived of as "variables" in the "objective function" used to measure the architecture's ability to address stakeholder concerns.

The second step in architecture design is *architectural synthesis*. This activity is the core of architecture design, as it moves from the problem space to the solution space (Hofmeister et al. 2007:113). The architect analyzes factors, explores the key architectural issues or challenges, and then develops strategies for addressing them (Hofmeister et al. 2007:109). In order to deliver a successful architecture, the architect needs to actively manage a multitude of concerns that may be articulated by a potentially large and diverse group of stakeholders (Lago et al. 2010:20, 22). Although many stakeholders often have similar concerns, some may hold unique concerns that conflict with those of others. Balancing conflicting concerns and finding tradeoffs acceptable to all can be challenging, albeit being a necessary and crucial task of the system architect (Rozanski & Woods 2011:22; also see Lago et al. 2010:22). The outcome of architectural synthesis is one or more "candidate architectural solutions." These candidates may present alternative or partial solutions (Hofmeister et al. 2007:113).

The last step in architecture design is *architectural evaluation*. This step measures candidate architectural solutions against architecturally significant requirements (Hofmeister et al. 2007:113). If candidate solutions sufficiently address the requirements, the architecture is validated. Otherwise, architecture analysis and synthesis may be carried out again. Validation is unlikely during the first round of evaluation. Most often multiple design and evaluation rounds are required, iteratively evolving candidate solutions until they become viable.

The previous have paragraphs presented a general model for the design of software architectures. Yet it is important to emphasize that this model derives from a more universal design model: the "function-behaviour-structure framework" initially developed by Gero (1990) and later refined by Gero & Kannengiesser (2004) (Hofmeister et al. 2007:112). The steps in the design process for software architectures resemble the generic design steps in that framework. Thus, we can conclude that although Hofmeister et al.'s (2007) model pertains to software architectures, it likely describes a design process that equally applies to solving design problems in other domains. In fact, in the following, we introduce a model that explains the structuring of organizations via a similar process.

4.3. The Structuring and Effectiveness of Organizations: Contingency and Strategic Choice Theory

Organizational theory attempts to describe and explain how organizations form, function, and survive with the explicit or implicit objective of supporting managers in guiding their organizations effectively (see e.g. Daft 2001:14; Scherer 2006:20). This particularly includes describing and explaining the process of how organizations become structured and making inferences about how organizations can achieve effectiveness, especially focusing on the relationship between organizational structure and effectiveness.

Organizational effectiveness can generally be defined as "the degree to which it [an organization] realizes its goals" (Etzioni 1964:8). While different approaches exist to measure organizational effectiveness (e.g. goal approach, systems approach, see Strasser et al. 1981), we use the *stakeholder approach*, or *multi-constituency approach* in this thesis. According to this approach, the effectiveness of an organization is determined by the extent to which "it satisfies the interests of one or more constituencies associated with the organization" (Tsui 1990:458). According to Tsui (1990:480), the multi-constituency approach is not only a viable framework to assess organizational effectiveness because it includes multiple criteria, but because it assesses effectiveness from the perspective of multiple evaluators (stakeholders). For example, from a customer's perspective, an organization may be effective if the quality of goods and services delivered is satisfactory according to their judgment, while owners of an organization deem an organization to be effective if it delivers a certain ROI.

In the following, we briefly introduce two theories that are relevant to the objectives of this thesis: (1) structural contingency theory, and (2) strategic choice theory (SCT).

(Structural) contingency theory posits that the structure of an organization exclusively depends on the manifestation of contextual factors to which the organization is exposed and that there must be some form of "fit" or "congruence" between the organization's structure and its environment if the organization is to survive or be effective (see e.g. Drazin & van de Ven 1985:514ff.). Contextual factors typically considered relevant include culture, size, technology, environmental complexity and dynamism (see e.g. Child 1972:3ff.; Drazin & van de Ven 1985; Pugh et al. 1969). Contingency theory rests on three fundamental premises (Schreyögg 1980:308f.): (a) for every specific contextual situation, there is one best structural "response"; (b) the organization has no control over or influence on its environment, which means that the environment is considered a given; and (c) organizations need to achieve a certain minimum level of economic performance to survive, and both the criteria for assessing performance and the required performance level are out of the organization's control.

SCT extends the contingency theory by introducing *human agency* as another critical factor (besides the environment) that influences organizational structure. More specifically, SCT posits that a "dominant coalition" has authority over decisions important for the organization as a whole, including its structure, which is why the structuring of organizations becomes a "political process," as it depends on the exercise of power (Child 1972:1f., 13f.). In other words, the dominant coalition acts as a "designer" or "architect" of the organization. The term dominant coalition does not necessarily refer to the holders of assigned formal authority, but refers to the group of individuals that actually wield authority over important organizational decisions, regardless of whether authority arises from formal or informal sources (Child 1972:13).

By introducing human agency as a critical determinant of organizational structure, SCT refutes the premises underlying contingency theory. More specifically, SCT refutes the "alleged" unidirectional deterministic relationship between the contextual factors of an organization and its internal structure by introducing human agency as another critical determining factor for organizational structuring and by regarding the relationships between environment and agency and between agency and structure to be dynamic (Child 1997:48). Moreover, the dominant coalition is assumed to have some degree of choice – that is, scope to maneuver according to its own preferences – because, compared with contingency theory, the extent to which contextual factors influence structure is weakened due to several factors, including the following (Child 1972:16f.). Organizational structure has only limited influence on achieved organizational performance. The dominant coalition is aware of the structure's limited influence, which is why contextual pressures pose only limited constraints on structural design. In addition, the dominant coalition has the power to influence contextual factors and to negotiate more easily attainable performance requirements with the organization's stakeholders, allowing them to maintain a structure adapted to their own preferences without significant performance penalties. Finally, certain combinations of contextual factors may create perceived conflicting implications for structure design, and thus create further leeway for the dominant coalition to make choices according to their preferences. To conclude, SCT proposes that the influence of contextual factors remains relevant but is weakened, which is the prerequisite for positing that an organization's dominant coalition has an influence on the organization's structure.

By using the term "strategic," SCT implies that choice pertains to all matters important to an organization in its entirety (Child 1997:48), including both its internal and external strategy. Yet, choice is not limited to the strategies of a single organization, but also extends to interorganizational relationships and forms of organization. Although the initial focus of SCT was primarily on single organizations, Child (1972:2, italics added) still defines the construct of organizational structure as "the formal allocation of work roles and the administrative mechanisms to control and integrate work activities *including those which cross formal organization boundaries*." Considering the recent proliferation of interorganizational networks and other collaborative arrangements (e.g. joint ventures, strategic alliances), Child (1997:54, 58) observes that simple interorganizational relationships have evolved into "sets of arrangements which are themselves organized," or, in other words, "fully operational form[s]" of organizing. He therefore conceives of the inclusion of these forms as an "appropriate contemporary extension of strategic choice analysis," and this extension requires paying "attention to the ways they [organizational actors] may seek to attain their objectives through mutual accommodation and collaboration with the parties within an existing environment" (Child 1997:54f.).

The structure of interorganizational forms depends on the agreed-upon terms of cooperation among the parties involved and the extent to which these parties can identify mutual complementarities (Child 1997:58). For example, with regard to interfirm networks, Grandori (1997b:900) remarks that involved actors have the discretion to define the type of game they intend to play in their cooperative venture, which includes its degree of how informationally complex or integrative the game will be. The crucial question for interorganizational structures pertains to the degree of authority that decision makers in one organization can exert over other legally autonomous organizations (Child 1972:9f.). It is important to stress that in an interorganizational context, actors usually do not wield any kind of formal authority over organizations with which they aim to collaborate. Hence, exercising strategic choice in an interorganizational setting can only be based on informal sources of authority. In this respect, Child (1997:57) stresses the importance of bargaining power (e.g. resulting from control over critical resources, clients, and reputation) as an important determinant for strategic choice (see Section C.V.2). The process of interorganizational design is thus a "political process" spanning organizational boundaries among actors belonging to different organizations.

Figure 32 depicts the basic process model for SCT. The process of exercising strategic choice consists of three steps (Child 1972:17f.; Child 1997:46ff.). The first step is an evaluation of the organization's situation, including its past performance, expectations for the organization formulated by its stakeholders, external trends relevant to the organization, and the congeniality of its current internal configuration. To some extent, this evaluation is colored by the prior ideology (e.g. values, experiences) of the dominant coalition. Formulating the organization's goals is the second step. In the third step, the organization defines its external and internal strategies for attaining their goals. Externally-oriented choices include negotiating organizational performance targets with stakeholders (who can impose sanctions on the organization or have authority over imposing sanctions) and moving into favorable markets (and products) or exiting less promising ones. Internally-oriented actions include implementing a configuration of personnel, technology, and organizational structure that is internally consistent and in harmony with the planned scale and nature of operations. Depending on the choices made, the organization achieves a certain level of operating efficiency and overall organizational effectiveness, which in turn feeds back into the evaluation of the organization's situation. Thus, a circular process is established, which may enable evolutionary organizational learning (Child 1997:58).

Although we have already applied the term "choice" several times, a formal conceptualization has not yet been provided. Choice essentially refers to human behavior in a situation in which an individual (a decision maker) can select between multiple alternatives (Schreyögg 1980:312). Schreyögg (1980:312ff.) proposes conceptualizing choice in an (1) intentional and (2) historical manner, which requires combining two levels of analysis: the actions of individuals and the overriding institutional nexus. Note that this conceptualization is very similar to the "realist perspective" of human behavior as developed by Whittington (1988).

Choice refers to *intentional* human behavior. This means that an individual who needs to choose one of multiple alternatives selects the one he or she believes to be most conducive to achieving the desired ends. In other words, an individual purposefully takes a step that he or she believes is necessary to reach the desired ends; without this step, the ends could not be achieved (Schreyögg 1980:313). Schreyögg (1980:313) proposes interpreting intentional human behavior as behavior that derives from specific meanings. Thus, to understand why an individual has behaved intentionally in a certain way, we need to reconstruct the network of meanings from which the individual derived his or her behavior in the first place (Schreyögg 1980:313). This network consists of two types of

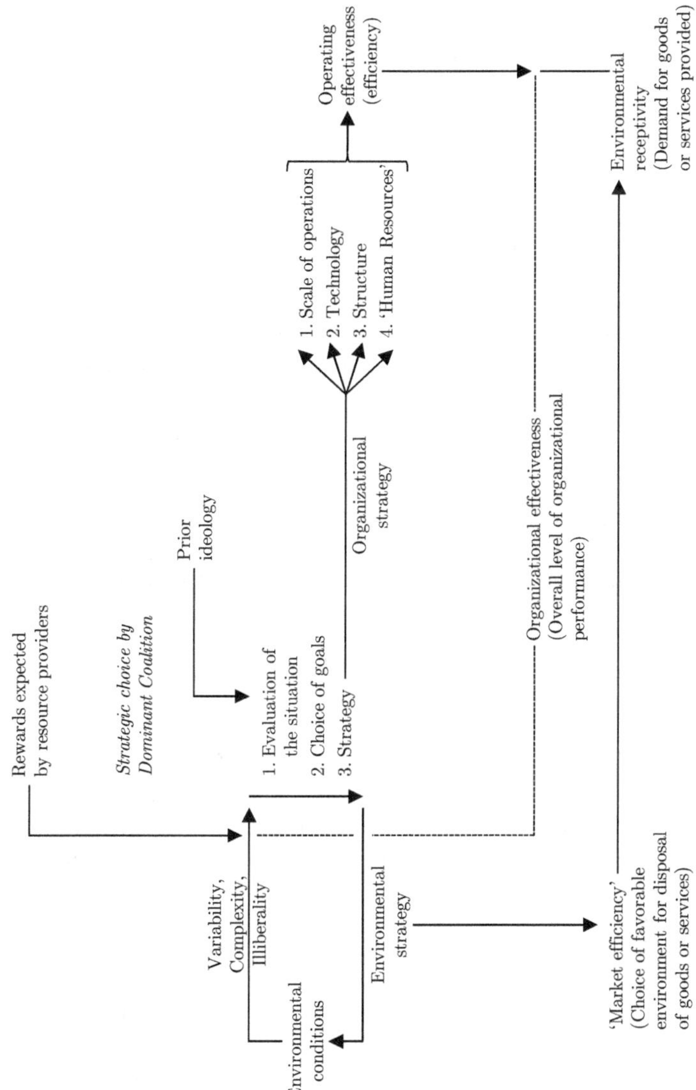

Figure 32.: The Role of Strategic Choice in the Theory of Organization (Child 1972:18)

meanings: (a) those internal to the individual (such as personal values and beliefs) and (b) those that manifest in the environmental structure (such as social norms and institutions) and therefore override the individual's internal meanings (Habermas 1970:182 as cited in Schreyögg 1980:313). Thus, to understand and explain intentional human behavior, we need to combine the level of analysis that focuses on the individual with the level that focuses on the overriding institutional frame.

The *historical* dimension of choice means that the network of meanings from which individuals derive behavior has developed over time (Schreyögg 1980:313f.). Individuals accumulate such meanings by reasoning with themselves or others and by acquiring knowledge and forming opinions about their environment (Schreyögg 1980:313). In addition, meanings that define the environmental structure are not externally given or fixed. Instead, they also result from intentional human action and are therefore changeable through human reasoning (see e.g. Bhaskar 1986 as cited in Whittington 1988:528f.; also see Schreyögg 1980:314). These (social) processes certainly take time, which establishes the historical dimension of choice.

The intentional and historical dimensions of choice cannot be understood independently of each other, but need to be considered together as they are closely intertwined. This is because meanings, especially those that belong to the (social) environmental structure, are simultaneously enablers for human agency and the result of it (Whittington 1988:528f.; Child 1997:56).

When choice is conceptualized as intentional human behavior, it fundamentally differs from human behavior that is explained in a stimulus-response manner – an explanation that underlies environmental determinism and (by definition) denies the existence of choice. Environmental determinism argues that environmental structure imposes such tight exogenous constraints on human behavior that, despite the existence of multiple alternatives, only one alternative is compatible with system survival (Whittington 1988:523ff.). However, denying the stimulus-response explanation of human behavior implies neither that environmental structure is irrelevant to made decisions; nor that having choice means having free choice, as suggested by interpretive voluntarism (Whittington 1988:523ff.). Instead, as argued above, it implies that the exercise of choice is subject to several constraints, including environmental structure, the internal (cognitive) properties of decision makers, and the availability of information about an organization's situation (Child 1997:49ff., 52). These constraints, however, do not determine choice to such an extreme extent that it is actually eradicated and that one specific solution must be selected because it is the only one compatible with organizational survival.

4.4. Similarities between Designing Software Architectures and the Structuring of Organizations

The preceding two subsections have introduced a process for software architecture design and a process for organizational structuring. Although the natures of software architectures and organizations are fundamentally different, comparing their design and structuring processes undeniably reveals similarities. We assess the similarities of both processes according to the three generic process steps of software architecture design.

In both cases, the design process is driven by a specified entity: the "architect" and the "dominant coalition," respectively. These entities analyze the status quo in the first step of the design process. Software architects analyze contextual factors and architectural concerns raised by stakeholders in order to identify those relevant for architectural choices. The dominant coalition analyzes the organization's situation, including past performance, its environment, and the reward expectations raised by its resource providers. Comparing the initial process steps between the design of software architecture and the structuring of organizations shows similarity in both the nature of activities and in the inputs to the process.

The second step relates to the actual design activity: moving from problem to solution space. Although the software architect and the dominant coalition have authority over design decisions, they are not entirely "free" in their decisions. In both cases, decision-making is constrained by external factors, including factors from the natural, economic, and social environments in which the software system and organization operate respectively. The architect needs to account for pressures resulting from contextual factors and needs to balance concerns raised by system stakeholders in order to develop an architecture that (sufficiently) satisfies their expectations. The dominant coalition similarly needs to choose an internal and external organizational strategy that achieves organizational effectiveness by aligning its strategies with prevailing environmental conditions and with the expectations raised by their resource providers.

Furthermore, the software architect and the dominant coalition face similarities in the nature of objectives they need to satisfy when making decisions: concerns raised by stakeholders. This is because, according to the multi-constituency approach to organizational effectiveness proposed by Tsui (1990), the aim of achieving organizational effectiveness can be understood as satisfying stakeholder concerns. The need for addressing and potentially balancing conflicting stakeholder concerns becomes particularly recognizable in an interorganizational context. For example, Gulati et al. (2012) describe the role of a "designer" who designs the structure of a network of legally autonomous firms or agents (referred to as "meta-organization"). Gulati et al. (2012:581) specifically describe the role of a meta-organization designer as:

> "A meta-organization's designer must recognize and accommodate members' organization-level strategies [...] and provide them a guiding economic logic for collective action while allowing appropriate flexibility for this logic to be realized."

Thus, the role of a meta-organization designer resembles that of a software architect. Both need to identify and balance concerns of different stakeholders in order to deliver a viable system.

Finally, the third step of the design process relates to some form of feedback loop. The software architect receives feedback when evaluating the candidate architecture(s) against architectural significant concerns. The dominant coalition receives feedback by measuring organizational performance against defined goals. If performance expectations are not met, both the software architect and dominant coalition need to adjust their proposed solutions. If goals are persistently not attained, the architect and dominant coalition may be "replaced" by pressures exerted by their principal stakeholders.

The preceding paragraphs show that although software architectures and organizations are different in nature, their design follows a very similar process and these processes are guided by environmental constraints and stakeholder concerns. Given these similarities, we conclude that the methods of organizational theory and design can be applied to designing a reference architecture for CLSs.

4.5. Making Architectural Design Decisions in Cloud Logistics Reference Architecture

In order to make architectural design decisions about the reference architecture of CLSs, we draw on contingency theory and on SCT. Thus, we analyze the environment of CLSs and, based on the (presumable) manifestation of environmental conditions, develop *propositions* for the design parameters of the CLS reference architecture so that we achieve congruence between the presumable environmental conditions and design parameters. In other words, propositions aim to achieve organizational effectiveness by addressing the interests of stakeholders, as in the multi-constituency approach. If needed, we complement these design decisions by exercising choice in an economically rational manner from the perspective of the dominant coalition in CLSs. The following paragraphs describe this design approach in detail.

While contingency theory and SCT are positivist in their nature, the process of designing organizational structures or a reference architecture of CLSs does not follow a positivist approach. In fact, organizational design requires normative reasoning. Hence, we face the critique of subjectivity regarding design decisions. Although absolute objective can never be reached (Rescher 1997:8), Rescher (1997:4 9) argues that

> "[o]bjectivity calls for not allowing the indications of reason, reasonableness, and good common sense to be deflected by 'purely subjective' whims, biases, prejudices, preferences, etc. Accordingly, objectivity always strives for the sensible resolution, while subjectivity gives rein to temper and lets personal inclination have its way. This does not require excluding values (how could humans ever achieve that?), but rather insists on not being deflected from the path of reason by rationally inappropriate prejudicial influences."

Thus, as long as our design decisions are guided by rationality and reason, the critique of subjectivity can be refuted. In order to achieve objectivity in making choices about organizational design (at least as much as we possibly can), Schreyögg (1980:314f.V) suggests following the principle of "transsubjective dialogue" because it is the only way to overcome our own subjectivity and thus allow for rational scientific reasoning on normative issues (Lorenzen 1969:82, *Konstruktive Philosophie* as cited by Schreyögg 1980:314f.). The principle of transsubjectivity requires a dialogue to be reasonable, symmetrical, and free of domination (*herrschaftsfrei*), which means that all arguments are taken into account, each participant in the dialogue is willing to challenge his or her own presuppositions and adapt to others' arguments if they are non-refutable, and arguments are neither accepted nor refuted through force or persuasion (Lorenzen 1969 as cited in Schreyögg 1980:315).

So far, the concept of CLSs only exists in our minds. This complicates designing a reference architecture for CLSs because there is no empirical knowledge about the "actual" environment of such systems, and such knowledge cannot be determined through observations. Design decisions can therefore only be based on the "presumable" environment of CLSs. We determine the conditions in this presumable environment through knowledge found in current cloud logistics sources. However, due to the scarcity of this knowledge, existing cloud logistics knowledge cannot provide a sufficient understanding of environmental conditions. Hence, we assume that the environment of CLSs largely resembles the typical environment of meso-logistics systems because, as suggested by current knowledge, CLSs represent meso-logistics systems, as they comprise a network of horizontally cooperating LSPs. This approach clearly represents a methodical limitation. However, due to the novelty of CLSs this limitation cannot be circumvented at this early stage of cloud logistics research.

SCT posits that the structuring of organizations depends on both constraints imposed by environmental conditions and choice exercised by the dominant coalition. If necessary throughout the process of reference architecture design, we exercise choice regarding architectural decisions from the perspective of the dominant coalition of a CLSs. Methodological questions about exercising choice in organizational design are extensively covered by Schreyögg (1980). In the following, we focus on selected issues particularly relevant to designing a reference architecture of CLSs: (1) critique of subjectivity, (2) selecting a norm of evaluation, (3) choosing between equally effective alternatives, and (4) having authority over structural choices.

Similar to making design decisions following the logic of contingency theory, making design choices in SCT equally faces the critique of subjectivity (Schreyögg 1980). Hence, architectural choices must be guided by rationality and reason.

In order to exercise choice, *norms of evaluation* are needed to determine whether, or the degree to which, a certain choice is desirable. When selecting an evaluation norm, Schreyögg (1980:314) argues that it must be proven that the norm facilitates choices that are desirable in the sense of being "justifiable." This is because organizational structures do not only enable collective action but also encroach upon individuals and impose constraints upon them (Schreyögg 1980:320). Hence, norms must be justifiable from a moral perspective. He points out that normative reasoning about values and norms in a scientific rational manner can only be accomplished by following the principle of the transsubjective dialogue, described above.

In order to identify a justifiable norm for exercising choice in organizational design, Schreyögg (1980) starts from the conceptualization of choice. Recall that choice derives from a network of meanings either internal to the individual or belonging to the overriding institutional frame, which override the individual's internal meanings (Subsection F.I.4.3). Due to the abstract nature of reference architectures, we limit our focus to meanings that belong to the institutional frame. Focusing on this frame shifts our attention to the market economy itself and thus to the principle of "economic rationality" – that is, choices that aim to maximize expected profit (Schreyögg 1980:315f.; also see Whittington 1988:532f.). Schreyögg (1980:316) argues that the principle of economic rationality can guide structural choice. This norm is also justifiable for evaluating structural choices when designing a CLS reference architecture for two reasons. First, economic rationality is deeply engrained in the logistics industry: LSPs face intensive price-competition and have a tendency to base logistical decisions on analytical models (e.g. network design). Second, many LSPs are publicly traded companies and thus subject to the frequent institutionalized monitoring of many economically rational behaving actors who maximize the expected profit of their investments. We thus conclude that economic rationality is a justifiable norm for making structural choices in the CLS reference architecture.

Given a justifiable evaluation norm, an organizational designer may face a situation in which he or she needs to *choose one of multiple equally effective alternatives*, as per the established norm. If so, Schreyögg (1980:321) argues: "among effective alternatives the one which imposes the least encroachments on those subjected to the structural arrangement has the best reasons on its side." Thus, choices are only justifiable from a moral perspective if they allow the greatest autonomy for those being constrained.

Finally, we face the methodological question regarding *who has authority* over structural choices. Schreyögg (1980:316) argues that structural choices reside with "those who are influential, namely, those who have the right of disposal according to the organizations constitution, or those to whom part of this right has been delegated." Thus, choice needs to be exercised from the perspective of the dominant coalition, taking its own interests into account and the interests of others. Yet, in order to exercise choice in the process of organizational design, the dominant coalition needs to first be determined.

5. Basis: The International Standard for Systems and Software Engineering — Architecture Description (ISO/IEC/IEEE 42010:2011)

The preceding section has focused on the process of architecture design. This section focuses on the outcome of this process. More specifically, it introduces a model for how the design outcome can be formally documented by means of an architecture description and how this model can form a basis for the design process itself.

5.1. Purpose, Evolution, and Application Scope

5.1.1. Purpose

The International Standard for Systems and Software Engineering — Architecture description (ISO/IEC/IEEE 42010:2011) establishes a core ontology for the description of architectures (ISO/IEC/IEEE 42010 2011b:v). Specifically, this core ontology establishes conventions about the content and the content's structure required for formally documenting architectures using architecture descriptions. This is achieved through an ontological model of architecture descriptions. By standardizing documentation, this model serves the "understanding of the system's essence and key properties pertaining to its behavior, composition and evolution" (ISO/IEC/IEEE 42010 2011b:v) and facilitates the communication between stakeholders (ISO/IEC/IEEE 42010 2011b:9f.).

Moreover, this model can serve as a "basis for system design and development activities" (ISO/IEC/IEEE 42010 2011b:8f.; see similar Rozanski & Woods 2011:38). This basis is established by specifying the concepts and their relationships necessary for documenting architectures. In this way, the model indirectly guides and structures the thought and design process of architects because they eventually need to document their result(s) in a manner compliant with the ontological model. In addition, the model's concepts and their relationships align with the logic of designing architectures (see Subsection F.I.4.2). For example, an architectural solution (documented in an "architecture view") must address stakeholder concerns, and the architecture description must include a rationale for why concerns can be addressed. Furthermore, architecture descriptions provide a basis for system design by requiring that the overall design problem is broken up into its fundamental sub-problems, where each has to be dealt with and documented separately, thus ensuring that "the right system gets built" (Rozanski & Woods 2011:38).

5.1.2. Evolution

The evolution of the ISO/IEC/IEEE 42010:2011 can be divided into three steps, as depicted in Figure 33. In 2000, the Institute of Electrical and Electronics Engineers (IEEE) published the initial version of the standard under the title "1471-2000 - IEEE Recommended Practice for Architectural Description for Software-Intensive Systems." In 2006, this standard was adopted by the ISO and the International Electrotechnical Commission (IEC) with the objective of creating a joint revision involving the ISO, IEC, and IEEE (Emery & Hilliard 2009:32). In 2007, these two organizations republished the initial standard with as slightly adapted title "Systems and software engineering — Recommended practice for architectural description of software-intensive systems (ISO/IEC 42010:2007)." In 2011, the ISO, IEC, and IEEE published a joint revision of the standard with the title "International Standard for Systems and Software Engineering — Architecture description."

Overall, the ISO/IEC/IEEE 42010:2011 and its preceding versions are well recognized in the field of systems and software development and widely followed in pertinent literature (see e.g. Rozanski & Woods 2011). For example, the ISO/IEC 42010:2007 serves as a basis for other standards in this field, such as the SOA-RAF (Laskey et al. 2012; also see Subsubsection D.III.4.3.3).

Figure 33.: Evolution of ISO/IEC/IEEE 42010:2011

5.1.3. Scope of Application

The scope of application specifies the "types" of systems whose architectures can be expressed by means of the ISO/IEC/IEEE 42010:2011 and for which systems the International Standard can serve as a basis for development activities. Throughout its evolution, the scope of application of the International Standard has expanded. This can be inferred from the changes in the titles of the different revisions of the standard. The IEEE's initial publication in 2000 and its re-publication in 2007 were both intended for architectures of "software-intense systems," as per their titles. Yet, the re-publication added "systems and software engineering" to the title, thus implying a broader scope of application, even though the standard itself was not changed. In 2011, the jointly revised standard completely dropped the term "software-intense systems" from the title, thus implying that the International Standard cannot be applied only to "software engineering" but also to other types of "systems" not necessarily in a computing context.

This wider application scope is also expressed in the International Standard. It applies to "software-intensive systems" and "software products and services" as well as to "systems" (ISO/IEC/IEEE 42010 2011b:3f.), where the International Standard, "takes no position on what constitutes a system" (ISO/IEC/IEEE 42010 2011b:4). To specify to what type of systems it applies, but not to constrain its application scope in any way, the International Standard lists several system definitions as examples. One examples specifies that the standard applies to

> "systems that are man-made and may be configured with one or more of the following: hardware, software, data, humans, processes (e.g., processes for providing service to users), procedures (e.g. operator instructions), facilities, materials and naturally occurring entities" (ISO/IEC/IEEE 42010 2011b:3).

One can argue that logistics systems are systems in the sense of this definition. In fact, given this definition and the definition of logistics systems (see Section B.II.1), we can define logistics systems as "man-made systems that serve the purpose to transfer objects in time and space and that may be configured with hardware, software, humans, processes, procedures, facilities, and materials." We thus conclude that the architecture of a logistics system can be expressed by means of the International Standard. Whether the International Standard also provides a "suitable" basis for designing and expressing a CLS reference architecture is investigated after the International Standard's ontological model is introduced in the next section.

5.2. Ontological Model of Architecture Descriptions

5.2.1. Model Overview

The ontological model of an architecture description is depicted in Figure 34. As pointed out, an architecture description expresses the architecture of a system. Hence, the architecture description identifies the system-of-interest, the system's stakeholders, and their concerns. In addition, the ontological model consists of the following concepts: architecture viewpoints and architecture views; model kinds and architecture models; architecture rationale; and correspondences and correspondence rules. While the terms "system" and "architecture" have been considered in sufficient detail above, stakeholders and their concerns require further attention along with the concepts that have not been considered so far.

5.2.2. Stakeholders and Stakeholder Concerns

Recall that *stakeholders* are parties with interests in a system, and these interests are referred to as *concerns*. Some stakeholders have concerns that are fundamental to the system architecture because they drive architectural choices. In other words, they have architecturally significant concerns. The International Standard requires that such stakeholders and their concerns are identified by and considered in an architecture description (ISO/IEC/IEEE 42010 2011b:20, 12).

In order to support system design, the International Standard also identifies several "generic" categories of concerns that, when applicable, are considered in architecture descriptions. These include (1) the system's purposes (e.g. what a system does and how it does it), (2) the architecture's suitability for achieving the purposes of the system, (3) the feasibility to construct and deploy the system (see "meta-systemic concerns" in Lago et al. (2010:22)), (4) the system's potential risks and impacts for its stakeholders, and (5) the ability to maintain and evolve the system (ISO/IEC/IEEE 42010 2011b:12; also see Maier et al. 2001:108f.). When designing and expressing the CLS reference architecture, we take the applicable concerns into account.

5.2.3. Architecture Views and Architecture Viewpoints

An *architecture view* expresses "the architecture of a system from the perspective of specific system concerns" (ISO/IEC/IEEE 42010 2011b:2). More specifically, a view specifies how a system architecture addresses – that is, how it deals with or aims to satisfy – one or more concerns relevant to one or more of its stakeholders (Rozanski & Woods 2011:34f.). Thus, a view describes the proposed solution of a design process.

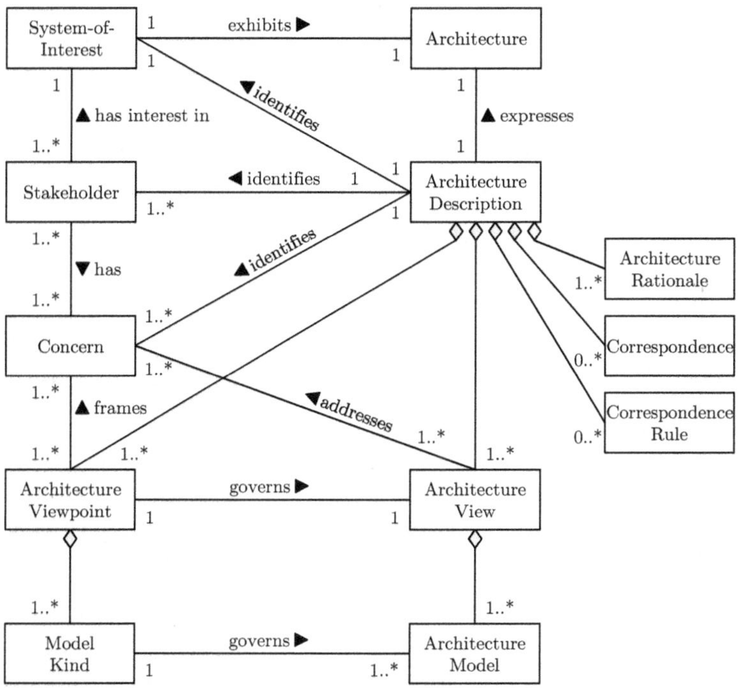

Figure 34.: Ontological Model of Architecture Descriptions (adopted from ISO/IEC/IEEE 42010 2011b:5) Legend: ◇ = is part of; 1...* = one or more; 0...* = zero or more; 1 = exactly one.

An *architecture viewpoint* (or perspective) establishes "the conventions for the construction, interpretation and use of architecture views to frame specific system concerns" (ISO/IEC/IEEE 42010 2011b:2). Framing concerns means that each viewpoint needs to identify the concerns that it encloses or covers (ISO/IEC/IEEE 42010 2011b:6). Each viewpoint also needs to frame at least one concern but may frame multiple, and each concern may be framed by more than one viewpoint (ISO/IEC/IEEE 42010 2011b:6). In order to establish conventions, a viewpoint needs to identify and specify the appropriate means that can be used for framing the set of selected concerns (Lago et al. 2010:22). Viewpoint conventions can therefore include "languages, notations, model kinds, design rules, and/or modeling methods, analysis techniques and other operations on views" (ISO/IEC/IEEE 42010 2011b:6; also see Lago et al. 2010:22).

It is important to note that viewpoints and views are closely related and sometimes even used synonymously, although they are distinct from one another: "A viewpoint is *a way* of looking at systems; a view is *the result of applying* a viewpoint to a particular system-of-interest" (ISO/IEC/IEEE 42010 2011b:20, italics added; see similar Hilliard 1999:4f.). Viewpoints can be applied to many different systems and can thus be reused in different architectural descriptions (see e.g. ISO/IEC/IEEE 42010 2011b:22). For instance, Clements & Bass (2010) introduce "The Business Goals Viewpoint," which generally frames concerns related to an organization's business goals in order to make systems more responsive to organizational needs, irrespective of a particular system-of-interest. Views, by contrast, are always specific to a particular system (see e.g. Maier et al. 2001:108). To illustrate the difference with a metaphor, one can say that the relationship between a viewpoint and a view is the same as that between a legend and a map (Emery & Hilliard 2009:34; ISO/IEC/IEEE 42010 2011b:21). In a nutshell, the relationship between viewpoints and views can be summarized as follows: A view is governed by its corresponding viewpoint (ISO/IEC/IEEE 42010 2011b:6), which is why the identification of viewpoints precedes the creation of associated views (Emery & Hilliard 2009:34). Note that architecture viewpoints and architecture views are related in a one-to-one manner: Each viewpoint governs exactly one view, and each view is governed by exactly one viewpoint. Hence, using multiple views to express an architecture implies the existence of an equal number of viewpoints being part of architecture descriptions.

System architectures are commonly expressed using multiple views (Hilliard 1999:1); the use of multiple views also represents a fundamental premise of the International Standard (ISO/IEC/IEEE 42010 2011b:21). Hence, architecture descriptions typically include multiple architectural views. The use of multiple views has been introduced as a construct to manage complexity in system design and documentation through the "*separation of concerns*," which means that different sets of concerns are addressed by different views (Hilliard 1999:1, italics original). Due to the complexity of many architectures, it has become common practice to use multiple views because it is most often not possible to describe and understand architectures through a single view, but architectures are best understood by using multiple (interrelated) views that collectively represent the architecture (see e.g. Clements et al. 2010:22; Kruchten 1995; Rozanski & Woods 2011:33; ISO/IEC/IEEE 42010 2011b:6). Clements et al. (2010:23) remark that it may be disconcerting that when an architecture cannot be described through a single view and that it can be perceived as inadequate to consider an architecture through several discrete views with potentially no or unclear relationships. However,

> "[t]he essence of architecture is the suppression of information not necessary to the task at hand, and so it is somehow fitting that the very nature of architecture is such that it never presents its whole self to us but only a facet or two at a time. This is its strength: Each view emphasizes certain aspects of the system while deemphasizing or ignoring other aspects, all in the interest of making the problem at hand tractable" (Clements et al. 2010:23; see similar Rozanski & Woods 2011:34f.)

While separation of concerns principle is applicable to many types of concerns, a separation of concerns may not always be possible. This is especially the case for concerns related to system properties, as

> "[p]roperties are harder to trace to specific elements of the design or the implementation because they are often cross-cutting concerns, or they affect too many elements." (Kruchten 2004:2)

In other words, concerns pertaining to system properties may need to be framed by and addressed in multiple viewpoints and views.

Although the use of multiple views and viewpoints is intriguing as it is associated with benefits like improved and tailored stakeholder communication, improved management of complexity, and improved developer focus (Rozanski & Woods 2011:38), it raises the tricky question of what the "right" views and viewpoints are to capture and describe an architecture. While many sets of views are proposed in literature (see e.g. Kruchten 1995; Laskey et al. 2012), there is no general or "easy answer to this problem, other than [...] an analysis of the most important concerns that affect your architecture" (Rozanski & Woods 2011:39). Likewise, the International Standard does not provide a list of views or viewpoints but "encourages the practice of defining or selecting viewpoints appropriate to the system-of-interest" (ISO/IEC/IEEE 42010 2011b:22).

In conclusion, when designing and expressing a reference architecture of CLSs, we inevitably encounter the question of what the most suitable views and viewpoints are for that particular system family.

5.2.4. Architecture Models and Model Kinds

The relation between model kinds and architecture models is the same as between viewpoints and views. A *model kind* refers to the "conventions for a type of modeling" while an *architecture model* uses these conventions to address concerns when constructing an architecture view of a system-of-interest (ISO/IEC/IEEE 42010 2011b:2, 6). A large variety of model kinds can be used in architecture viewpoints. The International Standard identifies several examples including "data flow diagrams, class diagrams, Petri nets, balance sheets, organization charts and state transition models" (ISO/IEC/IEEE 42010 2011b:2). It is remarkable that the International Standard gives examples within not only the domain of software development but also the fields of finance and organizational design. This again emphasizes the standard's wide application scope. To conclude, to design of a reference architecture for CLSs, we need to identify adequate model kinds for the design problems at hand.

5.2.5. Architecture Rationale and Correspondences

Architecture rationale "records explanation, justification or reasoning about architecture decisions that have been made" (ISO/IEC/IEEE 42010 2011b:7). This can include alternatives and tradeoffs considered as well as potential consequences of a decision. Although not every architecture decision can be documented for reasons of practicality, decisions deemed crucial for the architecture of a system-of-interest are recorded (ISO/IEC/IEEE 42010 2011b:15). Every architecture description thus includes a rationale for the selection of each viewpoint (ISO/IEC/IEEE 42010 2011b:15). The International Standard suggests several criteria that can be used to identify decisions about key architectural choices. These include decisions that are significant for architectural requirements; affect important or many stakeholders; are costly to change; necessitate non-obvious reasoning; are time-consuming to make, implement, or enforce; or cause capital expenditures (ISO/IEC/IEEE 42010 2011b:15).

A *correspondence* defines and expresses a relationship between two or more elements of an architecture description, for example between architecture viewpoints, model kinds, stakeholders, or concerns (ISO/IEC/IEEE 42010 2011b:7, 14). Typical uses of correspondence include the documentation of (in-)consistencies and dependencies between different parts of an architecture description (ISO/IEC/IEEE 42010 2011b:23). *Correspondence rules* can govern correspondence; that is, they can be used to express and enforce relationships between two or more architectural elements (ISO/IEC/IEEE 42010 2011b:7). Such rules are, however, not necessary when documenting the CLS reference architecture.

5.3. Interim Conclusion: The International Standard as a Basis for Reference Architecture Design of Cloud Logistics Systems

The preceding subsections have introduced the International Standard for architecture descriptions. This standard provides a suitable basis for designing and expressing a reference architecture for CLSs for three reasons: (1) unambiguous communication, (2) the use of multiple viewpoints, and (3) the separation of viewpoints and views.

By expressing the reference architecture of CLSs through a standardized ontological model comprising well-defined concepts and their interrelations, we can *communicate* the gist of CLSs unambiguously. This is crucial, given the lack of clarity in current cloud logistics knowledge about what CLSs may "look like;" the interdisciplinary nature of cloud logistics, which includes aspects of cloud computing, logistics, and the organization of interfirm networks (see Part E); and thus the different backgrounds of the scholars and practitioners involved.

The *use of multiple viewpoints* from which to design and to document architectures is particularly useful for designing the reference architecture of CLSs for two reasons. First, the use of multiple viewpoints allows the overall design of the reference architecture of CLSs to be broken down into several sub-problems, reducing design complexity, on the one hand. On the other hand, as all viewpoints are part of the same architecture description, they can be integrated into a single whole. Second, it allows the interdisciplinary nature of cloud logistics to be accounted for: Different design models and methods can be adopted in different viewpoints, where models and methods may originate from different domains. Thus, a viewpoint can be tailored to the specific design problem at hand. Moreover, this allows for explicitly anchoring the design of the reference architecture of CLSs in the different academic fields associated with cloud logistics.

Although the term "viewpoint" is not explicitly used in logistics literature, using multiple perspectives, with each potentially following a different methodological approach, is a cornerstone of logistics research for considering, describing, and designing logistics systems. As Delfmann et al. (2010:58) remarks, logistical inquiry adopts

> "a multiperspectival model [...], which means that logistics processes are illuminated from different perspectives with different methodological approaches".

We thus conclude that designing and documenting logistics systems from multiple viewpoints is a common logistical practice, even though the term "viewpoint" is not used. Hence, it is a suitable approach for CLSs as well.

The *separation of viewpoints and views* is particularly suitable for designing and expressing a reference architecture for CLSs because it allows for a clear distinction between the principles and concepts of cloud computing applied to logistics systems (viewpoints) and the results of applying them (views). Thus, we can clearly separate the contributions of this thesis from existing cloud logistics knowledge.

We thus conclude that the International Standard establishes a particularly suitable basis for designing and expressing a reference architecture of CLSs – and thus for achieving our research objective – due to its standardized form of expressing knowledge, use of multiple viewpoints, and separation between viewpoints and views.

6. The Process of Creating Architecture Descriptions

The preceding subsection has introduced an ontological model of architecture descriptions. However, this model does not provide any indication regarding the sequence in which the different concepts included in an architecture description are to be created. Following the methodological guidance from ISO on the creation of architecture descriptions (ISO/IEC/IEEE 42010 2011a; ISO/IEC/IEEE 42010 2011c), the development process can be intuitively structured into the following aggregated steps:

(1) Identify stakeholders and their concerns

(2) Select and define viewpoints; record the rationale for the selection

(3) Produce one view per viewpoint; record the rationale for architectural decisions

The sequence of these steps is intuitive: Identifying system stakeholders and their concerns represents the starting point for all design activities as concerns are the input to the design process. The selection of viewpoints defines the total scope of design activities. The viewpoints establish the methodological conventions of how these concerns can be framed. Next, architectural views are produced based on the conventions established by the corresponding viewpoints. Key design decisions and their rationales are recorded in these views.

These steps define the structure of the remainder of this part. Chapter F.II identifies the stakeholders and their concerns relative to CLSs. Chapter F.III identifies the viewpoints for designing a reference architecture of CLSs and provides a rationale for their selection. Chapter F.IV – Chapter F.XII provide a detailed definition of viewpoints and their associated views.

II. Stakeholders and Their Concerns

1. Abstract of Chapter

Stakeholders

Following existing knowledge, we derive three stakeholder classes for the CLSs reference architecture from the "SOA triangle:" (a) *LSPs* are firms that control logistics resources and have the capacity to deploy them to achieve logistics-related goals (logistics capabilities) and whose primary business purpose is to use these capabilities to achieve logistics-related goals of another entity against financial remuneration; (b) *logistics consumers* are any entity that requests and remunerates another entity to achieve its logistics-related goals on its behalf; and (c) *logistics IT platform and application provider (LITPAP)* is an IT vendor that uses its IT capabilities to make the cloud computing based platform available and that uses its logistics-related IT capabilities to make available the logistics-related functionality implemented on the platform (see Subsection F.II.3.1).

Stakeholder Concerns

Current cloud logistics knowledge does not provide a clear understanding of concerns held by the three identified stakeholder classes. Hence, concerns included the reference architecture draw on general logistics and organizational literature. They are structured into three interlinked categories: (a) *logistical concerns* (or requirements) represent any interest related to the achievement of logistical goals, including goals related to the planning, realization, control, and monitoring of logistics activities; (b) *economic concerns* represent any interest related to the achievement of pecuniary goals related to the achievement of logistical goals (similar to competitive advantage); and (c) *governance and organizational concerns* represent any interest related to the governance structure that regulates value exchanges between stakeholders and to the organizational structure that determines the division and coordination of labor among stakeholders related to the achievement of logistical goals (see Subsection F.II.4.1).

The five essential cloud characteristics are also interpreted as concerns and thus become integrated into the process of designing the reference architecture (see Subsection F.II.4.5). However, these concerns are not a separate category of concerns, but are related to logistical, economic, and organizational concerns. They further specify these concerns in the context of CLSs; moreover, they establish "minimum requirements" for how and the extent to which certain concerns need to be addressed in the reference architecture. In other words, these characteristics constrain the solution space for designing the reference architecture by specifying certain dimensions to be considered and by establishing constraints on these dimensions' values.

In a logistics context, *broad network access* means that logistics capabilities are offered and can be consumed at any geographic location. Following current cloud logistics knowledge, broad network access only pertains to administrative logistics processes, not to the physical transformation of logistics objects. Broad network access is self-evidently related to logistics functionality and specifically to *geographic coverage*.

In a logistics context, *on-demand* availability means that the fulfillment of specified logistical goal(s) begins shortly after the logistics consumer and respective LSPs have reached an agreement on the goals to be achieved and payments to be made. On-demand provisioning thus implies a short initial delay in the delivery of logistics capabilities that are – currently not provisioned – by LSP(s) for the specific logistics consumer. On-demand provisioning is related to *logistical effectiveness* because it specifies a temporal property with which logistical goals are to be achieved in CLSs.

In a logistics context, *rapid elasticity* means that the capacity of logistics capabilities – currently being provisioned – can be quickly scaled commensurate with demand by deploying additional or removing current resources thereby avoiding under- or overprovisioning, respectively. Thus, the capacity of logistics capabilities appears to be unlimited to consumers and can be appropriated at any time in any quantity. Similar to cloud computing, logistics elasticity refers to two core properties of the capacity reconfiguration process: speed and precision of scaling. Due to the physical tangibility of logistics resources and the large "unit capacity" of many logistics resources compared to a single consumer's actual demand, both speed and precision of scaling are lower than in cloud computing. Rapid elasticity is related to *logistical effectiveness* by aiming to reduce the likelihood of underprovisioning by requiring that any amount of additional capacity can be made available instantaneously and *logistical efficiency* by aiming to reduce logistics costs by "always" matching capacity provisioned with actual demand.

In a logistics context, *resource pooling* means that logistics capabilities are not dedicated to fulfilling the logistical goals of a single consumer, but the logistical goals of at least two consumers are fulfilled with deployed resources, either simultaneously or successively. Resource pooling is related to *logistical efficiency* because it refers to the degree to which deployed logistical resources are used.

In a logistics context, *pay-per-use* (or *measured service*) means that only the capacity of logistics capabilities actually consumed is charged to logistics consumers, and the amount of capacity consumed is made transparent to both consumers and LSPs. Pay-per-use is related to *economic concerns* because it specifies the underlying principle that determines any financial exchange between stakeholders.

In a logistics context, *self-service* means that logistics consumers can gain access to logistical capabilities without interacting with an LSP representative. Self-service does not apply to physical capabilities that transform goods in time and space, but only to the online procurement of logistics capabilities and administrative logistics capabilities related to the physical logistics capabilities. Self-service is related to *organizational concerns* because it specifies the mode logistics consumers use to interact with LSPs in order to access and manage logistical capabilities.

2. Abstract vs. Concrete Stakeholders and Concerns

Reference architectures are designed at a more abstract level than concrete system architectures (see Subsection F.I.2.2). We account for this abstraction by identifying and describing "classes of stakeholders" and "classes of their concerns," instead of concrete stakeholders and concerns. For example, LSPs are a stakeholder class, while DHL, a provider of logistics services, is a concrete stakeholder within that class. Similar, financial goals represent a class of concerns, while the goal of achieving a 5% gross profit margin represents a concrete concern.

3. Stakeholders

3.1. Identification of Stakeholders

Reference architectures are derived from existing architectures and knowledge: They extract and capture the most important aspects, usually at a higher level of abstraction (see Subsection F.I.2.2). Hence, we derive the stakeholder classes to be included in the reference architecture of CLSs from the deployment models found in existing cloud logistics knowledge.

The cloud logistics sources describe several deployment models for CLSs (see Subsection E.IV.5.4). Following the sources that derive their deployment models from SOA, we derive the fundamental stakeholder classes of CLSs from the "SOA triangle:" (1) LSPs, (2) logistics consumers, and (3) LITPAP. This triad of stakeholders is also supported by the sources that derive their deployment model from cloud computing. These sources equally distinguish between "providers" and "consumers," and they also mention some kind of "logistics platform provider" (see e.g. Delfmann & Jaekel 2012:20; Colajanni 2012:4).

The following three subsections briefly define and describe each stakeholder class.

3.2. Logistics Service Providers

An *LSP* is (a) a firm (b) that controls logistics resources and has the capacity to deploy them to achieve logistics-related goals (logistics capabilities) and (c) whose primary business purpose is to use these capabilities to achieve the logistics-related goals of another entity against financial remuneration.

The term "firm" shall imply that LSPs are profit-seeking entities. Note that LSPs do not necessarily need to own logistics resources, but it is sufficient if they exercise actual control over resources (e.g. through leasing contracts).

The understanding of "logistics resources" and "logistics capabilities" can be derived from the general definitions of resources and capabilities. *Resources* can be defined as

> "stocks of available factors that [...] are converted into final products or services by using a wide range of other firm assets and bonding mechanisms such as technology, management information systems, incentive systems, trust between management and labor, and more" (Amit & Schoemaker 1993:35; see similar Day 1994:38; Grant 1991:118ff.).

Examples of resources include physical assets, information, and human capital. Accordingly, *logistics resources* can be defined as any input factor necessary to support or carry out the planning, realization, control, and monitoring of logistics activities. By contrast, *capabilities*

> "refer to a firm's capacity to deploy *Resources*, usually in combination, using organizational processes, to effect a desired end" (Amit & Schoemaker 1993:35, italics and capitals original; see similar Day 1994:38; Grant 1991:119ff.).

Capabilities are "complex bundles of skills and accumulated knowledge, exercised through organizational processes" (Day 1994:38) that, unlike resources, cannot be traded or imitated as they are "so deeply embedded in the organizational routines and practices" (Dierickx & Cool 1989 as cited in Day 1994:38). Accordingly, *logistics capabilities* refer to the capacity of LSPs to deploy logistics resources in order to support or carry out the planning, realization, control, and monitoring of logistics activities to achieve desired logistics-related goals.

The specification that an LSP is a firm "whose primary business purpose is to utilize these capabilities to perform logistics activities on behalf of another entity in order to achieve the other entity's logistics-related goals" shall limit the types of firms with logistics capabilities that can act as LSPs to those that specialize in offering their logistics capabilities on the market (Pfohl 2010:16). Thus, firms that control logistics resources and have logistics capabilities but only pursue logistics goals as secondary goals in order to ultimately achieve their primary non-logistics business goals are not considered LSPs. Typical examples of such firms are large food retailers that usually operate large distribution networks.

The clause "against financial remuneration" shall imply that the relationship between LSPs and the entity on whose behalf they pursue logistics-related goals resembles a typical buyer-supplier relationship that involves transactions of services for money.

3.3. Logistics Consumers

A *logistics consumer* is (a) any entity that (b) requests and remunerates another entity to achieve its logistics-related goals on its behalf.

The term "any entity" shall specify "who" can act as a logistics consumer. We use this term to only impose the absolute minimum of restrictions on "who" can do so. This is reasonable given the abstract nature of reference architectures. Examples of logistics consumers include firms in the commercial trading sector or manufacturing industry that request and consume logistics capabilities in order to source and distribute physical goods (e.g. raw materials and semi-finished or finished goods) as part of their value creation processes. Moreover, logistics consumers can also include LSPs. This is the case if LSPs integrate the capabilities of multiple providers (which may include their own capabilities) in order to achieve (more complex) logistics-related goals for another entity, which the LSP may not be able to achieve in isolation.

The expression "requests and remunerates another entity to achieve its logistics-related goals on its behalf" shall indicate that logistics consumers create demand for logistics services within CLSs: They have a logistical need and are willing to pay another entity to satisfy this need.

3.4. Logistics IT Platform and Application Provider

A *LITPAP* is (a) an IT vendor (b) that uses its IT capabilities to make the cloud computing based platform available and (c) that uses its logistics-related IT capabilities to make available the logistics-related functionality implemented on the platform.

The term "IT vendor" shall specify the type of entity that can act as a LITPAP. Using the definition of LSPs, we define an IT vendor as a firm that controls IT resources, has the capacity to deploy them to achieve IT-related goals (IT capabilities), and whose primary business purpose is to use these capabilities to achieve IT-related goals on behalf of another entity for financial remuneration. *IT resources* include physical resources (e.g. servers, storage space, and networking equipment) and virtual resources (e.g. service description, virtual machines). *IT capabilities* include conceiving, developing, certifying proper implementation of, maintaining, and improving the architecture of a computing system, such as a cloud computing platform (see ISO/IEC/IEEE 42010 2011b:1). However, for an IT vendor to act as a LITPAP in a CLS, the vendor requires not only general IT capabilities but also *logistics-related IT capabilities.* These are necessary to implement logistics-related functionalities on the cloud-based platform, especially to support, carry out, and optimize the planning, realization, control, and monitoring of logistics activities and to enable the interactions and economic exchanges between LSPs and logistics consumers through the platform.

The clauses that the LITPAP uses "its IT capabilities to make available the cloud computing based platform" and "that utilizes its logistics-related IT capabilities to make available the logistics-related functionality implemented thereon" shall specify the two primary responsibilities of an IT vendor in a CLS. Accordingly, we refer to this stakeholder class as a LIT*PAP*. Several cloud logistics sources refer to this stakeholder class simply as a "platform provider," without clearly describing the responsibilities (see Subsection E.IV.5.4). Only one source clearly distinguishes between the platform provider and the providers that deploy their applications on the platform (see Subsection E.IV.3.5). In this reference architecture, we identify the two responsibilities separately, as they are fundamentally different from each other. At the same time, we consolidate them into a single stakeholder class, as we aim to develop a cloud *logistics* reference architecture rather than investigate the different responsibilities of IT vendors in such systems. In other words, the center of gravity of this thesis lies in the logistics domain, not IT.

While logistics consumers and LSPs are both "classic" stakeholder classes in logistics systems, the LITPAP is a comparatively "new" class. This stakeholder class is identified by several of the sources – usually without a clear rationale. We propose the following three reasons for supporting the inclusion of this stakeholder class in the reference architecture of CLSs. First, cloud computing is a technology commonly associated with the outsourcing to or procurement of IT resources and capabilities from third party IT vendors. As the platform in CLSs is cloud-based, one can assume that, in line with common industry practice, the platform is also made available by a third party IT vendor, instead of a logistics consumer or an LSP. In fact, cloud logistics sources that mention a "platform provider" commonly consider the platform and the software thereon to be made available by some form of a third party IT vendor (see Section E.IV.3; Section E.IV.4; Section E.IV.5).

Second, the undeniably high complexity of today's IT systems and the rapid speed of innovation in the IT industry, especially with regard to cloud computing, make it very unlikely for either logistics consumers or LSPs to independently realize, maintain, and evolve a cloud computing (or similar) technology-based logistics IT platform and the applications implemented thereon effectively and efficiently. In other words, only a specialized IT provider can achieve average industry costs and keep up with average innovation speed.

Third, the applications on the cloud-based platform store, maintain, and handle a large pool and flow of logistics-relevant information. On the one hand, this makes the platform and applications thereon an interesting target for any kind of cyber-attack and therefore requires professional security services, which most likely can only be offered by a professional IT vendor. On the other hand, LSPs and consumers may only be willing to store and maintain logistics-relevant data on the platform and use its applications if the LITPAP is neither a proximate nor distant competitor but a "neutral entity" that limits its business purpose on making the platform and its functionalities available. In this respect, Stinnes (2012:8) mentions a "neutral platform" and implies that it is made available by a third party, which we interpret as an IT vendor (see Subsection E.IV.5.4). Hence, similar to LSPs that act as "neutral arbitrators" between consignors and consignees (Zacharia et al. 2011:45), the LITPAP acts as another neutral entity between consumers and LSPs.

4. Stakeholder Concerns and Essential Cloud Characteristics

4.1. Identification and Model of Stakeholder Concerns

The preceding section has identified and described three stakeholder classes. This section identifies and describes the concerns these stakeholders are assumed to hold relative to CLSs. Recall that concerns are connected to influences originating in a system's environment and they represent the interests relative to a system that are held by one or more of its stakeholders; as such, they become the dimensions that need to be considered when designing a (reference) architecture (see Subsubsection F.I.5.2.2 and Subsection F.I.4.2). Due to the novelty of cloud logistics research, current cloud logistics knowledge does not provide a comprehensive understanding of stakeholder concerns. Hence, we draw on logistics and organizational literature to first establish a general categorization of concerns. We then use the existing cloud logistics knowledge related to the essential cloud characteristics to further detail this categorization.

Figure 35 depicts the proposed conceptual model of stakeholder concerns in the reference architecture of CLSs. Concerns are structured into three interlinked categories: (1) logistical, (2) economic, and (3) governance and organizational concerns.

First, *logistical concerns* represent any interest related to the achievement of logistical goals, specifically including goals related to the planning, realization, control, and monitoring of logistics activities. Logistical concerns are held by logistics consumers as well as by LSPs and the LITPAP, albeit from opposite perspectives. Consumers raise lo-

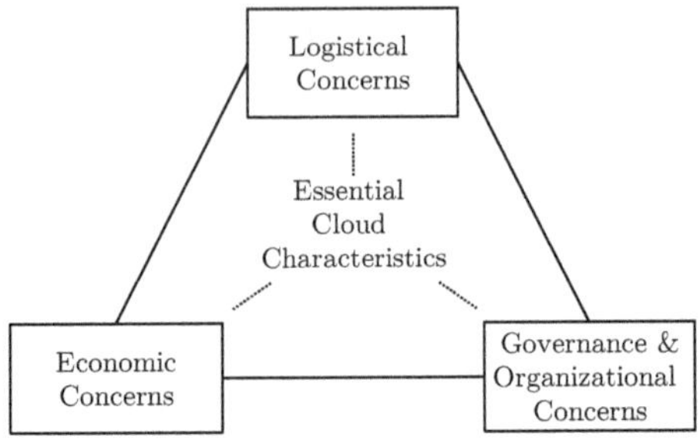

Figure 35.: Conceptual Model of Stakeholder Concerns and Relationships to Essential Cloud Characteristics

gistical concerns. LSPs and the LITPAP aim to address these concerns, although by different means due to the different resources and capabilities at their disposal. Logistical concerns can be interpreted as the *logistical requirements* of logistics consumers (see Subsection F.I.2.4; also see Rozanski & Woods 2011:22). Below, we use the terms "logistical concerns" and "requirements" interchangeably.

Second, *economic concerns* represent any interest related to the achievement of pecuniary goals in direct relation to the achievement of logistical goals. Addressing economic concerns equals the aim to achieve competitive advantage – that is, persistently earning (or having the potential to earn) a higher rate of profit than rivals (Grant 2010:211). Economic concerns are held by all stakeholder classes, as all ultimately aim to make a profit in the process of attaining logistical goals.

Third, *governance and organizational concerns* represent any interest related to the governance structure that regulates value exchanges between stakeholders and to the organizational structure that determines the division and coordination of labor among stakeholders in order to achieve logistical goals. Governance and organizational concerns are held by all stakeholders, as structure determines the ability of all stakeholders to achieve their ultimate pecuniary goals that they pursue in relation to logistical goals.

These three categories of concerns are interlinked. Governance and organizational concerns are linked to economic concerns because the governance structure determines the costs of making transactions, and the organizational structure determines the allocation of logistics tasks, costs, and revenue among stakeholders. Governance and organizational concerns are also linked to logistical concerns because the types of logistics capabilities exchanged critically influence transaction costs, and the organizational structure determines the ability of LSPs to achieve the logistical goals of consumers in an efficient and effective manner. Logistical and economic concerns are linked because the nature of logistical concerns determines the costs that accrue and the revenue that can be generated by achieving these logistical goals.

In addition to the three categories of concerns, Figure 35 also depicts the essential cloud characteristics (see Subsection E.IV.5.3). These characteristics are interpreted as concerns and thus become integrated into the process of designing the reference architecture of CLSs. However, they are not a separate category, but are related to logistical, economic, and organizational concerns, as shall be indicated the dotted lines. The essential characteristics further specify these concerns in the context of CLSs; moreover, they establish "minimum requirements" for how and the extent to which certain concerns need to be addressed by the cloud logistics reference architecture. In other words, these characteristics constrain the solution space for designing the reference architecture of CLSs by specifying certain dimensions to be considered and by establishing constraints on these dimensions' values in the reference architecture. The specific understanding of these essential characteristics in a cloud logistics context, and the minimum requirements for addressing these characteristics, are considered after the logistical, economical, governance and organizational concerns are introduced in detail.

4.2. Logistical Concerns

4.2.1. Model of Logistical Concerns

Logistical concerns are related to the logistical goals regarding the planning, realization, control, and monitoring of logistics activities. Logistical concerns are raised by logistics consumers and are addressed by LSPs and the LITPAP through their respective capabilities. As different logistics consumers may operate in different geographies, be part of different industries, offer products and services with various characteristics, and pursue different competitive strategies, logistical concerns can take a variety of forms – some of them common for many logistics consumers across industries, others unique to a single consumer's single product (for examples of varying logistical requirements see e.g. Cooper 1993;Salin & Jr 2003:918). This diversity of logistics requirements raises the questions of which and at which level of abstraction logistics concerns should be considered when designing the reference architecture of CLSs.

To design a reference architecture for CLSs, we propose structuring logistical concerns into two basic categories: (1) logistical functionality and (2) logistical performance. *Logistical functionality* represents any interest related to what kind of logistics capabilities a logistics system offers and thus what kind of logistical goals can be achieved. *Logistical performance* represents any interest related to the attributes with which logistics functionality can be made available. The fundamental categorization of logistical concerns into functionality and performance is motivated by existing literature on logistics and architecture design. Logistics literature commonly refers to logistics requirements by specifying the type of logistics process (see e.g. Pfohl 2010) and the "quality of service" (see e.g. Chow et al. 1994). In architecture design literature, stakeholder concerns are equally categorized into "functionality" (i.e. what a system does) and "system qualities" (i.e. how a system does it) (Maier et al. 2001:108f.; Lago et al. 2010:22).

Logistical functionality is further divided into three sub-categories: (1a) scope of logistical transformations, (1b) scope of freight, and (1c) geographic coverage. Performance is further divided into (2a) effectiveness, and (2b) efficiency. These sub-categories are considered in detail below.

4.2.2. Logistical Functionality

Logistical functionality refers to what kind of logistics capabilities a logistics system offers and thus what kind of logistical goals can be achieved. Functionality is further divided into (1a) scope of logistical transformations; (1b) scope of freight; and (1c) geographic coverage. These three sub-categories collectively specify the functionality of a logistics system by specifying what kind of logistical transformations can be achieved, for which scope of freight, and where.

Logistical concerns regarding the *scope of logistical transformations* represent any interest related to how a logistics system can initiate or manipulate the flow of physical goods and associated information. Logistics literature contains various categorizations of logistics activities (see e.g. Liu & Lyons 2011; Murphy & Poist 2000). However, these categorizations are often very detailed and therefore not abstract enough for a reference architecture. Hence, we draw on the basic categorization of core logistics processes and associated transformations, including storage, transportation, handling, packaging, load unitization, labeling, and order processing (see Section B.III.1). Regarding these transformations, logistical concerns can pertain to any or all of the following activities: planning, realization, control, and monitoring. In some cases, these logistical concerns may be connected to non-logistical concerns that relate to transforming goods in their form, such as final assembly. However, as our primary focus lies in designing a reference architecture of cloud *logistics* systems rather than cloud manufacturing, we do not further detail these concerns.

Logistical concerns regarding the *scope of freight* represent any interest related to the type and variety of freight for which logistical transformations can be performed. The scope of freight critically depends on the freight's economic and physical freight properties because physical logistical resources need to be designed and staff trained in different ways for handling freight with different properties (see Section B.III.2). As mentioned above, key economic and physical properties include value, size, weight, and susceptibility to damage.

Logistical concerns regarding *geographic coverage* represent any interest related to the spatial scope within which logistical transformations can be performed. The nature of geographic scope differs by type of logistics process. For logistics processes that transform goods in space (transportation), geographical coverage refers to the distance between or the areas within which goods can be moved. For example, inter-continental transportation implies large geographic coverage, while a local courier service implies small coverage. For any other type of logistics process, geographic coverage refers to the location(s) at which these transformations can be accomplished. For example, freight can be stored in a warehouse at a specific location.

4.2.3. Logistical Performance

Logistical performance refers to concerns related to attributes with which logistical functionality is made available and thus with which logistical goals can be attained. An analysis of logistics literature suggests a heterogeneous understanding of logistics performance: A plethora of "quality" attributes or criteria are suggested to specify logistical performance. Some scholars try to define logistics performance by means of various "rather specific" attributes, such as on-time delivery, low damage (Chow et al. 1994), order accuracy, and order release quantities (Mentzer et al. 1999). Other scholars try to define logistics performance by means of a few "more abstract" dimensions, such as operational performance, relational performance, cost performance (Stank et al. 2003), effectiveness, and efficiency (Mentzer 1991).

The large variety of attributes raises the question of how and at which level of abstraction logistics performance should be conceptualized in this reference architecture. Given the abstract nature of reference architectures, it is reasonable to adopt few more abstract dimensions instead of many very specific attributes. Hence, we follow Mentzer (1991:33f.) and define logistical performance along the two basic dimensions of (2a) logistical effectiveness, defined as "the extent to which goals are accomplished," and (2b) logistical efficiency, defined as "the measure of how well the resources expended are utilized."

4.3. Economic Concerns

Economic concerns are related to the achievement of pecuniary goals in direct relation to the achievement of logistical goals. These concerns are linked to achieving competitive advantage. Economic concerns are held by all stakeholder classes. For logistics consumers, economic concerns are, on the one hand, related to costs that accrue from tasking LSPs and the LITPAP with achieving or supporting the achievement of their logistical goals, respectively. On the other hand, they are related to the value they can derive from achieving logistics goals, which is equal to the money received for selling their products to their clients. For LSPs and the LITPAP, economic concerns are related to the revenue that can be generated and the costs that accrue for achieving the logistical goals on behalf of logistics consumers.

4.4. Governance and Organizational Concerns

Governance concerns represent any interest related to the governance structure that regulates value exchanges between logistics consumers, LSPs, and the LITPAP. Hence, concerns are especially related to safeguarding against opportunistic behavior. As the alignment of governance structures with transaction characteristics determines the amount of transaction costs, governance concerns are closely related to economic concerns.

Organizational concerns represent any interest related to the organizational structure that determines the division and coordination of labor in the process of achieving logistical goals. As logistical goals are achieved through some form of collective action in CLSs, involving a network of horizontally cooperating LSPs and the LITPAP (see Subsection E.IV.5.4), organizational concerns specifically relate to the *interorganizational structure* and whether this structure can enable effective attainment of logistical goals formulated by logistics consumers.

4.5. Essential Cloud Characteristics – Description and Relation to Concerns

4.5.1. Preliminary Considerations

The preceding subsections have introduced a general categorization of the concerns that are relevant to designing "any type" of logistics system. They are also relevant to designing a CLS. In addition to these, this subsection focuses on specific concerns within these general categories that are crucial for designing a *cloud* logistics system: the essential cloud characteristics (see Subsection E.IV.5.3). Existing cloud logistics knowledge refers to these characteristics without providing a clear definition or relationship to logistical, governance, organizational, and economic concerns. In other words, so far it has remained unclear whether these characteristics are entirely "new" types of concerns or whether they can be related to existing categories of concerns. Figure 36 depicts the proposed relationship between the essential cloud characteristics and the general categories of concerns. Establishing these relationships anchors the essential cloud characteristics in existing logistics literature, thus anchoring the design of CLSs within existing literature – our research objective. The essential characteristics detail these general categories by

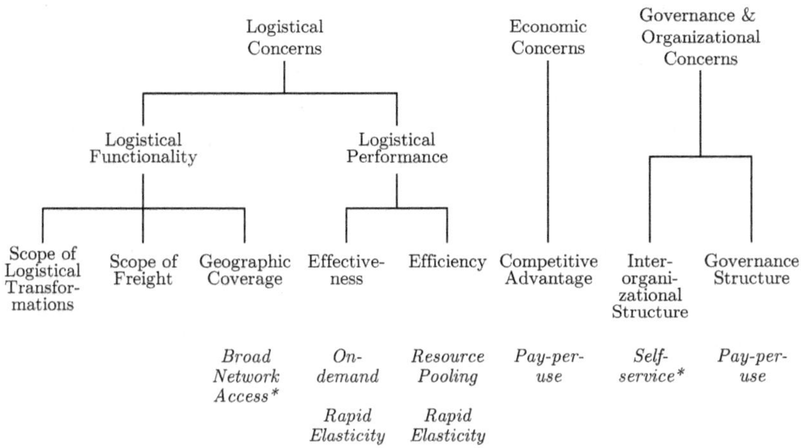

Figure 36.: Proposed Relationships between Essential Cloud Characteristics and Concerns (* = only for administrative logistics capabilities)

specifying the dimensions that are of particular importance for designing the reference architecture of CLSs. Moreover, these characteristics establish "minimum requirements" for how and for the degree to which these dimensions need to be achieved; they thus constrain the solution space for the cloud logistics reference architecture. However, the essential characteristics do not establish minimum requirements for all concerns. For example, the scopes of logistical transformations and freight remain unconstrained. Still, despite the lack of direct constraints, achieving the essential characteristics may nevertheless reduce the solution space for the remaining concerns. Hence, we encounter a question about what design choices would be "feasible" for those concerns that are not constrained by the essential characteristics. Specifically, the question of what logistical functionality can be offered by CLSs has not yet been answered.

The essential cloud characteristics are briefly defined below based on their meanings in cloud computing. Their relationships to the existing stakeholder concerns are explained as well, and their relevance for logistics consumers is emphasized. Emphasizing relevance is crucial because only concerns that are relevant to at least one stakeholder class can be incorporated into the design process. Yet we only emphasize relevance for logistics consumers because, if concerns are relevant for consumers, they are also relevant for LSPs and the LITPAP in a mediated manner, as they need to fulfill consumers' logistics requirements.

4.5.2. Broad Network Access

In a logistics context, *broad network access* means that logistics capabilities are offered and can be consumed at any geographic location; for example a transportation capability can connect any pair of locations. However, existing cloud logistics knowledge suggests that broad network access does not apply to capabilities that "physically" transform goods in time and space (see Subsection E.IV.5.3). Instead, based on current knowledge, we can argue that broad network access applies to geographic accessibility of administrative logistics capabilities only, which include the planning, control, and monitoring of physical logistics processes.

Broad network access is self-evidently related to *geographic coverage*; specifically, broad network access establishes a minimum requirement for the locations in which administrative logistics capabilities must be accessible in CLSs: ubiquitously.

Ubiquitous access to administrative logistics capabilities is relevant for logistics consumers because it can create management flexibility: logistical processes can be managed at anytime from anywhere. Thus, logistics consumers can respond to environmental uncertainties more quickly, which can contribute to supply chain agility.

4.5.3. On-demand availability

In a logistics context, *on-demand* availability means that the fulfillment of specified logistical goal(s) begins shortly after the logistics consumer and respective LSPs have reached an agreement on the goals to be achieved and payments to be made. On-demand provisioning thus implies a short initial delay in the delivery of logistics capabilities that are – currently not provisioned – by LSP(s) for the specific logistics consumer. Examples of on-demand provisioning include picking up freight at a new logistics consumer's location on the same day a contractual agreement has been reached.

On-demand provisioning is related to *logistical effectiveness* because it specifies a temporal property with which logistical goals are to be achieved in CLSs; moreover, it establishes a minimum requirement on this temporal property: Goal achievement must start briefly after an agreement has been concluded. In other words, logistics consumers must be able to access a logistics capability potential without initial delay (see Section B.III.3). However, note that on-demand provisioning establishes a minimum requirement for the beginning of the fulfillment process, but not the end. In fact, the time until completion of the fulfillment varies based on nature of the logistical goals, e.g. an intercontinental ocean transport requires more time than a local e-commerce shipment.

The relevance of on-demand provisioning for cloud logistics stakeholders is examined together with rapid elasticity, as both characteristics refer to the temporal characteristics with which logistics capabilities can be made available in CLSs.

4.5.4. Rapid Elasticity

In a logistics context, *rapid elasticity* means that the capacity of logistics capabilities – currently being provisioned – can be quickly scaled commensurate with demand by deploying additional or removing current resources in order to avoid under- or overprovisioning, respectively. As a result, the capacity of logistics capabilities appears to be unlimited to logistics consumers and can be appropriated at any time and in any quantity.

Like in cloud computing, elasticity in logistics is related to scalability. A scalable logistics system can maintain its performance level in the event of increasing freight volumes, provided that resources are added proportionally. Scalability of logistics systems may be limited in some cases, which means that resources can only be added up to a certain maximum point, which is given by the upper scalability bound. For example, additional flights can only be added to an airport up to the maximum capacity of flights per hour. Similar to cloud computing, logistics elasticity refers to two core properties of this adaptation (or reconfiguration) process: (1) speed of scaling, and (2) precision of scaling (see Subsection D.II.2.4). *Speed of scaling* refers to the time required for deploying additional logistics resources (or removing currently deployed logistics resources) in order to move from an underprovisioned state to an optimal or overprovisioned state (or move from an overprovisioned state to an optimal or underprovisioned state). *Precision of scaling* refers to the deviation between the capacity of currently deployed logistics resources and the capacity that is actually demanded.

The upper scalability bound and elasticity properties depend on the types of logistics resources, like in cloud computing (Herbst et al. 2013:24). They moreover depend on the differences between the "current physical state" of the resources to be adapted and their "target physical state," in which they become productive and contribute to the achievement of logistical goals. Due to the large variety of logistics resources and the infinite number of potential states logistics resources can occupy in relation to target states to achieve logistical goals, investigating the upper scalability bound and the elasticity properties for each conceivable combination is unfeasible. Hence, we investigate the properties of logistics elasticity on a more abstract level; we focus on the "nature of adaptation processes" for three assumptions regarding the upper scalability bound, independent of resource types, specific resource states, and logistical goals. Accordingly, three "elasticity types" are defined as: (1) intra-resource elasticity, (2) intra-LSP elasticity, and (3) inter-LSP elasticity. *Intra-resource elasticity* means that LSPs currently provisioning logistics capabilities allocate additional logistics resources or remove resources in order to scale the capacity of logistical capabilities commensurate with demand. For example, a logistics consumer is allocated additional storage capacity in a warehouse the consumer is already using. *Intra-LSP elasticity* means that LSPs currently provisioning logistics capabilities deploy additional or remove whole units of logistics resources in order to scale the capacity of logistics capabilities commensurate with demand. For example, an additional truck is deployed to handle an increase in freight volume. *Inter-LSP elasticity* means that LSP(s) currently *not* provisioning logistics capabilities deploy additional whole units of logistics resources in order to scale the capacity of logistics capabilities commensurate with demand. The properties of these elasticity types are summarized in Table 21.

The three elasticity types differ in their assumptions regarding the upper scalability bound. In the case of intra-resource elasticity this bound is determined by the maximum capacity of logistics resources currently deployed, for example, the maximum load capacity of a single truck. For intra-LSP elasticity it is determined by the aggregate logistical capacity controlled by the LSPs currently provisioning logistics capabilities, for example the truck fleets of active providers. Finally, for inter-LSP elasticity, it is determined by the aggregate logistical capacity controlled by LSPs that participate in a CLS but are currently *not* provisioning logistics capabilities. The upper scalability bound thus increases from intra-resource elasticity to inter-LSP elasticity, which implies that an increasing magnitude of demand surges can be compensated for.

The *adaptation precision* (or precision of scaling) is low in logistics, regardless of the upper scalability bound. This is due to the fact that logistics resources are only deployable in discrete units whose "unit capacity" often exceeds the capacity required to achieve the logistical goal(s) of a single consumer. For example, transporting a single pallet for a logistics consumer still requires dispatching a whole truck. Hence, in the case of demand surges, often only more capacity than is actually required can be added, while, in the case of demand declines, excess capacity of the resources currently deployed cannot be removed. To conclude, the large unit capacity of logistics resources increases the likelihood

Elasticity Type	**Upper Scalability Bound**	**Adaptation Precision**	**Adaptation Speed**
Intra-resource elasticity	Maximum capacity of resources currently deployed	low	high
Intra-LSP elasticity	Aggregate capacity controlled by those LSP(s) currently provisioning capabilities	low	low – high
Inter-LSP elasticity	Aggregate capacity controlled by those LSP(s) currently *not* provisioning capabilities	low	low – high

Table 21.: Types of Logistical Elasticity

of overprovisioning if logistical performance levels are maintained.

The three types of logistics elasticity differ in their *adaptation speed* (or speed of scaling). These differences result from the different types of adaptations necessary to add or remove logistical capacity. Intra-resource elasticity does not require any physical adaptation of resources already deployed. Hence, capacity can be allocated or de-allocated instantly, without any delay. For example, a truck that transports a single pallet today can immediately carry a second pallet.

Intra-LSP elasticity, by contrast, does require some form of physical resource adaptation. For example, additional trucks need to travel to the desired pick-up location, staff needs to be trained to handle a new type of cargo, or warehouse racks need to be adjusted to fit a new cargo type. The time required for carrying out these reconfigurations depends on the type of resources and on the difference between the current states of the resources to be adapted and their target states in which they will become productive for achieving a desired goal. For example, moving a truck to a desired pick-up location may take longer than moving a small van (resource type). However, the time required for moving the truck or van to the pick-up location also depends on their current locations (physical difference between current and target states of resources) (Leukel et al. 2011a:17). Furthermore, resource adaptations may be subject to certain external factors, such as physical or regulatory constraints, that prevent or delay the reconfiguration process. For example, railway wagons can only be added at a railway yard, additional flights can only be scheduled in the vacancies of an airport's existing flight schedule, and traffic regulations may constrain the routing of trucks. Hence, the speed of adaptations can vary significantly, ranging from low to high.

In addition to physical resource adaptations as described above, inter-LSP elasticity requires adaptations of (a) legal relations, e.g. negotiating a contract with another LSP; (b) organizational interfaces, e.g. setting up joint decision-making bodies to govern the delivery of logistics capabilities; and (c) technical interfaces, e.g. interfacing IT systems to exchange logistics-relevant information between stakeholders (see Leukel et al. 2011a:17). Hence, adaptation speed in inter-LSP elasticity can vary significantly, ranging from low to high.

To conclude, considering the speed of adaptations for each of these elasticity types, one can suggest that adaptation speed on average decreases from intra-resource to inter-LSP elasticity due to the increasing scope of the necessary adaptations. Furthermore, compared with cloud computing, the speed of scaling is generally lower due to the additional physical, organizational, and technical constraints (Leukel et al. 2011a:17).

Rapid elasticity is related to *logistical effectiveness* because it aims to reduce the likelihood of underprovisioning. In order to ensure that logistical goals are met effectively, rapid elasticity characterizes the temporal and quantitative aspects of how logistical capacity needs to be adjusted in case of demand surges in those logistics capabilities currently being provisioned[1]. Rapid elasticity establishes two minimum requirements in this respect: (a) additional capacity needs to be made available without delays for (b) demand surges of any magnitude (capacity appears to be unlimited). Considering the three elasticity types described above shows that the adaptation speed may, in some cases, decrease for an increasing upper scalability bound. In other words, the larger the requested capacity adjustment, the longer the adjustment will take in some cases. Hence, in some cases, the minimum requirement for scaling capacity without limits conflicts with the minimum requirement of scaling capacity rapidly. Therefore, the CLS reference architecture needs to identify design solutions that achieve rapid scaling without capacity limits. As elasticity properties vary for different resource types and for different initial resource states in relation to target states, finding such solutions means identifying a set of resource types, a set of initial resource states, and logistical goals such that capacity can be scaled sufficiently quickly for large demand surges.

Rapid elasticity is also related to *logistical efficiency* because it aims to reduce the likelihood of overprovisioning. In order to ensure that logistics resources are used efficiently, rapid elasticity establishes two minimum requirements: (a) idle capacity must be quickly removed (which allows LSPs to redeploy capacity to other consumers), and (b) the deployed capacity must match actual resource demand at any time (precision of scaling). The CLS reference architecture must find solutions that satisfy both requirements.

The previous paragraphs have provided a basic understanding of rapid logistics elasticity. The following investigates the relevance of on-demand and rapid elasticity for logistics consumers. The relevance of these characteristics is investigated jointly, as both relate to temporal characteristics with which logistics capabilities can be made available or logistics capacity can be adapted. They become relevant for logistics consumers through *logistical effectiveness*. As logistical capabilities represent crucial components of supply chains, supply chain scholars suggest the importance of on-demand and rapidly elastic logistics capabilities for supply chain performance and thus for the overall competitiveness of logistics consumers. Christopher & Peck (2004:10) argue that a logistics consumer's "ability to respond rapidly in terms of delivery and to be able to cope with short-term changes in volume and mix requirements" contributes to supply chain agility. Zhang et al. (2005:75) find that agility in logistical activities is necessary to be able to "adjust the inventory, packaging, warehousing, and transportation of physical products to meet customer needs, quickly and effectively" and thus achieve customer satisfaction. The relevance of rapid

[1]Note that rapid elasticity only refers to logistics capabilities currently being provisioned, while on-demand relates to capabilities currently not being provisioned.

elasticity and on-demand availability for logistics consumers manifests in the requirements they request from LSPs and their willingness to renew contracts. Jharkharia & Shankar (2007:279) identify LSPs' ability to offer "surge capacity" as a critical criterion for provider selection. Similarly, Lieb & Bentz (2005:7) find that logistics consumers consider the "scalability of the [logistics] solutions offered" to be an important determinant for whether consumers renew contracts with LSPs. We thus conclude that on-demand availability and rapid elasticity frame key logistical and economic concerns for logistics consumers, especially for those that operate in volatile and uncertain environments. These characteristics allow the consumers to quickly make available new capabilities or adjust the capacity of existing capabilities whenever needed, without delay, and at any amount, thus enabling them to compete effectively by "always" implementing the most adequate logistical response to changing conditions.

Additionally, rapid elasticity becomes relevant for logistics consumers through *logistical efficiency* by enabling them to reduce logistics costs by "always" matching the capacity provisioned with the actual resource demand.

4.5.5. Resource Pooling

In a logistics context, *resource pooling* means that logistics capabilities are not dedicated to fulfilling the logistical goals of a single consumer, but the logistical goals of at least two consumers are fulfilled with deployed resources, either simultaneously or successively. Doing so simultaneously means that capabilities serve multiple consumers at the same time, in a consolidated manner, through either temporal or spatial aggregation of logistics consumers' demands. Doing so successively means that capabilities are dynamically (re-)assigned to different consumers according to their demand, at different points in time, in a dedicated manner. Typical examples of pooled resources are multi-user warehouses or freight consolidation in LTL networks. As in cloud computing, there is some form of "location independence." Thus, logistics consumers cannot decide (or only on a higher level of abstraction) which resources are deployed to achieve their goals. For example, logistics consumers cannot request a specific routing of freight, but only that goods are not routed via country x.

Resource pooling is related to *logistical efficiency* because it refers to the degree to which deployed logistical resources are used. In fact, resource pooling establishes the minimum characteristic that logistics capabilities must be shared. The CLS reference architecture must make adequate design choices to enable resource pooling.

Resource pooling is highly relevant for logistics consumers. Pooling increases the utilization rate of resources and thus lowers unit production costs for each logistics consumer, while improving service levels (see e.g. Pooley & Stenger 1992:153; Tyan et al. 2003:55).

For example, pooling can increase delivery cycle time (Tyan et al. 2003:60) and thus contribute to on-demand availability.

4.5.6. Pay-per-use

In a logistics context, *pay-per-use* (or *measured service*) means that only the capacity of logistics capabilities actually consumed is charged to logistics consumers, and the amount of capacity consumed is made transparent to both consumers and LSPs.

Pay-per-use is related to *economic concerns* because it specifies the underlying principle that determines any financial exchange between stakeholders. Moreover, pay-per-use establishes a minimum requirement on the charging model for logistics capabilities in CLSs. The CLS reference architecture must be designed so that consumers only pay for actually consumed capabilities and have transparency of the capacity consumed.

While charging logistics capabilities based on actual consumption is an intuitive concept, metering the actual amount of capacity consumed can be complex. This is due to the question of which factor (or combination of factors) and metering scales should be used; moreover, these factors may need to be tailored to the particularities of logistical capabilities. Factors that can be used as basis for determining actual consumption include (a) physical properties of freight (e.g. weight, volume, density, dangerous goods), (b) economic properties of freight (e.g. security for high-value), (c) duration of use (e.g. storage duration, transit time and distance, carbon emissions), (d) frequency of use (e.g. number of handling steps per shipment, number of pick-ups), (e) point-in-time of use (e.g. weekend), and (f) external factors (e.g. fuel surcharges to pass fluctuations in crude oil prices on to consumers). Some of the above factors may be charged on a continuous (e.g. duration, carbon emissions) and others on a tiered scale (e.g. one to five pickups; 6 to 10 pick-ups) or binary scale, which means that surcharges or fees are levied only in certain situations (e.g. hazardous goods, weekend operation). If charging is based on a tiered or binary scale, the granularity of tiers and the situations for which fees are levied, respectively, need to match actual consumption. For example, a handling fee for dangerous goods should not be levied for handling dangerous goods, but for handling these goods at a specific number of locations. Otherwise, tiered or binary scales may not be considered pay-per-use.

Pay-per-use is relevant for logistics consumers because it creates financial flexibility: Logistics capabilities can be consumed without any up-front, long-term financial commitments or usage independent costs. Hence, pay-per-use appears especially relevant for logistics consumers that lack sufficient funds to make large investments or those that refrain from making long-term commitments or committing to usage independent costs due to a volatile and uncertain environment. Pay-per-use may also be relevant for logistics consumers that aim to improve balance sheet performance, as pay-per-use transforms capital expenditure into operational expenditures from an accounting perspective.

Moreover, the combination of flexible logistics expenses with rapid elasticity is especially intriguing to logistics consumers because they cannot only rapidly scale logistical capacity commensurate with demand, but can equally scale associated financial expenses at the same time. Consequently, logistics consumers neither suffer from excessive logistics costs due to overprovisioning (e.g. vacant warehouse space) nor from opportunity costs incurred from underprovisioning (e.g. lost revenue from stock-outs).

4.5.7. Self-service

In a logistics context, *self-service* means that logistics consumers can gain access to logistical capabilities without interacting with an LSP representative. Based on cloud logistics knowledge, self-service does not apply to physical capabilities that transform goods in time and space, but only to the online procurement of logistics capabilities (see Subsection E.IV.5.3) and the administrative logistics capabilities related to the physical logistics capabilities, including the planning, control, and monitoring.

Self-service is related to *organizational concerns* because it establishes a minimum requirement on the mode logistics consumers use to interact with LSPs in order to access and manage logistical capabilities. This mode must be reflected in the design of the CLS reference architecture.

Self-service is relevant to logistics consumers because it can increase their interaction flexibility by removing spatial and temporal interaction constraints between logistics consumers and LSPs. This can enable logistics consumers to respond to changing environmental conditions more quickly, thus contributing to supply chain agility.

5. Interim Summary

This chapter has introduced the stakeholders of CLSs and their concerns. Following existing cloud logistics knowledge, the CLS reference architecture includes three different classes of stakeholders: logistics consumers, LSPs, and the LITPAP. These stakeholders have logistical, economic, governance, and organizational concerns. Each of these categories has been defined generically based on existing literature.

Moreover, this chapter has proposed a relationship between the essential cloud characteristics and general categories of concerns that stakeholders can have regarding any type of logistics system. This relationship anchors the cloud characteristics in logistics and organizational literature, thus contributing to our second research objective. The cloud characteristics specify important dimensions in these general categories of concerns. They establish minimum requirements for how and the degree to which these dimensions need achieved; they thus constrain the solution space for the cloud logistics reference architecture. However, the essential characteristics do not establish minimum requirements for all concerns. Still, despite the lack of direct constraints, achieving the essential characteristics may nevertheless reduce the solution space for the remaining concerns. Hence, we encounter a question of what design choices would be "feasible" for those concerns that are not constrained by the essential characteristics. Specifically, the question of what logistical functionality can be offered by CLSs has not yet been answered.

III. Selection, Overview, and Outline of Viewpoints

1. Abstract of Chapter

Due to the novelty of cloud logistics, no viewpoints have been developed so far (see Chapter F.III). We propose to frame stakeholder concerns using nine viewpoints that are motivated and based on existing logistics research. Each viewpoint has a different perspective on CLSs, and perspectives differ in the level of granularity at which CLSs are considered. Viewpoint (1) considers CLSs from the most aggregate perspective by focusing on the value creation logic. Viewpoints (2) – (6) consider CLSs on a more granular level, each focusing on a distinct aspect related to the physical logistics infrastructure, structural governance, or organization of CLSs. Viewpoints (7) – (9) consider CLSs on an even more granular level by considering three specific mechanisms that belong to the characteristic mix of coordination mechanisms described in (6). More specifically, viewpoints (7) – (8) are related to adopting the principles of service-orientation for logistics system design. Viewpoint (9) focuses on market mechanisms, which are used to facilitate economic exchanges. Each viewpoint essentially frames logistics, economic, governance, and organizational concerns, either directly or indirectly. This is because these different types of concerns are interlinked. This is also because we model the essential cloud characteristics as concerns, and such "system characteristics" cannot be easily associated with a single system element or framed by a single viewpoint.

2. Selection and Overview of Viewpoints

Viewpoints identify "from where to look" by specifying the concerns to be framed and the appropriate means (e.g. models, design rules) to frame them (see Subsubsection F.I.5.2.3). While the International Standard provides clear guidance for what needs to be included in a viewpoint, it does not stipulate what viewpoints to select for designing and expressing the architecture of a particular system but "encourages the practice of defining or selecting viewpoints appropriate to the system-of-interest" (ISO/IEC/IEEE 42010 2011b:21; see

Subsubsection F.I.5.2.3). This section discusses the selection of viewpoints and provides an overview and rationale for this selection, as required by the International Standard.

Due to the novelty of cloud logistics, no viewpoints have been developed so far. We propose to frame stakeholder concerns using nine viewpoints, as depicted in Figure 37. Each viewpoint has a different perspective on CLSs, and perspectives differ in the level of granularity at which CLSs are considered. Viewpoint (1) considers CLSs from the most aggregate perspective. Viewpoints (2) – (6) consider CLSs on a more granular level, each focusing on a distinct aspect related to the infrastructure, structural governance, or organization of CLSs. Viewpoints (7) – (9) consider CLSs on an even more granular level by considering three specific mechanisms that belong to the characteristic mix of coordination mechanisms described in viewpoint (6).

This selection of viewpoints is motivated by and based on the perspectives typically adopted by logisticians for analyzing and designing logistics systems (see Subsection B.II.3.3). Viewpoint (1) creates an overarching understanding of logistics systems and therefore relates to all perspectives described above. Viewpoint (2) is based on the infrastructural and functional perspectives of logistics systems (see Chapter B.III). Additionally, this viewpoint draws on the construct of asset specificity and is thus related to TCE (see Chapter C.III). Viewpoint (3) is based on the institutional perspective on logistics systems (see Chapter B.IV; Section C.VI.3), so it draws on TCE (see Chapter C.III). Viewpoints (4) – (8) are based on the organizational perspective on logistics systems (see Section C.VI.4). They thus draw on the literature concerned with the organization of interfirm networks (see Chapter C.V). Viewpoint (9) is again based on the institutional perspective of logistics systems. This viewpoint additionally draws on TCE, game theory, and mechanism design (see Section C.III.5). In conclusion, by adopting a multiperspectival design approach comprising viewpoints rooted in different academic fields, we can anchor the design of the CLS reference architecture in current literature.

3. Outline of Viewpoints

Viewpoint (1) establishes the conventions for conceptualizing CLSs and their inherent value creation logic. This logic can be understood as the particular ways in which economic value is created, which value is exchanged through interactions, and thus how value flows among stakeholders in order address their concerns. To conceptualize CLSs and their inherent value creation logic, this viewpoint proposes to follow the abstract nature of reference architecture. Hence, CLSs can be conceptualized by abstracting context- or system-specific details contained in the cloud logistics sources, while keeping the most fundamental architectural elements and their relationships.

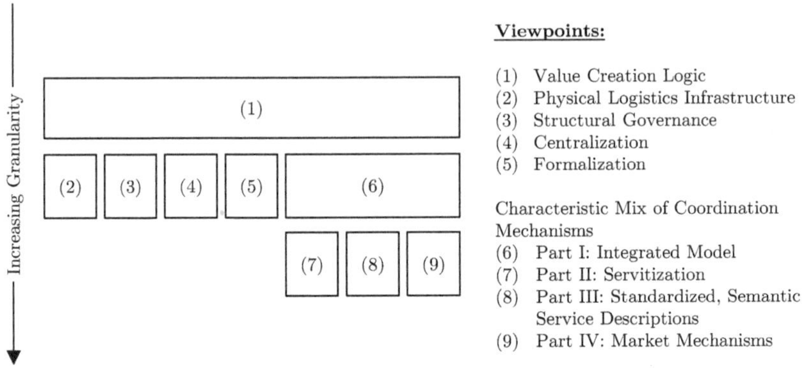

Figure 37.: Proposed Viewpoints of the Cloud Logistics Reference Architecture

Viewpoint (2) establishes the conventions for designing the *physical logistics infrastructure* of CLSs. Although infrastructural design is frequently considered in logistics research, current literature does not provide a model suitable for framing the logistical concerns held by stakeholders in CLSs, especially those related to the essential cloud characteristics, such as on-demand delivery and rapid elasticity. Hence, this viewpoint establishes a new design convention for logistical infrastructure in CLSs. This approach aims to balance concerns regarding cloud characteristics with economic concerns of LSPs and logistics consumers by using the construct of asset specificity.

Viewpoint (3) establishes the conventions for designing the *governance structure* of CLSs. Specifically, this viewpoint proposes determining the governance structure of CLSs as a function of both TCE and the RBV. While TCE provides a well-established framework for determining the governance structure by matching transaction characteristics with governance structure attributes such that transaction costs are minimized, the RBV is commonly not used for this purpose. Recall that the RBV is primarily concerned achieving competitive advantage as a function of the resources accessible to a firm. Despite this initial focus, scholars nevertheless have developed propositions for how the bundles of resources accessible to firms and contributed to an alliance can influence alliance governance structure. Hence, the governance structure of CLSs is determined based on TCE and the RBV.

Viewpoints (4) – (8) establish the conventions for designing the *interorganizational structure* of CLSs. These viewpoints propose conceptualizing the interorganizational structure of CLSs by adopting Grandori & Soda's (1995) tripartite conceptualization consisting of centralization, formalization, and a characteristic mix of coordination mechanisms (see Section C.V.1). Viewpoint (4) specifically establishes the conventions for determining the degree of *centralization*, viewpoint (5) for determining the degree of *formalization*, and viewpoint (6) for conceptualizing the *characteristic mix of coordination mechanisms.* Viewpoints (7) and (8) establish the conventions for designing two specific mechanisms within this characteristic mix. Viewpoint (9) is also part of this mix and establishes the conventions for designing a mechanism that enables market-based exchanges. To design these dimensions of the interorganizational structure of CLSs, we draw on the design logic of contingency theory and SCT (see Section F.I.4).

The use of multiple viewpoints and views generally aims to reduce design complexity through the separation of concerns (see e.g. Hilliard 1999; Rozanski & Woods 2011:38; see Subsubsection F.I.5.2.3). This means that different viewpoints ideally frame different types of concerns, thus reducing design complexity. However, the proposed selection of viewpoints can only partially separate logistical, economic, governance, and organizational concerns. In fact, each viewpoint essentially frames each category of concerns, either directly or indirectly. On the one hand, this is because these different types of concerns are interlinked (see Subsection F.II.4.1). On the other hand, this is because we model the essential cloud characteristics as concerns, and such "system characteristics" cannot be easily associated with a single system element. Again, we quote Kruchten's (2004:55) remark that

> "[p]roperties are harder to trace to specific elements of the design or the implementation because they are often cross-cutting concerns, or they affect too many elements."

Hence, although it is desirable, the essential cloud characteristics cannot be framed using single viewpoint. Due to the importance of the essential cloud characteristics for cloud logistics, we need to accept and manage this complexity in the following design process.

The following eight chapters each discuss one viewpoint and its associated view.

IV. Value Creation Logic

1. Abstract of Chapter

The value creation logic can be understood as the total sum of ways in which economic value is created, exchanged through interactions, and thus flowing among stakeholders in order to address their concerns.

The *architectural viewpoint* proposes to adopt a systemism perspective, thus conceptualizing the value creation logic through a set of components, their interrelations, and mechanisms embedded in an environment (see Section F.IV.2).

The *architectural view* summarizes the value creation logic via a formal definition which specifies the constituent features of CLSs (see Subsection F.IV.3.1). "Constituent" means that a "normal" logistics system becomes a "special type" of logistics system – a cloud logistics system – *if and only if* it exhibits all four features. Features can pertain to system components, to relations between components, or to mechanisms.

Proposition 1. *Cloud logistics systems are a special type of meso-logistics systems that comprise logistics consumers, LSPs, and a LITPAP and that offer Logistics-as-a-Service: Logistics consumers are provided with access to a shared, open pool of service-oriented physical and virtual logistics capabilities that can be made available in an on-demand, rapidly elastic, and pay-per-use manner with minimal provider interaction by means of formalized mechanisms via a universally accessible cloud computing technology based platform.*

This definition consists of four constituent features: (1) *a triad of stakeholder classes*: logistics consumers, LSPs, and a LITPAP, which mirrors the structure of stakeholders in SOA; (2) *a cloud-based platform*, which connects all stakeholders participating in a CLS and enables them to interact with each other and carry out value exchanges. In other words, the platform represents the "nexus for value creation"; (3) delivery of logistics capabilities with *essential cloud characteristics*; and (4) *computer-formalized mechanisms* to govern value creation processes including (a) interactions and value exchanges among stakeholders and (b) the planning, realization, control, and monitoring of logistics capabilities.

2. Viewpoint: Abstracting and Synthesizing Existing Cloud Logistics Sources

This first viewpoint of this reference architecture description establishes the conventions for conceptualizing the value creation logic of CLSs. This logic can be understood as the total sum of ways in which economic value is created, exchanged through interactions, and thus flowing among stakeholders in order to address their concerns.

In order to conceptualize the value creation logic, this viewpoint proposes to use a systemism perspective, thus conceptualizing the value creation logic through a set of components, their interrelations, and mechanisms embedded in an environment (see Section B.II.2). This viewpoint thus adopts an approach common in scientific logistics inquiry. Furthermore, the value creation logic can also be conceived of as the architecture of a CLS because it represents a fundamental conception of CLSs in our minds, embodied in its components and relationships, and the principles that govern its design and evolution (see Section F.I.2).

In accordance with the abstract nature of reference architectures, this viewpoint additionally proposes conceptualizing the value creation logic at a high level of abstraction, which abstracts system- and context-specific details provided by cloud logistics sources of the fourth meaning of cloud logistics (Logistics-as-a-Service) but keeps and synthesizes those details that relate to the most relevant architectural components and their interrelations (see Subsection F.I.2.2). This viewpoint thus conceptualizes the value creation logic at an aggregate level. This level is the most useful consolidating the existing cloud logistics knowledge in order to clear the fog around cloud logistics.

3. View: Formal Definition, Constituent Features, and Architectural Model

3.1. Proposition

The value creation logic is conceptualized through a (1) formal definition of CLSs and (2) a reference architecture model of CLSs, which illustrates the constituent features of CLSs included in its definition.

The systematic knowledge review of cloud logistics shows that no consensual definition of CLS has been proposed so far (see Subsection E.IV.5.1). Moreover, existing definitions have primarily defined cloud logistics by the *process* of designing logistics systems as per the concepts and principles of cloud computing, rather than by the *outcome* of this design process. Hence, the true notion of LaaS has remained cloudy so far. The lack of a shared understanding impedes research progress, especially as cloud logistics has attracted the attention of scholars from different academic fields. To overcome this research hurdle, we propose a formal definition for "cloud logistics systems." This definition is based on the NIST definition of cloud computing (see Chapter D.II) and abstracts and synthesizes existing definitions (see Subsection E.IV.5.1) and knowledge of the fourth meaning of cloud logistics as identified in our systematic review and summarized in a set of concepts and relationships (see Figure 29). We propose:

Proposition 1. *Cloud logistics systems are a special type of meso-logistics systems that comprise logistics consumers, LSPs, and a LITPAP and that offer Logistics-as-a-Service: Logistics consumers are provided with access to a shared, open pool of service-oriented physical and virtual logistics capabilities that can be made available in an on-demand, rapidly elastic, and pay-per-use manner with minimal provider interaction by means of formalized mechanisms via a universally accessible cloud computing technology based platform.*

This definition consists of four constituent features: (1) a triad of stakeholder classes: logistics consumers, LSPs, and a LITPAP; (2) a cloud computing technology based platform; (3) delivery of logistics capabilities with essential cloud characteristics; and (4) computer-formalized mechanisms to govern interactions and value exchanges. "Constituent" means that a "normal" logistics system becomes a "special type" of logistics system – a cloud logistics system – *if and only if* it exhibits all four features.

These constituent features specify the architectural components, relationships, mechanisms, or specific properties thereof, which are considered constituent to CLSs. They are depicted by the reference architecture model in Figure 38, exposes the *value creation logic* of CLSs by depicting how value is created (through the purposeful deployment of physical and virtual logistics resources), exchanged through interactions (virtual marketplace), and thus flowing among stakeholders (logistics capabilities and money flow between logistics consumers, LSPs, and the LITPAP). The model thus shows how stakeholders concerns are being addressed.

Before moving on, some further clarification is required regarding the types of logistics capabilities offered by CLSs in order to further specify the focus of this reference architecture. The proposed definition distinguishes between two types of logistics capabilities: (1) physical logistics capabilities; and (2) virtual logistics capabilities. *Physical logistics capabilities* are any physical capabilities necessary to realize or manipulate the physical flow of goods by transforming goods in time; space; quantity and composition; and in their storage, transportation, and handling properties (see Section B.III.1). Physical logistics capabilities are made available by LSPs. *Virtual logistics capabilities* are any logistics-related IT capabilities necessary to support the realization of physical logistics capabilities and to support or carry out the planning, control, and monitoring of physical logistics capabilities. This particularly includes capabilities to manipulate logistics-relevant information and to realize or manipulate the information flow, and thus to change the logistical determination of goods (see Section B.III.1). Virtual logistics capabilities are made available by the LITPAP, and they are requested and delivered via the cloud-based platform. Physical logistics capabilities are also requested through the cloud-based platform, but are delivered in the real world by LSPs (see below Subsection F.IV.3.4).

This first architectural view focuses on both virtual and physical logistics capabilities. However, the other viewpoints/views in the remainder of this reference architecture description center exclusively on physical cloud logistics capabilities, as this thesis ultimately aims to contribute to logistics research rather than IT.

The following subsections describe the four constituent features of CLSs and the reference architecture model in detail. Although it is not a constituent feature as such, we also describe the logistics-relevant environment of CLSs in order to provide a comprehensive

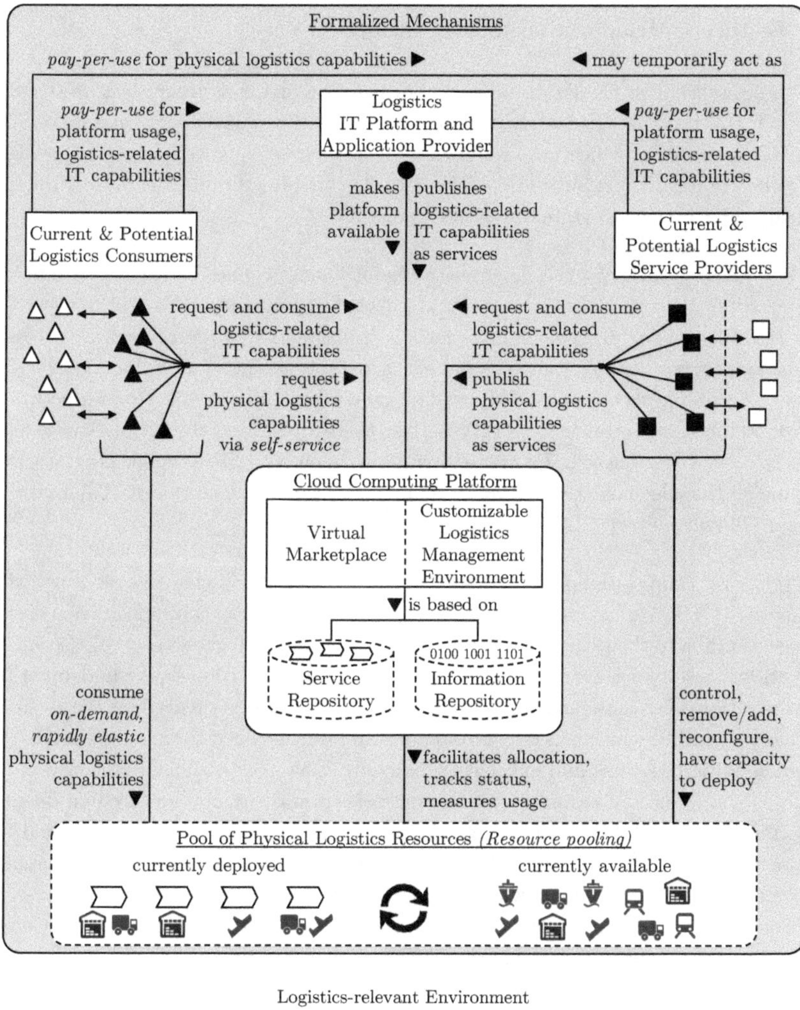

Figure 38.: Reference Architecture Model of Cloud Logistics Systems

understanding of the cloud logistics phenomenon and to form a basis for organizational design.

3.2. Feature 1: Triad of Stakeholder Classes

The three stakeholder classes introduced above are the first constituent feature (see Subsection F.II.3.1). This triad derives from and mirrors the stakeholder structure of SOAs (see "SOA triangle"), which has been used by several cloud logistics sources in this regard (see Subsection E.IV.5.2; Subsection E.IV.5.4). The architectural model depicts the triad of stakeholder classes as white boxes (see Figure 38).

While we have introduced the three abstract stakeholder classes above, we have not yet addressed the question of how many actual stakeholders exist in each class. In every CLS there are at least two, but usually many, logistics consumers. Two consumers are necessary to enable resource pooling – an essential characteristic of CLSs (see Subsection F.IV.3.4 below). According to current cloud logistics knowledge, LSPs cooperate in a networked horizontal manner (see Subsection E.IV.5.4), so there are at least three, but usually many, LSPs in every CLS. These LSPs are referred to as "member LSPs." Following the sources that model the stakeholder structure as in an SOA (see Subsubsection E.IV.5.4.2), there is one LITPAP in every CLS.

The temporal composition of actual stakeholders in CLSs may change over time, which means that CLSs are "open," to some extent. Consumers may join and leave, on short notice according to their current logistical needs. Some cloud logistics sources also suggest that LSPs can join or leave the CLS by joining or leaving the cooperative horizontal LSP network. However, these sources provide different conceptions regarding the degree of openness of a CLS, which can range from open and uncontrolled to more restricted, rule-based, conditional access and exit (see Subsection E.IV.5.4). Although openness is only suggested by a few sources and with varying conceptions, we consider system openness important for a CLS. Openness directly influences the upper scalability bound and thus the degree of elasticity – an essential cloud characteristic – with which logistics capabilities can be made available (see below Subsection F.IV.3.4). We conceive of the openness for LSPs as being rule-based. Following the terminology of Gulati et al. (2012:576), we refer to this rule-based openness in LSPs as a permeable system boundary, with the degree of permeability in CLSs yet to be determined.

The reference architecture model depicts actual stakeholders that belong to a CLS at a given point in time as triangles, circles, and squares, which are either "filled" meaning that the stakeholder currently participates in the CLS, or "unfilled," meaning that the stakeholder does not currently participate. The ability of logistics consumers to freely join or leave the system is indicated by double arrows. The constrained rule-based ability of LSPs to join or leave through a permeable system boundary is indicated by double arrows that pass through a dotted line.

3.3. Feature 2: Cloud Computing Technology based Platform

The second constituent feature is a cloud computing technology-based platform, hereafter referred to as a "cloud-based platform." This platform connects all stakeholders participating in a CLS and enables them to interact with each other and carry out value exchanges. In other words, the platform represents the "nexus for value creation" and is depicted in the middle of the architectural model (see Figure 38).

The cloud-based platform consists of four interrelated components to enable value creation: (1) a service repository, (2) an information repository, (3) a virtual marketplace, and (4) a customizable logistics management environment. These four components are adopted from the existing cloud logistics knowledge. The *service repository* is adopted from the fourth meaning of cloud logistics (LaaS) and specifically from the sources that suggest that the CLS deployment model resembles the SOA structure (see Subsection E.IV.5.4 and Subsection E.IV.5.6). The *information repository, customizable logistics management environment*, and *virtual marketplace* are adopted from the second meaning of cloud logistics, which refers to using a cloud-based platform for logistics IT systems (see Section E.IV.3). They each refer to one of the generic use cases of a cloud-based platform in logistics systems. The *virtual marketplace* is additionally adopted from sources of the fourth meaning of cloud logistics (see Subsection E.IV.5.4).

The integration of these four components into the cloud-based platform for LaaS emphasizes that the fourth meaning of cloud logistics is, to some extent, based on the second meaning. Moreover, these components anchor CLSs in the literature on SOC due to the service repository (see Section D.III.3) and in the literature on electronic markets due to the virtual marketplace (see Subsection C.III.4.1; Subsection C.VI.3.3).

The *service repository* lists the service descriptions of all physical and virtual cloud logistics capabilities that are published as services in the CLS. Member LSPs publish physical logistics capabilities as services; the LITPAP publishes virtual logistics capabilities as services. Logistics consumers browse the service repository to identify the physical and virtual logistics capabilities that can address their logistical and/or IT-related logistical concerns. Member LSPs browse the service repository to identify the virtual logistics capabilities that can address their IT-related logistical concerns. The composition of the service repository changes as member LSPs publish new descriptions or update, or remove existing ones, new LSPs join, or current LSPs exit the system.

The *information repository* stores all logistics-relevant information generated by consumers, member LSPs, and/or the LITPAP. This includes information related to the planning, realization, control, and monitoring of physical and/or virtual logistics capabilities, such as flight schedules, traffic information, and shipment information. Subject to access authorization, stakeholders can use this information for any operative or administrative activity related to physical or virtual logistics capabilities. The information in the repository is largely collected in an automated manner using IoT technologies (see below Subsection F.IV.3.5). The composition of the information repository continuously changes as information is updated, added, or removed during the planning, realization, control, and monitoring of physical and virtual cloud logistics services.

The *virtual marketplace* enables logistics consumers, member LSPs, and the LITPAP to carry out value exchanges using the price mechanism. Objects traded on the marketplace include both physical and virtual logistics capabilities previously published to the service repository. Physical cloud logistics capabilities are traded between logistics consumers and member LSPs. Virtual cloud logistics capabilities are traded between both the LITPAP and logistics consumers as well as between the LITPAP and member LSPs. Logistics consumers remunerate the member LSPs that deliver requested physical logistics capabilities. Logistics consumers and member LSPs remunerate the LITPAP for both making available the platform and delivering virtual logistics capabilities.

The *customizable logistics management environment* enables logistics consumers and member LSPs to deploy the virtual logistics capabilities previously purchased on the virtual marketplace and thus manage their logistics activities in the CLS using information stored in the information repository (in addition to any other private information they may possess). The platform offers each consumer and member LSP a dedicated logistics management environment (based on multi-tenancy) in which they can deploy a customized set of virtual logistics capabilities necessary to achieving their logistical goals. The platform thus replaces the need for logistics consumers and member LSPs to have their own IT system in order to manage logistics capabilities in CLSs.

3.4. Feature 3: Essential Cloud Characteristics

The third constituent feature of CLSs is the set of essential cloud characteristics with which logistics capabilities are made available. These characteristics have been described above for physical logistics capabilities (see Subsection F.II.4.5) and computing capabilities (see Section D.II.2). Based on current cloud logistics knowledge (see Subsection E.IV.5.3), we specify in the following *which logistics capabilities* can be made available with *which characteristics* in CLSs.

Based on the distinction between physical and virtual logistics capabilities, Table 22 specifies the essential characteristics of physical and virtual *cloud* logistics capabilities. Following cloud logistics knowledge, physical cloud logistics capabilities are characterized by on-demand, resource pooling, rapid elasticity, and measured service. Self-service and broad network access only apply to administrative logistics processes related to physical logistics capabilities (see Subsection E.IV.5.3). Virtual cloud logistics services possess the same characteristics as any other cloud computing capability, as they are based on cloud computing technology.

The essential cloud characteristics are depicted in the reference architecture model either by characterizing relationships between architectural components or as a separate architectural component (see Figure 38). However, not all essential characteristics are depicted: The model shows all essential characteristics of physical cloud logistics capabilities, but only selected ones of virtual cloud logistics capabilities. This is because the reference architecture focuses on the delivery of physical, not virtual, capabilities.

For virtual cloud logistic capabilities, only *pay-per-use* is depicted, in order to show how the LITPAP can address economic concerns. It specifies the relationship between architectural components: Logistics consumers and member LSPs may remunerate the LITPAP for platform usage and for the virtual cloud logistics capabilities actually consumed. *On-demand, rapid elasticity, self-service, broad network access*, and *resource pooling* are not shown in the model. Still, virtual cloud logistics capabilities – by definition – possess these characteristics as they are delivered via the cloud-based platform.

For physical cloud logistics capabilities, the essential characteristics of *on-demand, rapid elasticity, pay-per-use*, and *self-service* are depicted to specify the relationships between architectural components. Logistics consumers request physical capabilities via self-service on the virtual marketplace. Physical logistics capabilities are delivered to logistics consumers in an on-demand, rapidly elastic, pay-per-use manner.

Essential Characteristics	**Physical Cloud Logistics Capabilities**	**Virtual Cloud Logistics Capabilities**
On-demand	x	x
Self-service	(x)	x
Broad Network Access	(x)	x
Resource Pooling	x	x
Rapid Elasticity	x	x
Measured Service	x	x

Table 22.: Essential Characteristics of Cloud Logistics Capabilities;
(x) = only for administrative logistics processes

Unlike the other essential characteristics, *resource pooling* is depicted as a separate architectural component because pooling implies the existence of a shared collection of resources: a pool. In fact, within the reference architecture of CLSs, three such pools exist that are directly related to the delivery of physical cloud logistics services (see Figure 38): (1) a service repository, (2) in information repository, and (3) a pool of physical logistics resources. The *service repository* and *information repository* are pools of virtual logistics resources that are hosted on the cloud-based platform. They comprise service descriptions and logistics-relevant information, respectively (see Subsection F.IV.3.3). The *pool of physical logistics resources* comprises material resources that are controlled by member LSPs currently participating in the CLS and that they can deploy to address the logistical concerns of logistics consumers. Examples of physical logistics resources include warehouses, trucks, and airplanes. Resources may be substitutes or complements (Kersten et al. 2012:260ff.), and they may be geographically distributed, depending on the footprint of the CLS. From a conceptual perspective, this pool is divided into two subpools: (a) currently available and (b) currently deployed resources. If resources become deployed, they are associated with a service description from the service repository and are moved from the pool of currently available resources to currently deployed resources. After service delivery is completed, resources are moved back to the pool of available resources. This process of constant interchange between the pools is indicated by the circled arrows. The composition of the physical resource pool changes as member LSPs add, remove, or reconfigure resources, new LSPs join, or current LSPs leave.

The architectural model of CLSs depicts virtual and physical resource pools as boxes with dotted lines in order to emphasize their "openness," which means that their (spatial, qualitative, or quantitative) composition may change over time.

3.5. Feature 4: Computer-Formalized Mechanisms

The fourth and last constituent feature of CLSs is the use of formalized mechanisms to govern value creation processes including (a) interactions and value exchanges among stakeholders and (b) the planning, realization, control, and monitoring of logistics capabilities. Formalized mechanisms are depicted as the underlying gray box in the reference architecture model (see Figure 38), as they integrate the other architectural components into a functioning logistics system; they make a CLS "tick" (see Bunge 1997:410).

Although the understanding of a formalized mechanism might be intuitive, a formal definition is warranted to ensure terminological clarity of this constituent feature. A definition can be derived from combining the definitions of mechanisms and formalization. Recall that mechanisms are processes (or activities) that are able to cause or prevent change to individual system components or the system as a whole (see Section B.II.2). The regularities that govern these change processes are and remain hidden in some cases; in other cases, (causal) laws can be used to describe them (Machamer et al. 2000:7). If so, mechanisms can be interpreted as "laws of working" that explicitly formulate the rules by which change or its prevention is governed (Mackie 1974 as cited in Machamer et al. 2000:6f. and sources therein). Furthermore, recall that the degree of formalization principally refers to the degree of specification and the degree of codification of rules and procedures that prescribe behaviors and/or outputs (see Section C.V.3). Accordingly, formalized mechanisms can be defined as mechanisms for which the rules and procedures that prescribe behaviors and/or outputs *can be specified* and *are codified*, either in written form or in IT systems.

The central importance of formalized mechanisms for CLSs is suggested by cloud logistics knowledge. Our systematic knowledge review identifies several formalized mechanisms (see Section E.IV.5), which are implemented on the cloud-based platform. Identified mechanisms include (a) the process of virtualizing logistics resources; (b) service-orientation for enabling the coordination between multiple fragmented and distributed logistics processes toward an overall goal; (c) IoT technologies for connecting distributed resources with the cloud-based platform and for controlling and monitoring logistics processes; (d) algorithms for optimizing logistics processes (across organizational boundaries); and (e) resource allocation mechanisms for allocating service requests to several resource providers, such as market mechanisms and combinatorial auctions. Chapter F.IX – Chapter F.XII provide (detailed) designs of selected formalized mechanisms.

3.6. Environment: Limited Logistical Predictability and Hostility

Given that CLSs represent a special type of meso-logistics systems, their environment represents the union of all organizational environments of logistics consumers, member LSPs, and the LITPAP. Cloud logistics knowledge provides only an incomplete, fragmented description of the logistics-relevant environment within which CLSs are situated (see Subsection E.IV.5.3): No information on the legal and societal environment is provided and hardly any on the specific conditions in the political and ecological environment. Still, some information is provided regarding the technological and competitive environments, presumably the most important environmental dimensions. Regarding the former, several sources mention the IoT. This seems obvious, given that the IoT represents an underlying design concept of CLSs. Regarding the latter, current knowledge points toward environmental complexity, dynamism, and hostility.

We propose conceptualizing the competitive environment CLSs as dynamic and hostile. Regarding dynamism, we specifically propose that CLSs are exposed to dynamic logistical demand in terms of both quality and quantity. In other words, the environment is characterized by limited logistical predictability. Regarding hostility, we specifically propose that member LSPs are exposed to a hostile market environment, including competition among member LSPs in the CLS and competition with LSPs outside of the CLS.

We refrain from making any proposal regarding environmental complexity at this early stage of cloud logistics research, although some sources mention complexity. This is because the essential cloud characteristics make direct reference to the temporal properties with which physical logistics capabilities can be made available (on-demand availability, rapid elasticity) and to the temporal duration of financial commitments necessary by logistics consumers (pay-per-use). These characteristics clearly aim to enable effective logistical responses to dynamic demand, but they do not refer to complexity. We therefore conclude that, at this early stage of research, nothing limits the deployment of CLSs in environments of both low and high complexity.

3.7. Conclusions: The Outcome of Designing Logistics Systems in Accordance with the Principles and Concepts of Cloud Computing

The previous subsections have conceptualized CLSs and their inherent value creation logic through a formal definition, an architectural model, four constituent features, and their environment. This view thus provides the first comprehensive account of LaaS that goes beyond the frequent "procedural definition" which state that CLSs are systems designed in accordance with the principles and concepts associated with cloud computing. In fact, this view specifies the outcome of applying these principles and concepts on an aggregate level.

The formal definition, architectural model, and constituent features collectively expose the structure and types of stakeholders that participate in CLSs. They position a cloud-based platform at the center of a CLS. This platform represents the nexus of value creation by enabling the interaction and value exchanges among all stakeholders through computer-formalized mechanism. Physical logistics capabilities are delivered with cloud characteristics: in an on-demand and rapidly elastic manner based on a shared pool of logistics resources. Associated administrative capabilities are universally accessible and made available in a self-service manner through the cloud-based platform.

The following eight chapters of this reference architecture description each focus on specific components of CLSs and provide a detailed design of them.

V. Physical Logistics Infrastructure

1. Abstract of Chapter

The physical infrastructure of CLSs includes a *shared pool of logistics resources* and the capacity of member LSPs to deploy them to address logistical concerns (= logistical capabilities).

The *architectural viewpoint* establishes the conventions for designing the *physical logistical infrastructure* of CLSs such that physical logistical capabilities can be offered with cloud characteristics: on-demand availability, rapid elasticity, resource pooling, and pay-per-use (see Section F.V.2). We argue that not all types of physical logistics capabilities can be offered with these characteristics. Hence, this viewpoint establishes the conventions for determining which capabilities can (and which cannot) be offered by CLSs and thus, which logistics concerns can be addressed with cloud characteristics.

In order to deliver physical logistics capabilities with cloud characteristics, we argue that the production process of physical logistics capabilities includes three phases (see Subsubsection F.V.2.1.1) (a) in the *speculative pre-combination phase* LSPs establish a generic capability potential, where "speculative" means that LSPs create a potential without having received a concrete service request; (b) in the *tailoring phase* LSPs tailor the generic potential according to received consumer requirements; and (c) in the *delivery phase* logistics consumers access the tailored potential by handing over the logistics object to be transformed. Creating the capability potential in a two step approach is critical to achieving on-demand availability and rapid elasticity because the time necessary to tailor an established capability potential is – in many cases – lower than the time required to establish a capability potential "from scratch."

The scope of physical logistics capabilities that can be offered with cloud characteristics is ultimately constrained by the profit orientation (economic concerns) of logistics consumers and member LSPs in interaction with the speculative pre-combination, the tailoring phase, and the delivery phase. CLSs can only offer capabilities that are "economically feasible" for logistics consumers *and* member LSPs. From the perspective of consumers, two factors constrain the feasible set of logistical capabilities: (a) the costs associated with a certain degree of on-demand availability and rapid elasticity and (b) the requirement for a minimum degree of on-demand availability and rapid elasticity. From LSPs' perspective, two factors equally constrain the feasible set of logistical capabilities: (a) the willingness to invest speculatively and (b) the likelihood of (re-)deployment.

The construct of *asset specificity* is instructive to identify and design the economically feasible set of physical logistics capabilities in CLSs (see Subsubsection F.V.2.1.2). Asset specificity "has reference to the degree to which an asset can be redeployed to alternative uses and by alternative users without sacrifice of productive value" (Williamson 1988:70; Williamson 1991:281). Hence, asset specificity directly relates to capability costs and the likelihood of (re-)deployment. We therefore argue that *the lower the degree of asset specificity, the more likely that a logistics capability is offered by a CLS.* However, asset specificity does not need to be minimized, but only be "sufficiently low" in order to satisfy the minimum degree of (re-)deployment likelihood necessary to ensure that member LSPs are willing to invest speculatively.

The *architectural view* regarding the physical infrastructure of CLSs determines the feasible set of physical logistics capabilities by (1) assessing the degree of asset specificity of logistics resources and capabilities (see Subsection F.V.3.2) and (2) by then integrating the results from this assessment into "coherent bundles" of logistics resources and capabilities that can be offered with cloud characteristics, where "coherent" means that these bundles are likely of logistical value for logistics consumers (see Subsection F.V.3.3). These scenarios represent the "service models" of CLSs, similar to the service modes in cloud computing

The construct of asset specificity is multi-dimensional and therefore allows for a holistic analysis of logistics resources and capabilities. Four dimensions are relevant for our purposes: (a) temporal specificity (see Subsubsection F.V.3.2.1), (b) site specificity (see Subsubsection F.V.3.2.2), (c) physical asset specificity (see Subsubsection F.V.3.2.3), and (d) human and procedural asset specificity (see Subsubsection F.V.3.2.4). We derive the following propositions regarding these dimensions.

Proposition 2 (temporal specificity). *Cloud logistics systems will operate in an environment characterized by a particularly high degree of temporal specificity of logistics resources and capabilities.*

Proposition 3 (site specificity). *Cloud logistics systems will make use of warehouses and transshipment points that are located within or in the vicinity of areas of high economic intensity.*

Proposition 4 (site specificity). *Cloud logistics systems will make use of air and road transport.*

Proposition 5 (site specificity). *Cloud logistics systems will provide transportation links in, in the vicinity of, and between areas of intense economic activity.*

Proposition 6 (physical asset specificity). *Cloud logistics systems will make use of logistical resources that are compatible with either a standardized unit load or an industry-wide standardized handling interface.*

Proposition 7 (physical asset specificity). *Cloud logistics systems will offer transportation (spatial), storage (temporal), transshipment (quantity and composition), and industry-wide standardized value-added (non-)logistical transformations of logistics objects.*

Proposition 8 (human and procedural asset specificity). *Cloud logistics systems will use personnel with the capacity to deploy the resources required to achieve core logistical and industry-wide standardized, value-added, (non-)logistical transformations of logistics objects that use standardized unit-loads or an industry-wide standardized handling interface.*

These propositions show how the aim to deliver physical (non-)logistics capabilities with cloud characteristics limits the scope of capabilities that can be offered by CLSs. We integrate the results of the asset specificity assessment into five *logistical prototype scenarios* that can be differentiated along two dimensions (see Subsubsection F.V.3.3.1): (a) *capability scope*, which refers to the composition of logistics capabilities offered by the cooperative network of member LSPs, and it is measured in a three point scale: narrow (only one capability), selected compositions (few selected capabilities), and broad (any capability); and (b) *geographic scope*, which refers to the location(s) at which these capabilities are offered, and it is measured in a three point scale: localized (within a certain confined area), wide, networked (within certain geographic areas and links between them), and wide (any location). We summarize the above assessment in the following proposition:

Proposition 9. *The physical infrastructure of any cloud logistics system will match one of five logistical prototype scenarios, with each scenario's capability scope being either narrow or comprising selected compositions and its geographic scope either localized or wide, networked.*

Localized Transportation Capabilities as-a-Service (Scenario Ia) offers a narrow scope of logistics capabilities within a localized geography (see Subsubsection F.V.3.3.2). The capability scope covers end-to-end road transportation capabilities limited to goods compatible with standardized unit loads or an industry-wide handling standard. The geographic scope is confined to the vicinity of a single area of intense economic activity.

Localized Storage Capabilities as-a-Service (Scenario Ib) offers a narrow scope of logistics capabilities within a localized geography (see Subsubsection F.V.3.3.3). More specifically, the capability scope includes storage capabilities limited to standardized unit loads or an industry-wide handling standard. The geographic scope of these storage capabilities is limited to the vicinity of a single area of intense economic activity.

Networked Long-distance (end-to-end) Transportation Capabilities as-a-Service (Scenario II) offers a narrow scope of logistics capabilities within a wide, networked geography (see Subsubsection F.V.3.3.4). More specifically, the capability scope comprises transportation and transshipment capabilities for standardized unit loads or an industry-wide handling standard, thus enabling networked (or multi-leg) transportation. In addition, the capability scope can, but does not need to, include pick-up and last-mile delivery capabilities, thus enabling end-to-end transportation. The geographic scope of transportation capabilities is limited to the interconnection of multiple spatially distributed areas of intense economic activity. Transshipment, pick-up, and last-mile delivery are limited to the vicinity of these areas. The spatial distribution and distances between these interconnected areas are not confined, enabling long-distance (e.g. inter-regional or global) transportation.

Localized Storage and Distribution Capabilities with or without Standardized Value-added Capabilities as-a-Service (Scenario III) offers a composition of selected logistics capabilities in a localized geographic scope (see Subsubsection F.V.3.3.5). More specifically, the capability scope comprises a composition of storage capabilities integrated with (last-mile) distribution capabilities. These capabilities are limited to goods compatible with standardized unit loads or an industry-wide handling standard. In addition, it can, but does not need to, include standardized value-added capabilities (SVACs), such as picking-and-packing customized selections of goods for (last-mile) delivery. These SVACs thus enable fulfillment of customized client orders. The geographic scope is limited to the vicinity of a single area of intense economic activity. Due to the offering of last-mile distribution capabilities, this scenario is likely located in (dense) urban areas or industrial areas, rather than near locations of sole macro-logistics infrastructure.

Networked Spatially Distributed Storage Capabilities with or without Distribution and Standardized Value-added Capabilities as-a-Service (Scenario IV) offers a composition of selected logistics capabilities within a wide, networked geography (see Subsubsection F.V.3.3.6). More specifically, the capability scope comprises a composition of transportation, transshipment, and storage capabilities for goods compatible with standardized unit loads or an industry-wide handling standard. In addition, this composition can, but does not need to, include last-mile distribution and SVACs. It thus enables fulfillment of customized client orders. The geographic scope covers multiple areas of intense economic activity and the connections between these areas. Storage, last-mile distribution, and

SVACs are limited to the vicinity of these areas; transportation capabilities are limited to the connections between these areas, thus enabling networked storage capabilities. The spatial distribution and distances between these interconnected areas are not confined, thus enabling networked, spatially distributed capabilities.

The degree of logistical innovativeness varies across these scenarios. For Scenarios Ia – III, the degree ranges from low to medium because they represent modifications of existing types of logistics systems. By contrast, Scenario IV exhibits a high degree of innovativeness. To our knowledge, the capability and geographic scope offered by this scenario are not yet offered by any existing logistics system. In other words, this scenario creates a new amalgam of capability and geographic scope. More specifically, it integrates the concepts of (international) freight forwarding with last-mile distribution and with public warehouses that offer SVACs. All of these capabilities are provisioned with cloud characteristics and become accessible via a single cloud-based platform, even though they are made available by different LSPs.

Whether these prototype scenarios are of "value" for cloud consumers actually depends on their potential to create sources of competitive advantage (see Subsection F.V.3.4). Under simplifying assumptions a comparative assessment between cloud consumers and non-cloud consumers shows that the logistical prototype scenarios may in fact offer sources of competitive advantage. Due to on-demand, rapid elasticity, resource pooling, and pay-per-use, cloud consumers can adjust the degree of (vertical and/or horizontal) (de-)centralization of their logistics channels quicker as well as run (vertically and/or horizontally) more decentralized logistics channel(s) than non-cloud consumers. This enables them to respond quicker to emerging client demand, fulfill (customized) client orders with lower lead times, reduce inventory outages in case of demand peak, or postpone logistical investment decisions in case of new product launches.

2. Viewpoint: An Approach to Designing Physical Logistics Resources and Capabilities

2.1. Asset Specificity as Underlying Construct to Enable On-demand, Rapid Elasticity, Resource Pooling, and Pay-per-use

2.1.1. Production Process of Physical Cloud Logistics Capabilities and Constraints Imposed by Economic Concerns

Chapter B.III introduced the basic (non-)logistical transformations, their associated processes, and the underlying resources and capabilities. This viewpoint establishes the conventions for designing the *physical logistical infrastructure* of CLSs such that physical logistical capabilities can be offered with the cloud characteristics of on-demand availability, rapid elasticity, resource pooling, and pay-per-use (see Subsection F.IV.3.4). The physical infrastructure consists of the *shared pool of logistics resources* and the capacity of member LSPs to deploy them to address consumers' logistical concerns (logistical capabilities). Infrastructure design typically involves specifying the type(s), the spatial arrangement, and relationships between logistics resources and capabilities (see e.g. Klaas 2002:135; Delfmann et al. 2010:58). We argue that not all types of physical logistics capabilities can be offered with cloud characteristics. Hence, this viewpoint establishes the conventions for determining which physical logistical capabilities can (and which cannot) be offered by CLSs or, from a consumer perspective, which logistical concerns can be addressed with cloud characteristics.

In order to achieve cloud characteristics, we argue that the production process of physical cloud logistics capabilities needs to follow a certain structure. Recall that the production process of "regular" logistics capabilities can be structured into a pre-combination phase in which a capability potential is established (e.g. by means of pre-combining internal production factors, such as logistics personnel and facilities) and a delivery phase, in which this potential is accessed by the logistics consumer by handing over the logistics object to be transformed (see Section B.III.3).

In CLSs, the pre-combination phase is split into a "speculative pre-combination phase" and a "tailoring phase," as depicted in Figure 39. Member LSPs must establish a capability potential in a "speculative manner," which means that the pre-combination of logistical input factors needs to be achieved before receiving a concrete service request from logistics consumers. Once a service request is received, member LSPs "tailor" the capability potential so that the consumer's actual requirements can be addressed. The tailoring phase is depicted with a dashed outline because tailoring may not be required or feasible for every service request. Some consumers may request services that can be directly fulfilled by the established potential without tailoring, others may require tailoring,

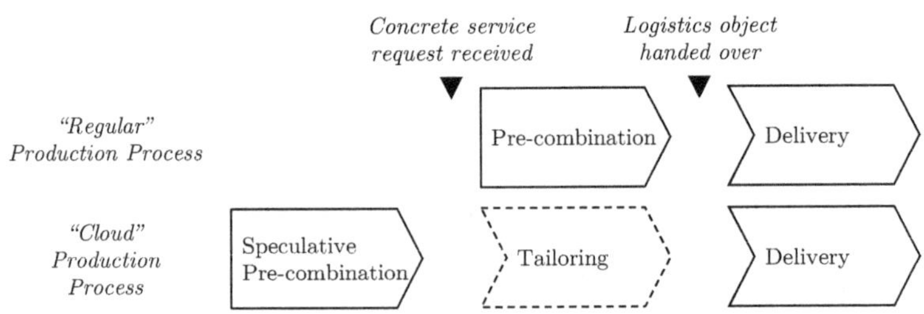

Figure 39.: Production Process of "Regular" and "Cloud" Logistics Capabilities

and still others cannot be delivered because tailoring is not feasible.

Splitting the pre-combination phase into a speculative pre-combination and a tailoring phase is critical to achieving on-demand availability and rapid elasticity. The time necessary to tailor an established capability potential is – in many cases – lower than the time required to establish a capability potential. Hence, speculatively establishing a capability potential does, in many cases, contribute to on-demand and rapidly elastic provisioning. A conceptual example of this production process are LTL networks: LSPs establish a capability potential (a set of transshipment points that are connected through frequent transports), which is tailored to consumer requirements by picking up their cargo at a location of their choice.

The scope of physical logistics capabilities that can be offered by CLSs is ultimately constrained by the profit orientation (economic concerns) of logistics consumers and member LSPs in interaction with the speculative pre-combination, the tailoring phase, and the delivery phase. CLSs can only offer capabilities that are "economically feasible" for logistics consumers *and* member LSPs. The following paragraphs elaborate on how the economic concerns of logistics consumers and LSPs limit the feasible scope of physical logistical capabilities.

From the perspective of logistics consumers, two factors constrain the feasible set of logistical capabilities: (a) the costs associated with a certain degree of on-demand availability and rapid elasticity and (b) the requirement for a minimum degree of on-demand availability and rapid elasticity, as depicted in Figure 40a. Note that on-demand availability and rapid elasticity are measured as the opposite of lead time. In principle, logistics consumers are willing to absorb higher costs for an increasing degree of on-demand and rapid elasticity, which is equal to a decreasing lead time. In addition, we assume that logistics consumers expect a minimum degree of on-demand and rapid elasticity because both are constituent characteristics of CLSs. Hence, if a speculatively established capability potential cannot be accessed with a sufficiently low lead time at sufficiently low costs, for example because tailoring would be too costly or take too long, consumers may not be interested in accessing this potential. In addition, consumers may not be interested if the capability potential cannot be tailored to adequately address their capabilities, for example, if a mismatch in their requirements and the capability offered incurs excessive production or opportunity costs. The curve has a negative slope with an increasingly negative gradient to emphasize that the costs logistics consumers accept for an additional unit of lead time reduction constantly decrease (diminishing returns). The curve represents an upper boundary for the willingness to absorb costs; hence, the set of economically feasible physical logistics capabilities lies below the curve. The curve has a greater y-axis than x-axis intercept to indicate the consumers' preferences for achieving on-demand availability and rapid elasticity over reducing costs. In other words, logistics consumers are assumed to prefer effectiveness over efficiency.

From an LSP perspective, also two factors constrain the feasible set of logistical capabilities: (a) the willingness to invest speculatively and (b) the likelihood of (re-)deployment, as depicted in Figure 40b. The need to speculatively invest in the setup of a capability potential exposes LSPs to a financial risk due to potentially underutilized capabilities: They must invest speculatively in the capability potential but are only remunerated if the potential is actually accessed by logistics consumers due to pay-per-use. In other words, LSPs must make large upfront investments and bear the risk of asset underutilization. Therefore, as profit-oriented entities, LSPs are only willing to speculatively establish capability potentials that have a high likelihood of actually being requested by consumers. In fact, they require a minimum degree of (re-)deployment likelihood, which corresponds to a certain level of expected resource utilization and profit. Reaching this minimum boundary is a necessary precondition for any CLS to become established. The willingness of member LSPs to invest speculatively increases with the likelihood of (re-)deployment. A higher degree of expected utilization increases the profit potential for LSPs. The curve in Figure 40b represents the upper boundary of willingness to invest speculatively for a given degree of likelihood. The feasible set of physical logistics capabilities thus lies below this curve. Due to uncertainty, the curve converges asymptotically against the likelihood of 1. We investigate the factors that influence the likelihood of (re-)deployment in the following subsubsection.

The preceding two paragraphs have investigated the economic concerns of logistics consumers and LSPs that limit the set of physical logistics capabilities that can be offered with cloud characteristics (on-demand, rapid elasticity, resource pooling, and pay-per-use). As argued above, CLSs can only offer physical logistics capabilities that are "economically feasible" for logistics consumers *and* LSPs. LSPs must be willing to invest speculatively

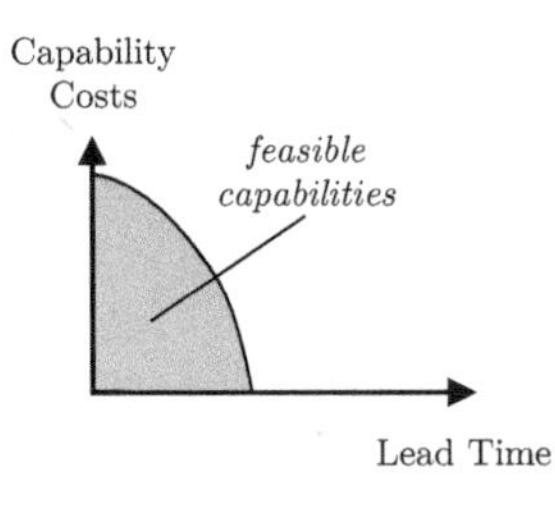

(a) Consumer Perspective

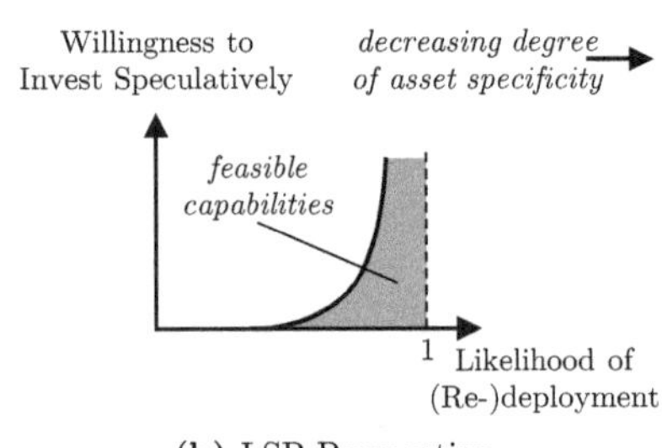

(b) LSP Perspective

Figure 40.: Constraints Imposed by Economic Concerns on the Feasible Set of Physical Logistics Capabilities

to establish a capability potential and, if required, be able to tailor this potential in a time and at a cost acceptable to logistics consumers' preferences for lead time and total capability costs.

2.1.2. Definition and Dimensions of Asset Specificity

Thus far, we have focused on the constraints that economic concerns of logistics consumers and member LSPs impose on the feasible set of physical cloud logistics capabilities. However, we have not yet attended to the question of *how* to identify and design this feasible set. To this end, we draw on the construct of *asset specificity.* While many definitions can be found in the literature (for an overview see de Vita et al. 2011:331), two are presented here to establish the understanding of asset specificity that guides the design of CLS physical infrastructure. First, asset specificity refers to

> "whether specialized investments are required to produce the service" (Brown & Potoski 2005:335).

A high degree of asset specificity means that significant specialized investments are required, while a low degree means that no specialized investments are required to produce a service. Second, asset specificity

> "has reference to the degree to which an asset can be redeployed to alternative uses and by alternative users without sacrifice of productive value" (Williamson 1988:70; Williamson 1991:281).

Thus, a high degree of specificity means that assets cannot be redeployed without a significant loss of productive value, while a low degree implies that assets can be redeployed without loss. More specifically, this definition identifies two distinct dimensions that determine the degree of asset specificity: redeployment to (1) alternative uses and by (2) alternative users. The degree to which an asset can be redeployed to *alternative uses* without a sacrifice in productive value depends on the degree to which an asset is tailored toward the unique requirements of its initial task (Williamson 1985 as cited in Erramilli & Rao 1993:21). The degree to which an asset can be redeployed to *alternative users* without a sacrifice in productive value depends on the availability of alternative users with similar requirements (see de Vita et al. 2010; Lyons 1995; as cited in de Vita et al. 2011:334). Sacrifices of productive value represent a deteriorated input/output ratio, and they typically manifest in (a) reconfiguration costs incurred by tailoring an asset toward fulfilling the requirements of an alternative task or an alternative user, and (b) higher production costs incurred by fewer economies of scale and scope (see Lyons 1995:432, 434) and/or by mismatches between the properties of an asset and the actual requirements of the alternative task and/or alternative user. In a logistics context, reconfiguration costs may be incurred by modifying a warehouse prior to redeployment; mismatches in requirements during service delivery may result from a sub-optimally positioned warehouse only reachable via a detour.

The degree of asset specificity directly relates to the likelihood of (re-)deployment: The higher the degree of asset specificity, the lower the likelihood of (re-)deployment. Figure 41 depicts the likelihood of (re-)deployment as a function of the number of alternative uses and the number of alternative users with similar requirements. Assets have a high likelihood of (re-)deployment if they can be (re-)deployed to many tasks and by many users without any loss in productive value. Conversely, they have a low likelihood of (re-)deployment if they can be (re-)deployed to few alternative tasks and by few alternative users without a loss in productive value. If assets can either be (re-)deployed to few (many) alternative tasks but by many (few) alternative users without a sacrifice of productive value, they have a medium likelihood of (re-)deployment.

Reverting to the initial question of which logistics capabilities can be offered by a CLS or, in other words, which logistics resources and capabilities are included in the physical infrastructure, we argue that *the lower the degree of asset specificity, the more likely that a logistics capability is offered by a CLS.* This is because the lower the degree of asset specificity, the higher the willingness of member LSPs to establish a capability potential in a speculative manner. However, asset specificity does not need to be minimized, but only be "sufficiently low" in order to satisfy the minimum degree of (re-)deployment likelihood necessary to ensure that member LSPs are willing to invest speculatively. The term "sufficiently low" is, of course, "slippery," as it does not state a precisely measurable sufficient condition. Yet, due to the abstract nature of reference architecture, specifying a precisely quantifiable condition does not seem feasible; too much depends on the particularities of the specific situation, for example on the degree of risk aversion of member LSPs. Hence, this reference architecture develops propositions that reasonably satisfy a sufficiently low degree of (re-)deployment likelihood. In other words, these propositions define what we believe to be a sufficiently low degree of asset specificity to enable the delivery of physical logistics capabilities with cloud characteristics. To exemplify these

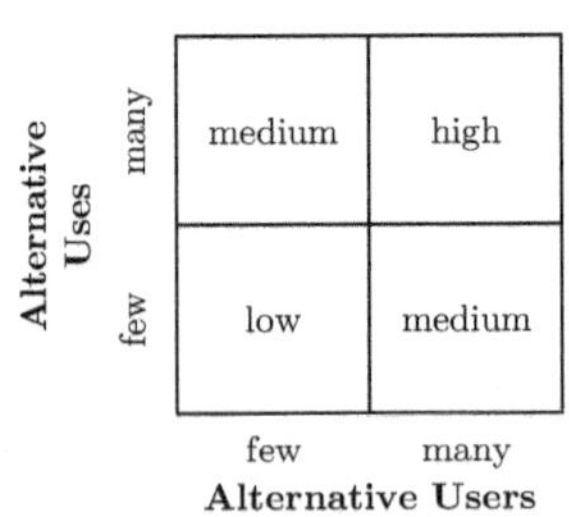

Figure 41.: Likelihood of (Re-)deployment as a Function of Alternative Uses and Users

propositions, we provide a few logistical prototype scenarios (see Subsection F.V.3.3).

In order to determine which capabilities can be offered by CLSs, we need to assess the degree of asset specificity of logistics resources and capabilities. While current logistics literature confirms the understanding of asset specificity as the degree of (re-)deployment to alternative uses and/or users without a sacrifice in productive value, it sheds little light on the drivers of asset specificity of logistics resources and capabilities. Skjoett-Larsen (2000:117) states that a dust-free warehouse is a "medium-specific asset," as the provider can use it to serve multiple clients (alternative users), thus improving its utilization. Yet, what exactly makes this warehouse exhibit medium asset specificity remains unclear – is it the dust-free outfitting or another factor? She further argues that warehouse trolleys, strapping machines, testing equipment and other warehouse equipment are equally "medium-specific" assets because the provider can use them to serve other shippers in the same line of business (alternative users) (Skjoett-Larsen 2000:122). Aertsen (1993:27) argues that special trucks built by one transacting party to carry and install printers of another transacting party are assets with high specificity (alternative uses). He also conceives of specialized equipment for performing fashion stock control activities as highly specific assets (alternative uses). Bourlakis & Bourlakis (2005:89) explain physical asset specificity as a situation in which a warehouse, owned by an LSP, can be redeployed to another consumer after some "minor modification." However, it remains unclear what exactly needs to be modified. Considering these examples, we conclude that the asset specificity of logistics resources and capabilities has not yet been analyzed systematically. Nevertheless, the underlying notion of asset specificity in these examples is consistent with the understanding adopted above.

To assess the degree of asset specificity of logistics resources and capabilities, we structure resources and capabilities according to a basic set of physical (non-)logistics transformations (see Section B.III.1) and assess their degree of specificity using the multidimensionality of the asset specificity construct. Seven distinct dimensions of asset specificity can be found in the literature (de Vita et al. 2011:332ff.): (a) site specificity, (b) physical asset specificity, (c) human asset specificity, (d) dedicated assets, (e) procedural asset specificity, (f) time or temporal specificity, and (g) brand name capital (Williamson 1985:55 (a-d); Williamson 1991:281f. (a-d, f-g); Malone et al. 1987:486 (f); Zaheer & Venkatraman 1994:553 (e)). In this reference architecture, we draw on five of these seven: (1) temporal specificity, (2) site specificity, (3) physical asset specificity, and (4) a combined construct of human and procedural asset specificity.

Dedicated assets and brand name capital are not included because they cannot provide insights into the assessment of logistics resources and capabilities for this thesis' purposes. *Dedicated assets* are large investments in general purpose production capabilities that are made contingent on a long-term supply agreement with a particular customer and lead to significant excess capacity in case of premature contract termination. (Williamson 1983:526; also see Williamson 1991:281; Williamson 2002:176). Physical logistics resources (e.g. warehouses) are cited as examples of dedicated assets (see e.g. Lohtia et al. 1994:268). According to this understanding, dedicated assets are those that have a high likelihood of (re-)deployment because they represent general purpose production capability which, by definition, can be (re-)deployed to alternative uses and by alternative users without a sacrifice in productive value. We thus follow Shelanski & Klein (1995:340) who argue that "[p]roxies such as capital intensity or fixed costs are very imperfect, and may not capture whether an investment has alternative value outside the transaction for which it was initially made." *Brand name capital* refers to investments in reputation, such as advertisement expenditures (Lohtia et al. 1994:268; de Vita et al. 2011:335). None of the physical (non-)logistical transformations discussed above refer to investments in reputation.

3. View: Five Logistical Prototype Scenarios

3.1. Assessing Asset Specificity and Deriving Logistical Prototype Scenarios

In order to determine the physical infrastructure of CLSs, we use a twostep approach: (1) separately assess asset specificity of logistics resources and capabilities, and (2) integrate the results from this assessment by proposing "coherent bundles" of logistics resources and capabilities that can be offered with cloud characteristics, where "coherent" means that these bundles are likely of logistical value for logistics consumers. These bundles are referred to as "logistical prototype scenarios," and they can be understood as the "service models" of CLSs (see Subsection E.IV.5.5), similar to the service models of cloud computing: IaaS, PaaS, and SaaS.

3.2. Propositions: Asset Specificity of Logistics Resources and Capabilities

3.2.1. Temporal Specificity

Temporal specificity, or time specificity, refers to the importance of timing between transacting parties (de Vita et al. 2011:335). As Malone et al. (1987:486) explain: "[a]n asset is time specific if its value is highly dependent on its reaching the user within a specified, relatively limited period of time." In a service context, time specificity depends on the degree of "service punctuality" necessary to prevent any decline in the quality of service (Brown & Potoski 2005 as cited in de Vita et al. 2011:335). Hence, an asset's degree of temporal specificity is not driven by its (physical) properties, but by the temporal requirements of the task for which the asset is an input factor.

As temporal specificity depends on the temporal requirements of a task, we do not assess the degree of temporal specificity for different types of logistical transformations and underlying logistics resources and capabilities separately, but do so on a more aggregate level without distinguishing between different types. We argue that logistics resources and capabilities exhibit a high degree of time specificity. This is due to simultaneous production and consumption as well as the perishability of logistics capabilities (see Section B.III.3). In addition, it is due to the frequent use of lean logistics concepts such as JIT, which raise the sensitivity of the temporal properties of logistics capabilities (see Chapter B.V). In CLSs, temporal specificity is particularly high due to the constituent characteristics of resource pooling, on-demand availability, and rapid elasticity. Pooling requires precise temporal coordination across multiple logistics consumers during service delivery. On-demand and rapid-elasticity imply logistics consumers' strong preference for temporal precision regarding the delivery of capabilities. We thus propose:

Proposition 2. *Cloud logistics systems will operate in an environment characterized by a particularly high degree of temporal specificity of logistics resources and capabilities.*

3.2.2. Site Specificity

Site specificity makes reference to "an asset immobility condition, which is to say that the setup and/or relocation costs are great" (Williamson 1985:95). An asset that is available at a certain location and can only be moved at significant costs is site specific (Malone et al. 1987:486). This understanding represents the starting point for assessing the degree of site specificity of logistics reassures. As site specificity makes reference to "asset (im-)mobility," one can separate (1) immobile logistics resources, which cannot be moved at all, such as warehouses and transshipment points (for reasons of argumentative simplicity, but without losing generality, we also include resources in this category that are typically deployed in these facilities, such as handling equipment and other equipment needed to accomplish value-added (non-)logistical transformations); from (2) mobile logistics resources, which are specifically designed to overcome physical distances, such as trucks, airplanes, vessels, and trains.

The degree of site specificity of *immobile logistics resources* is a function of the environment surrounding the resources' current locations precisely because they cannot be relocated at a reasonable cost and time. In other words, a capability potential consisting of immobile resources cannot be spatially tailored to suit the actual requirements of logistics consumers. Hence, the ability to deploy immobile resources to alternative users with no or little sacrifice of productive value depends on the availability of alternative users in the vicinity of the resources' current location. We therefore argue that site specificity of immobile logistics resources is a function of the *intensity of economic activity in the resource's vicinity.* Economic intensity is measured by the number of economic actors and the associated amount (or value of) goods and services produced and/or consumed in a specified area, including the logistics activities associated with these production and consumption processes (see "physical environment" in Hesse & Rodrigue 2004:181f.). Typical locations of high economic intensity are (dense) urban areas, industrial areas, and areas surrounding macro-logistics infrastructure, such as (arterial) roads and highways, airports, harbors, railway stations, and intermodal hubs. In areas of high economic intensity, it is more likely to find alternative users with similar logistical requirements, which means that it is more likely that resources can be redeployed with no or little loss in productive value. Moreover, due to spatial proximity, resources can be redeployed quickly to alternative users, thus enabling on-demand and rapid elasticity. At the same time, the spatial proximity of multiple alternative users enables resource pooling, either simultaneously or successively. We therefore argue that site specificity of immobile resources is high in low economic intensity and low in case of high intensity. Hence, member LSPs are more willing to speculatively establish a capability potential consisting of immobile logistics resources in areas of high economic intensity than low intensity. In other words, the higher the economic intensity in a resource's vicinity, the higher the likelihood that this resource will be included in the CLS infrastructure. With regard to the locations of immobile logistics resources, we thus propose:

Proposition 3. *Cloud logistics systems will make use of warehouses and transshipment points that are located within or in the vicinity of areas of high economic intensity.*

In conclusion, the degree of site specificity of immobile logistics resources limits the locations at which storage, transshipment, and other (non-)logistical transformations that may be carried out in immobile logistics assets can be offered by CLSs. This spatial limitation is necessary to offer logistics capabilities with cloud characteristics. By limiting locations to areas of high, intense economic activity, CLSs further enhance the trend of large-scale flows of goods being directed through major logistical hubs and gateways (e.g. ocean ports, airports) and through highway interactions with access to (nearby) markets (see Hesse & Rodrigue 2004:177).

The degree of site specificity of *mobile logistics resources* depends on both the environment surrounding a resource's location and the resource's ability to move and change its environment. Specifically, we argue that site specificity of mobile resources depends on three factors: (1) the costs of spatial reconfiguration and (2) the speed of spatial reconfiguration interacting with (3) the intensity of economic activity in the vicinity of transportation resources' (a) current locations and (b) the start-, end-point and intermediate handling locations of a transportation link.

Costs of spatial reconfiguration are the costs incurred by moving a transportation resource, such as fuel expenses, personnel costs, and depreciation. The higher these costs, the higher the degree of site specificity. For a pre-defined amount of money, a resource can be moved less far, thus reducing the number of alternative users with similar requirements that can be reached. Hence, the higher the costs of spatial reconfiguration, the lower the willingness of member LSPs to speculatively establish a capability potential consisting of this resource type. In addition, the upper boundary of consumers' willingness to absorb costs for tailoring resources to their requirements or adjusting the capacity of capabilities currently being provisioned is already reached for small spatial reconfigurations. We thus conclude that the higher the costs of spatial reconfiguration, the less likely that a resource is included in the physical infrastructure of CLSs.

Speed of spatial reconfiguration refers to the travel speed of mobile logistics resources. The higher a resource's travel speed, the lower its site specificity. For a given period of time, a resource can travel farther and therefore reach more potential alternative users with similar requirements. Hence, the higher the travel speed, the higher the willingness of member LSPs to speculatively establish a capability potential consisting of this type of transportation resource. Moreover, the higher the travel speed, the higher the feasible degree of on-demand availability and rapid elasticity, as resources can be quickly tailored spatially to meet new consumers' requirements or adjust the capacity of the capabilities currently being provisioned. Hence, the higher the travel speed, the higher logistics consumers' willingness of to accept the associated costs. We thus conclude that the higher the travel speed, the more likely that a resource is included in the physical infrastructure of CLSs.

As mobile logistics resources are specifically designed to overcome physical distances, spatial reconfiguration costs and speed directly correspond to the production costs. Table 23 depicts a comparative assessment of the reconfiguration and deployment costs and the speed of mobile logistics resources for different transportation modes. Following this assessment and the economic constraints of logistics consumers and LSPs described above

Transportation Mode	**Reconfiguration/ Costs**	**Deployment Speed**
Air	very high	very high
Road	high	high
Rail	medium	medium
Water	low	low

Table 23.: Comparative Assessment of Reconfiguration/ Deployment Costs and Speed (based on Bowersox & Closs 1996:326)

(see Section F.V.2), we propose:

Proposition 4. *Cloud logistics systems will make use of air and road transport.*

In conclusion, the degree of site specificity of mobile logistics resources limits the transportation modes that can offer transportation capabilities with cloud characteristics. This limitation manifests in the tradeoff between costs and speed in favor of the latter to meet the objective of achieving on-demand and rapid elasticity. In other words, despite the high costs of air and road transport, these modes are included in the physical infrastructure of CLSs due to their high travel speed. This tradeoff is already indicated in Figure 40a, as the feasible set of capabilities has a greater y-axis than x-axis intercept.

The aggregate degree of site specificity of mobile logistics resources depends on the resources' spatial reconfiguration properties (costs and time) in interaction with the *environment in which the resource is located.* This is because the higher the costs and time required for spatial reconfiguration, the less mobile a resource and, hence, the more important the resource's current environment. Using the argument of immobile resources, member LSPs are more willing to speculatively establish a transportation potential, in the vicinity of, or between areas of intense economic activity. Limiting the geographic scope in this way increases the likelihood for member LSPs to find alternative consumers with similar requirements in a short period of time or even to pool resources across multiple consumers (e.g. through consolidated transport). Furthermore, limiting the geographic scope fosters on-demand availability and rapid elasticity. Due to spatial proximity, the absolute time and costs required to spatially tailor transportation resources to either meet new consumers' requirements or adjust the capacity of capabilities currently being provisioned are more likely to align with the maximum acceptable reconfiguration costs and time stipulated by consumers. We thus propose:

Proposition 5. *Cloud logistics systems will provide transportation links in, in the vicinity of, and between areas of intense economic activity.*

In conclusion, the degree of site specificity of mobile logistics resources limits the geographic scope in which transportation capabilities can be offered if on-demand availability, rapid elasticity, and resource pooling are to be achieved. This limitation results in a focus on areas and links between areas of high economic intensity.

3.2.3. Physical Asset Specificity

Physical asset specificity refers to physical assets that add to a firm's special purpose production capability, which is necessary to fulfill other transacting parties' unique requirements (see Williamson 1983:526). Such assets have few alternative uses and users due to their specific (design) characteristics (see de Vita et al. 2011:334 and sources therein). Examples include specialized machine tools or complex computer systems designed for a single purpose (Malone et al. 1987:486).

The nature of the production capability of logistics systems depends on the composition of physical resources in their physical infrastructure. The nature of a logistical production capability manifests in (1) the variety of logistics objects that can be transformed; and (2) the variety of (non-)logistical transformations that can be accomplished by means of the resources in its infrastructure. Having a specialized production capability means that only certain types of logistics objects can be transformed, and/or only certain kinds of (non-)logistical transformations can be made. In other words, the greater the variety of logistics objects that can be transformed and the greater the variety of (non-)logistical transformations that can be accomplished, the lower the degree of the special purpose production capability of a logistics system. Our understanding of a special purpose production capability is motivated by Aertsen (1993:27), who defines physical asset specificity of transportation resources using the example of special trucks built by one of the transacting parties to move and install copiers for a transacting party. Hence, the special purpose production capability of logistics systems refers to a combination of logistics object types, (non-)logistical transformations, and underlying resources.

The *variety of logistics objects* than can be transformed by a logistics system depends on the logistics object's *physical properties* (e.g. size, weight, handling requirements) and *economic properties* (e.g. value), because these properties essentially determine the design of the underlying resources (see e.g. Coyle et al. 2003:333f.; Ballou 1999:457ff.; Bowersox & Closs 1996:433). When assessing the degree of physical asset specificity of (logistical) resources in order to determine the variety of logistics objects that can be transformed by CLSs, we do not distinguish between different types of resources because the line of reasoning presented in the following equally applies to all types. Regardless of the logistical resource type, two general approaches to logistical resource design can be differentiated: Resources can be specialized for compatibility with (a) a certain type of cargo or, in an extreme case, a single product of a single firm, or (b) a certain type of standardized unit load or handling standard. These design approaches can be illustrated by the following examples. Transportation resources can be tailored to a single type of cargo (e.g. oil tankers) or to standardized unit loads (e.g. container vessels). Likewise, the storage and materials handling system and the pick-and-pack equipment installed in a warehouse can be tailored to a specific type of cargo (e.g. truck tires) or a certain type of standardized unit load (e.g. pallets). The reconfiguration in the sense of making a resource compatible with another type of logistics object or another type of unit load usually incurs high costs and requires significant time. Hence, resource reconfiguration is likely not feasible in CLSs, as these cost and time requirements likely exceed logistics consumers' acceptable level of total capability costs and violate their temporal requirements regarding on-demand availability and rapid elasticity. In conclusion, any logistics resource that is designed for a specific task possesses a high degree of physical asset specificity, which means that it cannot be (re-)deployed to alternative uses.

Nevertheless, task-specific resources can still be (re-)deployed without a loss in productive value to alternative users with similar requirements. Therefore, the degree of physical asset specificity depends on the volume of logistics objects with similar physical and economic properties. As these properties are often relatively homogeneous in an industry, physical asset specificity refers to industry size. Lyons (1995:434) remarks in this respect: "[D]emand aggregation may be greatest when there is some industry specificity, but no firm specificity in the assets in question." Hence, member LSPs' willingness to speculatively establish a capability potential consisting of a certain resource type increases with the volume of logistics objects that are compatible with this resource type. We thus conclude that the greater the volume of objects with similar (physical and economic) properties, the more likely that a resource tailored to these requirements is included in the physical infrastructure of CLSs.

Standardized unit loads are a very common instrument to reduce the degree of physical asset specificity of logistics resources. Standardized unit loads remove interdependencies between the logistics object to be transformed and the physical resources necessary to achieve this transformation. On the one hand, standardized unit loads offer a standardized platform (e.g. pallets) or box (e.g. containers, ULDs) that can be flexibly loaded with objects with different physical and economic properties potentially dispatched by different firms of different industries. On the other hand, standardized unit loads offer standardized technical mounting devices which make them compatible with different types of logistics resources (Schulte 2005:175). Moreover, these standardized mounting devices enable the pooling of logistics resources across consumers: A single resource can be loaded with multiple unit loads, each carrying the cargo of another consumer. Thus, the use of standardized unit loads increases the variety of objects that can be transformed by a single logistics resource and the variety of logistics resources that can be used to transform a certain logistics object. Thus, tailoring logistics resources to standardized unit loads reduces their degree of physical asset specificity. In turn, this increases the willingness of member LSPs to speculatively establish a capability potential consisting of resource types compatible with standardized unit loads. Still, the willingness to invest speculatively also depends on the volume of logistics objects compatible with this type of standardized unit load. From a logistics consumer's perspective, the use of standardized unit loads contributes to achieving on-demand and rapid elasticity because logistics resources can be redeployed to alternative uses and users without physical reconfiguration. If standardized unit loads cannot accommodate idiosyncratic requirements of logistics objects, a time and cost-efficient approach is to develop a unit load with standardized external handling interfaces that is internally tailored to the properties of the logistics object, for example a temperature-controlled box. Hence, the use of standardized unit loads is likely to align with logistics consumers' willingness to absorb reconfiguration costs and their temporal requirements regarding lead time. We thus conclude that the greater the volume of logistics objects compatible with a certain type of standardized unit load or other type of standard handling interface, the more likely that resource types compatible with this type of unit load are included in the physical infrastructure of CLSs.

With regard to the type and variety of logistics objects that can be transformed by CLSs, we thus propose:

Proposition 6. *Cloud logistics systems will make use of logistical resources that are compatible with either a standardized unit load or an industry-wide standardized handling interface.*

In conclusion, the degree of physical asset specificity constrains the variety of logistics objects that can be transformed with cloud characteristics. This limitation represents the inevitable tradeoff between the capability scope of a logistics system and the aim of achieving the cloud characteristics. A remark is warranted about the "industry-wide standardized handling interface." By proposing that a standard is adopted industry-wide, we implicitly assume that an industry's volume of logistics objects with similar properties is large enough to ensure that member LSPs are willing to speculatively establish a capability potential. While this assumption will be satisfied for the majority of industries, some industries may be too small. Despite these exceptions, we believe this proposition is sufficiently precise for stating preconditions for establishing CLSs. Whether an industry is sufficiently large needs to be assessed on a case by case basis, as it cannot be determined at the level of abstraction of this reference architecture.

Before moving on, a remark regarding standardized unit loads is expedient. As argued above, standardized unit loads remove interdependencies between logistics objects and logistics resources. Their use can therefore be considered conceptually similar to the idea of hardware virtualization in cloud computing (see Section D.III.2), as depicted in Figure 42. Recall that hardware virtualization removes interdependencies between user applications and underlying physical computing resources by interposing an additional layer – a virtualization layer – between the user's operating platform and applications and the underlying physical computing hardware (see Section D.III.2). Using this idea, we conceive of standardized unit loads as a "logistical virtualization layer" that is being interposed between the logistics object and the underlying physical resource. Thus, by

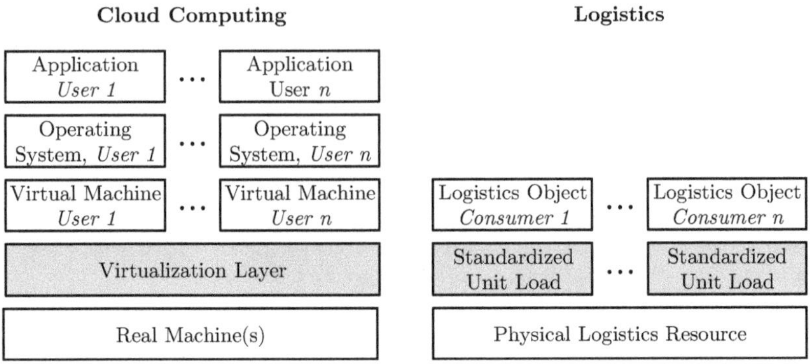

Figure 42.: Resource Virtualization in Cloud Computing and Logistics

using standardized unit loads, CLSs adopt the design principles of cloud computing.

The *variety of (non-)logistical transformations* that a logistics system can make depends on the composition of resources in its physical infrastructure. We conceive of a transformation being "specialized" if it can be made only by deploying a set of resources that possess a high degree of physical asset specificity, in the sense of having few alternative uses and/or users. When assessing physical asset specificity in the following, we focus on the availability of alternative logistics consumers with similar requirements, as retrofitting resources to alternative uses is likely to incur unacceptably high reconfiguration costs and time for logistics consumers, as argued above. Whether a certain kind of transformation becomes specialized depends on the extent to which the idiosyncratic requirements of logistics consumers regarding this kind of transformation – other than those requirements related to the physical and economic properties of the logistics object – necessitate the deployment of specialized resources. For example, a spatial transformation becomes specialized if the requirement for a very short transit time between continents necessitates the deployment of a "high speed" airplane. We argue that the lower the degree of physical asset specificity of deployed resources, the higher the willingness of member LSPs to speculatively establish a capability potential that can make this kind of transformation. Accordingly, the lower the degree of physical asset specificity of resources required for a transformation, the more likely that the (non-)logistical transformation is made by CLSs.

Due to the abstract nature of this reference architecture, we do not assess every conceivable (non-)logistical transformation in terms of whether or not it can only be made with or without specialized resources. Instead, we provide a comparative assessment between fundamental categories of transformations: (1) core logistical, (2) value-added logistical, and (3) value-added non-logistical. This assessment aims to provide fundamental guidance on the likelihood of transformations being made by CLSs.

The *core logistical transformations*, in terms of (a) spatial changes through transportation, (b) temporal changes through storage, and (c) quantity and composition changes through break-bulk/consolidation during transshipment, are most commonly made without specialized logistics resources. Resources for making these transformations are largely standardized across all industries, especially given the use of standardized unit loads. As Williamson (1985:54) remarks, "general purpose trucks and airplanes are likewise redeployable." This enables resource pooling, on-demand, and rapid elasticity.

Value-added logistical transformations, which change the logistics object(s) in terms of (a) storage, transportation, and handling properties through (re-)labeling and (re-)packaging and (b) quantity and composition through sorting and pick-and-packing, are more likely to require specialized resources. For example, logistics consumers may require certain labels that can only be created with a special printer, or they may require certain picking racks in order to streamline order fulfillment. Nevertheless, in some cases, these transformations can be relatively standardized across an industry. For example, product polybagging and shrink-wrapping resources may be used to package and mass-customize products (such as detergents) on-demand for different logistics consumers (Nexus Distribution 2016), thus enabling resource pooling, on-demand, and rapid elasticity.

Finally, *value-added non-logistical transformations* such as product design, marketing support, and final assembly (see Section B.IV.2) most often require specialized resources, ranging from basic mechanical tools to complex automated machinery, that cannot be redeployed to alternative consumers without a significant sacrifice of productive value. This is because these transformations are usually interwoven with logistics consumers' preceding or subsequent production activities. Still, some of these non-logistical transformations can be relatively standardized across an industry. For example, garment processing requires equipment for steam pressing, hanging, and folding, which is relatively standardized across the fashion industry, thus enabling resource pooling, on-demand availability, and rapid elasticity.

With regard to the scope of the transformations in CLSs, we thus propose:

Proposition 7. *Cloud logistics systems will offer transportation (spatial), storage (temporal), transshipment (quantity and composition), and industry-wide standardized value-added (non-)logistical transformations of logistics objects.*

In conclusion, the degree of physical asset specificity constrains the variety of (non-)logistical transformations that can be made with cloud characteristics. These limitations particularly manifest in the inability to make any type of value-added (non-)logistical transformation. Like the previous proposition, this one assumes implicitly that an industry's aggregated demand for a certain (non-)logistical transformation is sufficient to ensure that member LSPs are willing to speculatively establish a capability potential – if the requirements for this transformation are standardized across the industry.

3.2.4. Human and Procedural Asset Specificity

Human asset specificity refers to the degree to which a firm's personnel has acquired skills, knowledge, and experience that are specific to fulfilling another transacting party's requirements (Zaheer & Venkatraman 1995:377; also see Erramilli & Rao 1993:23). Human asset specificity refers to both physical and mental skills that cannot be readily put to work for alternative uses and users (Malone et al. 1987:486). Such specificity can arise from firm-specific training and/or learning-by-doing (Williamson 2002:176; Williamson 1991:281).

Using this definition, we conceive of human asset specificity as the capacity of deploying (logistics) resources in order to achieve a desired (non-)logistical transformation (capabilities). As purposeful deployment of logistics resources usually requires a well-defined sequence of activities, we propose to combine human and procedural asset specificity. *Procedural asset specificity* refers to organizational routines and workflows tailored to meet the unique requirements of the exchange partner (Malone et al. 1987 as cited in Zaheer & Venkatraman 1995:377). It depends on the degree to which these routines and workflows can be modified or redeployed without a loss in productive value (de Vita et al. 2011:335).

The assessment of human asset specificity is closely related to that of physical asset specificity. This is because humans ultimately deploy the resources necessary to achieve a desired (non-)logistical transformation of a logistics object. Hence, as with physical asset specificity, we do not assess the degree of human asset specificity regarding the deployment of every type of logistics resource, but we provide a comparative assessment between fundamental categories of transformations: (1) core logistical, (2) value-added logistical, and (3) value-added non-logistical. This assessment aims to provide fundamental guidance on the likelihood of transformations being made by a CLS.

In principle, we argue that the higher the degree of physical asset specificity of (logistics) resources, the higher the degree of human asset specificity. This seems reasonable because the more specialized a resource, the more specialized the capacity required to purposely deploy and operate it or handle a certain type of logistics object. The deployment of personnel to achieve *core logistical transformations* is not likely to require high levels of customer specific skills, knowledge, and experiences. The operation of transportation (e.g. airplanes) and certain handling resources (e.g. forklifts) are regulated by official driver's licenses. The deployment of *value-added logistics resources* is likely to involve a higher degree of human asset specificity. For example, warehouse personnel may acquire specific picking and packing skills. Likewise, the handling of logistics objects with unique requirements is likely to require personnel to internalize idiosyncratic procedures. Finally, the deployment of *non-logistics value-added resources* most likely involves the highest degree of human asset specificity. The operation of resources required to perform production-related transformations requires idiosyncratic skills and knowledge about the production technology of the specific logistics consumer. Still, the deployment of (non-)logistics value-added resources can be rather standardized across an industry for some transformations (see examples for physical asset specificity), thus enabling resource pooling, on-demand, and rapid elasticity. We thus conclude that the higher the degree of human asset specificity, the less likely that member LSPs establish a capability potential including this type of personnel. Hence, the more idiosyncratic the skills, knowledge, and experiences of personal necessary for a certain transformation, the less likely it is that this type of transformation will be made by CLSs.

With regard to the variety of logistics objects that can be transformed and the variety of transformations that can be made by CLSs, we thus propose:

Proposition 8. *Cloud logistics systems will use personnel with the capacity to deploy the resources required to achieve core logistical and industry-wide standardized, value-added, (non-)logistical transformations of logistics objects that use standardized unit-loads or an industry-wide standardized handling interface.*

In conclusion, this proposition reinforces the proposition made for physical asset specificity of logistics resources due to the relationship between the two specificity types. Likewise, the degree of human asset specificity limits the scope of logistics objects that can be transformed and the scope of (non-)logistical transformations that can be made.

3.3. Proposition: Five Logistical Prototype Scenarios

3.3.1. Proposition and Scenario Overview

The preceding subsection has assessed the degree of asset specificity of logistics resources and capabilities and derived propositions regarding which resources and capabilities are likely to be included in the physical infrastructure of CLSs and, hence, which (non-)logistical transformations can be made by CLSs. These propositions pertain to different dimensions of asset specificity and have not been integrated: Each imposes certain constraints on the physical infrastructure of CLSs, independent of the other propositions. This subsection develops "logistical prototype scenarios" that integrates them. Based on the propositions, each prototype scenario offers a "coherent bundle" of physical logistics capabilities with cloud characteristics, where "coherent" means likely to be of logistical value for logistics consumers.

In order to integrate the different propositions into logistical prototype scenarios, we draw on two dimensions included in the classification of horizontal LSP cooperation proposed by Schmoltzi & Wallenburg (2011:563) (also see Section C.VI.1): (1) *capability scope*, which refers to the composition of logistics capabilities offered by the cooperative network of member LSPs (this dimension is referred to as "service scope" in the original classification); and (2) *geographic scope*, which refers to the location(s) at which these capabilities are offered. These dimensions allow for the propositions derived in the preceding subsection to be integrated. Capability scope integrates physical asset specificity and human asset specificity; geographic scope depends on site specificity. Temporal specificity pertains to any capability independent of geography.

Capability scope and geographic scope are each measured on a three point scale. *Capability scope* ranges from (a) *narrow*, which means that capabilities offered are of one type only; to (b) *selected composition*, which means that certain combinations of a few different types of logistics capabilities are offered; to (c) *broad*, which means that any capability type or combination is are offered. *Geographic scope* ranges from (a) *localized*, which means that capabilities are offered within a certain confined area only; to (b) *wide, networked*, which means that capabilities are offered within certain geographic areas (nodes of a network) and geographic channels (links of a network) between these areas; to (c) *wide*, which means that capabilities are offered at any location without any spatial limits.

Based on the asset specificity assessment and the three point scales of capability and geographic scope, we propose:

Proposition 9. *The physical infrastructure of any cloud logistics system will match one of five logistical prototype scenarios, with each scenario's capability scope being either narrow or comprising selected compositions and its geographic scope either localized or wide, networked.*

Figure 43 depicts the proposed logistical prototype scenarios of CLSs. The figure shows that not any combination of capability and geographic scope can be offered by CLSs. In fact, the majority of combinations are not "economically feasible;" more specifically, any combination involving either a broad capability scope or a wide geographic scope cannot be offered because these scopes do not align with the economic concerns of member LSPs and logistics consumers. LSPs are not willing to speculatively establish these combinations of capability and geographic scope due to the low likelihood of redeployment. A broad capability scope likely involves resources and capabilities that are specific to the requirements of a single consumer or small group of them. A wide geographic scope likely

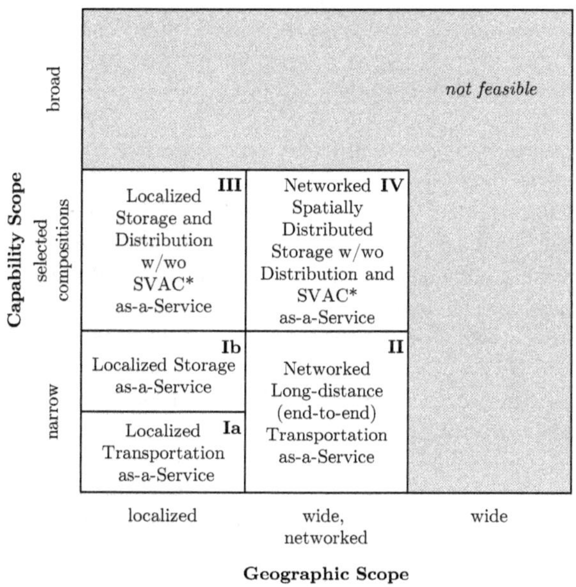

Figure 43.: Logistical Prototype Scenarios – Overview
(* = Standardized value-added (non-)logistics capabilities)

involves resources and capabilities at locations with few alternative logistics consumers. In addition, the reconfiguration of these resources and capabilities likely incurs high costs and requires significant time which is not acceptable for logistics consumers.

A brief description of each scenario is provided below. We specify the capability and geographic scope, explain how logistics consumers can access these capabilities, describe how each scenario achieves the cloud characteristics (resource pooling, on-demand availability, and rapid elasticity), assess the innovativeness compared to existing logistics systems, and provide empirical examples, if available. The subsequent subsection aims to identify potential sources of competitive advantage for logistics consumers that originate in consuming physical cloud logistics capabilities (seeSubsection F.V.3.4).

3.3.2. Scenario Ia: Localized Transportation Capabilities as-a-Service

Scenario Ia offers a narrow scope of logistics capabilities within a localized geography, as depicted in Figure 44. More specifically, the capability scope covers end-to-end road transportation capabilities limited to goods compatible with standardized unit loads or an industry-wide handling standard. Road transportation capabilities are realized through trucks and vans. Transportation links can be realized indirectly or directly, which means with or without transshipment, respectively. The geographic scope is confined to the vicinity of a single area of intense economic activity. Logistics consumers can physically access this capability potential by specifying a pick-up time and location. Goods are then picked up and delivered by providers with whom the consumers have concluded a service contract on the virtual marketplace.

Resource pooling, on-demand availability, and rapid elasticity are achieved by limiting the capability scope to transportation capabilities compatible with standardized unit loads or an industry-wide handling standard and by limiting the geographic scope to the same

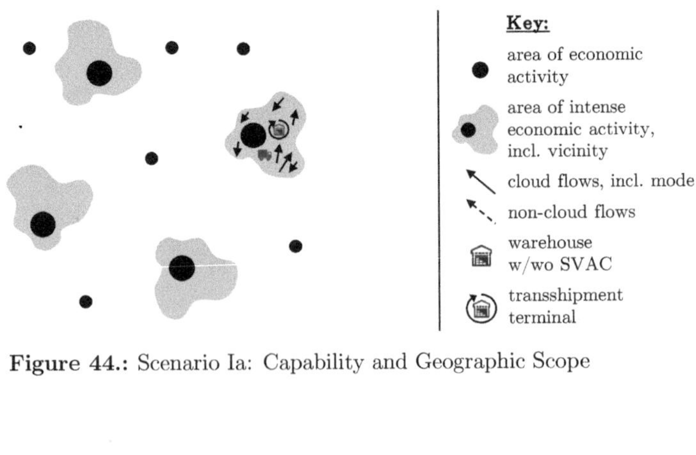

Figure 44.: Scenario Ia: Capability and Geographic Scope

area of intense economic activity. Limiting capabilities to a standard handling interface enables simultaneous and sequential use of transportation resources by different logistics consumers without the need for resource retrofitting. Limiting capabilities to the same area of intense economic activity enables on-demand availability and rapid elasticity as road transportation resources can be (re-)deployed with little time and costs due to spatial proximity between alternative logistics consumers. It also enables resource pooling due to spatial proximity between different consumers.

The degree of innovativeness in Scenario Ia is low. This scenario, in principle, resembles a freight exchange, which is limited to auctioning transportation capabilities within a single area of intense economic activity.

3.3.3. Scenario Ib: Localized Storage Capabilities as-a-Service

Scenario Ib offers a narrow scope of logistics capabilities within a localized geography, as depicted in Figure 45. More specifically, the capability scope includes storage capabilities limited to standardized unit loads or an industry-wide handling standard. The geographic scope of these storage capabilities is limited to the vicinity of a single area of intense economic activity. Logistics consumers can physically access this capability potential by delivering and/or retrieving their goods to and from the warehouse(s).

Scenario Ib achieves resource pooling, on-demand, and rapid elasticity by limiting the capability scope to storage capabilities for standardized unit loads or an industry-wide handling standard and by limiting the geographic scope to the vicinity of areas with intense economic activity. Different bays in a racking system compatible with a selected standard handling interface can be used by different consumers, thus enabling resource pooling. These bays can be quickly and effortlessly redeployed to alternative logistics consumers, thus enabling on-demand provisioning and rapid elasticity.

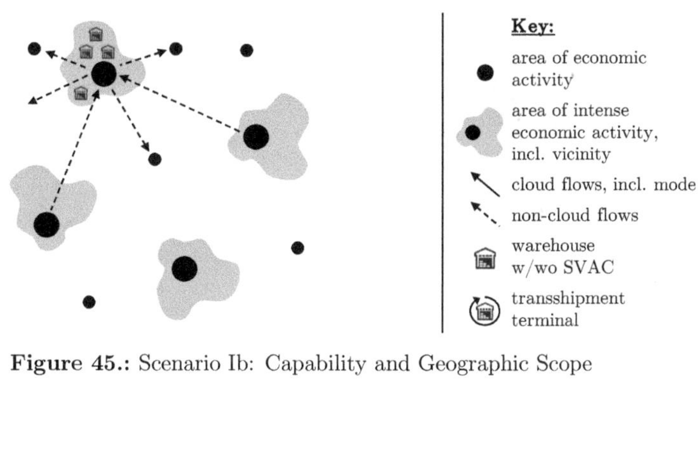

Figure 45.: Scenario Ib: Capability and Geographic Scope

The degree of innovativeness in Scenario Ib is low. This scenario in essence resembles the idea of public warehouses in which any logistics consumer can store goods loaded on a standardized handling device for a duration dependent fee. This scenario is slightly innovate because it connects multiple warehouses through a single cloud-based platform, providing access to a larger pool of storage capacity, which, in turn, increases the upper scalability bound. It furthermore innovates by using a market mechanism for matching storage demand with supply. Today, one can assume that most LSPs offer their storage capabilities in public warehouses for a fixed time-dependent fee.

3.3.4. Scenario II: Networked Long-distance (end-to-end) Transportation Capabilities as-a-Service

Scenario II offers a narrow scope of logistics capabilities within a wide, networked geography, as depicted in Figure 46. More specifically, the capability scope comprises transportation and transshipment capabilities for standardized unit loads or an industry-wide handling standard, thus enabling networked (or multi-leg) transportation. In addition, the capability scope can, but does not need to, include pick-up and last-mile delivery capabilities, thus enabling end-to-end transportation. The geographic scope of transportation capabilities is limited to the interconnection of multiple spatially distributed areas of intense economic activity. Transshipment, pick-up, and last-mile delivery are limited to the vicinity of these areas. The spatial distribution and distances between these interconnected areas are not confined, enabling long-distance (e.g. inter-regional or global) transportation. Depending on the distance between areas, transportation links are established by means of either air or road capabilities. Pick-up and last-mile delivery are always conducted through road capabilities. Logistics consumers can physically access this capability potential by either delivering or picking up their goods at any transshipment terminal that is part of the cloud, or they can make use of pick-up capabilities offered by the cloud, as long as the pick-up location lies within the vicinity of an area

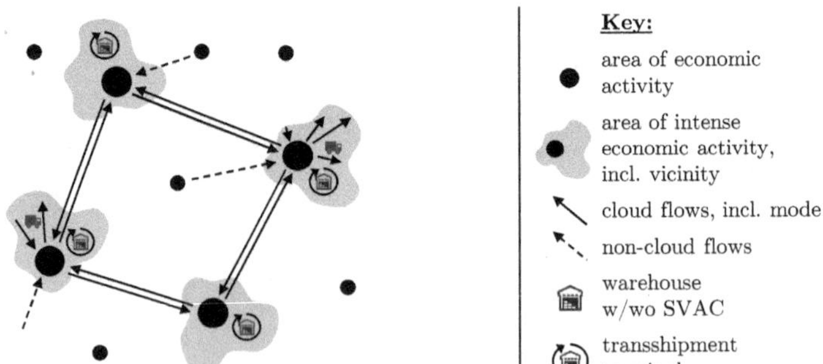

Figure 46.: Scenario II: Capability and Geographic Scope

with intense economic activity.

Scenario II achieves resource pooling, on-demand availability, and rapid elasticity by limiting the capability scope to goods compatible with standardized unit loads or an industry-wide handling standard and by limiting the geographic scope to transportation links between areas of intense economic activity or to the vicinity of such areas. Limiting transportation to a standard handling interface enables pooling in long-haul transportation and in pick-up and last-mile delivery as the goods of multiple consumers can be carried by the same transportation resource, either simultaneously or sequentially without reconfiguration. Limiting transportation to links between areas of intense economic activity enables on-demand and rapid elasticity. Such links can be operated with a higher frequency, and logistics resources can be redeployed to a large group of proximate alternative logistics consumers, thus minimizing lead times for delivering new capabilities (on-demand availability) and for expanding or reducing the capacity of capabilities currently being provisioned (rapid elasticity).

The innovativeness of this scenario is medium. It integrates different components of existing logistics systems in an innovative manner. The capability and geographic scope of this scenario resembles that of domestic and international express networks and of LTL networks – if pick-up and delivery locations are confined to areas with intense economic activity. However, while express networks are usually operated by a single LSP with centralized control, this cloud scenario is operated by a cooperative LSP network with decentralized control due to the use of market mechanisms. Yet, alliances with decentralized control are more frequently found in the freight forwarding segment in which auctions are used to allocate transportation capacity offered by (air/ocean) carriers to LSPs. However, while freight forwarding services are typically limited to main-haul transportation between macro-logistics infrastructure, this cloud scenario can (but does not need to) offer end-to-end transportation if pick-up and delivery are within areas of intense economic activity. We thus conclude that this cloud scenario adopts a capability and geographic scope typically associated with express and LTL networks and adopts an organizational form typically found in the freight forwarding segment. This scenario can thus be conceived of as a cloud-based platform freight exchange for networked, long-distance, end-to-end transportation links provisioned by a network of horizontally cooperating LSPs.

To our knowledge, there is no network of horizontally cooperating LSPs that resembles Scenario II. However, there are a few firms – especially IT providers of cloud-based logistics platforms – that are only "a few steps away" from realizing a logistics system that resembles this scenario. For example, the firms GT NEXUS (see GT Nexus 2014) and Cloud Logistics (see Cloud Logistics 2013) both operate a cloud-based platform that enables the different entities involved in transportation links to interact and exchange information and enables the management of (national, regional, and global) transportation links or entire supply chains. Today, both firms limit their business models to providing a cloud-based platform that can be used by others to manage their logistics activities. However, it is within their feasible scope of "strategic action" to introduce business logic to their platform in the form of market mechanisms, thus becoming the LITPAP of a horizontally cooperating network of LSPs. Consequently, they would not only operate the cloud-based platform, but moreover define standards for describing physical logistics capabilities as services, facilitate economic exchanges, and regulate the boundary of the cooperative LSP network.

Another example of a logistics system that partially resembles this prototype scenario is the cooperative LTL network Cloud Logistic (see IMA RWTH Aachen University 2011; Tummel et al. 2014), which we identified in our systematic knowledge search. This LTL network is operated by multiple horizontally cooperating LSPs that use a central cloud-based platform for coordination purposes. In its capability and geographic scope, this cooperative LTL network focuses on a "line-based logistics model" in Germany (Tummel et al. 2014:672), which means that shipments with similar origin and destination areas (zones) are picked up and distributed by one of the cooperating LSPs, thus achieving a higher utilization of transportation capabilities due to cross-consumer sharing. This capability and geographic scope would resemble that of this prototype scenario if the target and destination zones geographically corresponded with areas of intense economic activity.

3.3.5. Scenario III: Localized Storage and Distribution Capabilities with or without Standardized Value-added Capabilities as-a-Service

Scenario III offers a composition of selected logistics capabilities in a localized geographic scope, as depicted in Figure 47. More specifically, the capability scope comprises a com-

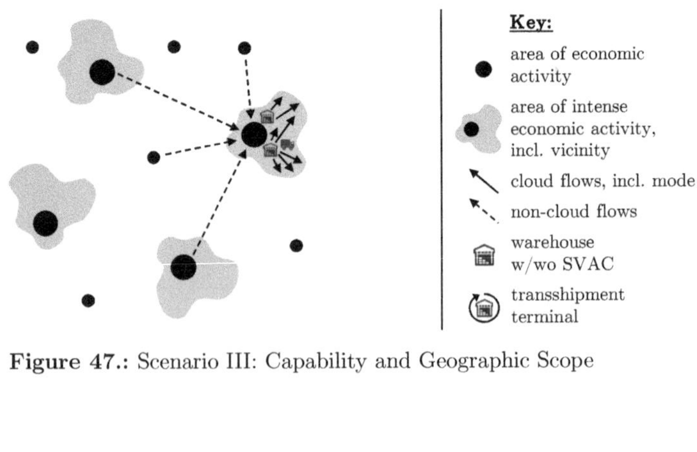

Figure 47.: Scenario III: Capability and Geographic Scope

position of storage capabilities integrated with (last-mile) distribution capabilities. These capabilities are limited to goods compatible with standardized unit loads or an industry-wide handling standard, most likely pallets and (modular) master cartons. In addition, it can, but does not need to, include SVACs. For example, this may include picking-and-packing customized selections of goods for (last-mile) delivery, (promotional) co-packaging and labeling, or garment processing. These SVACs thus enable fulfillment of customized client orders. The geographic scope is limited to the vicinity of a single area of intense economic activity. Due to the offering of last-mile distribution capabilities, this scenario is likely located in (dense) urban areas or industrial areas, rather than near locations of sole macro-logistics infrastructure. Logistics consumers can access this capability potential by delivering goods to the storage location(s) included in the cloud.

Scenario III achieves resource pooling, on-demand availability, and rapid elasticity by limiting the capability scope to goods compatible with standardized unit loads or an industry-wide handling standard and by limiting the geographic scope to a single area of intense economic activity. Limiting the capability scope to cargo compatible with a handling standard enables the sharing of storage and transportation capabilities, and limiting SVACs to those that are common across an industry enables sharing, on-demand availability, and rapid elasticity by reducing reconfiguration costs and time. Limiting the geographic scope to a single area of intense economic activity enables on-demand availability, rapid elasticity, and the pooling of capabilities due to spatial proximity between alternative logistics consumers.

The degree of innovativeness in this scenario is considered medium. The capability and geographic scope resemble the common idea of "city logistics," or out-of-town consolidation centers with storage capabilities attached. However, this scenario innovates by additionally integrating SVACs. They can thus be used as a final segment of a logistics channel or supply chain, rather than just the last segment in a distribution system.

To our knowledge, no cooperative logistics system exists that offers the capability and geographic scope of this scenario. However, within our systematic knowledge search, we identified one LSP whose capability and geographic scope match this scenario: "Cloud Pallet Storage" (offered by the LSP Cloud Storage and Logistics Limited, see Cloud Pallet Storage 2015) offers on-demand, rapidly elastic, pay-per-use storage capabilities and picking-and-packing capabilities in the vicinity of London. It seems within the feasible scope of "strategic action" of this firm to start offering additional standardized (non-)logistical value adding capabilities and to start cooperating with other LSPs in order to establish a storage and SVAC network around London, thus increasing the upper scalability bound.

3.3.6. Scenario IV: Networked Spatially Distributed Storage Capabilities with or without Distribution and Standardized Value-added Capabilities as-a-Service

Scenario IV offers a composition of selected logistics capabilities within a wide, networked geography, as depicted in Figure 48. More specifically, the capability scope comprises a composition of transportation, transshipment, and storage capabilities for goods compatible with standardized unit loads or an industry-wide handling standard, enabling networked storage capabilities. In addition, this composition can, but does not need to, include last-mile distribution and SVACs. It thus enables fulfillment of customized client orders. The geographic scope covers multiple areas of intense economic activity and the connections between these areas. Storage, last-mile distribution, and SVACs are limited to the vicinity of these areas; transportation capabilities are limited to the connections between these areas. The spatial distribution and distances between these interconnected areas are not confined, thus enabling networked, spatially distributed capabilities. Areas may be located in proximity to each other, in different countries, or even on different continents. Depending on the absolute distance, transportation links are either established by means of road or air transport. Logistics consumers can physically access this capability potential by delivering or picking up their goods to or from any location with storage or transshipment terminals which are part of the cloud.

This scenario achieves resource pooling, on-demand availability, and rapid elasticity by limiting the capability scope and geographic scope. Limiting transportation, transshipment, storage, and last-mile distribution capabilities to logistics objects compatible with standardized unit loads or an industry-wide handling standard enables quick redeployment and shared usage. Physical resource reconfiguration is not necessary. Limiting value-added capabilities to those that are common across an industry also enables quick redeployment and sharing. Personnel do not need to be (re-)trained and equipment necessary for capabilities can be used without reconfiguration. Limiting the geographic scope

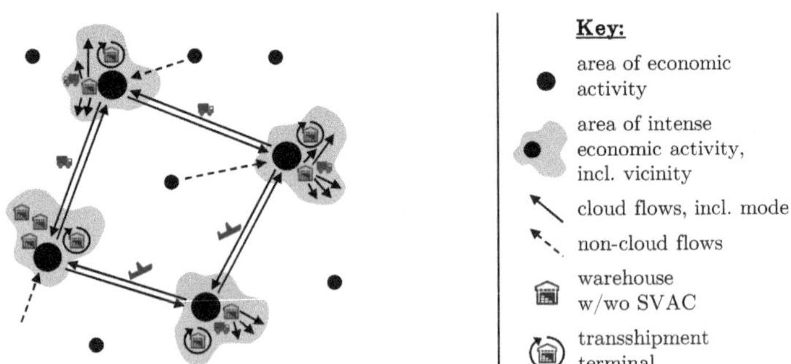

Figure 48.: Scenario IV: Capability and Geographic Scope

to areas of intense economic activity and the links between these areas enables the quick redeployment of resources due to a large number of proximate alternative consumers.

The degree of innovativeness in this prototype scenario is considered high. To our knowledge, the capability and geographic scope offered by this scenario are not yet offered by any existing logistics system. In other words, this scenario creates a new amalgam of capability and geographic scope. More specifically, it integrates the concepts of (international) freight forwarding with last-mile distribution and with public warehouses that offer SVACs. All of these capabilities are provisioned with cloud characteristics and become accessible via a single cloud-based platform, even though they are made available by different LSPs.

3.4. Potential Sources of Competitive Advantage for Logistics Consumers

The preceding section has proposed five logistical prototype scenarios that can offer physical logistics capabilities with cloud characteristics. This section investigates how logistics consumers can potentially gain competitive advantage by using the cloud logistics capabilities offered by these scenarios. Hence, this section aims to answer the question of whether CLSs may be of value for logistics consumers.

3.4.1. Preliminary Considerations: Logistics Channel Design

Each cloud logistics prototype scenario offers a unique set of physical logistics capabilities with cloud characteristics. In the following, we analyze how logistics consumers can integrate the cloud capabilities offered by these scenarios into their logistics channels and potentially gain competitive advantage over logistics consumers that do not utilize cloud capabilities. To specify how the use of cloud capabilities impacts on logistics channels, we describe channel design via three common parameters: (1) (de-)centralization, (2) logistics postponement and speculation, and (3) consolidation and separation.

(De-)centralization is the geographical distribution pattern of production and storage locations within an intra- or interfirm value creation system (see e.g. Klaas 2002:150f. and sources therein). Vertical (de-)centralization refers to number of sequential production and/or storage locations within a value creation system; horizontal refers to the number of geographically distributed production and/or storage locations at the same step of the value creation system. *Logistics postponement and speculation* refer to the geographical differentiation of inventory as a function of the existence of client orders (see e.g. Klaas 2002:152ff. and sources therein). Logistics postponement can be defined as deferred geographical differentiation of goods based on manifested client orders, while logistics speculation can be defined as the early geographical differentiation of goods based on

predicted client demand (see e.g. Klaas 2002:152 and sources therein). *Consolidation and separation* refer to the consolidation or separation of flows within a logistics system (see e.g. Klaas 2002:154f. and sources therein). These three parameters are however not independent of each other, but closely interrelated (Klaas 2002:157f.). As Klaas (2002:157f.) argues geographical de-centralization is associated with logistics speculation and the consolidation of flows, while geographical centralization is associated with logistics postponement and the separation of flows.

In the following discussion, we investigate potential sources of competitive advantage by means of a comparative assessment between "cloud consumers" and "non-cloud consumers." For comparability reasons, we make the simplifying assumption that both cloud consumers and non-cloud consumers have similar logistics requirements, for example in terms of clients, client locations, client volumes, and product characteristics. Moreover, we primarily focus on potential sources of competitive advantage that are related to the effectiveness of logistics responses in the sense of on-demand availability, rapid elasticity, and interaction with pay-per-use. We do not investigate potential sources of competitive advantage related to lower costs. This is because actual costs cannot be determined at the level of abstraction of this reference architecture as costs are determined through market mechanisms and are, *inter alia*, a function of the degree of resource pooling achieved at that particular time.

3.4.2. Localized Transportation Capabilities as-a-Service (Ia)

Scenario Ia allows cloud consumers located in an area of intense economic activity, to transport goods compatible with standardized unit loads or an industry-wide handling standard to clients located in the same area of intense economic activity.

Compared with non-cloud consumers, cloud consumers can potentially gain a competitive advantage due to *lower transportation lead times* that result from on-demand availability and due to the ability to achieve *higher delivery reliability* in case of demand peaks as a result of rapid elasticity.

3.4.3. Localized Storage Capabilities as-a-Service (Ib)

Scenario Ib allows cloud consumers to store goods compatible with standardized unit loads or an industry-wide handling standard in a set of shared warehouses, all located in the vicinity of the same area of intense economic activity. This scenario enables cloud consumers to temporarily or permanently increase the degree of decentralization by adding the cloud-warehouse to their logistics channels that "pass through" the area of intense economic activity covered by the cloud. The cloud-warehouse is usually added in the middle of a logistics channel as its capabilities are limited to storage.

Depending on the cloud consumers' current logistics setup, they can add the cloud-warehouse as an additional stage in their value creation or distribution system (vertical decentralization) or complement existing storage capabilities in this area (horizontal decentralization). The ability to increase the degree of decentralization of logistics channels can be a potential source of competitive advantage for cloud consumers over non-cloud consumers who operate a similar logistics channel because of *fewer inventory outages* in downstream warehouses in case of demand peaks. In the case of vertical decentralization, the cloud-warehouse can serve as an additional "more downstream" warehouse (logistics speculation) and thus experience fewer inventory outages due to quicker restocking. In the case of horizontal decentralization, the cloud-warehouse can provide additional capacity to hold "strategic stock" and thus experience fewer inventory outages due to more reliable restocking.

3.4.4. Networked Long-distance (end-to-end) Transportation Capabilities as-a-Service (II)

Scenario II allows cloud consumers to realize end-to-end transportation of goods compatible with standardized unit loads or an industry-wide handling standard if their clients are located in areas of intense economic activity covered by the cloud. This scenario corresponds to a centralized logistics channel setup, as the geographical differentiation of inventory can be deferred: Cloud consumers can hold inventory at a geographically "neutral" location as long as possible (logistics postponement) and dispatch goods to their clients through direct deliveries once orders are received (separation of flows).

Cloud consumers can potentially gain competitive advantage through *shorter delivery times* compared with non-cloud consumers who operate a logistics channel with a similar degree of decentralization and transportation modes. Shortened transit times result from more frequent pick-ups and last-mile deliveries in areas of intense economic activity and from more frequent transportation links between these areas. Cloud consumers can thus respond more quickly to demand peaks, which may give them the opportunity to increase market share compared with non-cloud consumers.

3.4.5. Localized Storage and Distribution Capabilities with or without Standardized Value-added Capabilities as-a-Service (III)

Scenario III allows logistics consumers to store goods compatible with standardized unit loads or an industry-wide handling standard in a set of shared warehouses located in the vicinity of the same area of intense economic activity. These storage capabilities are integrated with distribution capabilities to deliver goods to clients located in this area, and they may be further integrated with SVACs, allowing for the fulfillment of (customized) client orders.

Depending on the current setup of their logistics channel, cloud consumers can attach these capabilities as an additional “final segment” to their current logistics channel, thus increasing channel stratification (vertical decentralization), or complement already existing logistics capabilities in this area (horizontal decentralization). As geographical decentralization of logistics channels is associated with logistics speculation, cloud consumers can increase the geographical differentiation of their inventory. The ability to increase vertical or horizontal decentralization of a logistics channel (either temporarily or permanently) in an on-demand, rapidly elastic, and pay-per-use manner can be a potential source of competitive advantage for cloud consumers over non-cloud consumers.

Cloud consumers who do not have dedicated logistics capabilities in the area of intense economic activity covered by the CLS (e.g. due to a lack of cargo volume and clients in this area, which would economically justify dedicated capabilities, or due to a lack of financial resources to establish them) may gain competitive advantage over non-cloud consumers that do not have dedicated logistics capabilities in this area either. This advantage can occur in three general ways after adding cloud capabilities as a new final segment to the logistics channel (vertical decentralization).

First, due to on-demand availability, cloud consumers can *establish logistical access to the area of intense economic activity more quickly* than non-cloud consumers. They can start fulfilling (customized) orders of new clients in this area with low lead time more quickly than non-cloud consumers can. Hence, using cloud capabilities may secure a “first-mover” advantage for cloud consumers, especially in the event of unpredictable demand occurrences. Cloud consumers may thus be able to capture a larger market share in the long-run. From a different perspective, cloud consumers can quickly adjust their geographical logistical footprint.

Second, due to pay-per-use, on-demand availability, and rapid elasticity, cloud consumers may gain advantage by *fulfilling (customized) client orders with lower lead times* than non-cloud consumers. By adding cloud capabilities as final segment and scaling logistical capacity commensurate with demand, cloud consumers can operate a more decentralized logistics channel and adopt a better logistics speculation strategy than non-cloud consumers, regardless of cargo volume or financial power. Hence, this may be particularly relevant for small and medium-sized logistics consumers. Following the same line of reasoning, small and medium-sized cloud consumers may close the strategic disadvantage compared to (larger) competitors that have dedicated capabilities in this area: They can fulfill (customized) client orders with equally low lead times as their competitors, even without sufficient cargo volume or financial power. Hence, small or medium-sized cloud consumers are likely to use cloud capabilities as a permanent substitute for dedicated logistics capabilities.

Finally, due to pay-per-use, on-demand availability, and rapid elasticity, cloud consumers can potentially gain competitive advantage by *postponing logistical investment decisions* about the most suitable logistical setup for fulfilling (customized) client orders in an area of intense economic activity. This can be helpful for market entries or product launches. Cloud consumers can use cloud capabilities based on actual demand in an on-demand and pay-per-use manner during initial market growth and development, a period usually characterized by high demand uncertainty. Once market development is completed and demand more predictable, cloud consumers can determine their optimal (long-term) logistical setup in this area. Hence, by being able to postpone investment decisions, cloud consumers can potentially gain competitive advantage in the long-run through either *lower logistics costs* (by minimizing costs from idle capabilities due to over dimensioning) or *less lost revenue* (due to insufficient logistics capacity due to under dimensioning).

Furthermore, due to pay-per-use and rapid elasticity, cloud consumers that already have dedicated capabilities in an area covered by the CLS have a *higher reliability of maintaining a desired level of fulfilling (customized) client orders* during demand peaks. This can give them competitive advantage over non-cloud consumers that have dedicated logistics capabilities in the area covered by the CLS. While non-cloud consumers are likely unable to fulfill orders during demand peaks due to limited logistics capacity, cloud consumers can use rapid elasticity (horizontal decentralization) to quickly complement their existing capabilities with cloud capabilities commensurate with demand. They can thus offer a higher or unchanged service level as long as they can provide sufficient raw materials or inventory.

In conclusion, Scenario III enables cloud consumers to temporarily or permanently add or complement logistics capabilities in a single area of intense economic activity and thus potentially gain competitive advantage through quicker logistics access, lower lead times, postponed logistical decisions, and higher reliability in fulfilling (customized) client orders.

3.4.6. Networked Spatially Distributed Storage Capabilities with or without Distribution and Standardized Value-added Capabilities as-a-Service (IV)

Scenario IV allows cloud consumers to store goods in a set of spatially distributed, interconnected warehouses located in the vicinity of at least two different, but usually multiple, areas of intense economic activity. These distributed storage capabilities are integrated with and connected by transportation capabilities and, thus can be considered "networked storage capabilities." These capabilities can be further integrated with distribution capabilities to deliver goods to clients located in the area of intense economic activity, and with SVACs, thus enabling the fulfillment of (customized) client orders. In this sense, cloud consumers can use cloud capabilities as a temporary or permanent substitute for a dedicated, multi-staged, regional, or global distribution and fulfillment network.

Depending on their current setup, cloud consumers can add these cloud capabilities as an additional "final segment" to their logistics channels, increasing channel stratification (vertical decentralization), or complement existing logistics capabilities (horizontal decentralization), as in Scenario III. Yet, unlike in Scenario III, cloud consumers can realize vertical and horizontal decentralization concurrently – albeit in different areas of intense economic activity; while complementing their capabilities in one area, they may add cloud capabilities as a new final segment in another. As geographical decentralization is associated with logistics speculation, cloud consumers are also able to implement logistics speculation strategies. Compared with Scenario III, this strategy can be more complex due to the networked, multi-staged character of the storage capabilities. As a result, cloud consumers can increase the geographical differentiation of inventory not only in a single area of intense economic activity, but across multiple areas, and thus in different logistics channels. The ability to increase horizontal and vertical decentralization in an on-demand, rapidly elastic, and pay-per-use manner can be a potential source of competitive advantage for cloud consumers over non-cloud consumers.

The potential sources of competitive advantage in Scenario IV resemble those of Scenario III. This is because, if we were to consider each area of intense economic activity covered by Scenario IV in isolation, Scenario IV would equal Scenario III in terms of capability and geographic scope. However, the networked and geographically distributed character of cloud capabilities of in Scenario IV further strengthens these sources of competitive advantage.

Cloud consumers that do not have dedicated logistics capabilities in areas covered by the CLS can add these capabilities as final segments to their logistics channels (vertical decentralization). This provides cloud consumers a potential competitive advantage in three general ways compared to non-cloud consumers that do not have dedicated logistics capabilities in the areas covered by the CLS either.

First, because of on-demand availability, cloud consumers may gain competitive advantage by *establishing logistical access to more areas of intense economic activity more quickly* than non-cloud consumers. More specifically, cloud consumers may gain competitive advantage by being able to start fulfilling (customized) orders of new clients in multiple areas with low lead times more quickly than non-cloud consumers can. Using cloud capabilities may thus create a "first-mover" advantage for cloud consumers. From a different perspective, on-demand availability enables cloud consumers to adjust their logistical geographical footprint more quickly than non-cloud consumers by adding or removing final segments of their logistics channel in order to quickly respond to new demand in areas covered by the CLS.

Second, due to pay-per-use, on-demand availability, and rapid elasticity, cloud consumers can potentially gain competitive advantage by *fulfilling (customized) client orders with lower lead times* in *more areas* of intense economic activity than non-cloud consumers. Cloud consumers can cover more areas because, due to pay-per-use and rapid elasticity, they do not require a minimum cargo volume to economically justify an additional, more forward location in their logistics channels. Hence, cloud consumers can serve a larger number of clients with lower lead times, which may enable them to grow more quickly and capture more market share in the long-run. In addition, due to the networked character of cloud capabilities, cloud consumers can also benefit from a higher degree of consolidation in their flows: They can feed inventory into the cloud through one consolidated delivery to the most proximate cloud location. Inventory is then further distributed using the frequent and fast transportation links that connect the different cloud locations. This may result in additional cost advantages.

Third, due to pay-per-use and rapid elasticity, cloud consumers may gain competitive advantage by *postponing logistical investment decisions* regarding the most suitable logistical setup for fulfilling (customized) client orders in the areas covered by the CLS. This may be helpful to support a market entry or product launch in multiple areas of intense economic activity simultaneously. The distributed geographic coverage of cloud capabilities is especially important in a globalized economy in which more and more products are introduced on a regional or global scale, and clients expect the simultaneous availability of products across regions (e.g. mobile phones, tablets). Cloud consumers can use cloud capabilities based on actual demand in a pay-per-use manner during initial market growth and development, a period usually characterized by high demand uncertainty. Once market development is completed and demand more predictable, cloud consumers can determine their optimal (long-term) logistical setup in this area. Hence, by being able to postpone investment decisions, cloud consumers can potentially gain competitive advantage in the long-run through either *lower logistics costs* or *less lost revenue* due to overprovisioning or underprovisioning of logistics capacities, respectively.

Furthermore, due to pay-per-use and rapid elasticity, cloud consumers can achieve a *higher reliability of fulfilling (customized) client orders with a desired level of low lead times* during demand peaks. This may provide them with potential competitive advantage over non-cloud consumers that have a similar logistics channel setup. Due to the networked and geographically dispersed character of cloud capabilities, cloud consumers can simultaneously complement multiple stages of their regional or global distribution and fulfillment network. They can implement a sophisticated logistics speculation strategy that may help achieve a desired level of low lead times. Moreover, due to the networked capability character, cloud consumers can quickly re-supply cloud locations that experience demand surges. Inventory can be moved quickly between different cloud locations using the frequent and fast transportation links. In other words, cloud consumers can increase reliability in fulfilling (customized) orders by "balancing" inventory between different cloud locations commensurate with demand.

In conclusion, similar to Scenario III, Scenario IV enables cloud consumers to temporarily or permanently add or complement logistics capabilities in multiple areas of intense economic activity and thus to gain competitive advantage through quicker logistics access, lower lead times, postponed logistical decisions, and higher reliability in fulfilling (customized) client orders.

3.5. Conclusions: A Limited Capability and Geographic Scope due to Cloud Characteristics

The architectural view of the physical infrastructure of CLSs has presented five logistical prototype scenarios developed by using the construct of asset specificity. While all scenarios achieve resource pooling, on-demand availability, and rapid elasticity, each uses a unique combination of capability and geographic scope. The scenarios thus show that CLSs can address different sets of logistics concerns. However, the scenarios also show that not any set of logistics concerns can be addressed, but only certain "standardized" concerns, where "standardized" refers to the variety of logistical transformations, the variety of logistics objects to be transformed, and the geography in which the transformations can be made. In other words, the objective of attaining cloud characteristics constrains the set of addressable logistics concerns. We have also identified scenarios that resemble the capability and geographic scopes of existing logistics systems, such as public warehouses. While this is a sobering result – as it shows that, to some extent, CLSs already exist today – we still identified innovative compositions of capability and geographic scope. In this respect, Scenario IV is particularly interesting because it integrates storage, transportation, distribution, and SVAC capabilities across different geographies and makes them accessible through a single cloud-based platform.

Moreover, this architectural view also identifies several potential sources of competitive advantage that logistics consumers may realize by adding cloud capabilities to their current logistics channel setups compared to their competitors who are not using cloud capabilities. However, our comparative assessment is based on the simplifying assumption that both cloud consumers and non-cloud consumers have similar logistical requirements *and* that all these requirements can be addressed by cloud capabilities. In other words, we have confined our assessment to a relatively small scope of logistics requirements. This assumption is necessary to compare "apples to apples" and thus derive specific sources of competitive advantage. However, in reality, many logistics consumers may have idiosyncratic value creation processes and thus idiosyncratic logistical requirements that may not be readily addressable by CLSs. Therefore, logistics consumers face the choice of either adjusting their value creation processes in order use cloud capabilities or leave their value creation processes unchanged and use a tailored logistics solution. Whether such logistics consumers can also gain competitive advantage if they use cloud capabilities cannot be assessed universally: Different consumers suffer from different productive losses when adjusting their idiosyncratic value creation processes. Those that expect a net loss from complying with the standards imposed by CLSs are likely to opt for unique solutions, regardless of resource pooling, on-demand, and rapid elasticity. Those that expect a net gain from using cloud capabilities are likely to switch. Hence, future research should focus on identifying concrete logistical requirements for which a change in value creation processes and logistical requirements is worthwhile. This would provide valuable guidance for logistics consumers searching for the most suitable logistics setup.

VI. Structural Governance

1. Abstract of Chapter

The structural governance is concerned with the institutional framework within which logistics consumers, LSPs, and the LITPAP coordinate value exchanges.

The *architectural viewpoint* establishes the conventions for designing the governance structure (see Section F.VI.2). We adopt TCE to predict governance structure. Relevant contingency factors include asset specificity, transaction frequency, uncertainty, and complexity of product descriptions (see Subsubsection F.VI.2.2.1). We complement TCE with the RBV. Relevant resource-based contingency factors include the type, similarity of resource types contributed to the alliance, and resource complementarity of partnering firms (see Subsubsection F.VI.2.2.2).

The *architectural view* summarizes the assessment of contingency factors by proposing the following governance structure for CLSs (see Subsection F.VI.3.1).

Proposition 10. *Cloud logistics systems represent a unilateral contract-based alliance consisting of a LITPAP and a network of horizontally cooperating LSPs; value exchanges between logistics consumers and LSPs participating in a CLS are facilitated through a virtual marketplace via the price mechanism.*

Asset specificity is the primary contingency factor to determine the governance structure.

The governance structure in CLSs can be structured into two institutional frameworks: one for transactions between logistics consumers and providers (consumer transactions) and one for exchanges among providers (provider transactions). The low degree of *asset specificity* of cloud logistics capabilities (which we proposed as a prerequisite for achieving cloud characteristics) enables market-based economic exchanges for consumer transactions. The high powered incentives offered by markets also ensure that member LSPs commit a sufficient amount of resources to establishing a generic capability potential in a speculative manner. In other words, opportunism of LSPs to secretly reduce resource contributions to the alliance is deterred. Despite the cloud logistics capabilities' low degree of asset specificity, establishing a generic capability potential in a speculative manner represents a significant idiosyncratic investment for member LSPs to the alliance. As member LSPs often deliver capabilities in a collaborative manner, provider transactions need to be supported by a hybrid structure, a contract-based alliance.

2. Viewpoint: An Approach to Determining the Governance Structure

2.1. An Integrative Approach: TCE and RBV

This viewpoint establishes conventions to determine the structural governance mode of CLSs. This viewpoint thus considers and determines the institutional framework within which logistics consumers, LSPs, and the LITPAP coordinate value exchanges.

TCE is the most common theory used to explain and predict the structure that governs economic exchanges. Hence, we draw on TCE to predict the governance structure of CLSs. In addition, we also draw on the RBV. This may be surprising; after all, the RBV was initially proposed to explain and predict differential firm performance as a function of resources accessible to a firm. In other words, the RBV was not intended to support any inferences about structural governance choices of interorganizational relationships. Still, scholars have tried to explain and predict the choice of governance structures as a function of resources. Surprisingly, it seems that scholars have, in fact, been able to successfully establish this link and add another explanatory dimension to alliance governance choice (see e.g. Das & Teng 2000; Chen & Chen 2003b). The RBV thus poaches in the traditional realm of TCE. The following subsection introduces contingency factors for structural governance choice rooted in TCE and RBV.

2.2. Contingency Factors from TCE and RBV

2.2.1. Contingency Factors Rooted in TCE

The argumentative logic of TCE and the factors that determine structural governance choice have already been presented in detail (see Subsection C.III.3.2). Therefore, we only summarize the logic in a very condensed form and name the underlying factors here. Given opportunistically behaving and boundedly rational economic actors, structural choice depends on matching transaction characteristics with governance structure in a transaction-cost economizing manner (discriminating alignment hypothesis). Transactions can be characterized by (1) asset specificity, (2) frequency, (3) uncertainty, and (4) complexity of product description. The behavioral assumptions and the transaction characteristics represent contingency factors for governance choice.

2.2.2. Contingency Factors Rooted in RBV

Two resource-based contingency factors for structural governance choice are frequently mentioned in organizational literature: (1) type and similarity of resource types contributed to the alliance; and (2) resource complementarity.

The *type and similarity of resource types that are contributed to the alliance* by cooperating firms are posited as contingency factors for alliance governance structure (Das & Teng 2000). Structural choices result from firms balancing the issues of "*being able to procure valuable resources from another party without losing control of one's own resources.*" (Das & Teng 2000:44, italics original; see similar Chen & Chen 2003b:5). Das & Teng (2000:41f.) distinguish between two basic resource types, *property-based resources* and *knowledge-based resources*, which differ in terms of imperfect immobility, imperfect imitability, and imperfect substitutability. Examples of property-based resources include physical resources, patents, copyrights, and trademarks; knowledge-based resources include technological and managerial resources and organizational resources such as culture. These authors argue that property-based resources are well protected from unintended transfer by law, but knowledge-based resources are more difficult to confine within the boundaries of a firm once another firm has acquired access to them (Das & Teng 2000:43). Hence, appropriate safeguards need to be installed. Following this argumentative logic, they suggest four different structural governance choices depending on the resource type contributed to the alliance and the similarity between resource types contributed by partnering firms, as depicted in Table 24. They distinguish four types of structural governance choice: (1) equity joint venture, (2) minority equity position, (3) bilateral contract-based alliance, and (4) unilateral contract-based alliance (see Subsection C.III.3.1). These structural choices are viewed from the perspective of a focal firm that aims to set up an alliance

Focal Firm	Prospective Partner Firm	
	Property-based Resources	*Knowledge-based Resources*
Property-based Resources	Unilateral Contract-based Alliances	Equity Joint Ventures
Knowledge-based Resources	Minority Equity Alliances	Bilateral Contract-based Alliances

Table 24.: Resource Types and a Focal Firm's Structural Governance Preferences (adopted with minor changes from Das & Teng 2000:45)

with a prospective partner. Hence, they may not necessarily align with the interests of the prospective partner, especially if firms contribute resources of different types.

The focal firm prefers an *equity joint venture* if it primarily contributes property-based resources while the prospective partner primarily contributes knowledge-based resources (Das & Teng 2000:46, P2a). This governance mode allows for intimate interaction, creating the opportunity to (secretly) appropriate the partner's knowledge. At the same time, the focal firm's resources are well protected against unwanted appropriation due to well-defined property rights.

Taking a *minority equity position* in the prospective partner firm will be preferred by the focal firm, if the focal firm contributes knowledge-based resources and the prospective partner contributes property-based resources (Das & Teng 2000:46, P2b). The equity stake held by the focal firm offers sufficient safeguards against opportunistic knowledge appropriation due to its long term perspective and the possibility to hold it hostage. An equity joint venture is less preferable, as there are no knowledge-based resources that the focal firm could appropriate in turn while at the same time excessively exposing its knowledge resources to the partner.

The focal firm prefers a *bilateral contract-based alliance* if it and the prospective partner both contribute knowledge-based resources to the prospective alliance (Das & Teng 2000:47, P2c). As knowledge-based alliances are primarily about (mutual) learning processes, partnering firms usually conceive of them as "learning races." Once learning processes are completed, such alliances are typically terminated. Bilateral contract-based alliances are preferred to unilateral alliances because they offer better opportunities to work together closely and thus foster learning. Bilateral contract-based alliances are preferred to equity joint ventures and minority equity alliances because they are easier to terminate once the learning processes are completed or in the case of excessive (secret) knowledge appropriation by the partnering firm.

Finally, the focal firm prefers a *unilateral contract-based alliance* if it and the prospective partner contribute property-based resources to the prospective alliance (Das & Teng 2000:47, P2d). Property-based alliances are essentially equivalent to an exchange of property rights, and neither firm intends to secretly appropriate the partnering firm's tacit knowledge. Therefore, a light alliance engagement suffices, as neither a forum for extensive learning or knowledge exchange (as offered by bilateral contracts or equity based alliances) nor additional safeguards to curb opportunistic behavior (as offered by equity) are needed.

The second contingency factor for alliance governance structure is *resource complementarity* (Chen & Chen 2003b). These scholars find support for their hypothesis that the greater the complementarity between partnering firms, the more likely that a contract-based alliance is preferred over equity (Chen & Chen 2003b:4, H4). They assume that a higher degree of complementarity results in a lower likelihood of conflicting interest between partners, which thus encourages partners to choose flexibility over control (Chen & Chen 2003b:10). If partners complement each other, there is less direct competition. In addition, firms are mutually interdependent, which reduces the risk for opportunistic behavior.

To conclude, although the RBV initially set out to explain differential firm performance by focusing on resources accessible to firms, a few scholars expanded the explanatory power of this theory to the field of alliance governance structures.

3. View: Networked Horizontal LSP Cooperation with Market Coordination

3.1. Proposition

Current cloud logistics knowledge predominantly suggests that CLSs comprise a network of horizontally cooperating LSPs including the LITPAP (see Subsection E.IV.5.4). In addition, knowledge suggests that value exchanges among stakeholders are facilitated through an electronic marketplace (see Subsubsection E.IV.5.4.4). These suggestions clearly pertain to the governance structure of CLSs. However, the literature does not provide any theoretical or empirical support for these suggestions. Following the argumentative logic of TCE (see Subsection C.III.3.2) and the inferences by the RBV on structural governance choice (see Section F.VI.2), we support the existing cloud logistics knowledge and propose:

Proposition 10. *Cloud logistics systems represent a unilateral contract-based alliance consisting of a LITPAP and a network of horizontally cooperating LSPs; value exchanges between logistics consumers and LSPs participating in a CLS are facilitated through a* *virtual marketplace via the price mechanism.*

The following subsection provides a rationale for this proposition by assessing the presumable conditions of contingency factors rooted in TCE and the RBV. We again focus on the contract that structures the networked cooperation between LSPs and the LITPAP in more detail in the formalization viewpoint/ view due to the direct link between governance structure and contract formalization (see Chapter F.VIII).

3.2. Assessment of Contingency Factors

3.2.1. Contingency Factors Rooted in TCE

Following TCE, we start by assessing the underlying assumptions about the economic actors in terms of (1) bounded rationality and (2) opportunism. Thereafter, we assess the factors that characterize transactions' (3) asset specificity, (4) frequency, (5) uncertainty, and (6) complexity of product descriptions in order to support the above proposition.

Literature does not provide any specific insights into the degree of *bounded rationality* of cloud logistics stakeholders. However, regardless of whether these stakeholders are constrained by either small or significant cognitive limitations, incomplete contracting remains an issue in CLSs, thus making future contract adaptations inevitable.

In contrast, logistics literature provides significant insights into the degree of LSPs' *opportunistic behavior* in horizontal alliances. As described above, the risk of opportunistic behavior in horizontal LSP cooperation is especially high due to the concurrent existence of cooperation and competition, and the risk of such behavior is pervasive even in trustworthy relationships (see Subsection C.VI.3.1). In fact, given that CLSs require cooperating LSPs' logistics capabilities to overlap in type and geography to a significant degree in order to achieve cloud characteristics, the risk of opportunistic behavior seems even more prevalent in CLSs than in "regular" horizontal LSP alliances. Hence, the governance structure of CLS must provide strong safeguards to deter opportunism.

In order to support our proposition and the current cloud logistics knowledge regarding the governance structure, we distinguish between two types of transactions in CLSs: (a) *consumer transactions*, which refer to transactions between logistics consumers and member LSPs, and (b) *provider transactions*, which refer to transactions between member LSPs and the LITPAP. We assess the contingency factors for each transaction type.

The degree of *asset specificity* of logistics capabilities has been identified as a key factor for delivering physical logistics capabilities with cloud characteristics (see Chapter F.V).

In order to attain these characteristics, logistics capabilities must possess a sufficiently low degree of specificity. Concrete propositions have been developed to specify what a sufficiently low degree of specificity means. Considering these propositions, we believe that one can conclude that cloud logistics capability possess a (rather) low degree of specificity in absolute terms. This assessment primarily relates to *consumer transactions.* From the perspective of consumers, cloud logistics capabilities possess a low degree of asset specificity because there is limited ability to tailor them to potentially idiosyncratic requirements. From the perspective of member LSPs, the generic capability potential speculatively established possesses a low degree of asset specificity because it can be (re-)deployed for alternative uses and/or users with little or no tailoring. Following the argumentative logic of TCE, the goal of achieving a sufficiently low degree of asset specificity creates pressures toward market coordination (see Subsection C.III.3.2).

Provider transactions pertain to the commitment of member LSPs to speculatively invest in building up a logistics capability potential that enables the delivery of logistics capabilities with cloud characteristics, especially on-demand availability and rapid elasticity. Although cloud logistics capabilities have a rather low degree of asset specificity, speculatively establishing such a potential is a significant idiosyncratic investment from the perspective of a single LSP. The LSP speculatively commits resources to the CLS in anticipation of potential clients with matching requirements. These resources cannot be deployed to consumers outside of the CLS. According to TCE, idiosyncratic investments create pressures toward hierarchy (see Subsection C.III.3.2).

In fact, as is typical for hierarchies, CLSs are suggested to make use of extensive administrative controls, above all the IoT. However, the IoT represents a rather "operational" control that can effectively monitor the delivery of logistics capabilities (which reduces cheating by LSPs). Such controls, however, cannot deter opportunistic behavior of member LSPs to (secretly) reduce the amount of resources committed to the CLS in the case that consumer requests were allocated by the LITPAP in a hierarchical manner. If the LITPAP allocated consumer requests, single LSPs would behave opportunistically by reducing the amount of resources committed to the CLS and offering them outside of the CLS (known as shirking). If on-demand provisioning and rapid elasticity failed to materialize after being allocated a consumer request, LSPs would then invoke *force majeure* or other non-controllables to justify non-performance. The LITPAP would likely face difficulties in proving whether the LSP did, in fact, commit sufficient resources, especially due to the large number and potential geographical dispersion of LSPs, as in Scenarios II and IV. Hence, accurately monitoring the capability potentials established by all member LSPs not only seems difficult, but would also incur high monitoring costs in order to ensure on-demand and rapidly elastic delivery. This would reduce the likelihood for CLSs to form and actually deliver physical capabilities with cloud characteristics.

Thus, we argue that the concurrence of sufficiently low asset specificity of cloud capabilities and the need for LSPs to make significant idiosyncratic investments to deliver these capabilities likely causes market and hierarchical structures to fuse, as suggested by current cloud logistics literature and proposed above, into an interfirm network of horizontally cooperating LSPs (including the LITPAP) that coordinate value exchanges through an electronic marketplace. High-powered incentives and decentralized decision-making, as offered by markets, appear to be the only effective and efficient safeguards to deter opportunistic behavior among a large and potentially geographically dispersed group of member LSPs. On the one hand, LSPs are willing to commit enough resources speculatively because delivery (non-)performance has immediate consequences and can be clearly determined through the IoT. In addition, LSPs voluntarily commit enough resources because the single virtual marketplace aggregates demand from many potential logistics consumers, thus enabling the efficient use of speculatively committed resources. On the other hand, central monitoring efforts for the LITPAP decrease, and the LITPAP acts as a neutral arbitrator between logistics consumers and LSPs in case of disturbances. To conclude, the assessment of asset specificity supports the current literature and the above proposition.

The *frequency* of consumer transactions is likely to be low, while provider transactions are recurrent. *Consumer transactions* are likely to be infrequent as logistics consumers explicitly require pay-per-use, aiming for discreteness of transactions and the elimination of dependencies beyond the current transaction. Still, recurrent consumer transactions may happen, but are rather a coincidence of LSPs' and consumers' congruent preferences regarding performance and price. According to TCE, a low frequency does not warrant a complex governance structure, as fixed costs for setting up such structures are unlikely to be recovered from a single transaction, thus creating pressure for market coordination. In addition to transaction discreteness, logistics consumers also require that the delivery of cloud logistics capabilities is monitored on a granular level for purposes of transparency into the capacity actually consumed, which forms a basis for pay-per-use. This requires setting up complex governance structures involving elaborate administrative controls, especially including the IoT. However, according to TCE, such structures are only economically justifiable if transaction frequency is high enough to recover fixed set-up costs. This would be the case in a hierarchy. Hence, the need for pay-per-use delivery imposes conflicting requirements on the governance structure. We argue that these conflicting requirements can be reconciled from a TCE perspective if the complex governance structure can be used to support similar kinds of transactions for many different logistics consumers, thus allowing fixed set-up costs to be recovered from a scale perspective. Under this assumption, we conclude that market coordination may facilitate consumer transactions in a transaction cost economizing manner. This assumptions is reasonably met in a CLS because of the homogeneity of capabilities it offers, thus supporting the current cloud logistics knowledge and the above proposition.

Provider transactions are recurrent, as member LSPs continuously commit logistics resources to the CLS in order to establish a capability potential suitable for on-demand and rapidly elastic delivery. Although LSPs may not be able to predict which specific other member LSPs they will work with, they plan for recurrent transactions with other member LSPs. This warrants a complex governance structure, as fixed costs incurred from erecting the structure can be recovered through recurring transactions, thus creating pressure for hierarchical coordination. More specifically, this may warrant a hybrid governance mode – structured through a cooperation contract that specifies administrative and operational procedures to enable the collective delivery of cloud logistics capabilities. These considerations also support the current cloud logistics knowledge and the above proposition that CLSs comprise a network of horizontally cooperating LSPs.

Uncertainty is higher in consumer transactions and lower in provider transactions. *Consumer transactions* are likely to be exposed to frequent disturbances, which result from the dynamic (business) environment of logistics consumers. These disturbances usually impact the demand for logistics capabilities in terms of geography, type, and capacity. According to TCE, this high frequency of disturbances creates pressures for either market or hierarchical coordination of consumer transactions (see Subsection C.III.3.2). Given the sufficiently low degree of asset specificity from a consumer perspective, this high level of uncertainty creates pressures for market coordination, thus supporting the current cloud logistics knowledge and the above proposition

Provider transactions are likely to be exposed to fewer disturbances than consumer transactions, although LSPs are exposed to the dynamic (business) environment of logistics consumers in a mediated way. However, not every disturbance in consumer transactions immediately translates into the need for adapting the structure that governs provider transactions. For example, a single consumer who would like to renegotiate rates due to unexpected additional volume is unlikely to affect the cooperation structure among LSPs. We thus conclude that the frequency of provider transactions is likely to be of medium degree. Following TCE, market, hybrid, or hierarchical coordination may therefore be cost-efficient structures to facilitate provider transactions. Given the idiosyncratic investments by LSPs, a medium level of uncertainty suggests a hybrid governance mode as a cost-efficient choice, thus supporting the current cloud logistics knowledge and the above proposition regarding the existence of a network of horizontally cooperating LSPs.

Finally, the *complexity of product descriptions* is high in CLSs because of the use of standardized, semantic service descriptions as an underlying coordination mechanism (see Chapter F.IX; Chapter F.XI below). Highly complex descriptions generally create pressure toward hierarchy (see Subsection C.III.4.2). However, cloud logistics capabilities are expressed by means of descriptions that are standardized and make use of well-defined semantics. Standardization, and especially well-defined semantics, aim to enable the automatic processing of these descriptions by computers, reducing coordination costs. In other words, CLSs explicitly aim to exploit the communication effect of IT (see Subsection C.III.4.2). This creates strong pressures toward market coordination. Given the sufficiently low degree of asset specificity from a consumer perspective, market coordination seems like a cost-efficient governance mode, based on Malone et al.'s (1987) argument. This supports the existing cloud logistics knowledge and the above proposition.

Based on the assessment of contingency factors associated with TCE, we conclude that a synthesized governance mode combining market coordination with a network of horizontally cooperating LSPs is a cost-efficient structure for CLSs. This aligns with current knowledge and provides first theoretical support for the suggestions therein.

3.2.2. Contingency Factors Rooted in the RBV

Two resource-based contingency factors relevant to governance choice are (1) type and similarity of resource types contributed to the alliance and (2) resource complementarity.

In order to determine the *type and similarity of resource types* contributed to a CLS, we refer to the cloud logistics prototype scenarios (see Subsection F.V.3.3). They suggest that all LSPs and the LITPAP primarily contribute property-based resources. LSPs contribute physical logistics resources, and the LITPAP contributes physical and virtual logistics-related IT resources. Based on the propositions of Das & Teng (2000), similar contributions of property-based resources make unilateral contract-based alliances a preferable choice for cooperating firms.

Resource complementarity varies across logistical prototype scenarios. Scenarios Ia and Ib do not exhibit any complementarity because LSPs contribute similar resources in a similar geographic scope. This creates pressure toward equity joint ventures. However, all other prototype scenarios involve some degree of complementarity, thus creating pressures toward contract-based alliances. We thus conclude that the resource complementarity factor provides conflicting pressures for different prototype scenarios. Therefore, making a general prediction based on this factor is not feasible.

The two preceding paragraphs have assessed two contingency factors for structural governance choice rooted in the RBV. For prototype scenarios II, III, and IV, both factors partially match their predictions: Both factors identify pressures toward contract-based alliances. However, the factor of *resource complementarity* also identifies pressures toward equity joint ventures for these scenarios. For prototype scenarios Ia and Ib, the two factors provide conflicting results. While the factor of *type and similarity in resource types* suggests a contract-based alliance, *resource complementarity* identifies pressure toward equity, primarily due to the need for deterring opportunism resulting from direct competition among partnering firms. Whether these pressures are strong enough to actually cause an equity joint venture to emerge cannot be determined based on these two factors alone. In fact, although the risk of opportunism is high, structures other than equity joint ventures may be available to deter opportunistic behavior in all scenarios.

3.3. Conclusions: Synthesizing Results from TCE and RBV

The previous two subsections have assessed contingency factors rooted in TCE and RBV relevant to structural governance choice. Despite considering the choice from fundamentally different perspectives, the results closely match and complement each other. While TCE suggests the existence of a hybrid governance structure among LSPs and the LIT-PAP, the RBV provides insight into the position of this hybrid structure between markets and hierarchies. The network of horizontally cooperating LSPs represents a unilateral contract-based cooperation, positioned closer to markets. This rather light cooperative form also aligns well with the requirement of a permeable system boundary that allows LSPs to leave or join according to a set of pre-defined rules. Furthermore, TCE suggests that transactions are facilitated through an electronic market, thus deterring opportunistic behavior. In fact, this marketplace represents a critical factor for offsetting pressures toward equity joint ventures, which we have identified based on the contingency factor of resource complementarity. Opportunistic behavior resulting from resource similarity can be controlled through markets' high-powered incentives. We thus conclude that contract-based cooperation is feasible for all scenarios.

In conclusion, the theoretical investigation of the structural governance of CLSs supports the conceptions of current cloud logistics knowledge and supports the above proposition that CLSs represent a unilateral contract-based cooperation that facilitates value exchanges through an electronic market.

VII. Centralization

1. Abstract of Chapter

Centralization is concerned with the distribution of authority between network members over decisions that affect either the whole or parts of the network and, thus, all or a subset of its members (see Section C.V.2).

The *architectural viewpoint* establishes conventions to determine the degree of centralization in CLSs (see Section F.VII.2). Based on the model by Provan & Kenis (2008), we distinguish between four potentially effective distributions of authority in CLSs, referred to as "governance modes:" (a) *shared governance* refers to the case in which authority over network-level decisions is widely dispersed among the network of horizontally cooperating LSPs and the LITPAP; (b) *lead LITPAP governance* refers to a situation in which the LITPAP has exclusive authority over network-level decisions; (c) *lead LSP governance* refers to a situation in which a single LSP in the network of horizontally cooperating LSPs has exclusive authority over network-level decisions. This leading LSP tasks and monitors the LITPAP to establish, maintain, and evolve the cloud-based platform and applications on its behalf; and (d) *NAO governance* refers to a situation in which a subset of LSPs in the horizontal LSP network voluntarily erect an NAO. These member LSPs monitor the activities of the NAO and become involved in strategic network-level decisions through board structures. Operational decisions reside with the NAO, which tasks and monitors the LITPAP to establish, maintain, and evolve the cloud-based platform and applications on its behalf.

Following the logic of contingency theory, effective organization of an interorganizational network depends on four contingency factors: (1) network size, (2) network-level goal consensus, (3) need for network-level competencies, and (4) distribution of trust (see Subsection F.VII.2.2).

The *architectural view* summarizes the assessment of contingency factors by proposing the following distribution of authority to be effective for CLSs (see Subsection F.VII.3.1).

Proposition 11. *Lead LITPAP governance will be an effective distribution of authority for network-level decisions in cloud logistics systems.*

In other words, we propose that CLSs cannot be effectively governed by one or more LSPs but only by an IT vendor that is capable of establishing, maintaining, and evolving a cloud-based platform with logistics-related applications.

Large *network size* is a presumed prerequisite in order to deliver logistics capabilities with cloud characteristics, thus making all governance modes effective except of shared governance as that mode requires that all members interact frequently in a direct manner (see Subsubsection F.VII.3.2.1).

Network goal consensus among horizontally cooperating LSPs and the LITPAP is presumed to be moderate because member LSPs and the LITPAP, on the one hand, need to cooperate to deliver logistics capabilities with cloud characteristics in a profitable manner, but, on the other hand, compete when allocating the joint profit. This makes all modes effective except of shared governance, as that modes requires the alignment of member goals (see Subsubsection F.VII.3.2.2).

The need for *network-level competencies* is presumably high due to the interdependent nature of logistics activities (internal factor) and the permeable system boundary (external factor). This makes LITPAP governance and NAO governance effective (see Subsubsection F.VII.3.2.3).

The *distribution of trust* is the pivotal contingency factor that makes lead LITPAP governance the only mode effective (see Subsubsection F.VII.3.2.4). The distribution of trust is presumed to be low among member LSPs as they are distant or proximate competitors, and it is presumed to be moderate to high between member LSPs and the LITPAP. As a consequence, only the LITPAP is able to establish, maintain, and evolve the cloud-based platform and its applications because of its neutrality. Neutrality is critical because all logistics-relevant data are exchanged or stored on the platform. NAO governance is not effective as member LSPs fear that the small group of member LSPs (which controls and monitors the NAO through board structures) will use its power to influence the LITPAP (which is contracted by the NAO) with the objective to gain advantage over other network members. The same argument applies to lead LSP governance, thus making this mode ineffective as well.

2. Viewpoint: An Approach to Identifying the System Architect

2.1. Potential Governance Modes for Cloud Logistics Systems

This viewpoint establishes conventions to determine the degree of centralization in CLSs. With regard to interfirm networks, we have defined centralization to be concerned with the distribution of authority between network members over decisions that affect either the whole or parts of the network and, thus, all or a subset of its members (see Section C.V.2). Centralization is closely related to the concept of the "dominant coalition" as described in SCT (Subsection F.I.4.3) or the "system architect," as described in software architecture design (see Subsection F.I.4.2). In fact, determining the degree of centralization can be considered conceptually similar to determining the dominant coalition or the system architect in an interorganizational network.

To identify the system architect in a CLS, we adopt Provan & Kenis's (2008) model for network governance modes. Recall and observe that these "governance modes" are not similar to "governance structures" or "governance modes" as used in TCE (see Chapter C.III) and the structural governance viewpoint/view (see Chapter F.VI). Provan & Kenis (2008) use this term to refer to different distributions of decision-making authority rather than different institutional frameworks in which contracts are initiated, negotiated, monitored, adapted, enforced, and terminated. Despite varying meanings, these terms are related to some extent (which may explain these authors' terminological choice), as the distribution of authority influences how contract adaptations can be made in interfirm networks.

Provan & Kenis (2008) distinguish between three governance modes in their model (see Section C.V.2): (1) shared governance, (2) lead organization governance, and (3) NAO governance, either voluntarily-erected or mandated. Each of these generic modes refers to a different distribution of authority over network-level decisions, which means that different organizations act as the system architect. In shared governance, the collective of network members acts as architect. In lead organization governance and NAO, a single member or a small group of them acts as the architect, respectively.

In order to determine the system architect in CLSs, we use these three generic modes of authority distribution to the CLS context. Hence, we specify (1) the set of network-level decisions to be made and (2) which cloud logistics stakeholder classes assume authority over network-level decisions in each of these generic modes.

Motivated by SCT (Subsection F.I.4.3), critical *network-level decisions* in CLSs include specifying (a) the logistical capabilities, especially in terms of type and geographic scope, that can be delivered with the essential cloud characteristics (see Chapter F.V) and (b) the interorganizational structure which enables member LSPs and logistics consumers to interact with each other and enables member LSPs to collectively deliver logistics capabilities with cloud characteristics (see Chapter F.VII – Chapter F.XII). This especially includes establishing, maintaining, and evolving a cloud-based platform and its applications, as the platform represents the nexus of value creation in CLSs (see Subsection F.IV.3.3).

Shared governance in CLSs refers to the case in which authority over network-level decisions is widely dispersed among the network of horizontally cooperating LSPs and the LITPAP.

We distinguish between two types of *lead organization governance* in CLSs. First, *lead LITPAP governance* refers to a situation in which the LITPAP has exclusive authority over network-level decisions. *Lead LSP governance* refers to a situation in which a single LSP in the network of horizontally cooperating LSPs has exclusive authority over network-level decisions. This leading LSP tasks and monitors the LITPAP to establish, maintain, and evolve the cloud-based platform and applications on its behalf. As argued above, LSPs generally lack IT capabilities to establish, maintain, and evolve a cloud-based platform cost-efficiently and effectively (see Subsection F.II.3.4). Hence, the leading LSP needs to contract with the LITPAP.

NAO governance refers to a situation in which a subset of LSPs in the horizontal LSP network voluntarily erect an NAO. These member LSPs monitor the activities of the NAO and become involved in strategic network-level decisions through board structures. Operational decisions reside with the NAO. And the NAO tasks and monitors the LITPAP to establish, maintain, and evolve the cloud-based platform and applications on its behalf. Note that we do not consider a case in which an NAO is mandated externally, as cloud logistics knowledge does not provide any indications in this respect, although governments may support establishing a cloud-based logistics platform (Li et al. 2014b).

2.2. Contingency Factors

2.2.1. Overview

Following the logic of *contingency theory*, Provan & Kenis (2008:241, P1) contend that an effective mode for organizing an interorganizational network depends on the fit between four contingency factors and the governance mode; otherwise, the network may suffer severe economic penalties, which lead to network ineffectiveness, dissolution, or change in governance mode. The four contingency factors they propose are (1) network size, (2) network-level goal consensus, (3) need for network-level competencies, and (4) distribution of trust. Table 25 depicts these contingency factors and shows which governance mode will be effective in each condition. These four factors and their proposed relevance for organizational effectiveness are discussed below.

2.2.2. Network Size

Network size refers to the number of network members. Network size critically determines the complexity of network governance in terms of facilitating decision-making on the network-level, accommodating members' needs, and coordinating their activities. This is because network size determines the number of potential relationships among network members that need to be governed: The greater the number of relationships, the higher the governance complexity (Provan & Kenis 2008:238).

Shared governance is effective in the case of few network members. Face-to-face meetings involving all members are effective for making network-level decisions, raising and balancing member needs, and coordinating their activities (Provan & Kenis 2008:238). However, shared governance becomes highly inefficient if network size increases; after all, the number of potential relationships increases exponentially in the number of network members (Provan & Kenis 2008:238). *Lead organization governance* and *NAO governance* are both effective for organizing networks with a moderate to large number of members (Provan & Kenis 2008:238f.). Both modes can effectively govern larger networks by centralizing au-

Generic Network Governance Modes	Network Size	Network Goal Consensus	Need for Network-Level Competencies	Trust Distribution
Shared governance	Few	High	Low	High density
Lead organization governance	Moderate	Moderately low	Moderate	Low density, highly centralized
Network administrative organization governance	Moderate to many	Moderately high	High	Moderate density, NAO monitored by members

Table 25.: Key Predictors of Effectiveness of Network Governance Forms (Provan & Kenis 2008:237)

thority in a single entity, thus reducing the involvement of members to a minimum (Provan & Kenis 2008:238f.). Moreover, centralized authority reduces the number of relationships to be governed, as members no longer need to interact with each other, but each member directly interacts with the lead organization or NAO (Provan & Kenis 2008:238f.).

These general considerations are supported by empirical evidence from the logistics industry. Albers & Klaas-Wissing (2012:192) argue that the larger the number of member LSPs in a horizontal LSP network, the higher the degree of centralization. In line with this, but from the opposite perspective, Midoro & Pitto (2000:36f.) argue that organizational complexity increases with a larger number of members. Hence, they argue that shared governance is not suitable for alliances with many members in ocean transport, which is why such alliances should aim to either move to a more centralized mode or reduce the number of members in order to achieve a "fit" between the decision-making mode and alliance size.

2.2.3. Network-level Goal Consensus

Network-level goal consensus refers to the network members' agreement on network-level goals (Provan & Kenis 2008:239f.). In other words, it refers to the degree of similarity between network-level goals and individual organizational goals. Organizational behavior of network members is thus guided by both individual goals and goals formulated on the network level (Provan & Kenis 2008:239). Network members' commitment to and involvement in the network depends on the degree to which network involvement supports their own goals (Provan & Kenis 2008:239).

Shared governance is effective in the case of high goal consensus among network members. Shared goals enable members to cooperate without conflict because each member's contributions are conducive to attaining individual and network-level goals concurrently (Provan & Kenis 2008:240). *Lead organization governance* is effective in the case of moderately low goal consensus. Network members are only partially committed to network-level goals and therefore unlikely to cooperate without conflict, let alone resolve conflict on their own (Provan & Kenis 2008:240). The lead organization can maintain an overall network-level view, make network-level decisions, and resolve conflict among members (Provan & Kenis 2008:240). *NAO governance* is effective in the case of moderately high goal consensus, at least among a subset of network members (Provan & Kenis 2008:240). These members are highly committed to network-level goals and become strategically involved in the network (Provan & Kenis 2008:240). At the same time, the NAO works actively with members that have a lower goal consensus in order to resolve possible conflicts among them and enhance their commitment to the network (Provan & Kenis 2008:240).

2.2.4. Need for Network-level Competencies

The need for network-level competencies is the necessity for having skills to carry out certain tasks on the network-level in order to attain network-level goals (Provan & Kenis 2008:240). The need for such competencies primarily arises from (1) the nature of tasks to be performed (internal factors) and (2) the need for buffering and bridging (external factors). The *nature of tasks* refers to the degree of task interdependence among network members. The greater the degree of task interdependence, the greater the need for network level competencies in the form of "network-level coordinating skills and task-specific competencies" to organize interdependent actions (Provan & Kenis 2008:240f.). *Buffering* refers to protecting the network against environmental shocks, such as changes in regulations, and *bridging* refers to linking network members with external stakeholders. Typical tasks include lobbying, building external legitimacy, acquiring funding, and selecting new members. The greater the need for bridging and buffering, the greater the need for network-level competencies (Provan & Kenis 2008:241). The crucial question of effective network governance relates to how network members, or a subset thereof, can provide network-level skills necessary for collective action.

Shared governance is effective in the case of a low need for network-level competencies. Conversely, *lead organization* and *NAO governance* are effective in the case of a moderate or high need for network level competencies. In the case of high task interdependency, shared governance is not effective because it may require network members to carry out tasks that require skills they do not possess (Provan & Kenis 2008:241). By contrast, brokered forms of governance are effective as they can build the required skills centrally. In the case of a high need for buffering and bridging, shared governance is not effective, as it cannot facilitate centralized actions across the whole network in response to environmental shocks or other external demands (Provan & Kenis 2008:241). Conversely, brokered governance can act as a focal point for interacting with external stakeholders and facilitate a network-wide response. Compared to lead organization governance, NAO governance is more effective for high needs of network-level competencies because it is the NAO's primary purpose to develop skills necessary for facilitating network-level actions (Provan & Kenis 2008:241). A lead organization may not possess and may be reluctant to invest in skills necessary to govern the network (Provan & Kenis 2008:241).

The importance of central coordination as a driver for network-level competencies is also supported by empirical logistics evidence. Albers & Klaas-Wissing (2012:192) argue that the need to achieve efficiency in operational logistics activities in an LTL network results in the creation of a centralized management unit.

2.2.5. Distribution of Trust

The (structural) distribution of trust refers to the degree to which trust is reciprocated among network members (Provan & Kenis 2008:238). A wide distribution (high density) denotes a case in which trust is pervasive among many network members: Many members trust one another, thus forming "a dense web of trust-based ties" (Provan & Kenis 2008:238). A narrow distribution (low density) describes a situation in which trust is reciprocated only differentially, within individual dyadic relationships or cliques.

Shared governance is effective if trust is widely distributed and reciprocated among network members (Provan & Kenis 2008:238). A dense web of trust-based ties builds a basis for collaboration and for collectively making network-level decisions. However, if trust is less pervasive in the network, more centralized governance modes are effective. On the one hand, if trust primarily occurs in a dyadic relationship centered around a single organization (low density, highly centralized), *lead organization governance* is most effective (Provan & Kenis 2008:238). On the other hand, if trust is distributed in a moderately dense manner among a subset of members, *NAO governance* is most effective because these members monitor the brokering organization and can become involved in strategic decision-making (Provan & Kenis 2008:238).

In line with these general considerations, Albers & Klaas-Wissing (2012:192f.) argue that in the case of high centralization in a cooperative LTL network, trusting relationships among members are not necessary for network stability; members only need to trust the central management unit.

3. View: Lead LITPAP Governance

3.1. Proposition

After having considered the potential governance modes of CLSs and relevant contingency factors, we now attend to the question of which of these potential modes are effective for CLSs. In principle, we propose:

Proposition 11. *Lead LITPAP governance will be an effective distribution of authority for network-level decisions in cloud logistics systems.*

We thus propose that CLSs cannot be effectively governed by one or more LSPs but only by an IT vendor that is capable of establishing, maintaining, and evolving a cloud-based platform with logistics-related applications. The rationale for this proposition is based on determining the conditions that presumably prevail in CLSs for each contingency factor and then comparing these conditions with Provan & Kenis's (2008) predictions about effective network governance modes, as depicted in Table 25. The assessment of contingency factors is summarized in Table 26. The distribution of trust is the pivotal contingency factor for determining the governance mode of CLSs because this factor only allows a single effective mode to emerge from among multiple potential modes. The next subsection provides a detailed assessment of the contingency factors.

3.2. Assessment of Contingency Factors

3.2.1. Network Size

Network size is presumably moderate to large, due to the goal of delivering logistics capabilities in an on-demand and rapidly elastic manner. The number of cooperating providers has been identified in cloud computing literature (Subsection D.II.4.2) and cloud logistics knowledge (see Subsubsection E.IV.5.4.3) as a critical factor for achieving rapid elasticity. In fact, the goal of achieving rapid elasticity is positioned as a primary motive for interfirm cooperation, both in cloud computing and cloud logistics. This is because the larger the shared resource pool, the higher the upper scalability bound and thus the larger the demand surges that can be handled effectively (capacity appears unlimited). Moreover, the larger the resource pool, the higher the likelihood of having (currently idle) resources in states in which little reconfiguration is required to bring them into target states in which they become productive in addressing the desired concerns. For example, the larger the resource pool, the larger the likelihood that a truck is currently idle in the vicinity of a new consumer or in the vicinity of an already established transportation

Potential Cloud Logistics Governance Forms	**Network Size**	**Network Goal Consensus**	**Need for Network-Level Competencies**	**Trust Distribution**
Shared governance				
Lead LITPAP governance	x	x	x	x
Lead LSP governance	x	x		
NAO governance	x	x	x	

Table 26.: Assessment of Centralization Contingency Factors; x = presumable condition in CLSs matches the condition proposed by Provan & Kenis (2008)

link. Thus, the larger the resource pool, the higher the likelihood of achieving on-demand availability and rapid elasticity.

Following the reasoning of Provan & Kenis (2008), the presumed moderate to large network size makes *lead LITPAP organization*, *lead LSP organization*, and *NAO governance* effective modes of governance.

3.2.2. Network-level Goal Consensus

Network-level goal consensus among horizontally cooperating LSPs and the LITPAP is presumed to be moderate due to (1) partial misalignment between financial goals and (2) potential misalignment between strategic priorities. The CLSs' network-level goals can be derived from the profit seeking nature of LSPs and the LITPAP: delivering logistics capabilities with cloud characteristics to make a profit. This conception of network-level goals is supported by Kersten et al. (2012:259), who argue that the network of horizontally cooperating LSPs focuses on the "common overall objective of customer satisfaction and service lead time," where customer satisfaction is understood as a proxy for financial performance and service lead time as a synonym for on-demand delivery.

Network-level financial goals are partially misaligned with the financial goals of individual member LSPs and the LITPAP. On the one hand, goals are aligned because profit earned on the network level can only be earned through cooperation. On the other hand, goals are misaligned, as network-level profit must be distributed among its members (see Subsection C.IV.3.3). From the perspective of a single network member, any profit distributed to other members reduces the attainment of its own financial goals and *vice versa*. Hence, the degree to which goals can be aligned depends on the distribution mechanism. Note that goals can never be fully aligned. The proposed distribution mechanism (an auction) is considered in detail in Chapter F.XII.

Furthermore, misalignment of network-level and individual goals can arise from different "strategic priorities" pursued on the network-level and priorities pursued by individual network members. For example, the network may aim for member growth, but current member LSPs may try to limit access to new LSPs to avoid increasing internal competition. Misalignment of goals may also be likely regarding the offered logistics capabilities. The cloud may require member LSPs to build up certain logistics capabilities to strengthen the position of the CLS; however, these capabilities may not necessarily fit with member LSPs' portfolios and priorities.

Following the reasoning of Provan & Kenis (2008), the presumed moderate degree of goal consensus makes lead LITPAP governance, lead LSP governance, and NAO governance effective.

3.2.3. Need for Network-level Competencies

The need for network-level competencies is presumably high due to (1) the interdependent nature of logistics activities (internal factor) and (2) the permeable system boundary (external factor). The sequential nature of logistics activities and the temporal and/or spatial consolidation of cargo creates high *coordination interdependence* among LSPs that are involved in collectively delivering logistics capabilities. These coordination interdependencies are addressed through a central cloud-based platform on the network-level, which enables all stakeholders to interact with each other and carry out value exchanges (see Subsection F.IV.3.3). Hence, establishing, maintaining, and evolving this platform and its functionality is a critical component of network-level competencies. The *permeable system boundary* also raises demand for network-level competencies: Constantly joining and leaving LSPs require active member management on the network level in order to maintain and evolve cloud logistics capabilities in line with network-level goals and logistical consumers' requirements.

Following the reasoning of Provan & Kenis (2008), the presumed high need for network-level competencies makes *NAO governance* the only effective mode. The NAO builds up control and monitoring skills in order to task the LITPAP with establishing, maintaining, and evolving the cloud-based platform. Recall that the NAO is not able to establish, maintain, and evolve the cloud-based platform efficiently and effectively, which is why contracting with an external party is inevitable (see Subsection F.II.3.4). In addition, the NAO builds up member management skills to control the permeable system boundary.

In addition to NAO governance, we argue that *lead LITPAP governance* is also effective, although Provan & Kenis (2008) consider lead organization governance to be effective only in the case of a moderate, not high, need for network-level competencies. Provan & Kenis (2008:241) argue that, compared with the NAO, lead organizations are only suitable for a moderate degree of network-level competencies because they may be more reluctant to make sufficient investments in building up necessary network-level competencies. However, the LITPAP *already has* the (logistics-related) IT capabilities necessary to establish, maintain, and evolve a cloud-based logistics platform and applications. In fact, these represent the core competencies of the LITPAP. Hence, the LITPAP only needs to invest in building up skills for active member management, which is a moderate investment of resources compared with the cloud-based platform.

3.2.4. Distribution of Trust

The distribution of trust is presumed to be (1) low among member LSPs and (2) moderate to high between member LSPs and the LITPAP. The *low distribution of trust among member LSPs* directly results from the high risk of opportunistic behavior prevalent in horizontal LSP alliances, which remains real even in trustworthy relationships (see Subsection C.VI.3.1). This risk is especially high in CLSs because of the need for capabilities to overlap in type and geography in order to achieve cloud characteristics. Moreover, trust is also unlikely be built among LSPs due to the permeable system boundary. With current members potentially leaving and new members joining, there may often be too few exchange episodes among LSPs to establish trust.

The *distribution of trust between member LSPs and the LITPAP* is moderate to high. This is because member LSPs and the LITPAP are complementors rather than competitors. Hence, the risk of opportunistic behavior is less pronounced. Moreover, trust-based ties between LSPs and the LITPAP are likely to emerge as the LITPAP establishes, maintains, and evolves the cloud-based platform in a neutral manner. Neutrality is critical because all logistics-relevant data are exchanged or stored on the platform. Improper data handling by the LITPAP or unauthorized access to these data could jeopardize the LSPs' competitive position. Hence, the level of trust depends on the LITPAP's ability to protect data from unauthorized external access resulting from hacker attacks and from unauthorized internal access resulting from improper handling or insufficient access rights management. The level of trust may also depend on the transparency the LITPAP provides LSPs regarding the platform's security features.

This presumed distribution of trust is likely to make *LITPAP lead governance* an effective mode due to the low density of trust among LSPs and moderately to highly centralized density of trust between LSPs and the LITPAP.

Lead LSP governance and NAO governance are effective governance modes because trust-based ties between member LSPs and the leading LSP or NAO are very unlikely to emerge.

For the sake of the argument, suppose that trust-based ties emerge among a (small) group of LSPs, for example due to personal relationships among the senior management teams. This group of LSPs can erect an NAO, which then tasks and monitors the LITPAP. However, other member LSPs would not perceive the LITPAP as a neutral entity; after all, the LITPAP is be tasked and monitored by the NAO, which, in turn, is monitored by the small group of LSPs that are either proximate or distant competitors. Other member LSPs would specifically fear that the NAO is using its authority to gain access to other LSPs' logistics-relevant data and to bias the design of the cloud-based platform to its advantage for improving its competitive positioning. This makes NAO governance ineffective. The same argument applies to lead LSP governance, presumably making this mode ineffective as well.

3.3. Conclusions: A Change of Power in the Logistics Industry?!

Based on the model by Provan & Kenis (2008), we proposed that the LITPAP acts as the lead organization and thus has decision-making authority over network-level decisions in CLSs. The LITPAP is assumed to have sufficient logistics-related IT capabilities to establish, maintain, and evolve a cloud-based platform that enables stakeholders to interact and to coordinate value exchanges. In particular, this includes the coordination of (operative) interdependencies between member LSPs in the delivery of logistics capabilities. Moreover, due to the LITPAP's neutral role, member LSPs are willing to use the cloud-based platform to exchange and store logistics-relevant data.

Although the LITPAP has authority over network-level decisions, it is important to recognize that authority is not unlimited. This is because authority derives from informal sources that may erode in certain situations, such as data security and functionality issues on the cloud-based platform. Such conditions may trigger member LSPs or logistics consumers to leave a CLS, thus diminishing informal authority of the LITPAP as its authority largely depends on network effects associated with the cloud-based platform's role as a two-sided market between many logistics consumers and many LSPs. Hence, the LITPAP must recognize and accommodate the strategic needs of logistics consumers and member LSPs. It must also offer LSPs a guiding economic logic that enables the cooperative delivery of logistics capabilities, while giving them enough flexibility to realize this logic (see Gulati et al. 2012:581). If it does so, member LSPs may be willing to accept the loss of control as a *quid pro quo* for the opportunity to earn an additional profit by gaining access to many potential consumers via the virtual marketplace.

Comparing the proposed distribution of authority in CLSs with the distribution of authority typically observed in logistics systems today suggests a major shift of power. Today, IT vendors are largely suppliers of LSPs, albeit being important ones due to IT' relevance in managing logistics processes. In the above proposition, IT vendors become "active" players. In fact, they assume a critical intermediary role between LSPs and logistics consumers: They own the "virtual" customer interface. This may represent a critical implementation hurdle for CLSs, as LSPs may not be willing to become "faceless" resource providers on a platform controlled by a third party. We therefore conclude that LSPs must evaluate whether the encroachments imposed by the LITPAP's authority and central role in CLS will be or will not be acceptable for member LSPs regarding their ability to achieve their (long-term) organizational objectives.

VIII. Formalization

1. Abstract of Chapter

Formalization in interorganizational networks refers the degree to which contingencies and rules and procedures that prescribe network members' behaviors and/or outputs are specified and/or adhered to and the degree to which these rules, procedures, and contingencies are codified in the cooperation contract and/or IT systems used in the cooperation (see Section C.V.3).

The *architectural viewpoint* establishes conventions for determining the degree of formalization for the governance and interorganizational structure of CLSs (see Subsection F.VIII.2.1). Structural formalization depends on contingency factors and managerial choice (Vlaar et al. 2007:437). We consider four contingency factors: (1) network size and scope, (2) type of network goals, (3) behavioral uncertainty, and (4) external environment. To introduce managerial choice, we consider tradeoffs between its functions and dysfunctions, with functions defined as consequences conducive to achieving desired ends, and dysfunctions obstructing the achievement of goals (Vlaar et al. 2007).

The *architectural view* summarizes the assessment of the contingency factors and tradeoff decisions in the following proposition (see Subsection F.VIII.3.1).

Proposition 12. *Cloud logistics systems will formalize (a) the relationship between the LITPAP and LSPs that intend to offer logistics capabilities on the virtual marketplace by means of a written "framework" cooperation contract and (b) the characteristic mix of coordination mechanisms in the applications implemented on the cloud-based platform (computer-formalized mechanisms).*

As indicated by the term "framework," the written cooperation contract possesses a comparatively low level of formalization by only codifying fundamental duties and obligations of partnering firms. This especially includes the exclusive authority of the LITPAP to make adaptations in the mix of mechanisms implemented on the cloud-based platform. The characteristic mix of coordination mechanisms, by contrast, possesses a comparatively high level of formalization, as mechanisms are implemented with a high level of detail in the applications on the cloud-based platform.

Different contingency factors create opposing pressures on the degree of formalization. Especially the presumable large *network size* and the high *behavioral uncertainty* among LSPs create very strong pressures toward formalization. Other contingency factors, such as a *dynamic environment*, create pressures toward medium to low formalization. We argue that these opposing pressures can be reconciled through a written "framework" cooperation contract and a mix of computer-formalized coordination mechanisms. Due to the low degree of contractual formalization, only very few disturbances are so significant that they require the LITPAP and member LSPs to bilaterally adapt the fundamental clauses of the cooperation contract, thus making a framework contract effective in dynamic environments. Yet, despite the low degree of contractual formalization, opportunistic behavior can still be deterred sufficiently as highly formalized coordination mechanisms are a substitute to contractual formalization in horizontal LSP alliances as long as logistics alliances are concerned with the exploitation of tangible resources – a condition clearly satisfied by CLSs (Raue & Wieland 2015:414; also see Subsection C.VI.4.3). If environmental disturbances require adapting coordination mechanisms, adaptations can be achieved quickly as the LITPAP has exclusive authority over these mechanisms.

Trading of the functions and dysfunctions of formalization in an economically rational manner from the perspective of the LITPAP supports our above proposition (see Subsection F.VIII.3.3). Due to the intertwined character of formalization with other structural dimensions, the choices of the LITPAP are considered with regard to centralization and the characteristic mix of coordination mechanisms.

Regarding the degree of centralization, the LITPAP aims to *legitimize its authority holding position* by means of a written contract because formal contracts offer a legitimate source of managerial intervention (Sitkin & Bies 1993 as cited in Vlaar et al. 2007:444). The LITPAP thus transforms informal authority over LSPs into formal authority. This is critical to assume exclusive authority over the mix of coordination mechanisms. In addition to legitimizing the LITPAP's authority, formalization is equally important to *legitimize the (business) relationships and interactions* among LSPs – after all, they remain proximate or distant competitors. Legitimizing such relationships is important to increasing stakeholders' commitment toward goals (Vlaar et al. 2007:444), especially if the relationships are exposed to a dynamic and discontinuous environment (Kale et al. 2001 as cited in Vlaar et al. 2007:444), as presumed in CLSs.

Regarding the characteristic mix of coordination mechanisms, the LITPAP aims to formalize the mix of mechanisms in order to deliver physical logistics capabilities with cloud characteristics. The LITPAP presumably chooses to formalize coordination mechanisms in order to achieve *on-demand availability* and *rapid elasticity*. Formalization fuels interaction processes and thus reduces the interaction time before service delivery. The LITPAP also chooses to formalize coordination mechanisms in order to enable *self-service*. Formalization establishes a common language and serves as signaling devices which reduces the need for personal interaction. Finally, the LITPAP presumably chooses to formalize coordination mechanisms in order to enable *pay-per-use*. Formalization enables the monitoring of progress and the identification of deviation from objectives which are prerequisites for determining the actual amount of capacity consumed.

While the functions of formalization are conducive to achieving cloud characteristics, its dysfunctions do not obstruct their achievement. Hence, we argue that it is an economically rational tradeoff decision for the LITPAP to formalize the authority relationships and the mix of coordination mechanisms.

2. Viewpoint: An Approach to Determining the Structural Formalization

2.1. An Integrative Approach: Contingency Factors and Choice

This viewpoint establishes conventions for determining the formalization level for the governance and interorganizational structure of CLSs. With regard to interfirm networks, we have defined formalization as the degree to which contingencies and rules and procedures that prescribe network members' behaviors and/or outputs are specified and/or adhered to and the degree to which these rules, procedures, and contingencies are codified in the cooperation contract and/or IT systems used in the cooperation.

Following Vlaar et al. (2007:437), we consider the level of structural formalization a function of *contingency factors* and *managerial choice*. To introduce managerial choice into the structuring of interorganizational relationships and thus complement contingency theory, Vlaar et al. (2007) suggest a "dialectic perspective" on the role of formalization in interorganizational relationships, as depicted in Figure 49. This perspective recognizes the duality inherent in formalization, which manifests in (a) *tradeoffs between its functions and dysfunctions* and (b) *dialectic tensions that arise from the coexistence of these functions and dysfunctions* (Vlaar et al. 2007:442), with functions defined as consequences conducive to achieving desired ends, and dysfunctions as consequences obstructing the achievement of goals. Vlaar et al. (2007) argue that managers make tradeoffs regarding formalization based on anticipated functions and dysfunctions. The extent to which the desired ends of an interorganizational relationship can be achieved depends on the tradeoffs made and on how well managers cope with the dialectic tensions that result from the

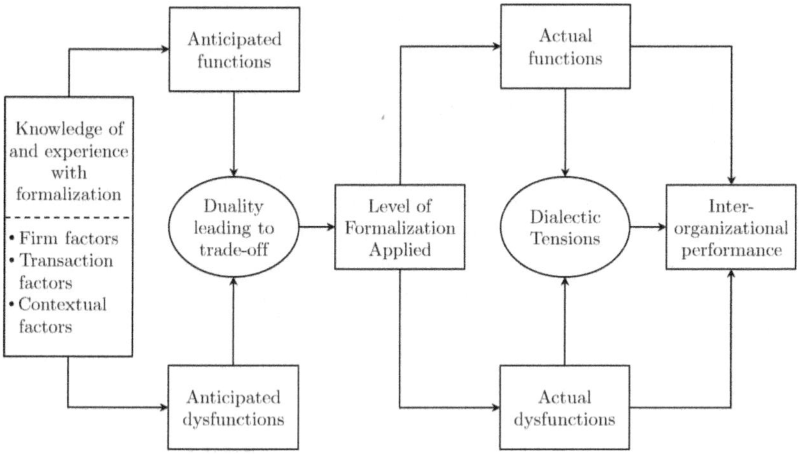

Figure 49.: A Dialectic Perspective on the Role of Formalization in Interorganizational Relationships (adopted from Vlaar et al. 2007:441)

actual functions and dysfunctions of the level of formalization applied.

Using this approach, we determine the level of formalization in CLSs by considering the constraints imposed by contingency factors and by exercising choice in an economically rational manner from the perspective of the LITPAP (see Subsection F.I.4.5), which acts as the system architect (see Chapter F.VII). More specifically, we determine the "level of applied formalization" (see Figure 49). The resulting actual interorganizational performance cannot be determined as part of this reference architecture, as we do not focus on a specific implementation, but on the abstract conceptual design of CLSs.

The dialectic perspective on formalization is particularly conducive to designing a reference architecture of CLSs. The functions and dysfunctions inherent in formalization are used to directly frame concerns related to the cloud characteristics. Specifically, these functions and dysfunctions show how formalization can support or prevent the delivery of physical logistics capabilities with cloud characteristics.

The following subsection introduces the contingency factors of formalization. Recall that the functions and dysfunctions of formalization have already been introduced above (see Section C.V.3).

2.2. Contingency Factors

Organizational literature provides rich knowledge about contingency factors that influence the degree of formalization in interorganizational relationships. Based on a comprehensive review of literature, Albers (2010) identifies four contingency factors that characterize the environment of an alliance: (1) network size and scope, (2) type of network goals, (3) behavioral uncertainty, and (4) external environment (also see Albers 2005:184ff.). Due to the abstract nature of this reference architecture, we do not consider factors related to specific characteristics of cooperating firms, such as firm size, alliance history, or culture. In the following, we present a condensed summary of Albers's (2010) propositions and underlying rationale for how these factors influence the degree of formalization in interorganizational relationships.

2.2.1. Network Size and Scope

Network size refers to the number of partnering firms and network scope to the domains covered by the alliance, such as products and geographies (Albers 2010:211). Despite being separate factors, they are discussed together, as they exert the same pressures on formalization. The greater the alliance's size and scope, the higher the degree of formalization (Albers 2010:211). An increase in size and scope entails a greater variety of members' interests, goals, and strategies, thus reducing the likelihood of members organizing to achieve network-level goals solely based on implicit agreements and mutual understanding (Phillips 1960:607f.). Hence, to achieve efficient coordination and control of behaviors and outputs, member rights and obligations need to be encoded in a way easily accessible and similar for all network members (Albers 2010:211; Phillips 1960:607f.).

2.2.2. Type of Network Goals

Network goals can be categorized as efficiency-oriented (pursued by scale alliances) or growth-oriented (pursued by complementary alliances) (Garette & Dussauge 2000 as cited in Albers 2010:212). The stronger the focus on efficiency-oriented goals, the greater the degree of formalization (Albers 2010:212). Efficiency-oriented goals require a greater rationalization of decision-making processes than growth-oriented goals; this is why efficiency-related goals are better achieved through an acquisition rather than an alliance, because alliances involve lengthy bargaining processes (Garette & Dussauge 2000:65f. as cited in Albers 2010:212). To overcome this structural disadvantage, Albers (2010:212) argues that efficiency-oriented alliances must replicate an acquisition's more hierarchical and formalized structures and mechanisms.

2.2.3. Behavioral Uncertainty (Opportunistic Behavior)

Behavioral uncertainty is the "probability and consequences that a partner firm does not commit itself to the alliance in the desired manner;" such behavior thus makes direct reference to opportunistic behavior (Das & Teng 2001:6) and is the opposite of trust. The greater the behavioral uncertainty associated with one or more network member, the higher the degree of formalization (Albers 2010:212). Formalization is a means of countering behavioral uncertainty by unambiguously fixing rules and procedures and consequences in case of deviating behavior in the cooperation contract (Albers 2010:212; see Schmoltzi & Wallenburg 2011:66f.; also see Subsection C.VI.4.3), which can ultimately be enforced by courts (Parkhe 1993). Formalization thus makes interactions among partnering firms more predictable and ensures that joint decisions are made by rules rather than by exception (Gulati & Singh 1998:786).

The relationship between behavioral uncertainty and formalization directly derives from TCE. Pointing towards proponents of TCE, Grandori (1997b:40) argues that the extent of formalization depends on the degree to which agreements and contracts are self-enforcing or must be enforced externally by a third party, either internally through hierarchical authorities or externally through courts. Likewise, Gulati & Singh (1998:781f.) argue that hierarchical controls in alliances – such as explicit command structures, standard operating procedures, and procedures for dispute resolution – are considered efficient means of governing exchanges in the case of high behavioral uncertainty among transacting parties because of their "ability to assert control by fiat, provide monitoring, and align incentives." The use of contractual safeguards (which clearly prescribe remedies and consequences for non-performance) and their legalistic interpretation can deter behavioral uncertainty in market transactions (see Subsection C.III.3.1).

2.2.4. Environmental Dynamism and Complexity

Recall that the external environment can be conceptualized along dynamism and complexity, resulting in four types of environments (see Chapter B.V). Alliance formalization is higher in stable/simple environments than in dynamic/complex environments (Albers 2010:212f.). Following the general argument of Mintzberg (1979:270f.), Albers (2010:212f.) argues that it is increasingly difficult and potentially inadequate to *ex-ante* formalize work and identify contingencies and adequate responses to frequent and varying kinds of environmental changes, thus making alliance formalization less effective in dynamic/complex environments than in stable/simple ones (also see Albers 2005:245f.).

3. View: Written "Framework" Cooperation Contract and Computer-Formalized Mechanisms

3.1. Proposition

The level at which interorganizational relationships are formalized depends on contingency factors and on tradeoff choices made by the system architect between the anticipated functions and dysfunctions of formalization (Vlaar et al. 2007). After reviewing relevant contingency factors in the last section and introducing functions and dysfunctions of formalization (see Section C.V.3), we propose:

Proposition 12. *Cloud logistics systems will formalize (a) the relationship between the LITPAP and LSPs that intend to offer logistics capabilities on the virtual marketplace by means of a written "framework" cooperation contract and (b) the characteristic mix of coordination mechanisms in the applications implemented on the cloud-based platform (computer-formalized mechanisms).*

This proposition directly refers to both the governance and the interorganizational structure of CLSs, which emphasizes the intertwined nature of formalization and other structural dimensions. The written framework contract directly refers to the governance structure (see Chapter F.VI). The characteristic mix of coordination mechanisms refers to both governance and interorganizational structure, as it includes mechanisms that govern value exchanges among stakeholders (market exchanges on the virtual marketplace) and mechanisms to support or carry out the planning, realization, control, and monitoring of physical logistics capabilities (see Subsection F.IV.3.5). Before providing a rationale to support this proposition, we briefly provide some further clarification regarding its meaning.

A written "framework" cooperation contract is understood as a contract that specifies only the fundamental rights and obligations of the LITPAP and member LSPs and responses to critical contingencies. Fundamental rights of the LITPAP include the exclusive authority over network-level decisions, especially the design and evolution of the characteristic mix of coordination mechanisms. Obligations include assuming a neutral position in the process of resolving disputes among member LSPs as well as in establishing, maintaining, and evolving the cloud-based platform and its applications. Critical contingencies include clauses to define and handle *force majeure* and persistent non-performance of member LSPs. Fundamental rights of LSPs include access to the cloud-based platform, especially the right to offer and sell logistics capabilities via the virtual marketplace. Their obligations include offering and delivering logistics capabilities with cloud characteristics. Cloud characteristics may be specified by threshold values, for example that every pick-up must happen within a specified lead time (on-demand). Logistics consumers and LSPs may, of course, negotiate "stricter" requirements.

The framework cooperation contract and the characteristic mix of mechanisms have different levels of formalization. As indicated by the term "framework," the written cooperation contract possesses a comparatively low level of formalization by only codifying fundamental aspects of the cooperation. By contrast, the characteristic mix of coordination mechanisms possesses a comparatively high level of formalization, as mechanisms are implemented with a high level of detail in the applications of the cloud-based platform. In other words, the governance structure possesses a lower level of formalization than the interorganizational structure. This dichotomous approach to formalization aims to reconcile the opposing pressures from contextual factors and aligns with economically rational choices from the perspective of the LITPAP. It specifically leverages the empirical evidence collected by Raue & Wieland (2015:414), who find that contractual formalization and operational process formalization (which includes coordination mechanisms) are substitutes in horizontal LSP alliances – as long as logistics alliances are concerned with the exploitation of tangible resources, a condition clearly satisfied by CLSs (see Subsection C.VI.4.3).

The following two subsections provide a rationale to support this proposition by assessing (1) the conditions associated with contingency factors, and (2) the tradeoffs associated with the anticipated functions and dysfunctions of formalization, especially in relation to cloud characteristics.

3.2. Assessment of Contingency Factors

Four contingency factors that influence the level of organizational formalization have been identified: (1) network size and scope, (2) type of network goals, (3) behavioral uncertainty, and (4) environmental dynamism and complexity (see Subsection F.VIII.2.2). The presumable conditions that prevail in CLSs regarding these factors are assessed below.

As elaborated above, *network size* is presumably moderate to large due to the necessity of creating a sufficiently large resource pool that enables rapid elasticity and on-demand provisioning (see Subsubsection F.VII.3.2.1). This creates strong pressures for formalizing the organizational structure.

Network scope is presumably limited to medium, depending on the logistical prototype scenario. The capability scope offered by CLSs ranges from narrow to a composition of a few selected capabilities, which can be made available in a localized or wide, networked geography (see Subsection F.V.3.3). However, a broad capability scope in a wide geography cannot be offered. This creates pressures toward a medium to low degree of formalization.

The *type of network goals* pursued by CLSs are primarily efficiency-oriented. This is because resource pooling is an essential characteristic of CLSs. Still, in addition to pursuing efficiency-oriented goals, CLSs may also pursue growth-oriented goals in prototype scenarios II, III, and IV if individual LSPs contribute complementary logistics capabilities, where complementarity can manifest in capability type and/or geography. The simultaneous existence of efficiency- and growth-oriented goals thus creates conflicting pressures toward and away from organizational formalization, respectively.

Behavioral uncertainty is presumably high among LSPs due to the inherent risk of opportunistic behavior (see Subsection C.VI.3.1). This creates strong pressures toward a highly formalized organizational structure and a formalized cooperation contract, which specifies rights, obligations, contingencies, and responses to a high degree of detail.

The *external environment* of CLSs is presumed to be highly dynamic (see Subsection F.IV.3.6), which creates pressures toward a low degree of formalization.

The assessment of contingency factors reveals that CLSs are presumably exposed to partially opposing pressures. While the presumed large network size, high behavioral uncertainty, and efficiency-oriented goals lead to a high degree of formalization, the presumable dynamic environment and the limited to medium scope of CLSs lead to a low to medium degree of formalization. We argue that these opposing pressures can be reconciled through a written "framework" cooperation contract and a mix of computer-formalized coordination mechanisms. Specifically, the interaction between behavioral uncertainty, environmental dynamism, and network size are critical in this respect.

The moderate to large network size and the high degree of behavioral uncertainty create strong pressures toward formalizing the cooperation contract and the mix of coordination mechanisms. However, instead of formalizing both, only the mix of coordination mechanisms is formalized to a high degree. The cooperation contract possesses a comparatively low degree of formalization by only fixing the most fundamental cooperation aspects. Hence, this contract develops little vigor to discourage opportunistic behavior. In spite of this, we argue that opportunistic behavior can still be deterred sufficiently. On the other hand, value exchanges between logistics consumers and LSPs are coordinated through an electronic marketplace, thus discouraging opportunistic behavior through high-powered incentives for LSPs. On the other hand, Raue & Wieland (2015) find empirical evidence in horizontal LSP alliances that contractual safeguards and operational process formalization (which includes coordination mechanisms) are substitutes in cooperations that use tangible resources. CLSs clearly aim to exploit tangible resources. Hence, formalizing coordination mechanisms to a high degree can (at least partially) offset the lower degree of contractual formalization.

A highly dynamic environment creates pressures toward a low degree of formalization of both the cooperation contract and the mix of coordination mechanisms. The cooperation contract succumbs to these pressures by only codifying the most fundamental rights, obligations, contingencies, and appropriate responses. Focusing on these most critical aspects reduces the need for adaptations because very few disturbances are so significant that they necessitate adapting these fundamental contract clauses. In this way, the need for bilateral adaptations of the cooperation contract through bargaining between the LITPAP and member LSPs is reduced. This is important, as such bargaining processes may take time, in which transacting parties suffer from costs incurred by maladaptations, especially in the case of a highly dynamic environment. Although the LITPAP may leverage its significant informal authority over member LSPs to speed up bargaining processes (see Chapter F.VII), bilateral adaptations are very likely to take longer than unilateral adaptations.

Due to the moderate to large network size, the degree of formalization of coordination mechanisms cannot be reduced without significant penalties, even in a highly dynamic environment. Hence, adaptations of mechanisms are inevitable and may occur frequently. In order to handle these adaptations quickly and efficiently, the framework cooperation contract gives the LITPAP exclusive authority over network-level decisions, especially the right to adapt the coordination mechanisms unilaterally. This enables the quick adaptation of coordination mechanisms, which helps avoid penalties from maladaptations, also in the event of frequent disturbances.

To conclude, a dichotomous approach toward formalization, which manifests in a written framework cooperation contract and a mix of highly formalized mechanisms implemented on the cloud-based platform is likely to enable effective and efficient governance and organization of CLSs.

3.3. Assessment of Tradeoffs between Functions and Dysfunctions

Formalization is intertwined with the other structural dimensions because it states their form (see Section C.V.3). Hence, we consider the LITPAP's choices regarding formalization in relation to (1) centralization and (2) the characteristic mix of coordination mechanisms. Specifically, we argue in an economically rational manner how the LITPAP presumably chooses to formalize the structure of CLSs. Choices are based on a tradeoff between the functions and dysfunctions inherent in formalization (see Table 10), especially focusing on concerns related to the essential cloud characteristics. We specifically account for the constraints on the LITPAP's choices imposed by the contingency factors, especially the external environment.

With regard to *centralization*, the LITPAP presumably chooses to sign a written framework cooperation contract with LSPs that intend to offer logistics capabilities on the virtual marketplace because contractual formalization represents a means to (1) establish and legitimize formal control over partnering firms and (2) increase legitimacy of interorganizational relationships among partnering firms (see Table 10).

The LITPAP aims to *legitimize its authority holding position* by means of a written contract because formal contracts offer a legitimate source of managerial intervention (Sitkin & Bies 1993 as cited in Vlaar et al. 2007:444). The LITPAP thus transforms informal authority over LSPs into formal authority. Being aware of the highly dynamic environment and the resulting high frequency of disturbances and adaptation needs, the LITPAP does *not* choose to govern the large network of horizontally cooperating LSPs by means of a comprehensive cooperation contract. Costs incurred from frequent and lengthy bargaining processes with many LSPs would be excessive. Instead, the LITPAP deliberately leverages the finding that contractual safeguards and procedural formalization are substitutes (Raue & Wieland 2015) by agreeing a framework contract and establishing a highly formalized mix of coordination mechanisms. In this way, overall governance costs can be reduced by, on one hand, reducing the need for bilateral contract adaptations while allowing for unilateral adaptations of coordination mechanisms and, on the other hand, by discouraging opportunistic behavior due to formalized mechanisms. However, note that having exclusive authority over the design and evolution of coordination mechanisms does not necessarily result in the LITPAP directing the behaviors and outputs of member LSPs in a hierarchical manner. In fact, the latitude of member LSPs is critically determined by latitude granted by the coordination mechanisms deployed, not necessarily by the authority over the design and evolution of these mechanisms. This argument becomes clearer when the design of market mechanisms is discussed (see Chapter F.XII).

In addition to legitimizing the LITPAP's authority, formalization is equally important to *legitimize the (business) relationships and interactions* among LSPs – after all, they remain proximate or distant competitors. Legitimizing such relationships is important to increasing stakeholders' commitment toward goals (Vlaar et al. 2007:444), especially if the relationships are exposed to a dynamic and discontinuous environment (Kale et al. 2001 as cited in Vlaar et al. 2007:444). Provan & Kenis (2008:243) argue in this respect:

> "legitimizing interactions among other organizations, some of which may be actual or potential competitors, is a critical function of network governance. If participants do not see interactions and coordinated efforts as being a legitimate way of conducting business, with potential benefits from these interac-

> tions (either social or economic), then the network is likely to exist in name only with little real commitment by participants to network-level goals and outcomes."

We thus conclude that a written framework contract and a mix of highly formalized coordination mechanisms represent economically rational choices for the LITPAP to legitimately enable the collective action of a network of horizontally cooperating LSPs embedded in a dynamic environment.

With regard to *coordination mechanisms*, the LITPAP presumably chooses to formalize the mix of mechanisms because this can achieve concerted action of partnering firms (coordination) and restrain and direct their behaviors (control) (see Vlaar et al. 2007:442f.). Formalization can support addressing organizational concerns by coordinating and directing the large network of horizontally cooperating and competing LSPs toward the logistical goals of logistics consumers. In addition, formalization enables control from a distance which is especially important for logistical prototype scenarios with a wide, networked geographic coverage. Moreover, formalizing the mix of coordination mechanisms is conducive to delivering logistics capabilities with cloud characteristics. Below, we investigate how the functions of formalization can contribute to achieving these characteristics, specifically focusing on on-demand availability, rapid elasticity, resource pooling, and pay-per-use.

The LITPAP presumably chooses to formalize coordination mechanisms in order to achieve *on-demand availability* and *rapid elasticity* because formalization fuels interaction processes (see Table 10). The instant implementation of new types of logistical capabilities (on-demand availability) and the rapid capacity adjustment of capabilities already being provisioned (rapid elasticity) requires not only physical reconfiguration of resources, but also interaction either between logistics consumers and LSPs or among LSPs. Hence, reducing the time necessary for these interactions contributes to achieving on-demand availability and rapid elasticity from an organizational perspective.

The LITPAP presumably chooses to formalize coordination mechanisms to achieve *resource pooling* because formalization (a) enables control from a distance and (b) fuels interaction processes (see Table 10). Exercising control from a distance is necessary to coordinate resource pooling among a spatially distributed network of horizontally cooperating LSPs that deliver geography spanning logistics capabilities. Fueling interaction processes is necessary for the pooling of physical logistics resources because the time windows are continually decreasing within which pooling needs to be coordinated among LSPs and resources need to reconfigured. This is due to an increasingly dynamic environment that reduces time to plan ahead.

The LITPAP presumably chooses to formalize coordination mechanisms in order to enable *self-service* because formalization (a) establishes a common language and (b) serves as signaling device (see Table 10). A common formalized language is a prerequisite for logistics consumers to interact with the LITPAP and LSPs via a cloud-based platform without any personal assistance from provider staff. A signaling device is necessary for logistics consumers to communicate their logistical concerns. In fact, formalization can help them break complex logistical problems down into clear and understandable terms (Singh 1997 as cited in Vlaar et al. 2007). A signaling device is also necessary for the LITPAP and LSPs to demonstrate their logistical capabilities (see Vlaar et al. 2007:444) by publishing them as services to the service repository. Thus, formalization can align the expectations between consumers and LSPs without the need for personal interactions.

The LITPAP presumably chooses to formalize coordination mechanisms in order to enable *pay-per-use* because formalization enables (a) the monitoring of progress and (b) the identification of deviation from objectives (see Table 10). Monitoring the delivery progress of logistics capabilities and identifying any discrepancies between the expected and actually realized objectives are prerequisites for determining the actual amount of capacity consumed and thus for charging in a pay-per-use manner.

In addition to these functions, formalization also contributes to dysfunctions (see Table 10). For example, formalization inhibits flexibility and innovation, reduces commitment and aspirations (Vlaar et al. 2007:445f.), and creates conditions for organizational strife (Vlaar et al. 2007:445f.; Wallenburg & Raue 2011; see Section C.V.3). Although these are dysfunctions and are thus undesirable, none of them directly reduces the ability to deliver logistics capabilities with cloud characteristics. We therefore argue that a mix of computer-formalized coordination mechanisms combined with a less formalized framework cooperation contract represents an economically rational design choice from the perspective of the LITPAP, which aims to enable the delivery of physical logistics capabilities with cloud characteristics.

3.4. Conclusions: The Need for Social Interaction vs. Computer-formalized Mechanisms

Based on contingency theory and on economically rational choices made by the LITPAP, we proposed that the structure of CLSs is formalized through a written framework cooperation contract and a mix of computer-formalized coordination mechanisms. Both design approaches yield similar results, thus providing strong support for the proposition.

While these design results appear reasonable from a governance and organizational perspective, they may still face major obstacles in actual implementation, especially the mix of highly formalized coordination mechanisms. This is because humans and human-to-human interactions are critical components of logistics systems, which is why a social perspective plays a decisive role in logistical inquiry (Delfmann et al. 2010:58f.; Krupp & Klaus 2012:72). Given mankind's inherent need for social interactions and autonomy, the question of whether human logistical actors are willing to exclusively coordinate the delivery of logistical capabilities through a mix of highly formalized mechanisms becomes highly relevant. Due to the central role of these formalized mechanisms in CLSs, the potential lack of "human willingness" may prevent CLSs from becoming a reality.

Moreover, the exclusive reliance on formalized mechanisms may also translate into a competitive disadvantage for member LSPs *vis-à-vis* competitors outside of the CLS. Panayides & So (2005) find evidence that LSPs can gain competitive advantage by proactively developing and maintaining relationships with clients (relationship-orientation). Given the exclusive reliance on formalized mechanisms, however, member LSPs are largely barred from developing personal relationships with consumers in CLSs. This may reduce LSPs' willingness to join, especially if these disadvantages are unlikely to be offset by potential benefits resulting from offering physical logistics capabilities with cloud characteristics.

IX. Characteristic Mix of Coordination Mechanisms – Part I: Integrated Model

1. Abstract of Chapter

Coordination mechanisms enable integrating the efforts of various actors to achieve a common goal. Different types of interfirm networks use different *characteristic mixes of coordination mechanisms* in order to coordinate their members' efforts toward network-level goals (Grandori & Soda 1995; see Section C.V.1).

The *architectural viewpoint* establishes the conventions for identifying and designing the characteristic mix of CLS coordination mechanisms (see Section F.IX.2). To identify the complete set of mechanisms that makes up the characteristic mix, we draw on the "life cycle of web services" (see Subsection F.IX.2.1). This cycle specifies all phases of a web service's existence and thus the fundamental activities and exchanges that stakeholders need to carry out in pursuit of their goals related to an SOA. As CLSs adopt service-orientation as an underlying design principle (see Subsection E.IV.5.2), we argue that "service-oriented physical cloud logistics capabilities" pass through the same life cycle phases as web services. Hence, CLSs must also include a set of mechanisms to coordinate the fundamental activities and exchanges along this life cycle, just as any other SOA (see Subsection D.III.4.1). In order to determine what kinds of mechanisms can effectively coordinate activities and exchanges along the life cycle, we use contingency theory.

Due to the abstract nature of the reference architecture of CLSs, we adopt an aggregate perspective on the web service lifecycle consisting of six phases (see Subsection F.IX.2.2): (a) *service development*: LSPs express their physical cloud logistics capabilities by means of formal service descriptions (Ludwig 2014:51f.); (b) *publication phase*: LSPs become admitted to CLSs, where they can publish their service descriptions to the service repository; (c) *design phase*: logistics consumers express their logistical concerns by using the service descriptions available in the service repository (Ludwig 2014:52); (d) *negotiation and selection phase*: logistics consumers and LSPs negotiate the conditions and constraints of requested services, especially price; (e) *delivery phase*: LSPs deploy resources to realize the agreed logistical transformations (Ludwig 2014:52f.); and (f) *unpublication phase*: service descriptions that are no longer in demand or that require refinement are removed from the service repository. In some cases, this may also include removing the associated LSPs from the CLS.

Based on current literature, five contingency factors critically influence the effectiveness of coordination mechanisms: number of actors (size) and their geographic dispersion, task complexity, external environment (complexity and dynamism including hostility), task interdependence, and task modularity (see Subsection F.IX.2.3).

The *architectural view* summarizes the assessment of the contingency factors of coordination mechanisms along this life cycle in the following proposition (see Subsection F.IX.3.1).

Proposition 13. *Cloud logistics systems will use a characteristic mix of formalized coordination mechanisms implemented on a cloud-based platform and consisting of programmed routines, system-supported skills, system-supported supervision, implicit coordination, price, and service-orientation all structured along the life cycle of cloud logistics services, where service-orientation is an underlying mechanism that supports all life cycle phases.*

In the following, we provide a brief description of each mechanism and also highlight which contingency factors critically influence the effectiveness of each mechanism to coordinate the activities and exchanges carried out in the respective life cycle phase.

The *cloud-based platform* is the underlying infrastructure that enables coordination along the entire life cycle of cloud logistics services by creating horizontal communication links among stakeholders and by providing a computing platform on which the characteristic mix of mechanisms can be implemented (see Subsection F.IX.3.2). Specifically, the cloud-based platform facilitates coordination by hosting (a) a *service repository*, which is a structured database and contains the service descriptions of all logistics capabilities currently being offered in the CLS and thus achieves coordination through the mechanism *implicit coordination*; (b) an *information repository*, which is a structured database and contains all logistics-relevant information necessary for delivering physical logistics capabilities in the CLS and thus achieves coordination through the mechanism *implicit coordination*; and (c) a *virtual marketplace*, which represents a virtual forum for stakeholders to interact and trade physical logistics capabilities and thus achieves coordination through the *price* mechanism. Critical contingency factors for coordination effectiveness include network size, geographic dispersion of stakeholders, and environmental dynamism (see Subsubsection F.IX.3.2.2).

Standardized, semantic service descriptions encode the capabilities, conditions, and constraints of using the respective physical cloud logistics service via well-defined semantics, which means that they are understandable by humans and machine-processable (see Subsubsection F.IX.3.3.1). These descriptions are developed according to the design principles of *service-orientation.* They are standardized by using a "logistics-domain specific ontology-based description language," where logistics-domain specific means that the description language's expressiveness and inference mechanisms are tailored to capture the particularities of the logistics domain, and "ontology-based" means that semantic descriptions are created through semantic markup: The language provides a generic "top-level" ontology to encode capabilities, conditions, and constraints of use. Following the SOA-RAF model of service descriptions (se Subsubsection D.III.4.3.3), proposed mandatory concepts of cloud logistics service descriptions are: service functionality, service policies, service metrics, service interface description, and service reachability. Chapter F.XI provides a more detailed design of cloud logistics service descriptions, especially with regard to the standardized top-level ontology. Critical contingency factors for coordination effectiveness include task complexity, task modularity, and the external environment (see Subsubsection F.IX.3.3.2).

Servitization refers to a *programmed routine* that specifies the exact steps LSPs need to perform to develop formal descriptions of physical cloud logistics capabilities (see Subsubsection F.IX.3.4.1). This routine consists of four steps: (a) *identification*, (b) *resource and capability modeling*, (c) *encapsulation*, and (d) *description.* Conditions that are likely to critically influence the effectiveness of servitization relate to the following contingency factors: task complexity and task interdependence (see details Subsubsection F.IX.3.4.2).

Boundary permeability refers to a *programmed routine* implemented on the cloud-based platform that specifies the exact steps that the LITPAP performs to admit new LSPs to or exclude member LSPs from the networked LSP cooperation within a CLS (see Subsubsection F.IX.3.5.1). The degree of boundary permeability – that is, the degree of openness – is determined by the boundary arrangements between the LITPAP and candidate member LSPs. These arrangements are conceptualized along three dimensions (1) who chooses member; (2) criteria for membership; (3) duration and exclusivity of membership (Gulati et al. 2012:576). First, the LITPAP has exclusive authority over the degree of boundary permeability. This involves specifying criteria for assessing candidate member LSPs and setting minimum thresholds for these criteria. It also includes carrying out the assessment process. Second, the degree of permeability (criteria for membership) depends on three conditions: (a) current degree of redundancy in the shared resource pool, (b) how a candidate member LSP influences the degree of redundancy, and (c) logistics market demand. Continuously assessing the degree of resource redundancy is critical for the LITPAP because it directly determines the ability to deliver capabilities with cloud characteristics, especially in an on-demand and rapidly elastic manner by being able to absorb unanticipated demand occurrences or surges. Moreover, the degree of redundancy determines the degree of internal competition among member LSPs. This, in turn, determines the extent to which member LSPs can address their economic concerns. A higher degree of redundancy increases the ability to achieve cloud characteristics but also increases internal competition. Hence, the LITPAP needs to balance the ability to achieve cloud characteristics with the economic concerns of member LSPs. Third, membership exclusivity is limited to the capability and geographic scope covered by the CLS. In other words, member LSPs may engage in other alliances as long as these alliances do not jeopardize the strategic position of the CLS. Membership duration is generally unlimited, but includes a minimum period of engagement to ensure member LSPs' commitment (deter opportunism). Conditions that are likely to critically influence the effectiveness of boundary permeability relate to the following contingency factors: environmental dynamism, task and environmental complexity, and the degree of centralization interacting with behavioral uncertainty of the LITPAP.

Self-service refers to *system-supported skills* that let logistics consumers express their logistical concerns in a formal service request using the service descriptions available in the service repository (see Subsubsection F.IX.3.6.1). System-supported skills is a software program that specifically supports logistics consumers in identifying logistics capabilities that match their concerns according to the information accessible in the service description; evaluating these candidate services; comparing alternative candidate services; and, if needed, combining multiple services into more complex solutions via an iterative, dynamic process without the support of provider personnel (self-service). System-supported skills leverage the use of well-defined semantics in service descriptions to automate the discovery, evaluation, composition of services. Conditions presumed to influence the effectiveness of system-supported skills to coordinate service design are related to the contingency factors of task complexity, and environmental complexity and dynamism (see Subsubsection F.IX.3.6.2).

Market mechanisms are a *programmed routine* that specifies the exact steps for logistics consumers and LSPs to carry out negotiations about formally requested services on the virtual marketplace in order to reach a formal contract (see Subsubsection F.IX.3.7.1). These contracts specify (a) which LSP is awarded which requested service in the case of composite services or multiple providers of candidate services, (b) the conditions and constraints of service use, and (c) the payments to be made by the logistics consumer to LSPs of awarded services. Market mechanisms represent a decentralized mechanism, which means that economic decisions (allocation of services and payments to be made) are not determined by the LITPAP in a hierarchical manner but are exclusively determined based on the information provided by consumers and LSPs during the negotiation phase. Conditions that presumably influence coordination effectiveness are related to the following contingency factors: size, task complexity, environmental dynamism and hostility, task interdependence, behavioral uncertainty, and the interaction of selected factors with the degree of centralization (see Subsubsection F.IX.3.7.2).

The *IoT* relates to *system-supported supervision*, which enables the direction and monitoring of logistics resources which are deployed to address agreed-upon logistical concerns through IoT technologies, such as RFID (see Subsubsection F.IX.3.8.1). Specifically, physical resources are equipped with sensors that capture relevant data. IoT technologies associate these resources with their respective virtual service descriptions, thus real-world logistics resources can be sensed, identified, monitored, and controlled by virtual services. Collected information is stored in the information repository on the cloud-based platform and is made available to (a) LSPs involved in delivering services as operationally required during delivery phase and (b) the respective logistics consumers to ensure that the delivered services comply with the services that were agreed to. By combining cloud computing technology with the IoT, the "full potential" of both technologies can be realized (Gubbi et al. 2013:1651). This is because the IoT is a concept that enables the automated and ubiquitous collection of data, while cloud computing is a technology that allows for the systematic, large-scale economic exploitation of large amounts of information. In this respect, the combined use of IoT and cloud computing relates to the idea of "big data." Conditions that presumably influence the effectiveness of coordination during the delivery phase are related to the following factors: size and geographic dispersion, task complexity, environmental complexity and dynamism, and task interdependence (see Subsubsection F.IX.3.8.2).

2. Viewpoint: An Approach to Identifying and Designing the Characteristic Mix of Coordination Mechanisms

2.1. The Life Cycle of Web Services as a Model to Structure the Characteristic Mix of Coordination Mechanisms

Coordination mechanisms enable integrating the efforts of various actors to achieve a common goal. A general taxonomy of coordination mechanisms, including computer-enabled mechanisms, has already been introduced (see Subsection C.V.4.1). Different types of interfirm networks use different *characteristic mixes of coordination mechanisms* in order to coordinate their members' efforts toward network-level goals (Grandori & Soda 1995; see Section C.V.1).

Cloud logistics knowledge identifies several coordination mechanisms that are formalized as they are implemented on the cloud-based platform (see Subsection F.IV.3.5). However, beyond merely "listing" these mechanisms, current knowledge provides hardly any information regarding the characteristic mix of mechanisms in CLSs. Four issues are of particular relevance in this respect. Current cloud logistics knowledge does not provide (a) any conceptual proof of the "completeness" of the identified mechanisms; (b) any insights into the structure or "interplay" of these mechanisms, which currently appear to exist independently rather than forming a coherent mix; (c) any anchoring of the identified list in organizational theory, which is why it remains unclear whether these mechanisms can in fact achieve effective coordination; nor (d) a concrete design proposal for any of the identified mechanisms.

This viewpoint establishes the conventions for identifying and designing the characteristic mix of CLS coordination mechanisms. To identify the complete set of mechanisms that makes up the characteristic mix, we draw on the "life cycle of web services" (see Subsection D.III.4.1). This cycle specifies all phases of a web service's existence and thus the fundamental activities and exchanges that stakeholders need to carry out in pursuit of their goals related to an SOA. As CLSs adopt service-orientation as an underlying design principle (see Subsection E.IV.5.2), we argue that "service-oriented physical cloud logistics capabilities" pass through the same life cycle phases as web services. Hence, CLSs must also include a set of mechanisms to coordinate the fundamental activities and exchanges along this life cycle, just as any other SOA (see Subsection D.III.4.1). This mix of mechanisms lets cloud logistics stakeholders address their logistical and economic concerns in CLSs. In other words, the life cycle of web services provides a supporting model to identify the mix of coordination mechanisms of CLSs and expose their interplay. In order to determine what kinds of mechanisms can effectively coordinate activities and exchanges along the life cycle, we use *contingency theory*.

The following subsection discusses the life cycle of cloud logistics and links it to stakeholder concerns. The subsequent subsection introduces contingency factors that critically influence the effectiveness of coordination mechanisms.

2.2. Life Cycle of Cloud Logistics Services and Link to Stakeholder Concerns

The life cycle of web services has already been introduced at its most granular level with a discussion of its nine distinct phases (see Subsection D.III.4.1). Due to the abstract nature of reference architectures, however, we adopt a more aggregate version of the life cycle here, aggregating these nine phases into six. Specifically, service discovery and composition are combined into the design phase (see Erl 2008:413), and invocation, execution, and monitoring are combined into delivery phase (see "enactment phase" Burstein et al. 2005:74). This aggregated life cycle is the "life cycle of cloud logistics services," which is depicted in Figure 50 along with each phase's final milestone. These phases largely align with the life cycle of cloud logistics services proposed by Ludwig (2014:51ff.), who structures the life cycle into servitization (≈ development), service development (≈ design), service delivery (≈ delivery), and termination (≈ unpublication). Furthermore, these phases generally align with the procurement process of logistics services as conceptualized by Andersson & Norrman (2002), who identify a service definition and specification phase, a Request for Proposal (RfP) phase (≈ service design), and a negotiations and contracting phase (≈ negotiation and selection). Moreover, by adopting the life cycle of web services as a basis for the CLS coordination model, we can anchor the model within current literature.

During the service development phase, LSPs express their physical cloud logistics capabilities by means of formal service descriptions (Ludwig 2014:51f.). During the publication phase, LSPs become admitted to CLSs, where they can publish their service descriptions to the service repository. During the design phase, logistics consumers express their logistical concerns by using the service descriptions available in the service repository (Ludwig 2014:52). During the negotiation and selection phase, logistics consumers and LSPs negotiate the conditions and constraints of requested services, especially price. During the delivery phase, LSPs deploy resources to realize the agreed logistical transformations (Lud-

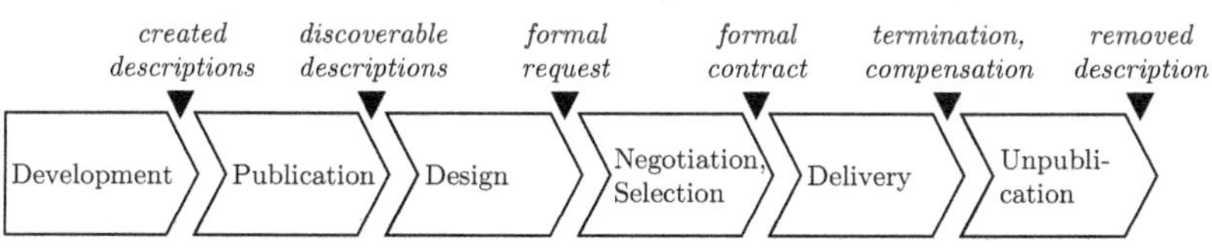

Figure 50.: Life Cycle of Cloud Logistics Services with Milestones

wig 2014:52f.). During the unpublication phase, service descriptions that are no longer in demand or that require refinement are removed from the service repository. In some cases, this may also include removing the associated LSPs from the CLS.

The life cycle of web services is a model suitable for framing stakeholder concerns. All life cycle phases directly frame organizational concerns, as this viewpoint focuses on coordination mechanisms in the life cycle. In addition, however, this viewpoint also frames logistical and economic concerns. The development, publication, and design phases frame logistical concerns. They allow, on one hand, LSPs to offer their capabilities as services and, on the other, logistics consumers to specify their concerns by using the descriptions of published services. The negotiation and selection phase frames logistical and economic concerns, as the negotiation outcome exactly specifies the logistical concerns to be addressed and the payments to be made. *Delivery* also frames logistical and economic concerns as logistical concerns are addressed through logistical transformations and payments are made.

2.3. Contingency Factors

2.3.1. Overview

This subsection attends to the question of which mechanisms can effectively coordinate work under which conditions. Based on current literature, five contingency factors critically influence the effectiveness of coordination mechanisms: (1) number of actors (size) and their geographic dispersion, (2) task complexity, (3) external environment (complexity and dynamism including hostility), (4) task interdependence, and (5) task modularity. Task interdependence and modularity are discussed last, as both refer to a task's structure and external environment.

2.3.2. Size and Geographic Dispersion

Literature suggests that the effectiveness of a mechanism to coordinate work depends on the number of actors to be coordinated (see e.g. Kieser & Walgenbach 2007:328ff.; Pugh et al. 1969) and their geographic dispersion (see e.g. Grinter et al. 1999). Despite being distinct factors, size and geographic dispersion are discussed together, as they have similar effects on the effectiveness of coordination mechanisms.

Size refers to the number of actors who need to be coordinated to accomplish a task. Note that size, with regard to coordination, differs from the absolute number of network members (network size) introduced above: A network can be large, but a specific task only requires coordination among a subset of its members. *Geographic dispersion* refers to (a) the (average) physical distance between actors and (b) the number of sites at which they are located (O'Leary & Cummings 2007:434). A mechanism's capacity to effectively coordinate groups of different sizes and geographic dispersions depends on its ability to handle different communication, information processing, and information storage requirements associated with different sizes and different degrees of geographic dispersion. Figure 51 provides an overview of the effectiveness of non-technology-enabled and computer-dependent coordination mechanisms along these two dimensions. Computer-dependent mechanisms are all implemented on a computer platform (depicted as a box with a dashed border), which establishes communication links between the actors to be coordinated.

Non-technology enabled mechanisms (in their pure forms) exclusively rely on the human capacity to communicate, process, and memorize information. Thus, natural constraints limit our ability to coordinate work, both in the number of actors and in their geographic dispersion. Groth (1999:77) remarks in this respect:

> "The fact that unaided human information exchange can only take place locally and in real time puts severe constraints on the possibilities for building and sustaining large organizations. The only means of communication over distance is then the dispatch of messengers, and the messengers have to rely on their memory to ensure that the message reaches its destination uncorrupted."

Mutual adjustment imposes high communication costs as it essentially "requires everyone to communicate with everyone else," which makes it effective for coordinating only a few actors (Groth 1999:51). In addition, actors need to be geographically clustered due to the ongoing need for informal face-to-face communication. *Direct supervision* is effective for coordinating groups of moderate size, as coordination is achieved through a single supervisor who issues orders and monitors the progress of work (Mintzberg 1979:7f.). Without delegation, group size is constrained by the supervisor's capacity to communicate with each worker and to process and memorize the information required to direct and monitor employees. The group needs to be geographically clustered, although to a lesser degree than for mutual adjustment, as the supervisor must coordinate work through frequent, personal, direct communication. *Standardization of work processes, -outputs, and -skills* allows for the effective coordination of many actors through the use of programs (Mintzberg 1979:5ff.). Standards imposed on processes, outputs, and skills need to be defined only once and can subsequently be internalized and applied by many actors. With regard to the standardization of work processes and output, actors need to be reasonably geographically clustered, so the supervisor can monitor behaviors and outputs, respectively. The standardization of skills allows geographically dispersed actors to be coordinated as actors only need to be clustered during the indoctrination process (which is administered and monitored by the supervisor), but not during the actual transfor-

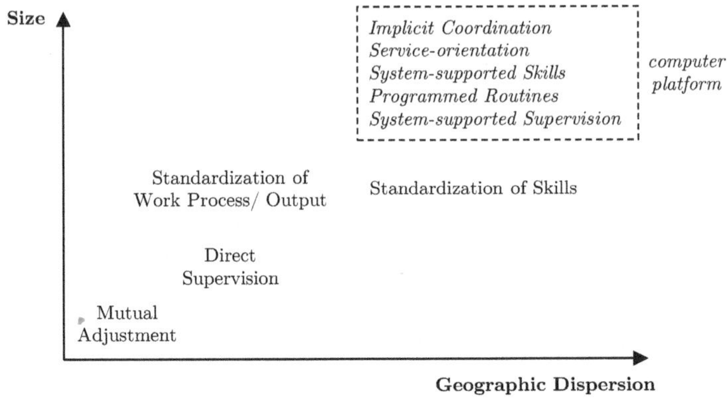

Figure 51.: Effective Coordination Mechanisms by Size and Geographic Dispersion

mation processes. Hence, work can be carried out at geographically dispersed locations. To illustrate, Mintzberg (1979:6) provides the example of a king who trusts his remote governors because he has trained them in person beforehand.

The use of IT can enhance coordination (see Subsection C.V.4.1). Any technology (e.g. email, phone, video conferencing) that improves the communication or handling of information and feedback relations allows for (a) a higher degree of control if distance is held constant or (b) a greater distance if the degree of control is held constant (Groth 1999:327). Hence, computer-dependent coordination mechanisms can effectively coordinate significantly more actors that can be more geographically dispersed than their non-computer-dependent correspondents. Each computer-dependent mechanism is implemented on a computer platform, which creates communication links between the actors to be coordinated. *Implicit coordination* can effectively coordinate many (likely unlimited) highly geographically dispersed actors due to the database's wide geographic reach enabled by the internet, the database's capacity to connect to many actors, and the swiftness with which information can be stored or retrieved from the database (see Subsection C.V.4.2; Groth 1999:304ff.). *System-supported supervision* allows for the effective coordination of many geographically dispersed actors. If supervision is handled by a single manager, the number of actors to be coordinated ultimately depends on the manager's coordination capacity. However, if supervision can be automated, e.g. through a network of sensors such as GPS tags, there are no limits to the number of actors, other than computer's technical capacity constraints. *Programmed routines*, *hyperautomation*, and *system-supported skills* can be installed on the computers of many geographically dispersed actors and thus effectively coordinate their efforts. Monitoring the usage of routines and software suites can generally be automated through another software application; if so, there are no general limits to the number of actors that can be coordinated, other than technical constraints. *Service-orientation* can effectively coordinate many, geographically dispersed actors. Different service modules can be assigned to actors at different locations, and these modules can be integrated to accomplish a common task through standardized module interfaces. If the outputs of service modules can be monitored electronically, there are no limits to the number of actors, other than technical capacity constraints.

2.3.3. Task Complexity

Task complexity has long been a focus of organizational scholars. Definitions of task complexity can be classified into the following approaches: (a) complexity as a psychological experience, (b) complexity as a task-person interaction, and (c) complexity as a function of objective characteristics (Campbell 1988:44). To investigate how task complexity affects coordination, we focus on complexity as a function of objective characteristics. In this sense, task complexity is determined by the information processing requirements associated with a task (Campbell 1988:43):

> "any objective task characteristic that implies an increase in information load, information diversity, or rate of information change can be considered a contributor to complexity."

Based on a literature review, Campbell (1988:44) suggests four objective characteristics that determine complexity. Any single or combined presence of these characteristics contributes to objective task complexity by increasing information load, diversity, or rate of change. The four characteristics are (a) multiple potential paths to arriving at a desired outcome, (b) multiple desired outcomes, (c) conflicting interdependence among paths to multiple outcomes, and (d) uncertain or probabilistic links among path and outcomes.

Mintzberg (1979:7ff.) argues that different mechanisms can coordinate tasks of differing complexity with different levels of effectiveness, as depicted in Figure 52. *Mutual adjustment* can effectively coordinate the most simple and the most complex tasks: In-

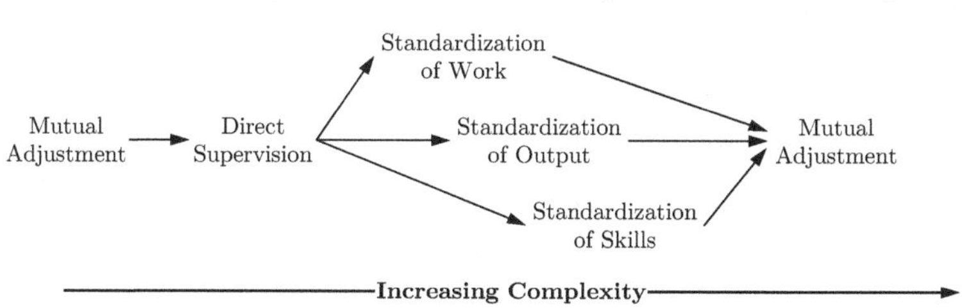

Figure 52.: The Coordinating Mechanisms: A Rough Continuum of Complexity (adopted from Mintzberg 1979:7)

dividuals can communicate informally throughout the transformation process and adapt their activities to the unfolding task conditions, regardless of how simple or complex these might be (Mintzberg 1979:3, 7). According to the complexity continuum, *direct supervision* enables effective coordination of tasks that are slightly more complex than in mutual adjustment (compared with the most simple tasks that are coordinated via mutual adjustment). However, Mintzberg (1979:7ff.) makes no direct reference to complexity to explain this positioning but argues that direct supervision can coordinate a larger group of individuals. Hence, he refers to size rather than task complexity. Nevertheless, direct supervision may be effective in coordinating more complex tasks, as a larger group collectively has a greater information processing capacity, and hence may be able to accomplish a more complex task, e.g. by quickly identifying the only feasible path for achieving a desired outcome. However, the degree of task complexity that can be coordinated ultimately depends on the single manager's qualification to direct and monitor the tasks performed by subordinates (see Kieser & Walgenbach 2007:110f.). *Standardization of work processes* can effectively coordinate tasks that are "simple and routine" (Mintzberg 1979:8). Effective coordination requires that the path of how a task can be accomplished can be understood and clearly specified *ex ante.* This, however, may not be achievable in cases of complex tasks with unclear means-ends relationships. Therefore, if tasks are more complex, through multiple potential paths and uncertain or probabilistic relationships between paths and outcomes, *standardization of output* is an effective mechanism. Instead of specifying how a task must be carried out, this mechanism defaults to specifying *ex ante* only what needs to be achieved (Mintzberg 1979:8). If tasks are even more complex and neither the transformation process nor the outcome can be specified, *standardization of skills* is an effective mechanism if the skills necessary for carrying out the task can be taught before the work begins (Groth 1999:53). As Mintzberg (1980:334) remarks, "[c]omplexity demands the use of skills and knowledge that can be learned only in extensive training programs." Finally, if tasks are so complex that the paths to task accomplishment, outputs, and skills cannot be standardized, *mutual adjustment* is an effective mechanism again, as described above.

Just as the use of IT enhances the capacity of mechanisms to coordinate many actors that are geographically dispersed, IT can also enhance the coordination of complex tasks. The same three characteristics are crucial: (a) immense and loss-less storage capacity, (b) quick and less costly communication, and (c) immense and less costly information processing capacity (Groth 1999:144; also see Argyres 1999; see Subsection C.V.4.1). These properties in combination with the programmability of computers can significantly enhance the coordination of complex tasks:

> "Computer-based systems *process* information, they *store* information, and they *communicate.* And, the key to the power of computers, to all of their capabilities, is their *programmability*—the possibility to have immensely com-

plex sets of logical operations executed automatically" (Groth 1999:215, italics original).

With regard to the objective task characteristics driving complexity, introduced above, the capacity of computers to process large amounts of data quickly and cheaply contributes to (a) quickly evaluating potential paths and identifying feasible ones, (b) pursuing multiple outcomes in parallel, (c) making "optimal" tradeoffs in case of conflicting interdependence among paths and associated outcomes, and (d) unveiling statistically significant relationships between paths and outcomes. Following this general argument, we propose to position the computer-dependent mechanisms higher on the complexity axis than their non-technology enabled correspondents, as depicted in Figure 53. With regard to *system-supported supervision*, Groth (1999:320f.) argues that computers can capture and analyze more information and thus enable the coordination of more complex tasks. *Programmed routines* can be more numerous and diverse than explicit routines and accomplish computationally complex tasks that humans cannot not easily perform (Groth 1999:277f.). *System-supported skills* enable the coordination of complex tasks, as computer programs can support decision-making through pre-defined decision rules and quick data analysis in situations of cognitive overload (artificial intelligence) (Groth 1999:258f.,260). With regard to *implicit coordination*, structured databases enable the exchange of very simple and of complex information sets, including various attributes and formats, that would be difficult to transmit in a face-to-face discussion. Hence, implicit coordination is the mechanism that also coordinates the most complex tasks. In line with this, Argyres (1999:163) finds evidence that a database was pivotal in enabling intra- and interfirm coordination in the project to develop the B-2 "Stealth" bomber:

> "The database eventually contained nearly all the information necessary to build the highly complex and innovative aircraft—information that was directly used both in advanced structural analyses and to create machine tools and control machines to manufacture all major aircraft sections."

Likewise, Malone et al. (1987:489) argue that databases and modern high-bandwidth communication technologies are more suitable than traditional ways of communication to handling and exchanging complex, multidimensional product descriptions. Finally, with regard to *service-orientation*, electronic service descriptions can be made much more complex by including more detailed output specifications and more input and output variables than paper-written descriptions could reasonably include.

2.3.4. External Environment

Organizational scholars suggest that conditions in the external environment of an organization determine the effectiveness of mechanisms to coordinate work. Mintzberg (1979:286) states that mechanism effectiveness varies based on the degree of environmen-

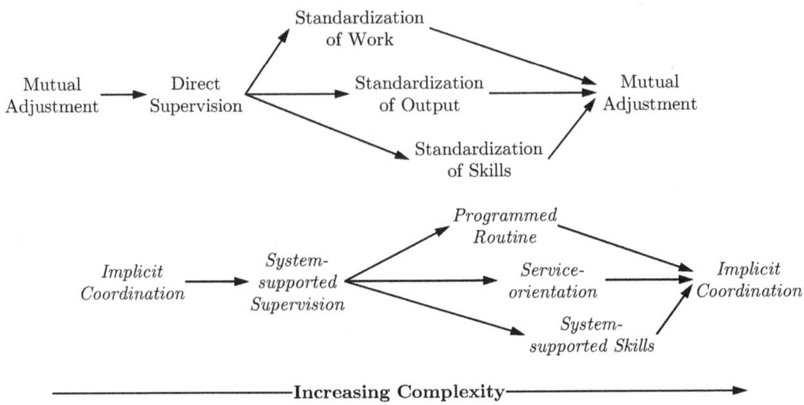

Figure 53.: Classic and Computer-dependent Coordination Mechanisms along the Complexity Continuum

tal complexity and stability, as depicted in Table 27. Essentially the same arguments apply to environmental complexity as to task complexity; from a coordination perspective, there is no difference in whether complexity originates in the task or the environment. Hence, we focus on the degree of dynamism in the following.

The degree to which mechanisms can achieve effective coordination in a dynamic environment depends on their ability to respond to unpredicted changes in a timely and economic manner. There are two fundamental mechanism categories: (1) coordination by feedback and (2) coordination by program. *Coordination by feedback* allows for adequate responses to external disturbances due to real-time coordination; on the other hand, *coordination by program* is less flexible and is only successful if the future turns out as predicted/assumed in programmed processes, outputs, or skills (Kieser & Walgenbach 2007:122). *Mutual adjustment* and *direct supervision* can effectively coordinate work in

Environmental Dimensions	Stable	Dynamic
Complex	Standardization of skills	Mutual adjustment
Simple	Standardization of output Standardization of work processes	Direct supervision

Table 27.: Effective Coordination Mechanisms per Organization Environment (adopted with minor changes from Mintzberg 1979:286)

dynamic environments as both achieve coordination through feedback. New information about changes in the environment can be incorporated into activities and directions in an *ad hoc* manner (Thompson 1967:56; Kieser & Walgenbach 2007:110). By contrast, *standardization of work processes, -outputs, and skills* can effectively coordinate work in stable environments, as these mechanisms achieve coordination by program. Because of the stable environment, new information relevant to a task is very unlikely to surface during the transformation process, which is why work processes, outputs, and skills can be adequately specified before the work is carried out.

So far, we have focused on the suitability of non-computer-dependent mechanisms to coordinate work as a function of environmental dynamism. In the following, we investigate the suitability of computer-dependent mechanisms to coordinate work in environments of different dynamism relative to their classic correspondents. As depicted in Figure 54, no general statement seems feasible regarding whether computer-dependent mechanisms are more or less suited to effectively coordinating work in a dynamic environment. Hence, the mechanisms must be investigated separately. With regard to *implicit coordination* and *system-supported supervision*, we propose that both are less effective than their classic correspondents for coordinating work in dynamic environments. Adapting databases and supervision applications likely takes more time than simply adapting informal communication processes (mutual adjustment) or the methods of issuing orders verbally and supervising workers in person (direct supervision). Yet, both computer-dependent mechanisms can effectively absorb rapid capacity fluctuations that occur *within* the boundaries of the task model programmed into the computer. This is because the capacity of computers can be scaled more quickly by adding and removing hardware (especially when using cloud computing technology) than adding and removing personnel, which might require initial training.

With regard to *programmed routines*, *hyperautomation*, and *system-supported skills*, we refrain from making a general proposal about whether these computer-dependent mechanisms are more or less effective than their classic counterparts for coordinating work in more or less dynamic environments. Too much seems to depend on the particularities of the situation, such as the magnitude of adjustments needed, the programming language initially used, the documentation of source code, and the personnel's cognitive capacity to internalize adapted processes and skills through training. As Groth (1999:287) remarks with regard to changing internalized work processes: "The process of renewing or changing routines is also difficult, because it requires workers to actively forget the old routines and thoroughly learn the new ones." Likewise Groth (1999:259) points to the difficulty of changing IT systems "because of the limits of the systems they reside in, and simply because nobody may have a complete knowledge of the systems involved." However, we propose that *service-orientation* is effective for coordinating work in more dynamic

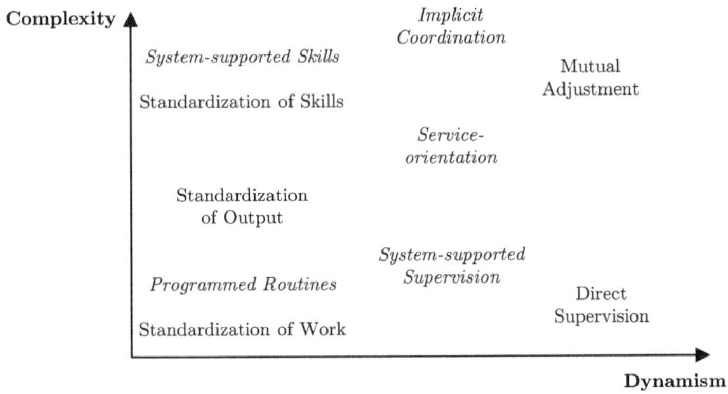

Figure 54.: Effectiveness of Classic and Computer-dependent Coordination Mechanisms by Complexity and Dynamism

environments *vis-à-vis* standardization of output, due to the inherent modularization of service descriptions. In response to environmental changes, services can be combined into new configurations, new services can be added, or affected services can be changed. Thus, due to modularization, time and costs associated with making adjustments become minimized.

2.3.5. Task Interdependence

Task interdependence refers to the degree to which actors need to work together with others outside of their organization in order to achieve their goals (Tushman 1979:88). In interorganizational networks, it is the degree to which different network members need to work together in order to achieve their own and network-level goals. Interdependence necessitates "coordination, joint decision-making, and problem solving" among the interdependent actors, and the greater the degree of interdependence, the greater the need for coordination, joint decision-making, and problem solving (Tushman 1979:85).

Different types of interdependencies can be effectively coordinated by different types of mechanisms (March & Simon 1958 as cited in Grandori 1997a:901). Based on literature, Grandori (1997a:901ff.) distinguishes four interdependency types: (1) *pooled*, where network members are interdependent due to their joint use of resources such as buildings, competencies, and technologies; (2) *intensive*, where network members are interdependent due to their joint application of complementary resources to a common activity; (3) *sequential*, where network members are interdependent because the outputs of one network member represent the inputs for another member; and (4) *reciprocal*, where network members are interdependent because the outputs of one network member represent inputs for another member, and these outputs are tailored to the requirements of the receiving member, which is why they can only be produced after the receiving member has provided some form of input, such as information or materials. The former two types refer to interdependencies due to collective action and the latter two to interdependencies due to transactions.

Table 28 summarizes Grandori's (1997a) argument. With regard to *pooled interdependencies*, Grandori (1997a:901f.) suggests *communication and decision procedures* (standardization of work processes) as effective mechanisms. These rules prescribe some actions and forbid others, thus ensuring the efficient use of pooled resources. Likewise, Nassimbeni (1998:548) argues that standardized work processes are suited to coordinating activities that possess process interdependencies, which means that network members' tasks are not independent of each other *per se*, but build on a common set of activities (which can be conceived of as a set of pooled resources). If members have incentives for free-riding by reducing their contributions, *mutual monitoring* (mutual adjustment) or the *supervisory hierarchy* of a third party (direct supervision) can effectively coordinate work as long as activities are not complex and differentiated, and are thus observable.

With regard to *intensive interdependencies*, Grandori (1997a:902) suggests *group decision-making* (mutual adjustment) and *social monitoring or property-rights sharing* as effective coordination mechanisms. The need for mutual adjustment arises from the complexity of the tasks involved and the need to coordinate and respond in real time. However, if activities are difficult to monitor, network-members' incentives must be aligned either through social norms or the pooling of property rights associated with given activities.

With regard to *sequential interdependencies*, Grandori (1997a:903) proposes *cross-activity programming* (standardization of work processes) and *hierarchical decision-making* (direct supervision) as effective mechanisms, thus ensuring a synchronized flux of goods and full

Pooled	**Intensive**
Communication and decision procedures Mutual monitoring or supervisory hierarchy	Group decision-making Social monitoring or property-rights sharing
Sequential	**Reciprocal**
Cross-activity programming Hierarchical decision-making for inter-unit adjustment	Integration and liaison roles Authority by exception and residual arbitration

Table 28.: Types of Interdependence and Effective Coordination Mechanisms (adopted from Grandori 1997a:909)

utilization of production capacity in subsequent production stages, where each stage is controlled by a different network member. Nassimbeni (1998:547f.) also proposes direct supervision as an effective means of coordinating this kind of interdependency (see "flow interdependencies" Nassimbeni 1998:547f.).

For *reciprocal interdependencies*, Grandori (1997a:903) proposes different mechanisms depending on the degree of environmental dynamism and cognitive complexity. If the external environment is stable, *extensive rules and programs* (standardization of work processes) are effective, just as for sequential interdependencies because complexity does not result from the direction of flows, as such (Grandori 1997a:903f.). If, however, reciprocal interdependencies are embedded in a dynamic environment, *authority by exception and residual arbitration* (direct supervision) are proposed to be effective, thus enabling quick responses to changing conditions (Grandori 1997a:903). Finally, if capabilities are diffused among network members and uncertainty arises from cognitive complexity associated with the interdependent activities rather than environmental dynamism, *lateral relations, liaison roles, and integration units* (mutual adjustment) are proposed to be effective (Grandori 1997a:903), providing each member the opportunity to influence and monitor the given activities.

2.3.6. Task Modularity

Thus far, the considerations regarding the effectiveness of service-orientation as a coordination mechanism have focused on the ability to reasonably specify the output(s) of tasks *ex ante*. However, to determine the conditions under which *service-orientation* can effectively coordinate work, an additional factor needs to be considered. This is because service-orientation comprises not only design principles concerned with specifying and codifying a service's capabilities (e.g. standardized service descriptions), but also those that determine the structure of the underlying task (reusability and composability). Service-orientation, therefore, strongly refers to the concept of *modularity*, and, hence, one can argue that service-orientation is only an effective coordination mechanism if the structure of the underlying task exhibits some degree of modularity. Hence, the factors that determine the degree of modularity need to be investigated.

Schilling (2000) proposes a general model for explaining whether a system will migrate toward an increasing degree of modularity or in the opposite direction – toward integration (see Figure 55). Her model is rooted in general systems theory and is therefore reasonably abstract and applicable for explaining (and predicting) the degree of *task modularity* or *process modularity* (see Section B.IV.3). The model posits four factors that pressure a system (in our case, a specific task) to migrate to modularity or in the opposite direction. The influencing factors are: (1) synergistic specificity, (2) heterogeneity of inputs, (3)

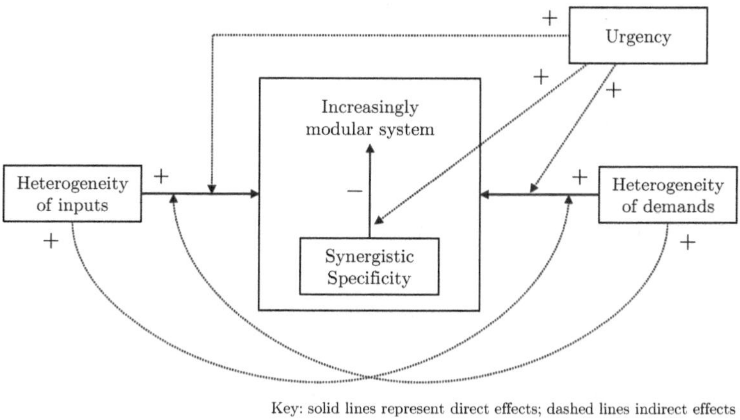

Figure 55.: Modular Systems – Factors Influencing the Migration toward Increasing or Decreasing Degree of Modularity (adopted from Schilling 2000:319)

heterogeneity of demands, and (4) urgency. Synergistic specificity is a factor that is internal to a system; the others originate in the system's context.

Synergistic specificity moves a system away from modularity. It pertains to the degree to which a system achieves superior functionality and/or performance by having components specific to a particular configuration or function (Schilling 2000:315f.). In other words, a system achieves higher gains from an integral structure than from a modular one (see Campagnolo & Camuffo 2010:260). In this respect, systems with a high degree of synergistic specificity trade off recombinability against potential synergies arising from component specificity (Schilling 2000:316). *Heterogeneity of inputs* refers to the number and variety of available inputs that can be combined in a system (Schilling & Steensma 2001:1153). The larger the number and variety of inputs available in the environment, the more system configurations are possible, leading to a higher degree of modularity. *Heterogeneity of demands* also moves a system toward modularity. It refers to the variety of demands that are placed on the system, such as hetegreonous customer demands (Schilling 2000:317). Hence, the hetereogeniety of demands directly relates to the degree of environmental complexity. The more complex the environment, the higher the degree of modularity. Finally, *urgency* acts as a catalyzing force towards a higher degree of modularity. Urgency, among other things, results from environmental factors such as time constraints causing "rapid obsolescence of products and processes caused by technological change" (Schilling 2000:318). Hence, urgency is directly related to the degree of environmental dynamism: The more dynamic the environment, the higher the degree of modularity.

The relevance of these factors for explaining and prediciting modularity in logistics processes is corroborated by empirical logistics research. With regard to synergistic specificity, Pekkarinen & Ulkuniemi (2008:99) remark:

> "managers need to know, for which services modular design and production is appropriate. If it is not possible to share or separate the independent elements [e.g. warehousing space] of a service without losing some of its functionality, modular services are not the right answer. For example, in the case of full load road transportation from origin A to destination B, it is difficult to see any benefits arising from the use of modular platform planning. For a firm with several service portfolios, however, such as our LSP, full load direct shipments are just one of the service offerings within a modular service platform."

For the other factors, LSPs report that the adoption of modularity will become more important in the future in service production and design because it allows them to respond flexibly and cost-efficiently to increased competition, technological change, and heterogeneous customer demand through concurrent standardization and customization (Rajahonka 2013: 45f.; see similar Lin & Pekkarinen 2011:353; Pekkarinen & Ulkuniemi 2008:98).

To address the initial question of task modularity as a contingency factor for service-orientation, we propose that if outputs can be reasonably specified *ex ante*, then the higher the degree of modularity inherent in the task to be coordinated, the more effective the mechanism of service-orientation.

3. View: An Integrated Model of the Characteristic Mix of Coordination Mechanisms

3.1. Proposition and Mechanism Overview

Above we have introduced a taxonomy of coordination mechanisms (see Subsection C.V.4.1) and described conditions in which they are considered to be effective coordination instruments (see Subsection F.IX.2.3). In addition, we have adopted the life cycle of web services as an underlying model to identify and structure the value adding activities and exchanges among logistics consumers, LSPs, and the LITPAP (see Subsection F.IX.2.1). Based on these inputs, we derive the characteristic mix of coordination mechanisms of CLSs. We propose:

Proposition 13. *Cloud logistics systems will use a characteristic mix of formalized coordination mechanisms implemented on a cloud-based platform and consisting of programmed routines, system-supported skills, system-supported supervision, implicit coordi-*

nation, price, and service-orientation all structured along the life cycle of cloud logistics services, where service-orientation is an underlying mechanism that supports all life cycle phases.

Figure 56 depicts the proposed model of the characteristic mix of coordination mechanisms for the CLS reference architecture. This model reveals the complete set of mechanisms, including their interplay and structure along the life cycle of cloud logistics services. The model's contribution is twofold. First, it integrates the set of formalized mechanisms that we have identified from current cloud logistics knowledge into a single model through the service life cycle. The model specifies the primary coordination mechanism during each phase of the life cycle and the interplay between mechanisms. Second, the model anchors the characteristic mix of coordination mechanisms in existing literature, thus contributing to our second research objective. This anchoring is achieved by associating the mechanisms identified in cloud logistics knowledge (see Subsection F.IV.3.5) with those included in the taxonomy of coordination mechanisms initially proposed by Mintzberg (1979) and later extended by Groth (1999).

In the remainder of this subsection, we provide a brief description of each mechanism and explain how it fits into the proposed model. In the subsequent subsections (Subsection F.IX.3.2 – Subsection F.IX.3.8), a high-level design for each mechanism is provided. Moreover, an assessment of the conditions presumed to prevail in each of the life cycle phases is provided in order to validate that the proposed mechanisms can effectively coordinate according to the logic of *contingency theory*. However, a detailed design for each mechanism is beyond the scope of this thesis. Nevertheless, we provide a more detailed design for three mechanisms in this reference architecture; these three mechanisms are critical when adopting the design principles that underlie cloud computing for the logistics domain because they enable the functioning of CLSs. These include servitization (Chapter F.X); standardized, semantic service descriptions (Chapter F.XI); and market mechanisms (Chapter F.XII).

The cloud-based platform is the underlying infrastructure that enables coordination along the entire life cycle of cloud logistics services by creating horizontal communication links among stakeholders and by providing a computing platform on which the characteristic mix of mechanisms can be implemented. Specifically, the cloud-based platform facilitates coordination by hosting (a) a *service repository* and (b) an *information repository*, which both achieve coordination through the mechanism *implicit coordination*, and (c) a *virtual marketplace*, which achieves coordination through the *price* mechanism. The platform is depicted as a gray box that underlies the other coordination mechanisms and stretches over the entire life cycle to indicate its infrastructural, enabling character (see Figure 56).

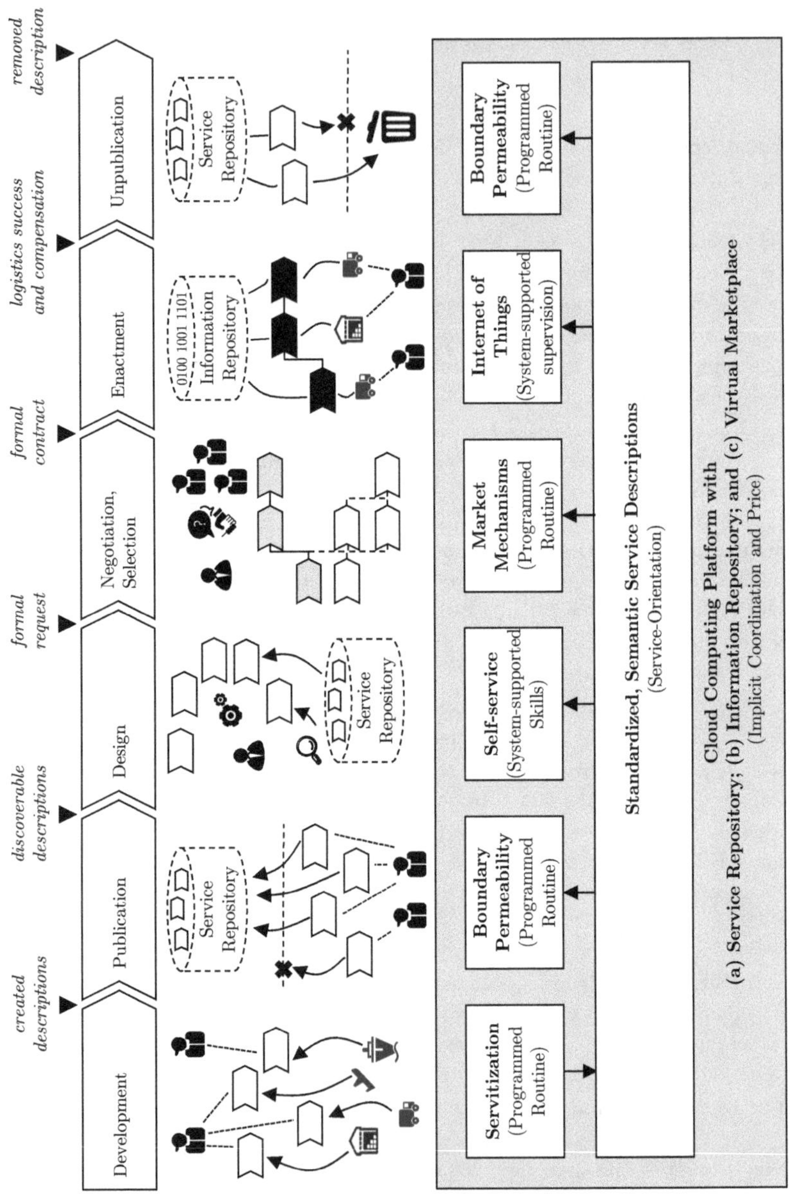

Figure 56.: Cloud Logistics – Characteristic Mix of Coordination Mechanisms

Standardized, semantic service descriptions relate to *service-orientation* that specifies the formal requirements for physical logistics capabilities to be expressed as services. These requirements are based on the design principles of service-orientation. Standardized, semantic service descriptions support coordination along the entire life cycle. Hence, they are depicted as a box stretching over the entire life cycle. The interplay between the different mechanisms is depicted by the arrows that connect them: Standardized, semantic service descriptions are the output of the Servitization, and they represent the input for each mechanism in subsequent phases (see Figure 56).

Servitization refers to a *programmed routine* that specifies the exact steps LSPs need to perform to develop formal descriptions of physical cloud logistics capabilities. The specific steps in the programmed routine are guided by the design principles of *service-orientation.* Servitization is depicted as arrows beginning at physical logistics resources and pointing toward formal service descriptions to emphasize the transformational process in the service development phase (see Figure 56). Different service descriptions are created by different LSPs.

Boundary permeability refers to a *programmed routine* that specifies the exact steps that (a) LSPs need to perform to publish a new service description or update or remove an existing one in the service repository and (b) the LITPAP performs to admit new LSPs to or exclude member LSPs from the CLS. While "regular" SOA systems focus on the technical routine(s) of publishing new descriptions, updating, or removing them, the CLS reference architecture focuses on the organizational aspects of admitting and excluding LSPs from the networked LSP cooperation. In particular, this includes evaluating the candidate and current member LSPs against a set of pre-defined (performance) criteria and rules in order to decide on their admittance or removal from the CLS. During the publication phase, the permeable boundary is depicted as a dashed line that prevents the admittance of certain LSPs by preventing them from announcing their descriptions to the service repository. During the unpublication phase, the permeable boundary is also depicted as a dashed line to illustrate the rule-based exit of member LSPs. Descriptions are then removed from the repository and moved in to the recycle bin.

Self-service refers to *system-supported skills* that let logistics consumers express their logistical concerns in a formal service request using the service descriptions available in the service repository. System-supported skills specifically support logistics consumers in identifying logistics capabilities that match their concerns according to the information accessible in the service description; evaluating these candidate services; comparing alternative candidate services; and, if needed, combining multiple services into more complex solutions via an iterative, dynamic process without the support of provider personnel. Self-service is depicted as magnifying glasses and toothed wheels to emphasize the search and discovery functions for services in the repository and the composition of complex services (see Figure 56).

Market mechanisms are a *programmed routine* that specifies the exact steps for logistics consumers and LSPs to carry out negotiations about formally requested services on the virtual marketplace in order to reach a formal agreement (contract). These contracts specify (a) which LSP is awarded which requested service in the case of composite services or multiple providers of candidate services, (b) the conditions and constraints of service use, and (c) the payments to be made by the logistics consumer to LSPs of awarded services. Market mechanisms are depicted as multiple paths running through a network of candidate services, with only the "selected" path connected via a solid line. The negotiation process and potential signing of a contract between consumers and LSPs is illustrated as a callout with a question mark and a handshake.

The *IoT* relates to *system-supported supervision*, which enables the direction and monitoring of logistics resources deployed to address agreed-upon logistical concerns through IoT technologies, such as RFID. Collected information is stored in the information repository on the cloud-based platform and made available to (a) LSPs involved in delivering services as operationally required during delivery phase and (b) the respective logistics consumers to ensure that the delivered services comply with the services that were agreed to. The IoT is depicted as solid lines that connect deployed resources with the service description and the information repository (see Figure 56).

3.2. Cloud-based Platform with Service Repository, Information Repository, and Virtual Marketplace (Implicit Coordination, Price)

3.2.1. Description

The cloud-based platform represents the foundational computing infrastructure that enables coordination along the entire life cycle of cloud logistics services. The platform generally enables coordination by (1) establishing horizontal communication links among all stakeholders and providing the infrastructure for the characteristic mix of coordination mechanisms to be implemented on. Specifically, the platform facilitates coordination by hosting (2) a service repository, (3) an information repository, and (4) a virtual marketplace. Recall that the cloud-based platform also hosts a customizable logistics management environment (see Subsection F.IV.3.3); however, this environment is not part of the characteristic mix of coordination mechanisms because it does not enable coordination between stakeholders, but only supports logistics consumers and member LSPs in managing their respective logistics activities in CLSs.

The cloud-based platform establishes *horizontal communication links* among all stakeholders, including logistics consumers, LSPs, and the LITPAP. These links enable interactions through messages and, thus interfirm coordination. We have argued above that the cloud-based platform represents the "nexus of value creation" because it interconnects all stakeholders (Subsection F.IV.3.3); in fact, it is because the platform enables coordination among the stakeholders. In addition, the cloud-based platform also represents the infrastructure on which all other computer-dependent mechanisms are deployed (see Laskey et al. 2012:72). This specifically includes the service and information repository and the virtual marketplace.

The *service repository* is a structured database that enables the coordination of demand and supply of logistics capabilities through the mechanism of *implicit coordination* (see Subsection C.V.4.2). The repository contains all relevant information about the physical cloud logistics capabilities currently offered in the CLS. The repository can thus be conceived of as a "logistics service map" that structures logistics capabilities, for example by the type of logistical transformation (Ludwig 2014). Coordination is achieved indirectly, or in other words, through "mediated awareness" between consumers and providers (Laskey et al. 2012:59ff.), where "awareness" refers to a state in which one stakeholder has knowledge about the existence of another stakeholder (MacKenzie et al. 2006:28). On the one hand, logistics consumers can browse the repository to gain transparency into the current supply of capabilities and request those that match their requirements. On the other hand, LSPs can observe which services are being searched for and requested by consumers and can thus adjust their offerings accordingly. Thus, coordination between demand and supply becomes possible.

The *information repository* is a structured database that enables the coordination of physical cloud logistics capabilities during the delivery phase through *implicit coordination*. Accordingly, the repository unfolds its coordinative power by storing and making available logistics-relevant information to logistics consumers and LSPs. In other words, the coordinative power results from synchronizing information among stakeholders. This can help to, among other things, coordinate (a) the handover of physical goods between logistics consumers and LSPs and among LSPs and (b) timely responses to unpredictable events in the natural, social, or economic environment that impact the delivery of physical logistics capabilities. Coordination is achieved because stakeholders can adapt their delivery of capabilities based on the information in the database.

The *virtual marketplace* is the forum for trading physical cloud logistics capabilities via the *price* mechanism. The virtual marketplace enables interactions among stakeholders through messages – or so-called proposal exchanges – regarding the constraints and conditions of requested capabilities, especially price. The virtual marketplace additionally provides functionality for stakeholders to sign formal contracts that document conditions and constraints and agreed-to prices. Thus, demand for and supply of logistics capabilities is coordinated.

3.2.2. Conditions

The cloud-based platform represents the fundamental computing infrastructure that enables coordination along the entire life cycle of cloud logistics services. Fundamental conditions that presumably influence the effectiveness of coordination along this life cycle are related to the following factors: (1) the number of actors to be coordinated and their degree of geographic dispersion, (2) task complexity, and (3) environmental dynamism.

Network size is presumed to be moderate to large (see Subsubsection F.VII.3.2.1). Although not all stakeholders need to communicate with each other at all times, communication among them may be necessary at some point in the cloud logistics service life cycle and therefore must be technically possible.

Geographic dispersion of stakeholders varies based on the logistical prototype scenario (see Subsection F.V.3.3). Geographic dispersion is high in scenarios that have a wide, networked geographic coverage. However, even in localized scenarios, distances between stakeholders and the number of their locations can be high, considering the size of dense urban areas or megacities, for example. In fact, geographic dispersion will always be present in CLSs, as overcoming geographical distances to connect geographically dispersed locations represents a fundamental logistical transformation. Hence, geographic dispersion is presumed to be at least medium, but often high.

Regardless of the specific task to be coordinated in each phase of the cloud logistics service life cycle, *task complexity* is presumably medium to high due to the significant information processing requirements resulting from the mechanisms, standardized, semantic service descriptions, and system-supported supervision. Standardized, semantic service descriptions increase information processing requirements (information load) due to many diverse attributes used to specify physical logistical capabilities. System-supported supervision raises information processing requirements (information load) due to the diversity and rate of information change resulting from real-time monitoring and control of logistics capabilities during delivery via the IoT.

Environmental dynamism is presumably high. Hence, logistics-relevant information becomes obsolete rapidly and changes frequently and unpredictably. This further raises the information processing requirements along the cloud logistics service life cycle.

Considering the presumed conditions described above, we conclude that effective coordination can be achieved through a common IT infrastructure that establishes communication links among all stakeholders and that hosts structured databases for service descriptions and logistics-relevant information. Thus, large amounts of information can be exchanged instantaneously among a large number of potentially geographically dispersed stakeholders. In fact, a common IT infrastructure is a foundational enabler for achieving on-demand and rapidly elastic provisioning in a dynamic environment by allowing for quick communication and thus coordination. The effectiveness of price coordination via the virtual marketplace is assessed later, together with market mechanisms (see Subsection F.IX.3.7).

3.3. Standardized, Semantic Service Descriptions (Service-orientation)

3.3.1. Description

Standardized, semantic service descriptions are the proposed foundational coordination mechanism in the life cycle of cloud logistics services. They are the output of servitization and the input to mechanisms in subsequent life cycle phases. During the publication phase, descriptions are published to the service repository. During the design phase, logistics consumers use the descriptions as a basic unit for creating individual logistics solutions through composition. Service descriptions also provide the starting point for negotiation and selection. In fact, the content of service descriptions, especially the conditions and constraints of use, is the primary object of negotiation. During service delivery, descriptions specify what kind of physical cloud logistics capabilities need to be delivered by LSPs. Finally, service descriptions are removed from the service repository.

Standardized, semantic service descriptions are service descriptions that are developed according to the design principles of *service-orientation* (see Subsection D.III.3.2). They encode the capabilities, conditions, and constraints of the respective physical cloud logistics service by using well-defined semantics, which means that they are understandable by humans and machine-processable (see Subsection D.III.4.2). Descriptions are standardized with a "logistics-domain specific ontology-based description language" (see "semantic markup" in Subsection D.III.4.3). "Logistics-domain specific" means that the language's expressiveness and inference mechanisms are tailored to capture the particularities of the logistics domain. This requirement is reasonable because of the limited applicability of current web service standards to adequately describe logistics services (see Subsection D.III.4.4). "Ontology-based" means that semantic descriptions are created through semantic markup: The language provides a generic "top-level" ontology with which services are described. Service descriptions use these concepts and relationships to encode capabilities, conditions, and constraints of use. Following the SOA-RAF model of service descriptions (se Subsubsection D.III.4.3.3), mandatory concepts of cloud logistics service descriptions are: service functionality, service policies, service metrics, service interface description, and service reachability. Chapter F.XI provides a more detailed design of cloud logistics service descriptions, especially with regard to the standardized top-level ontology.

3.3.2. Conditions

Conditions that are likely to influence the effectiveness of standardized, semantic service descriptions as an underlying coordination mechanism in the entire life cycle are related to the following contingency factors: (1) task complexity, (2) task modularity, and (3) environmental dynamism.

Task complexity varies significantly for different physical logistics capabilities; for example, road transportation of general cargo is less complex than pick-and-pack activities in e-commerce fulfillment centers. However, considering the nature of the physical cloud logistics capabilities offered by the different logistical prototype scenarios (see Subsection F.V.3.3), task complexity is more likely to be low rather than high. After all, cloud logistics scenarios do not include any "complex services" such as logistics management capabilities or the engineering of unique, integrated solutions (see Andersson & Norrman 2002:4). A low degree of task complexity – if considered in combination with the high number of actors to be coordinated – generally makes coordination via *standardized work processes* effective. In fact, standard operating procedures are pervasive in the logistics industry. However, coordinating the operative delivery of physical logistics capabilities by means of standardized work processes is presumably ineffective in CLSs. Different member LSPs are likely to use different standard procedures. Overriding these procedures with a standard procedure for cloud logistics capabilities seems infeasible; after all, LSPs may participate in a CLS for a certain period of time only. In fact, this may result in service "over specification," which can reduce the LSPs' ability to optimize their own processes and thus lead to higher costs or lower quality (see Andersson & Norrman 2002:9). Hence, CLSs are likely to default to achieving effective coordination through *standardization of output* and more specifically through *service-orientation*. This gives each LSP the freedom to deliver capabilities according to their unique procedures, while only specifying what logistical transformation needs to be achieved by when. Giving LSPs sufficient leeway for delivering physical cloud logistics capabilities is also critical for enabling resource pooling across various customers, an essential characteristic of CLSs. LSPs can share resources in various ways across consumers, as long as the agreed-upon conditions and constraints of service use are met.

However, using standardized, semantic service descriptions requires outcomes of logistics capabilities and their conditions and constraints of use to be specified to a sufficient level of detail *ex ante*. In this respect, Andersson & Norrman (see 2002:8f.) argue that task complexity influences the ability to specify services: Complex services are more difficult to specify than basic services due to their higher degree of "intangibility" (see similar Lindberg & Nordin 2008:293f., 297). However, unlike other services (e.g. the training of personnel), physical cloud logistics capabilities presumably possess a very low degree of intangibility. This is because logistics capabilities such as freight transport and warehousing are "physical acts to owned objects" and hence include tangible (i.e. specifiable) processes with tangible (either ephemeral or permanent) outcomes (Lovelock & Gummesson 2004:27, 31). The performance of such services can be determined by means of an immediate before-after comparison (Fitzsimmons et al. 1998:374). Hence, the outcomes of physical cloud logistics capabilities delivered by CLSs can presumably be described to a sufficient degree *ex ante*.

The degree of *task modularity* has been identified as a critical factor that influences the effectiveness of service-orientation. Hence, to determine whether or not service-orientation can be an effective coordination mechanism, we need to determine the degree of modularity of the physical cloud logistics capabilities offered by CLSs. The concept of modularity is generally applicable to the logistics domain. For example, several scholars choose logistics as an "example domain" for researching process or service modularity in a more general context (Pekkarinen & Ulkuniemi 2008:90; Bask et al. 2010:368; Bask et al. 2011:381). Moreover, its applicability is also corroborated by empirical evidence from the logistics industry (Rajahonka 2013:45f.; see similar Lin & Pekkarinen 2011:353; Pekkarinen & Ulkuniemi 2008:98). For example, Rajahonka (2013:41, 46) find in an empirical study of 25 Finnish LSPs that

> "[a]lmost all of the interviewees expressed the feeling that modularity is a suitable approach for the logistics industry [...] and the use of modularity will increase in the future."

According to the interviewees, logistics modularity manifests in the fact that services can be sold separately or combined into unique service solutions (Rajahonka 2013:42f.; see similar Pekkarinen & Ulkuniemi 2008; Corsten & Gössinger 2007). For example, LSPs may offer highly standardized core logistics services that they supplement with customized value-adding service modules to fulfill heterogeneous demand (Pekkarinen & Ulkuniemi 2008:92). Logistics processes are customized in the beginning and at the end (e.g. pick-up and delivery), but are standardized in the middle (e.g. main haul) (Rajahonka 2013:48).

As argued above, the degree of (task) modularity is a function of four factors: heterogeneity inputs and demands, synergistic specificity, and urgency (see Subsubsection F.IX.2.3.6). In CLSs, task modularity is presumably high due to the low degree of synergistic specificity, high urgency, and high heterogeneity of demands. Low synergistic specificity results from the unique design approach of logistics resources and capabilities in cloud logistics, which aims to reduce asset specificity to a sufficiently low level (see Section F.V.2). Thus, resources and capabilities are deliberately designed to be shared across multiple consumers and (re-)deployed and reconfigured quickly for rapid responses to newly emerging or changing consumer demand. Low synergistic specificity creates pressures toward modularity. The heterogeneity of inputs is low because the variety of resources and capabilities is limited in their type and geographic coverage, considering the logistical prototype scenarios. This creates pressure for moving away from modularity. Heterogeneity of demands is high, as every service request is unique with regard to the combination of time and space, creating pressures toward modularity. High urgency results from the dynamic external environment, creating pressures toward modularity. In sum, three of four factors create pressures toward modularity. These pressures likely outweigh the pressure exerted by the low heterogeneity of inputs, especially due to the low synergistic specificity. We therefore conclude that – in aggregate – these factors create pressures toward a high degree of modularity in CLSs.

The *environment* is presumably dynamic (see Subsection F.IV.3.6). Such environments require quick responses. Semantic service descriptions enable quick responses by enhancing the speed of coordination along the life cycle through automation. Coordination can be automated because well-defined semantics are understandable by humans and machine-processable. Well-defined semantics' ability to increase the speed of service discovery, design, and negotiation through automation is well recognized in the IT domain (see Subsection D.III.4.3). In fact, it is also recognized by logisticians. Hoxha et al. (2010:73) argue, for example, that

> "the semantic representation of logistic services [...] enables automated and intelligent techniques for discovery, ranking, execution and efficient composition of services into more complex and flexible logistics processes."

The use of semantics substantially "reduces time" for service discovery and selection logistics services (Preist et al. 2005:999f.) and enables "to accelerate matching demand with supply of logistics services" and "to rapidly and flexibly integrate logistics services" (Scheuermann & Hoxha 2012:108).

In addition to well-defined semantics, the service descriptions of physical cloud logistics capabilities use an ontology-based description language that defines a standardized set of ontological concepts and relationships used to describe the capabilities. This standardization further enhances coordination speed, as it minimizes the risk of inconsistencies in ontologies, which are more likely to occur if semantic service descriptions are created through semantic annotation rather than semantic markup. Moreover, standardized, semantic service descriptions enable flexible responses to quickly changing conditions through different service compositions. Well-defined semantics allow modular logistics services to flexibly interact with other services and thus be dynamically (re-)combined into more complex solutions (Hoxha et al. 2010:73). In addition, semantic service descriptions let LSPs unambiguously express their logistical capabilities (Scheuermann & Hoxha 2012:108, 106). This enables *self-service* provisioning: Logistics consumers can understand available capabilities without further assistance and they can unilaterally judge whether services are suited to their logistical needs. As a result, logistics consumers can quickly respond to changing conditions in a dynamic environment due to increased interaction flexibility. We thus conclude that using well-defined semantics to describe cloud logistics capabilities contributes to enabling adequate responses in a dynamic environment through on-demand, rapidly elastic provisioning and self-service.

Considering the presumable conditions described in the previous paragraphs, we conclude that standardized, semantic service descriptions represent an underlying mechanism that can enable effective coordination in the life cycle of cloud logistics services. Despite presumed rather low degree of task complexity associated with physical cloud logistics capabilities, standardized, semantic service descriptions are effective because they give LSPs sufficient freedom to pool resources across consumers by only focusing on describing the outcomes rather than the process itself. In addition, the physical logistics capabilities offered by CLSs are sufficiently modular, thus enabling the effective use of standardized, semantic service descriptions. Moreover, standardized, semantic service descriptions contribute to on-demand and rapidly elastic provisioning of physical logistics capabilities by enhancing the speed of coordination along the life cycle. Finally, they also enable self-service from an organizational perspective by making offered physical logistics capabilities unambiguously comprehensible to logistics consumers.

3.4. Servitization (Programmed Routine)

3.4.1. Description

During the development phase, LSPs encode their physical cloud logistics capabilities, and conditions and constraints of their use, into standardized, semantic service descriptions.

Servitization is the proposed coordination mechanism during the development phase. It is a *programmed routine* that specifies the steps for creating formal service descriptions as per the design principles of *service-orientation*. Consolidating existing literature suggests that this routine consists of four steps: (1) identification, (2) resource and capability modeling, (3) encapsulation, and (4) description (see the literature quoted in Chapter F.X): In *identification* each physical resource is tagged with an identifier that can be unambiguously resolved within the system. CLSs use IoT technologies to link physical logistics resources with the cloud-based platform. *Resource and capability modeling* involves expressing the functional and non-functional features of heterogeneous physical resources and capabilities in an isomorphic manner, usually by means of a resource expression model. *Encapsulation* is selecting (logical) resources and capabilities to be included in the service to be created. *Description* involves encoding the new service's functionality, conditions, and constraints of use with a formal service description language. As argued above, CLSs use an ontology-based description language (semantic markup). Each of these phases is based on one or more service-orientation design principles. Chapter F.X provides a more detailed description of these phases and shows how the design principles of service-orientation guide this programmed routine.

Before assessing the conditions that presumably determine the effectiveness of coordination during service development, a terminological remark is helpful to explain why we refer to this programmed routine as "servitization" rather than "resource virtualization," as is often found in existing literature (see e.g. Li et al. 2013; Liu et al. 2014; Liu & Li 2012; Liu et al. 2011b). We believe that the term "resource virtualization" is too narrow because it does not intuitively capture the fact that this routine eventually creates *service descriptions.* In fact, virtualization only "refers to abstraction of logical resources from their underlying physical characteristics" (He & Xu 2014:5f.), and these virtualized resources are then packaged into cloud services (see Liu et al. 2014) and ultimately described as a service. In other words, resource virtualization only refers to a subset of activities necessary to create cloud services, namely step two identified above: resource and capability modeling. Hence, in order to use a single term to describe the entirety of activities necessary for creating service-oriented capabilities, we use the term "servitization" (as proposed by Glöckner et al. 2014:187; Ludwig 2014:51f.). This term specifies the process' final output – service description – more explicitly. Of course, servitization involves creating virtualized resources, but does not stop at that point.

3.4.2. Conditions

Conditions that are likely to influence the effectiveness of servitization relate to the following contingency factors: (1) task complexity, (2) environmental dynamism, and (3) task interdependence.

Task complexity is presumably medium. On one hand, the exact process steps necessary for creating semantic service descriptions can be clearly specified *ex ante*, which means that a clear relationship between paths and outcomes can be established, creating pressures toward a low degree of complexity. On the other hand, the varying design principles included in service-orientation have conflicting goals, for example loose coupling and composability (see Subsection D.III.3.2). These conflicts will increase complexity. Furthermore, the degree of complexity likely varies based on the types of encoded logistics resources and capabilities because of differences in their functional and non-functional features. The variety and nature of these features determine the information processing requirements for servitization, and reduce or increase complexity accordingly. Hence, at the level of abstraction for this reference architecture, no general statement for "all" logistics resources and capabilities can be made. However, considering the logistical prototype scenarios (see Subsection F.V.3.3), one can argue that the resources and capabilities necessary to manipulate the flow of physical goods are likely to be more simple than complex, thus following the same argument as for standardized, semantic service descriptions. This lowers the degree of task complexity. Still, developing reasonable modules that can be reused and combined into complex solutions is a rather complex task. As Baldwin & Clark (1997:86) argue in a general context,

> "modular systems are much more difficult to design than comparable interconnected systems. The designers of modular systems must know a great deal about the inner workings of the overall product or process in order to develop the visible design rules necessary to make the modules function as a whole."

We thus conclude that – in aggregate – these factors lead to a medium degree of task complexity when creating service descriptions.

The process of creating service descriptions is presumably exposed to a low degree of *environmental dynamism*, independent of the high degree of dynamism in logistics consumers' and LSPs' business environment. This is because changes in the business environment are very unlikely to change the fundamental design principles used to develop service descriptions. Nevertheless, dynamic changes in the business environment are likely to change the type of services created.

Task interdependence between the service development processes of different member LSPs is high. This is because the service descriptions created by different LSPs become part of the same shared pool and must be compatible with each other to ensure composability. This interdependence is well described by Baldwin & Clark (1997:86), who argue,

> "[W]hile [the] designs at the modular level are proceeding independently, it may seem that all is going well; problems with incomplete or imperfect modularization tend to appear only when the modules come together and work poorly as an integrated whole."

Hence, development processes across member LSPs are subject to interdependencies related to specialization (Mintzberg 1979:122). Following Nassimbeni (1998:547) such interdependencies can be effectively coordinated by standardizing work processes: Common specifications for processes are adopted to ensure uniformity of outputs. In line with this Erl (2008:131), a computer scientist, argues:

> "Formal processes need to be introduced to ensure that services are modeled and designed consistently, incorporating accepted design principles, conventions, and standards."

The preceding paragraphs have assessed the presumable conditions that influence the process of creating standardized, semantic service descriptions. These conditions presumably make coordination via a programmed routine effective. The medium degree of task complexity can be effectively handled by a programmed routine, especially because of the clear relationship between means and ends can be established. In addition, the stable task environment ensures that the routine remains effective over a longer period of time. However, considering the degree of task complexity and environmental stability, one can argue that the creation of standardized, semantic service descriptions can be coordinated equally effectively by means of standardization of output (see Table 27). After all, whether service descriptions comply with formal requirements of the service description language can be easily validated through an algorithm. However, standardization of output cannot ensure that the bundle of resources and capabilities encapsulated by each service is "practical." Service modules may contain "large chunks" of resources and capabilities that cannot be easily recombined with other modules into reasonable logistics solutions, thus violating the principle of service reusability. This necessitates pre-defining the routine within which the design principles of service-orientation are applied in order to ensure its correct application. Specifically, consistent application can be ensured by implementing this routine on the cloud-based platform so that it is similar for all member LSPs. In conclusion, a programmed routine is presumed to be an effective instrument to coordinate the creation of standardized, semantic service descriptions.

3.5. Boundary Permeability (Programmed Routine)

3.5.1. Description

In the publication phase, LSPs become admitted to the CLS and can then publish their service descriptions to the service repository. In the unpublication phase, LSPs become excluded from the CLS, and their service descriptions are removed from the repository. As already mentioned, we focus on the mechanism that determines how LSPs become admitted to or excluded from the CLS rather than on the technical mechanism of how service descriptions are (un-)published to the repository.

Boundary permeability is the proposed mechanism for the publication and unpublication phases and thus refers to two phases in the life cycle. We consolidate both of these phases into a single mechanism, as LSP admittance and removal are closely linked conceptually. Boundary permeability refers to a *programmed routine* that specifies the process of assessing candidate member LSPs against a set of membership criteria. Boundary permeability can be considered a funnel process: Candidate LSPs need to clear different stages of assessment. The assessment eventuates in candidate member LSPs being either offered a membership contract or not. This programmed routine can be considered a sub-process of a wider set of activities that are carried out to identify, select, and retain member LSPs, such as sensing potential partners in the marketplace (Albers et al. 2013).

The degree of boundary permeability – that is, the degree of openness – is determined by the boundary arrangements between the LITPAP and candidate member LSPs. Following Gulati et al. (2012:576), such organizational arrangements can be conceptualized along three dimensions:

> "(1) who chooses members; (2) criteria for membership (i.e., the attributes members possess and the degree of redundancy between them); and (3) duration and exclusivity of membership (i.e., whether members can belong to more than one meta-organization)."

We propose adopting these three dimensions to conceptualize the CLS openness. The following paragraphs consider them in detail.

First, the *authority over choosing members* resides with the LITPAP, as it has authority over strategic decisions that pertain to the network as a whole (see Section F.VII.3). In fact, member identification, selection, and retention are network-level competencies. The LITPAP will decide on membership offers through a programmed routine implemented on the cloud-based platform and transparent to member LSPs rather than through informal communication processes. The LITPAP's authority focuses on specifying the steps of this routine, the assessment criteria, and their minimum thresholds for these criteria and carrying out the routine.

Second, cloud logistics knowledge provides heterogeneous suggestions for the degree of boundary permeability, ranging from highly open, unconditional access to rule-based admission (see Subsubsection E.IV.5.4.2). Yet, cloud logistics knowledge does not provide any insight into the specific *membership criteria* for controlling the degree of permeability. Nevertheless, the identification of criteria for partner selection has long been a focus of management scholars (see e.g. Hitt et al. 2000; Das & Teng 2003). With regard to the selection of member LSPs for CLSs, one criterion that frequently recurs in management literature is of particular importance and is considered below: the similarity of resources in terms of type and geography (see e.g. Das & Teng 2003:287). Resource similarity is important because it directly relates to the degree to which a CLS can frame stakeholder concerns regarding resource pooling, on-demand availability, and rapid elasticity. In fact, LSP resources need to be similar in type and geography in order to create a pool sufficiently large to achieve these cloud characteristics.

Hence, we propose that the LITPAP controls the degree of permeability as a function of (a) the amount of current resources of a certain type in a certain area relative to the average demand for this type of resource in this area (*resource redundancy*) and (b) the amount of resources of a certain type in a certain area that a candidate member LSP intends to add relative to the current degree of resource redundancy. Choosing the degree of resource redundancy is a critical choice for the LITPAP because it determines both the extent to which cloud characteristics can be achieved and the degree of internal competition among member LSPs. This, in turn, determines the extent to which member LSPs can address their economic concerns. A more dynamic environment presumably leads to a higher degree of redundancy in order to enable responses to unpredictable demand surges. At all times, redundancy should be larger than 1, which means that resource supply exceeds demand, in order to rapidly absorb demand surges and offer on-demand availability. If redundancy is insufficient for achieving a desired level of on-demand availability and rapid elasticity, new LSPs are admitted until the desired level is achieved. Conversely, if resource redundancy is sufficient, no new LSPs are admitted. If redundancy significantly exceeds the degree necessary to achieve a desired level of on-demand availability and rapid elasticity (e.g. demand has declined persistently), resources may consistently run idle. Thus, LSPs may decide to withdraw resources or leave the CLS because they see better opportunities elsewhere. In an extreme situation, the LITPAP may suspend or terminate the membership with selected member LSPs to balance demand and supply, for example by taking into account membership tenure and past member performance. Hence, the degree of permeability is not static in CLSs but dynamic, depending on internal redundancy, how candidate LSPs influence redundancy, and external logistics demand.

Third, regarding the *exclusivity of membership*, the LITPAP presumably forbids LSPs from joining other alliances that directly compete with the CLS by offering a similar set of logistics capabilities in a similar geography. By doing so, the LITPAP aims to secure access to strategic resources critical for delivering logistics capabilities with cloud characteristics. Securing exclusive access to resources is especially critical for the LITPAP because of the dynamic and competitive logistics market environment. As Gulati et al. (2012:577) remark in this respect:

> "Especially in highly competitive market environments, exclusive contributions from members may be vital to strategically differentiate [...] products and services."

However, the LITPAP presumably does not constrain member LSPs' ability to join alliances that offer either different capabilities in the same geographies or similar or different capabilities in different geographies. On the one hand, the LITPAP presumably lacks sufficient informal authority to encode restraints for member LSPs into the membership contract that go beyond the scope of the CLS. On the other hand, whether or not member LSPs ally with other LSPs on a combination of capabilities and geographies that does not overlap with the CLS presumably poses little strategic risk to the CLS. In sum, membership is exclusive with regard to the scope of capabilities in the geography covered by a given CLS, but not beyond.

Finally, regarding the *duration of membership*, the LITPAP presumably requires LSPs to agree to a minimum period of engagement. Establishing exit barriers builds trust among partnering firms and fosters commitment to the alliance (see Gulati et al. 2008:150), ultimately contributing to alliance success (see e.g. Gulati et al. 2008:153; Schmoltzi & Wallenburg 2011:63). Increasing commitment in this way is critical, as the LITPAP lacks other instruments to secure commitment, such as participatory decision-making (Schmoltzi & Wallenburg 2011:63). If a member LSP breaches the membership contract, for example due to significant operational underperformance or engaging in forbidden partnerships, the LITPAP may terminate its membership, regardless of the minimum engagement time. There is presumably no general upper limit on membership duration.

To summarize, the LITPAP has authority over the degree of boundary permeability. This involves specifying criteria for assessing candidate member LSPs and setting minimum thresholds for these criteria. Candidate member LSPs are evaluated based on the criteria through a programmed routine implemented on the cloud-based platform. The degree of permeability depends on three conditions: (a) current degree of resource redundancy in the CLS, (b) how a candidate member LSP influences this degree, and (c) logistics market demand. Membership exclusivity is limited to the capability and geographic scope covered by the CLS. Membership duration is generally unlimited, but includes a minimum period of engagement to ensure member LSP commitment.

3.5.2. Conditions

Conditions that presumably influence the effectiveness of boundary permeability are related to the following factors: (1) environmental dynamism, (2) task and environmental complexity, and (3) the degree of centralization interacting with behavioral uncertainty of the LITPAP.

The *task environment* of the member selection process is presumably stable. The assessment process and criteria presumably remain unchanged even though LSPs and the LITPAP are embedded in a dynamic business environment. This is because environmental dynamism does not affect the objective that guides the assessment process, which is: admitting or excluding member LSPs to deliver logistics capabilities with cloud characteristics.

Complexity in the candidate assessment process is presumably medium. Complexity arises from both the task itself and the environment. *Environmental complexity* is driven by the variety of factors that need to be considered to estimate logistics demand and thus determine the amount of resources necessary to maintain or achieve a desired level of redundancy and thus on-demand availability and rapid elasticity. *Task complexity* is driven by the variety of factors that need to be considered to assess the "fit" of candidate member LSPs. This especially includes assessing the amount and type of resources candidate member LSPs can add to a CLS. Given the plethora of analytical models in management literature for selecting "optimal" partners, one can argue that task programmability is high. The ability to establish a clear relationship between paths (the assessment process) and outcomes (the offer of a membership contract) lowers task complexity. We thus conclude that, in aggregate, complexity associated with the assessment of candidate members is medium. Moreover, the analytical complexity of the assessment process and the variety of factors, presumably exceed human information processing capacity, thus requiring the memory and processing power of computers.

In sum, the stable task environment combined with the medium degree of complexity presumably makes a programmed routine an effective coordination mechanism.

However, in addition to the factors discussed above, boundary permeability is influenced by the interaction between the *degree of centralization* and *behavioral uncertainty*. These factors create pressures to formalize the assessment process on a cloud-based platform and to make the process and assessment status of candidate member LSPs transparent to current member LSPs. As argued above, the LITPAP holds exclusive authority over the member selection process. However, this process crucially encroaches on member LSPs's ability to address their economic concerns in the CLS; admitting additional LSPs increases resource redundancy and thus competition, which, in turn, reduces the potential to make a profit – if logistics demand remains constant. Barring member LSPs from participating in the member selection process can reduce their commitment to the CLS and thus adversely impact overall cloud performance. The LITPAP must therefore continuously legitimize its member selection actions, as member LSPs are autonomous organizations and free to leave the CLS (at least after the minimum engagement period).

Moreover, given the importance of the member selection process, (some) member LSPs may fear that the LITPAP's assessment may not comply with standards encoded in the membership contract. For example, some member LSPs may fear that other member LSPs try to influence the LITPAP during the assessment in order to improve their competitive positions in the CLS. On the other hand, member LSPs may not trust the LITPAP's ability to carry out the process according to the agreed-upon standards, thus adversely impacting overall cloud competitiveness, which, in turn, can reduce their ability to make a profit by participating in the cloud. To ensure legitimacy and counter the risk of behavioral uncertainty, effective coordination can presumably be achieved by formalizing the assessment process in a programmed routine that is transparent to member LSPs, as formalization offers the functions to foster legitimacy and commitment (Vlaar et al. 2007:443f.). They can thus monitor the LITPAP's behavior, making the LITPAP's choices transparent and predictable and thus acceptable.

3.6. Self-service (System-supported Skills)

3.6.1. Description

During the design phase, logistics consumers express their logistical concerns by using the descriptions of physical cloud logistics capabilities available in the service repository as a basic unit of design. Depending on the complexity of their concerns, logistics consumers either select a single service or combine multiple services in a complex solution. The design phase ends once logistics consumers have forwarded the desired (composite) service as a formal request to the LSPs that offer the selected candidate services.

Self-service is the proposed coordination mechanism for the design phase. It refers to *system-supported skills* and is realized through a software program that offers the functionality to enable and support logistics consumers in specifying their logistical concerns in a self-service manner. Service design can be done by logistics consumers unilaterally because the software application on the cloud-based platform has knowledge, functions, and rules embedded in it to support and structure design activities. This specifically includes mechanisms to support the discovery, comparison, and composition of candidate services. A semantic search engine enables logistics consumers to either automate discovery or efficiently navigate through the available logistics capabilities in a structured manner (Dong et al. 2008:820). Search can be furthermore enhanced by using "template-based service composition," which means that domain knowledge is incorporated into the search and design process (Leukel & Kirn 2011:296f.). Consumers then need to select a service for each "slot" of the template. Automated algorithms aggregate the conditions and constraints of use for all the candidate services that are part of a service composition, creating an end-to-end performance perspective. The use of well-defined semantics can thus contribute to improving the quality of logistical decisions (Hoxha et al. 2010:73). The software includes functionality to intuitively juxtapose individual candidate services or entire compositions with specific requirements, for example by means of graphs or color-coded tables. This lets consumers quickly determine if (or the extent to which) currently identified candidate services can satisfy their requirements. If current candidates are not deemed adequate, consumers may dynamically switch back and forth between the discovery, composition, and comparison functionalities, expressing their concerns unilaterally, in an iterative manner.

3.6.2. Conditions

Conditions presumed to influence the effectiveness of system-supported skills to coordinate service design are related to the contingency factors of: (1) task complexity, and (2) environmental complexity and dynamism.

Task complexity during service design is presumably high and represents the critical contingency factor for service design. Complexity arises from multiple conditions. First, complexity is driven by the potentially large number of attributes used to specify each service's functionality and conditions and constraints of and the potentially large number of services with similar functionalities but different conditions and constraints of use. Consumers need to understand the attributes of each service in order to judge whether services can adequately address their concerns. Second, logistics consumers usually aim to achieve multiple objectives simultaneously, some of them most likely conflicting. This further increases complexity. Complexity also increases due to unclear relationships between paths and outcomes: There may be various compositions of physical logistics services equally suited to addressing a consumer's concerns.

In addition to task complexity, *environmental complexity* is presumably high as well. This is because logistics consumers may take into account a variety of factors in their business environment when expressing their logistics concerns, such as the anticipated demand for their products in different customer segments and geographies. Conversely, *environmental dynamism* in service design presumably low despite the dynamic overall business environment of LSPs and logistics consumers. This is because unpredictable changes in the business, natural, or social environment do not change the processes for discovering and comparing physical logistics capabilities.

Considering the conditions in service design, we conclude that system-supported skills are presumably effective mechanisms due to the high complexity embedded in a stable task environment.

3.7. Market Mechanisms (Programmed Routine)

3.7.1. Description

After a logistics consumer has specified his or her requirements through a formal service request, the request is published on the virtual marketplace and forwarded to the providers of candidate services to initiate the negotiation and selection process. During the negotiation and selection phase, consumers interact with LSPs to determine which LSP will deliver which component of a composite service, the conditions and constraints of service use, and the payments the LSPs will receive from the consumer. The phase ends once logistics consumers and LSPs have signed a service contract that reconciles their preferences. If no agreement can be reached, negotiations are aborted.

The cloud logistics knowledge provides a heterogeneous conception regarding the allocation of resources, ranging from centralized to decentralized mechanism (see Subsubsection E.IV.5.4.4). We argue that CLSs make use of decentralized mechanisms. Thus, *market mechanisms* are the proposed coordination mechanism for the negotiation and selection phase. A market mechanism refers to a *programmed routine* that specifies the interaction process for reaching economic decisions in the virtual marketplace. The programmed routine enables logistics consumers and LSPs to negotiate any of the attributes used to specify the conditions and constraints of logistical capabilities.

In general, market mechanisms can be designed using three fundamental components: a bidding language, an allocation function, and a payment function (see e.g. Vries & Vohra 2003; Schnizler et al. 2008). Yet, for designing a market mechanism for the reference architecture for CLSs, we use none of the above. Instead, we design market mechanisms from a more abstract perspective, which exclusively focuses on the mechanism's economic properties of allocative efficiency, budget balance, inventive compatibility, and individual rationality (see Subsection C.III.5.4). This gives us the possibility to design market mechanisms independently of a specific solution or implementation, as required by a reference architecture. More specifically, we put forth the proposition that exchanges of physical cloud logistics capabilities can be effectively coordinated through a multi-attribute combinatorial auction that favors individual rationality and weak budget balance over incentive compatibility and allocative efficiency. The design of this market mechanism is covered in detail in Section F.XII.3.

3.7.2. Conditions

The structural governance viewpoint/view has suggested that exchanges between logistics consumers and member LSPs are mediated through markets (see Chapter F.VI). That argument has been based on TCE. In the following, we discuss further conditions that influence the effectiveness of market-based exchanges in the service negotiation and selection phase. These conditions complement TCE. The argument is structured in two parts. The first argues that exchanges can be effectively coordinated through a programmed routine and the second that this routine is a market mechanism based on non-cooperative game theory. Conditions that presumably influence coordination effectiveness are related to the following contingency factors: (1) size, (2) task complexity, (3) environmental dynamism and hostility, (4) task interdependence, (5) behavioral uncertainty, and (6) the interaction of selected factors with the degree of centralization.

The *number* of involved actors in the negotiation and selection phase is presumably high. For each service request, a single logistics consumer can negotiate with multiple LSPs. This is because of the inherent resource redundancy. Hence, for each service request, multiple LSPs should be able to offer requested capabilities. The number of LSPs further increases if consumers request composite services. If so, for each component of the composite service, multiple LSPs should be able to offer the requested capabilities.

Task complexity is presumably high during the negotiation and selection phase. Complexity arises from three conditions. First, it results from high information load due to various attributes included in service descriptions. Logistics consumers and LSPs need to process and may negotiate the value of each attributes. Second, consumers likely aim to achieve several of these attributes simultaneously, thus increasing complexity. However, some of them are likely to conflict, such as price and transit time, further driving complexity. Finally, complexity is increased because of rich preference structures of consumers and LSPs in the sense of "I only want A, if I also get B."

The degree of *environmental dynamism* is presumed to be low in terms of the nature of the task; regardless of changes in the quality or quantity of demand, requested services need to be allocated to member LSPs and payments determined. However, due to the dynamic business environment of logistics consumers and LSPs, their preferences for certain capabilities and conditions and constraints of use are likely to change frequently. For example, depending on the nature of the logistics demand logistics consumers face, they may change their preferences for functionality and costs.

The degree of *environmental hostility* is presumably high among member LSPs. On one hand, low margins are common in many segments of the logistics industry. On the other hand, competition is particularly intense among member LSPs due to the low asset specificity of physical cloud logistics capabilities combined with resource redundancy, which is necessary for achieving a desired level of on-demand availability and rapid elasticity. Competition among member LSPs is further increased due to the transparency of the logistics capabilities and associated conditions and constraints, which are published to the service repository.

Task interdependence presumably high. The incoming stream of service requests can be considered a pool of resources that is collectively exploited by the network of horizontally cooperating member LSPs and the LITPAP. The allocation of service requests, associated revenue, and costs to member LSPs represents pooled interdependence due to the finite number of service requests for a given period of time.

Behavioral uncertainty is presumably high. Although the LITPAP formally assures in the cooperation contract that it behaves neutrally, member LSPs may still fear behavioral uncertainty associated with the allocation of service requests to LSPs and the determination of payments to be made. Specifically, member LSPs may fear that other member LSPs may bias the allocation to their own advantage by lobbying or even bribing the LIT-PAP. These fears are particularly strong due to the direct impact of the allocation on the revenue and costs of member LSPs in a highly hostile environment.

The preceding paragraphs have assessed the conditions that presumably prevail in the negotiation and selection phase. We argue that the large number of involved stakeholders in a stable task environment makes a programmed routine effective despite a high degree of task complexity. This is because the complexity sources do not affect the programmability of the process. In fact, the processing power and storage capacity of IT systems can support handling the large number of attributes included in service descriptions. Complexities resulting from conflicting consumer goals and complex preference structures do not reduce process programmability, but only complicate the ability of stakeholders to make decisions that are "optimal" according to their preferences. Programmed routines are furthermore considered effective in coordinating situations of pooled interdependence. Finally, the high risk of opportunistic behavior and the direct impact on economic concerns creates strong pressures toward formalization, as it can increase behavioral predictability and legitimize interactions among horizontally cooperating LSPs.

After having argued that a programmed routine enables effective coordination, we now move on to the second part of the argument and discuss why this routine is a market mechanism. The pivotal conditions for making market mechanisms more effective than other types of allocation mechanisms are related to the interactions between environmental dynamism, environmental hostility, behavioral uncertainty, and the high degree of centralization. Behavioral uncertainty in this context specifically refers to member LSPs' incentives to lie about their true "type" in terms of capabilities and preferences. In other words, LSPs behave strategically to influence the outcome of the negotiation and selection process to their advantage. The propensity to reveal private information untruthfully likely increases with the degree of environmental hostility as they need to use every possible means to earn a profit.

Aside from the question of which LSP delivers which component of a composite service, cooperating LSPs inevitably encounter the question of how to divide the revenue collected from a logistics consumer. A common and generally perceived as fair approach is to allocate gains according to some form of "weight" proportional to each firm's contribution (Grandori 1997a:907f.). Such approaches are also very common in horizontal logistics alliances, such as a rule that uses the number of drop points to determine the contributions made (see Cruijssen et al. 2007c:31f.). Although such basic rules can be intriguing to LSPs at the alliance's outset due to their simplicity, they may eventually become a source of frustration and may lead to alliance disintegration if partner's contributions are not valued fairly (Cruijssen et al. 2007c:31f.). Considering the variety of attributes included in the description of cloud logistics services, specifying a rule that fairly determines the contributions of member LSPs as a function of these attributes is not easy. Moreover, even if such rules could be specified, LSPs' preferences may quickly change due to varying environmental conditions, making them inadequate again. However, if allocation rules do not reward the "true" contributions of member LSPs, rapid frustration and potential network disintegration are likely because member LSPs have no direct influence on the design of the allocation mechanism, as this authority resides with the LITPAP. To overcome the issue of fair contributions, Cruijssen et al. (2007c:32) suggest allocating gains based on each member's *marginal* contribution, which can be determined via cooperative game theory.

In principle, *cooperative game theory* is concerned with predicting the formation of a coalition (subset of all actors in the game) and the payoffs that participating actors receive by participating in the coalition (Mas-Colell et al. 1995:673f.). In the context of the problem at hand, cooperative game theory is concerned with determining the subset of member LSPs that would deliver a requested service and the payoffs they would receive. Cooperative game theory provides the tools to distribute value in a coalition according to the marginal contributions of its members (see "Shapley Value" Mas-Colell et al. 1995:679ff.), thus enabling sustainable cooperation. However, one critical issue remains related to cooperative game theory. It assumes that decision relevant information – especially information about each actor's type – is available to the decision-maker. The decision-maker uses this information to calculate the marginal value created by each actor in each possible coalition and to determine which coalition forms and what payoffs are received. However, in a real-world situation, actors may not be willing to truthfully reveal their private information to the decision-maker. In fact, they may have incentives to behave opportunistically – that is, to lie – about their true type in order to influence the formation of coalitions and thus their payoff. Due to the high degree of competition, which results from the inherent resource redundancy and the hostile environment in CLSs, member LSPs have strong incentives to untruthfully reveal their types to the LITPAP. We thus conclude that cooperative game theory is not suitable for achieving effective coordination in the negotiation and selection phase in CLSs.

These strong incentives for behaving opportunistically shift our attention to market-based exchanges and, more specifically, to *mechanism design*, a sub-field of *non-cooperative game theory*. Compared with cooperative game theory, non-cooperative game theory adopts a "procedural approach" that focuses on how decisions are reached, rather than on what outcomes are likely to be achieved. As discussed above, mechanism design is concerned with reaching economic decisions as a function of the information that is privately held by actors (see Subsection C.III.5.3). Market mechanisms are likely to achieve effective coordination in the negotiation and selection phase in CLSs for three reasons. First, market mechanisms explicitly account for the fact that member LSPs have private information and incentives for behaving strategically by revealing information untruthfully. This is crucial, as the LITPAP has no knowledge about the types of member LSPs, especially as their memberships may last only for a limited period of time. Second, market mechanisms remain effective in dynamic environments. Environmental changes may influence the types and preferences of member LSPs. However, as the decision for each service request is made based on LSPs's private information available at the time of the service request, member LSPs have the ability to adapt to changing conditions immediately and autonomously. In other words, the LITPAP does not need to adapt the mechanism in case of environmental changes. Finally, market mechanisms impose the least encroachments on the autonomy of member LSPs; although the LITPAP defines the mechanism hierarchically, economic decisions are reached in a decentralized manner, only depending member LSPs' actions in the sense of revealing private information to the LITPAP. In other words, the extent to which stakeholders can address their economic and logistical concerns depends on their own actions and preferences. By reducing the hierarchical intervention to a minimum and formalizing the interaction process, market mechanisms are likely to be accepted as a legitimate exchange mode.

We thus conclude that the conditions prevailing in the negotiation and selection phase in CLSs can make market mechanisms an effective way to achieve coordination. CLSs thus follow the concept of hybrid clouds in cloud computing, which are also considered to allocate computing resources and capabilities based on market mechanisms (see Subsection D.II.4.2).

3.8. Internet of Things (System-supported Supervision)

3.8.1. Description

The delivery phase is initiated by the logistics consumer according to the conditions and constraints specified in the service contract. The logistics consumer initiates delivery by notifying the LSP about the time, location, and condition in which the logistics objects shall be handed over. If the logistics object has already been given to the LSP, the consumer notifies the LSP to start with the logistical transformation. LSPs deploy resources to achieve the agreed-on logistical transformation. Hence, LSPs associate physical resources with virtual service descriptions at the beginning of delivery. The type of algorithm used determines the extent to which LSPs can achieve resource pooling across various logistics consumers; several algorithms are available in literature (see e.g. Li et al. 2013; He & Xu 2014). These are not part of the following discussion, as the characteristic mix of coordination mechanisms focuses on interfirm coordination rather than on the coordination within member LSPs. The delivery phase ends once the agreed-upon logistical transformation is completed and associated payments are made. In the case of service failure, compensating financial transactions conclude the delivery phase, if agreed to.

The IoT is the proposed coordination mechanism during the delivery phase. The IoT provides *system-supported supervision* that enables the real-time control and monitoring of physical cloud logistics capabilities. To this end, physical resources are connected to the cloud-based platform. More specifically, they are connected to the service descriptions included in the service contract by means of IoT technologies (see Atzori et al. 2010:2791; Guinard et al. 2010). Depending on the types of logistical resources, different sensing technologies may be used. For example, GPS is deployed for tracking transportation services and near field communication (NFC) and RFID for indoor material handling processes. Figure 57 depicts how physical resources can be linked with (composite) services by means of the IoT. Physical resources are equipped with sensors that capture relevant data. IoT technologies associate these resources with their respective virtual descriptions (see "resource and capability modelling" in Subsubsection F.IX.3.4.1). These virtual resources are encapsulated by services (see "encapsulation" in Subsubsection F.IX.3.4.1), which may be combined into more complex solutions. In this way, real-world logistics resources can be sensed, identified, monitored, and controlled by virtual services. System-supported supervision is therefore similar to the perception layer as described by Wang et al. (2012b:561).

The information gathered through IoT sensors is analyzed, aggregated, and stored in the information repository on the cloud-based platform. Thus, system-supported supervision is linked to implicit coordination. The gathered information is used to determine the extent to which conditions and constraints agreed in service contracts are satisfied during service delivery. In particular, the information is used to determine the amount of logistical resources consumed, enabling pay-per-use. Furthermore, consumers and LSPs

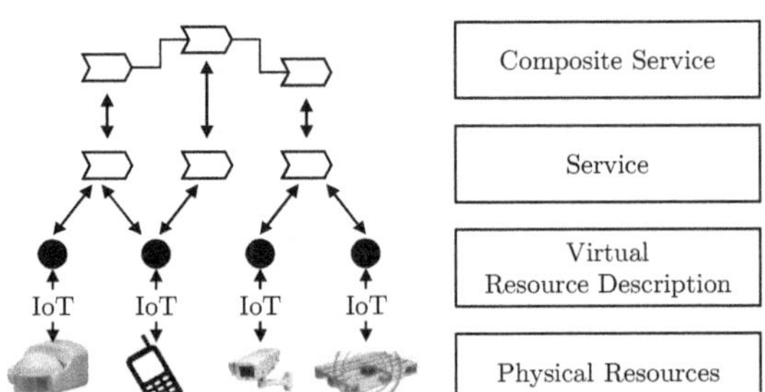

Figure 57.: Linking Smart Physical Resources and Capabilities to Services (motivated by Atzori et al. 2010:2792)

can use this information to manage logistics processes within the CLSs, for example by re-routing their trucks in case of congestion. By combining cloud computing technology with the IoT, the "full potential" of both technologies can be realized (Gubbi et al. 2013:1651). This is because the IoT is a concept that enables the automated and ubiquitous collection of data, while cloud computing represents a technology that allows for the systematic, large-scale economic exploitation of large amounts of information. In this respect, the combined use of IoT and cloud computing relates to the idea of "big data." This is because, just like in big data, combining the IoT with cloud computing involves significantly increasing the amount of available logistics data (volume), the speed with which data become available (velocity), and the number of different sources from which data are collected (variety) (see McAfee & Brynjolfsson 2012:62f.). Moreover, just as in the general concept of big data, the collected logistics data can enable better predictions, smarter decisions, and more effective interventions (see McAfee & Brynjolfsson 2012:62).

3.8.2. Conditions

Conditions that presumably influence the effectiveness of coordination during the delivery phase are related to the following factors: (1) size and geographic dispersion, (2) task complexity, (3) environmental complexity and dynamism, and (4) task interdependence.

The *number* of LSPs involved in delivering a requested (composite) service can vary. At one extreme, a (composite) service can be delivered by a single LSP. At the other extreme, a composite service can be delivered by as many LSPs as the number of services that are part of the composite service. Yet, the actual number of actors that needs to be coordinated during service delivery is usually much larger than the number of LSPs as each physical cloud logistics capability is commonly realized by multiple individuals deploying various physical resources. Hence, the number of actors to be coordinated during delivery is presumably high.

As argued above, *geographic dispersion* varies by prototype scenario and is presumably at least medium, but most often high (see Subsubsection F.IX.3.2.2)

As argued above, *task complexity* is most likely low, as cloud logistics capabilities offered by the prototype scenarios are limited to physical logistical transformations and do not include any "complex" logistical management activities (see Subsubsection F.IX.3.3.2). However, the complexity of controlling and monitoring the delivery of physical logistics capabilities is medium. This is due to the significant number and variety of data that are collected in an automated manner through the IoT. These data must be processed and integrated to enable effective logistics control.

Environmental complexity associated with executing, controlling, and monitoring physical logistics capabilities is presumably medium, as the effective delivery of logistics capabilities requires accounting for various environmental factors, such as traffic and weather.

The *environmental dynamism* during the delivery of physical cloud logistics capabilities is presumably high in CLSs. This is due to the fact that logistics capabilities are exposed to, above all, the natural and economic environments. Hence, logistics capabilities need to be adjusted in response to changing conditions in the natural environment (e.g. a disaster) or changing demand for goods already being transformed logistically (e.g. varying geographic demand).

In case of composite services, *sequential task interdependencies* between multiple cloud logistics services are inevitable, as logistics objects must be passed between different services.

The preceding paragraphs have assessed the presumable conditions during the delivery of physical cloud logistics capabilities. We argue that these conditions, in aggregate, enable effective coordination through system-supported supervision, which uses a sensor network to coordinate many geographically dispersed actors. As physical logistics capabilities presumably exhibit a rather low complexity, the information necessary to direct and monitor them can be exchanged and stored electronically. Despite a dynamic environment, system-supported supervision is an effective coordination instrument, as the task environment for directing and monitoring physical logistics capabilities is stable. Environmental changes are very unlikely to change the type of information to be exchanged, such as the about time, location, or condition of shipments. Moreover, using an electronic sensor networks provides LSPs and logistics consumers with real-time transparency of the status of capabilities currently being delivered, thus enabling them to rapidly respond to environmental changes. Sequential dependencies prevail between different cloud logistics services. These interdependencies can be effectively coordinated by combining the IoT with a central database. This database provides transparency of the status of each capability at all times, thus enabling LSPs to coordinate the handover of logistics objects in time and space, especially if actual delivery deviates from what was agreed to in the service contract. In conclusion, these conditions make system-supported supervision an effective coordination mechanism in the service delivery phase.

3.9. Conclusion: A Computer-Formalized Mix along the Cloud Logistics Service Life Cycle

Based on contingency theory, we have proposed a characteristic mix of coordination mechanisms for CLSs. This mix comprises a set of computer-formalized mechanisms structured along the life cycle of cloud logistics services. Table 29 summarizes the presumed conditions for the identified contingency factors and the proposed mechanism per life cycle phase. By adopting the web service life cycle as the underlying model to structure the characteristic mix of coordination mechanisms of CLSs, we transfer the design principles and concepts of cloud computing to the logistics domain, thus contributing to our research objective.

X. Characteristic Mix of Coordination Mechanisms – Part II: Servitization

1. Abstract of Chapter

Service-orientation is a design paradigm that underlies cloud computing and comprises multiple design principles that collectively shape the design of services (see Section D.III.3).

The *architectural viewpoint* establishes the conventions for adopting service-orientation to the logistics domain, thus creating *service-oriented physical cloud logistics capabilities* (see Section F.X.2). Specifically, this viewpoint argues that following the design principles of service-orientation, contributes to delivering physical logistics capabilities with cloud characteristics – from a coordination perspective. Following these principles (a) shortens the time span between the emergence of a logistical concern and the satisfaction of this concern in various ways, especially through automating activities along the lifecycle of cloud logistics services (on-demand and rapid elasticity); (b) provides flexibility for LSPs to pool resources across various consumers (resource pooling); (c) enables consumers to unilaterally interpret service descriptions (self-service); and (d) enables consumers to determine actual resource consumption (pay-per-use).

The *architectural view* summarizes the detailed design of the mechanism servitization, which coordinates the creation of formal service descriptions of physical cloud logistics capabilities in accordance with the design principles that underlie cloud computing, in the following proposition (see Subsection F.X.3.1):

Proposition 14. *Servitization refers to a programmed routine that consists of four steps: (1) identification, (2) resource and capability modeling, (3) encapsulation, and (4) description. These steps incorporate the IoT and are guided by the design principles of service-orientation.*

In the *identification* step, physical logistics resources are associated with at least one identifier (a sequence of characters) that can be resolved unambiguously within the sys-

Life Cycle Phase	All	All	Development	(Un-) publication	Design	Negotiation Selection	Delivery
Proposed Mechanism	Cloud-based Platform	Stan., Sem. Service Description	Servitization	Boundary Permeability	Self-service	Market Mecha-nisms	Internet-of-Things
	Implicit Coordination, Price	Service-orientation	Programmed Routine	Programmed Routine	System-supported Skills	Programmed Routine	System-supported Supervision
Contingency Factors							
Size	High					High	High
Geographic Dispersion	Medium or Higher						Medium or Higher
Task Complexity	Medium or Higher	Low	Medium	Medium	High	High	Low / Medium
Env. Dynamism	High	High	Low	Low	Low	Low	High
Env. Hostility						High	
Env. Complexity				Medium	High		Medium
Task Interdepen.			High				High
Task Modularity		High					
Behav. Uncertainty				High		High	
Centralization				High		High	

Table 29.: Characteristic Mix of Coordination Mechanisms – Summary of Factors, Conditions, and Mechanisms

tem at any point in time, such identifiers are typically referred to as universal resource identifiers (URIs) (see Laskey et al. 2012:31; Li et al. 2013:1698; Shi et al. 2007:173; see Subsection F.X.3.2). This first step is associated with the concept of IoT.

The *resource and capability modeling* step involves isomorphically expressing the functional and nonfunctional features of heterogeneous physical resources and capabilities, usually by means of a *resource expression model* (Li et al. 2013:1697; synonyms include "virtual specification" Liu et al. 2011b; "virtual description model" Liu & Li 2012; Liu et al. 2014:72; Shi et al. 2007:173; see Subsection F.X.3.3). This step is associated with the principles of (a) service abstraction because descriptions inevitably need to abstract from reality in order to describe resources and capabilities with a finite number of dimensions and (b) loose coupling because descriptions remove a one-to-one mapping between resources and descriptions.

The *encapsulation of virtual resources and capabilities* step focuses on the question of what kind of functionality, resources, capabilities, and information are packaged into a single service and are thus contained within the service boundary (Erl 2008:235, 43; see Subsection F.X.3.4). Encapsulation is related to the design principles of (a) service reusability because the potential to reuse a service depends on the extent to which a specific selection of resources and capabilities is of value for different consumers and (b) service composability because the greater the potential for reuse, the greater the potential for the same service to become a member of several compositions (Erl 2008:280).

The *description* step focuses on the development of service descriptions once the selection is fixed for what virtual logistical resources and capabilities will be encapsulated by a service (see Subsection F.X.3.5). Service descriptions are created by expressing a service's functionality (which derive from the encapsulated virtual resources and capabilities) and its conditions and constraints of use by means of a *service description language* (Xu 2012:79f.; Li et al. 2013:1698). The choice of the service description language is critical because the properties of service descriptions in terms of automatic discovery, composition, and delivery depend on the language chosen. Standardized, semantic service descriptions of cloud logistics services are created by means of a logistics-domain specific, ontology-based description language. This enables adequately capturing and expressing the particularities of the logistics domain and automating coordination along the life cycle of cloud logistics, thus contributing to achieving on-demand availability, rapid elasticity, and self-service. The creation of service description is related to the design principles (a) service abstraction because the description determines the amount of information to be exposed to or hidden from stakeholders; (b) discoverability and (c) composability because, depending on the description language, service descriptions exhibit varying properties regarding the degree of automatic discovery and composition; and (d) standardized service description because all descriptions must use (at least) the same description language, which requires in the case of semantic markup that descriptions are created by using the same generic top-level ontology.

2. Viewpoint: An Approach to Creating Service-oriented Physical Cloud Logistics Capabilities

Service-orientation is a design paradigm that underlies cloud computing and comprises multiple design principles that collectively shape the design of services (see Section D.III.3). Our systematic cloud logistics knowledge review has identified the adoption of service-orientation for the logistics domain as a central element of the cloud paradigm and therefore as a key element of LaaS (see Subsection E.IV.5.7). However, our review also revealed that, although many sources consider service-orientation as a key element, very few pay attention to how it can actually be applied to logistics (Subsection E.IV.5.2).

This viewpoint establishes the conventions for adopting service-orientation to the logistics domain, thus addressing our research objective. Specifically, this viewpoint establishes the conventions for designing *service-oriented physical cloud logistics capabilities*. The design of these logistical capabilities is anchored in existing literature by explicitly applying the design principles associated with service-orientation to the logistics domain.

Service-orientation comprises six design principles: standardized service descriptions, service abstraction, service loose coupling, service discoverability, reusability, and composability (see Subsection D.III.3.2). However, it has not yet been shown whether designing physical logistics capabilities in accordance with these principles can actually contribute to delivering physical logistics capabilities with cloud characteristics. Hence, this viewpoint investigates whether following these design principles can frame concerns related to the essential cloud characteristics. We argue that following these principles can in fact frame such concerns, but only from a coordination perspective. This means that service-oriented design is not a sufficient precondition for delivering physical logistics capabilities with cloud characteristics. In fact, Chapter F.V has identified the physical infrastructure of CLSs as another key factor that determines the degree to which cloud characteristics can be achieved. The following paragraphs show how each design principle of service-orientation can contribute to delivering physical logistics capabilities with cloud characteristics.

The principle of *standardized service descriptions* aims to express a service's functionality as well as its conditions and constraints of use in some standardized form. Using standardized service descriptions allows for the framing of concerns related to (a) on-demand availability, (b) self-service, and (c) pay-per-use from a coordination perspective. Concerns regarding on-demand availability and self-service can be framed because consumers can interpret und understand standardized descriptions quickly and unilaterally (see Erl 2008:130), reducing the time between the emergence of a logistical need and when it begins to be addressed. Concerns regarding pay-per-use can be framed because describing the bundle of logistics resources and capabilities encapsulated by a service in a standardized manner is a precondition for measuring the actual usage of deployed resources and for determining the financial valuation of this usage (see Leukel & Scheuermann 2014:42).

Service abstraction aims to expose only information to consumers via the service description that they require to effectively use the service. Service abstraction allows for the framing of concerns regarding (1) on-demand availability and (2) self-service from a coordination perspective. These concerns can be framed because reducing the amount of information available to consumers enables them to quickly and unilaterally determine whether (or the extent to which) a service is suited to addressing their logistical needs.

Service loose coupling aims to reduce the interdependencies between the service description and the real-world implementation of a service. Adopting this principle allows for the framing concerns related to (1) on-demand availability, (2) rapid elasticity, and (3) resource pooling from a coordination perspective, because loose coupling grants LSPs a considerable degree of latitude in how to implement a logistics service. With regard to on-demand availability and rapid elasticity, LSPs can use this latitude to coordinate initial implementation and capacity adjustments in whatever way is necessary to achieve the desired temporal requirements. With regard to resource pooling, LSPs may use this latitude to share physical resources across different logistics consumers.

Service discoverability aims to design services so that their functionality can be easily located and interpreted by consumers. Hence, this principle allows for the framing of concerns related to (1) on-demand availability and (2) self-service provisioning from a coordination perspective. This is because concise descriptions can be located and interpreted by consumers quickly and unilaterally, reducing the time between the emergence of a logistical need and when it begins to be addressed.

Reusability aims to design services so that their functionality can be used for multiple purposes and/or consumers. Applying this principle allows for the framing of concerns related to (1) on-demand availability, (2) rapid elasticity, and (3) resource pooling from a coordination perspective. Concerns regarding on-demand availability and rapid elasticity can be framed because LSPs, which frequently deliver similar capabilities, are likely to gain experience in coordinating deliveries with low lead time and rapidly adjusting capacity commensurate with demand. Concerns regarding resource pooling can be framed because reusability enables LSPs to pool their physical logistics resources across consumers sequentially or simultaneously.

Composability is the ultimate goal of service-oriented logistics capabilities. Specifically, this principle aims to design services so that consumers can satisfy their logistics concerns by combining two or more services into customized solutions, with each service potentially offered by a different LSP. Composability allows for the framing of concerns related to (1) on-demand availability and (2) self-service from a coordination perspective. This is because consumers can quickly and unilaterally specify their complex logistics concerns by drawing on a set of existing services, reducing the time between the emergence of a logistical need and when it begins to be addressed.

The preceding paragraphs suggest that the design principles associated with service-orientation allow for the framing of concerns regarding cloud characteristics from a coordination perspective. Table 30 summarizes which design principle frames which cloud characteristic. We conclude that applying these principles to the design of logistics capabilities is reasonable from a coordination perspective for creating physical logistics capabilities that, in fact, possess cloud characteristics.

SOC Design Principle	**On-demand**	**Self-service**	**Rapid Elasticity**	**Resource Pooling**	**Pay-per-use**
Standardized Service Descriptions	x	x			x
Service Abstraction	x	x			
Service Loose Coupling	x		x	x	
Service Discoverability	x	x			
Service Reusability	x		x	x	
Service Composability	x	x			

Table 30.: Cloud Characteristics Framed by Service-orientation Design Principles

3. View: A Four Step Process Model for Creating Service-oriented Physical Cloud Logistics Capabilities

3.1. Proposition

Servitization refers to a programmed routine implemented on a cloud-based platform that specifies the steps LSPs need to carry out in order to create formal service descriptions of physical cloud logistics capabilities according to the design principles that underlie cloud computing (see Subsection D.III.3.2). We propose:

Proposition 14. *Servitization refers to a programmed routine that consists of four steps: (1) identification, (2) resource and capability modeling, (3) encapsulation, and (4) description. These steps incorporate the IoT and are guided by the design principles of service-orientation.*

Figure 58 depicts the four proposed process steps and maps them to the design principles and concepts that underlie cloud logistics: IoT and service-orientation (see Subsection E.IV.5.2). More specifically, the process model maps the IoT to the initial process step and the design principles of service-orientation to subsequent steps. Thus, the process of creating service-oriented physical cloud logistics capabilities is guided by the design principles of service-orientation. The proposed split of activities into four steps aims to structure the process according to the fundamental (sub-)activities that need to be carried out to create service-oriented physical logistics capabilities.

This process model is derived from existing literature on (1) cloud logistics and (2) cloud manufacturing, an idea conceptually related. We draw on cloud manufacturing literature, on one hand, because of the scarcity of cloud logistics knowledge and, on the other hand, because cloud manufacturing essentially faces the same problem but is more advanced in this field. Hence, cloud manufacturing results can reasonably provide input for cloud logistics. Existing literature in both fields reveals only a partial consensus on the structure of activities for creating service-oriented capabilities. There are several, albeit related, conceptualizations of this process, as depicted in Figure 59.

With regard to *cloud logistics*, Li et al. (2013:1697) point out that a resource virtualization process consists of two tasks: (i) developing a resource expression model through the analysis of resources' features; and (ii) encapsulating this information into services by building a service description using web service technology. Hu & Zhang (2015:128f.) describe a three step process: (i) identifying scattered resources using IoT technologies; (ii) creating virtual resources and uploading them into a virtual resource pool hosted on the cloud-based platform; and (iii) encapsulating, releasing, and registering virtual resources to become cloud services. Wang et al. (2015b:322) describe four step process. However, only three are directly related to the creation of service descriptions: (i) perception, (ii) virtualization, and (iii) encapsulation. The fourth step, "release and registration," relates to publishing the service in a service repository.

With regard to *cloud manufacturing*, Liu & Li (2012) describe the process of resource virtualization as comprising two steps: (i) manufacturing resource modeling; and (ii) manufacturing cloud service encapsulation (see similar Liu et al. 2011b). Xu (2012:79)

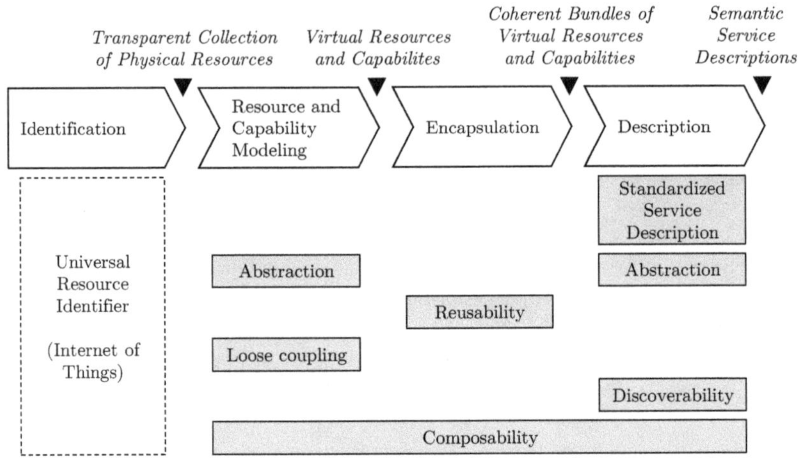

Figure 58.: Process Model for Creating Service-oriented Cloud Logistical Capabilities

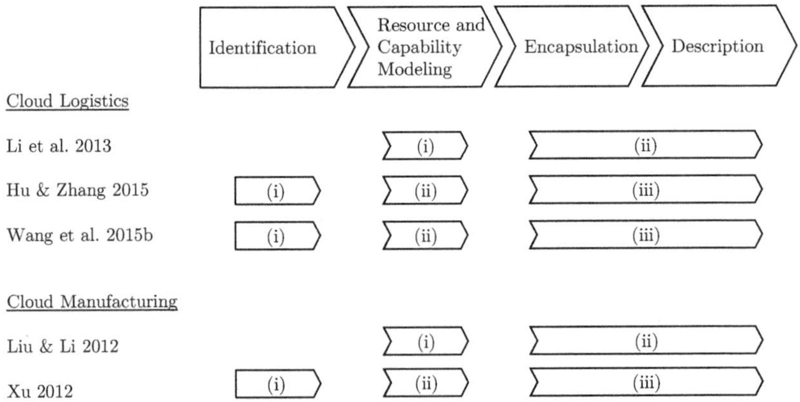

Figure 59.: Process Model for Creating Service-oriented Logistical Capabilities

identifies three steps: (i) identifying manufacturing resources and capabilities, (ii) virtualizing them, and (iii) packaging them into cloud manufacturing services. Despite different conceptions about the number and naming of process steps, all of these processes are conceptually comparable because they accomplish the same task: transforming heterogeneous physical resources and capabilities into cloud services (see Liu & Lyons 2011:1003). However, none of these models explicitly refer to the design principles of service-orientation. Hence, they lack anchoring in the existing literature on SOC.

Comparing the proposed process model for creating service-oriented physical cloud logistics capabilities with process models found in literature reveals that existing literature does not differentiate between encapsulation and description. We believe that this is due to a misconception regarding the meaning of the term "encapsulation" with regard to different software design paradigms. In *service-oriented design*, encapsulation refers to what is enclosed in a service. However, in *object-oriented design*, another paradigm, encapsulation is about hiding information; it means that objects are only accessible via published interfaces (which are described by means of a language) while the data and their functions and the object's implementation all remain hidden (Erl 2008:458; Rouse 2007). Hence, in object-oriented design, encapsulation is related to the design principle of service abstraction and standardized service descriptions as defined in service-orientation.

Considering current cloud logistics and cloud manufacturing literature, it seems that, although it relates to service-oriented design, encapsulation is understood as it is in object-oriented design. For example, Li et al. (2013:1698) say: "the other [step] is to encapsulate resource information in services by building a service description method with web service technologies." Likewise, Liu & Li (2012:46) say, "[t]he OWL-S [a service description language] is adapted to an upper level ontology model, according to which manufacturing resources & capabilities are encapsulated into manufacturing cloud service." Similarly, Xu (2012:79) says, "[t]he next step is to package the virtualized manufacturing resources to become cloud manufacturing services. To do this, resource description protocols and service description languages can be used." All three quotations use the term "encapsulation" to exclusively refer to the description of services and their interfaces, not to specify coherent bundles of underlying resources and capabilities. In order to avoid any confusion between "what is being enclosed within a service" and "how this service is described," we propose to separate the steps of encapsulation and description. We thus follow Erl (2008:235), an SOC guru, who states with regard to service-oriented design,

> "A service, in its entirety, is comprised of a contract [= service description] and what it encapsulates."

The following subsections describe the process steps of servitization in detail. We specifically point out how the design principles of service-orientation guide the creation of service-oriented logistics capabilities. We thus anchor the process of servitization in current literature. Due to the abstract nature of this reference architecture, we do not apply any of these principles to an actual example or create an actual service description, but only focus on pointing out how they can guide the creation process.

3.2. Step 1: Identification

In identification, physical logistics resources are associated with at least one identifier (a sequence of characters) that can be resolved unambiguously within the system at any point in time, such identifiers are typically referred to as URIs (see Laskey et al. 2012:31; Li et al. 2013:1698; Shi et al. 2007:173). IoT technologies suited to identifying (or tagging) physical resources include RFID, GPS, 2D codes, and WSNs (see e.g. Xu 2012:79; Li et al. 2013:1698; Liu et al. 2014:72; He & Xu 2014:5). Thus, the process of identification lets resources become connected to the cloud-based platform and thus synchronize the virtual and physical worlds. Unambiguous identifiers are a crucial means to gaining transparency of the current status of resources and thus enable their effective management. Unambiguity in identification is furthermore important to determining the rights and authorization of resource use, to understand the functions performed by a resource, and ensuring repeatability of outcomes (Laskey et al. 2012:31). Note that "capabilities" cannot be tagged physically. This is because they refer to the capacity of an organization to deploy resources toward a desired end, and this capacity may not be physically tangible. For example, the capacity to operate a truck cannot be tagged, but the truck and the truck driver can be.

3.3. Step 2: Resource and Capability Modeling

Resource and capability modeling is the second step in the process of creating service-oriented logistics capabilities. This step involves isomorphically expressing the functional and nonfunctional features of heterogeneous physical resources and capabilities, usually by means of a *resource expression model* (Li et al. 2013:1697; synonyms include "virtual specification" Liu et al. 2011b; "virtual description model" Liu & Li 2012; Liu et al. 2014:72; Shi et al. 2007:173). This model specifies (a finite number of) dimensions along which resources and capabilities are described (Liu et al. 2011b:1004; Shi et al. 2007:173) and thus provides a universal description of a resource or capability (Li et al. 2013:1696). In other words, the model abstracts from the heterogeneity of physical resources and capabilities by establishing a unified logical view on heterogeneous resource and capabilities (Li et al. 2013:1697). In comparison with computing resources, however, Li et al. (2013:1696) argue that the virtualization of logistics resources is more complex because

"Logistics resources have the characteristics of variability, geographical distri-

> bution, heterogeneity, morphological diversity and self-governing zone." (for similar arguments regarding cloud manufacturing see Liu et al. 2014:72f.; Liu & Li 2012:46f.)

These particularities of logistics resources and capabilities must be accounted for when virtualizing them, for example by including optional dimensions that can capture relevant information about functional and non-functional characteristics.

As different types of physical resources and capabilities exhibit different features, different expression models become necessary to adequately capture the information relevant for each resource and capability type (see Xu 2012:79; He & Xu 2014:6). It is useful to develop (hierarchical) classifications of physical resources and capabilities as a basis for identifying the different types of expression models needed (for logistics resources see Li et al. 2013:1699; Wang et al. 2012b:560f.; for manufacturing resources see Shi et al. 2007:172). Resources and capabilities of the same type are represented by the same model, and resources and capabilities of different types are represented by different models. For example, trucks with different loading capacities are of the same type and hence are represented by the same model, even though they are filled with different values, such as 7.5t and 40t. By contrast, trucks and forklifts are of different types and are represented by different models. An example of a logistics resource expression model can be found in Li et al. (2013:1698ff.).

Resource and capability modeling is related to the design principles of (1) service abstraction and (2) loose coupling (see Section F.X.2). It is related to *service abstraction* because the resource expression model must necessarily abstract the underlying heterogeneous physical resources and capabilities by describing them with a selected number of dimensions. By neglecting some information, a uniform view can be created of the underlying set of heterogeneous resources and capabilities, which fosters agility, enhances flexibility, and reduces costs (He & Xu 2014:5f.). LSPs that face the task of creating expression models for logistics resources and capabilities need to carefully decide which dimensions to include and which to neglect in order to capture the "essential" functional and non-functional features of resources and capabilities. Using too many dimensions makes the expression model difficult to understand and may unnecessarily constrain the types and number of physical resources and capabilities that it can represent (Liu et al. 2014). For example, creating a resource expression model for trucks should include dimensions that capture information about (a) the dimensionality of the loading space, (b) the maximum loadable weight, (c) the types of licenses necessary for operating the truck, and (d) any specific transportation capabilities such as installed appliances for refrigerated goods or hanging garments. However, the model should not include any dimensions for capturing information about a truck's manufacturer or age.

Resource and capability modeling is related to *loose coupling* because describing logistics resources and capabilities by means of generic expression models removes any one-to-one relationships between a physical resource and capability and its virtual expression model. Consequently, many physical resources and capabilities may exist, each possessing the functional and non-functional features described in the expression model. In other words, many different physical resources and capabilities can be mapped to the same virtual model (He & Xu 2014:6; also see Liu et al. 2014).

Before moving on to the third process step, a terminological remark is expedient. Resource and capability modeling is often associated with the terms "virtualization" and "virtualized resources" (see e.g. Xu 2012; Liu et al. 2014), thus inferring some relationship with the concept of hardware virtualization. In fact, the relationship is twofold. First, hardware virtualization and resource and capability modeling both create *virtual resources*. Hardware virtualization creates virtual machines (on which users can deploy an operating system and applications of their choice) by inserting an additional software layer (the VMM) between the operating system and the physical computer hardware (see Section D.III.2). Resource and capability modeling creates resource expression models that, by themselves, represent virtual resources (see Laskey et al. 2012:45), as they are stored and managed on and by computer systems and can be reused by different stakeholders. Second, hardware virtualization and resource and capability modeling both remove one-to-one dependencies between virtual and physical resources and capabilities. Inserting a VMM is conceptually similar to representing physical resources and capabilities through resource expression models because both – the VMM and the expression model – abstract the heterogeneity of physical underlying resources and capabilities by establishing a uniform view of them, thus removing interdependencies. To conclude, these terminological and conceptual relationships re-emphasize the importance of abstraction and loose coupling for this second step of servitization.

3.4. Step 3: Encapsulation of Virtual Resources and Capabilities

Encapsulation is the third step in creating service-oriented physical cloud logistics capabilities. Generally speaking, "[t]o encapsulate means to enclose something in a container" (Erl 2008:458). With regard to service-oriented design, encapsulation refers to the question of "*what* is being enclosed within (encapsulated by) the container" (Erl 2008:459, italics original), where "container" refers to a service. In other words, encapsulation refers to the question of what kind of functionality, resources, capabilities, and information are packaged into a single service and are thus contained within the service boundary (Erl 2008:235, 43). Thus, the process step of encapsulation involves determining the combination of logical logistics resources and capabilities necessary to perform a set of tasks that from a coherent functionality. Some services may encapsulate only a single resource and capability, others may comprise multiple that collectively provide a specific functionality (Liu et al. 2014:72). Encapsulation may therefore be understood as a process that maps virtual resources and capabilities onto services (Xu 2012:80). Each service comprises and controls a coherent bundle of resources and capabilities to deliver some coherent functionality (see Subsection D.III.3.1). We use the term "coherent" to imply that the bundle of resources and capabilities must fit together; for example, a truck would be combined with a truck driver rather than a pilot.

Encapsulation is related to the design principles of (1) service reusability and (2) service composability (see Section F.X.2). The resources and capabilities encapsulated by a service determine the degree to which it can be *reused* (Erl 2008:43). Services that encapsulate bundles of resources and capabilities of value for many consumers exhibit a higher degree of reusability than those that can only address the unique concerns of a single consumer. Encapsulation is indirectly related to *composability* because composability can be understood as a form of reusability (Erl 2008:280). The greater the potential for reuse, the greater the potential for the same service to become a member of several compositions (Erl 2008:280).

Through its relationships with reusability and composability, encapsulation directly refers to the concept of (process) modularity. In fact, selecting a set of virtual resources and capabilities to be included in a single service as per the principles of reusability and composability is conceptually similar to building process modules: selecting bundles of virtual resources and capabilities in such a way that services (modules) can be easily mixed and matched (possibly into new configuration) with no or little loss of functional performance (see Langlois 1992; Sanchez 1995 as cited in Schilling 2000:315).

Current logistics literature has developed a few approaches for how to modularize logistics processes and systems (see Section B.IV.3). These methods appear generally applicable to modularization in a cloud logistics context with the following caveat: All methods start the modularization process based on specific customer requirements or a current state of the logistics systems. In CLSs, however, LSPs need to start modularization based on "anticipated customer requirements," as service modules must be published to the repository before the reception of any specific service request. Hence, LSPs need to thoroughly understand the current logistics demand from current and potential customers (see "customer markets" Pekkarinen & Ulkuniemi 2008:99) and anticipate furture demand. We believe that Schilling's (2000) model for explaining and predicting the degree of modularity in systems provides useful guidance for the general factors that LSPs need to account for when designing logistics modules: heterogeniety of inputs and outputs, synergestic specificity, and urgency. Based on these factors, they can develop a general understanding of the size of the needed modules.

3.5. Step 4: Description

The last step in creating service-oriented cloud physical logistics capabilities is concerned with the actual creation of service descriptions (see Liu et al. 2014:72). The development of service descriptions begins once the selection is fixed for what virtual logistical resources and capabilities will be encapsulated by a service. Service descriptions represent crucial information sources for potential consumers to determine whether (or the extent to which) a given service can address their logistical concerns because they contain all information necessary for using, deploying, managing, or otherwise controlling a service (see Subsection D.III.3.2; Laskey et al. 2012:43). This typically includes information about a service's functionalities and its conditions and constraints of use.

Service descriptions are created by expressing a service's functionality (which derive from the encapsulated virtual resources and capabilities) and its conditions and constraints of use by means of a *service description language* (Xu 2012:79f.; Li et al. 2013:1698; see Subsection D.III.4.3). Choosing a service description language is a critical design choice in the process of creating service-oriented physical cloud logistics capabilities because the properties of service descriptions in terms of automatic discovery, composition, and delivery depend on the description language chosen (see Subsection D.III.4.3). As argued above, standardized, semantic service descriptions of cloud logistics services are created by means of a logistics-domain specific, ontology-based description language (see Subsection F.IX.3.3). This enables adequately capturing and expressing the particularities of the logistics domain and automating coordination along the life cycle of cloud logistics, thus contributing to achieving on-demand availability, rapid elasticity, and self-service. However, the current status of applying semantic theory to the logistics domain (see Subsection D.III.4.4) and current cloud logistics knowledge (see Li et al. 2013; Hu & Zhang 2015) reveal that so far no logistics-domain specific, ontology-based description language has been developed. The development of such a language, however, represents a critical research avenue towards the realization of CLSs. The following chapter provides an initial proposal for a "top-level" ontology for such a description language, thus establishing a foundation for future research along this avenue (see Chapter F.XI).

The creation of service description is related to the design principles (1) service abstraction, (2) service discoverability, (3) service composability, and (4) standardized service description. It is related to *service abstraction* because the description determines the amount of information to be exposed to or hidden from stakeholders. It is related to *discoverability* and *composability* because, depending on the description language, service descriptions exhibit varying properties regarding the degree of automatic discovery, composition, and delivery. Finally, it is related to the principle of *standardized service description* because this principle requires that all descriptions use (at least) the same description language. With regard to semantic markup, standardized descriptions are created by using the same generic top-level ontology.

3.6. Conclusions: Implementation Hurdles and a Transformation of LSPs

The previous subsections proposed a mechanism for the creation of service-oriented physical cloud logistics capabilities. This mechanism structures the design principles associated with service-orientation along four process steps. Applying these design principles contributes to achieving cloud-like provisioning of physical logistics capabilities.

The introduction of service-orientation to logistics imposes a significant documentation burden on LSPs. They need to describe physical logistics resources and capabilities by means of expression models and describe coherent bundles of virtual logistics resources and capabilities by means of standardized semantic service descriptions. Such extensive documentation needs may be an implementation hurdle for any service-oriented system, especially if organizations lack a history of using standards (Erl 2008:131). Empirical evidence from the logistics domain suggests a heterogeneous application of standardized documentation. On the one hand, Rajahonka (2013:44) finds evidence that many LSPs do not have comprehensive descriptions of their processes and that descriptions appear not to follow any kind of industry standard. Similarly, Arnold (2014a:21) provides (anecdotal) evidence that logistics practitioners show considerable resistance when asked to adjust their IT systems to make them compatible with a standardized ontology for freight documents. On the other hand, Pekkarinen & Ulkuniemi (2008:93) interview a globally operating LSP that "offers highly standardized services including over fifty different, carefully defined service modules." Despite this heterogeneous evidence on the current degree of formalization in the logistics industry, formalization may likely increase in future. Regulatory pressures drive process standardization and formalization, especially in international, but increasingly also in domestic trade (Rajahonka 2013:43). We thus conclude that introducing service-orientation to logistics seems possible today, albeit only among a subset of LSPs. These LSPs may therefore be promising candidates for participating in an early prototype realization of a CLS. If the concept of cloud logistics becomes more common, LSPs will be more used to extensive documentation, thus enabling more and more LSPs to join.

Furthermore, the introduction of service-orientation to logistics is likely to extend LSPs' responsibilities. LSPs are likely to assume a more active and guiding role in the development of services than today. Instead of responding to service requests from logistics consumers, LSPs must publish their capabilities as services to a repository before any interaction with potential consumers has taken place, similar to providers in cloud manufacturing systems (see He & Xu 2014:5). This aligns well with the prediction of Andersson

& Norrman (2002:11) that, in future, LSPs will assume a more active role in the development of services in terms of resources, processes, and outputs.

The introduction of service-orientation may also create pressures for LSPs to adjust their organizational structures to align with the structural needs resulting from creating and delivering modular service-oriented logistics capabilities. As indicated by the modular logistics service platform developed by Lin & Pekkarinen (2011) (see similar Pekkarinen & Ulkuniemi 2008), logistics services, processes, and activities are associated with organizational modules. Hence, LSPs need to define services not only in terms of required physical resources and capabilities but also in terms of the organizational resources and capacities required to deliver these physical capabilities. The need for an adjusted (more modularized) organizational structure is also considered if IT vendors deliver value through a "regular" SOA (see e.g. Cherbakov et al. 2005:658; Bieberstein et al. 2005:696ff.).

XI. Characteristic Mix of Coordination Mechanisms – Part III: Standardized, Semantic Service Descriptions

1. Abstract of Chapter

The last step in the servitization mechanism – description – refers to the actual creation of service descriptions.

The *architectural viewpoint* establishes the convention for creating service descriptions of physical cloud logistics capabilities in accordance with the design principles of (1) standardized service description, (2) abstraction, (3) discoverability, and (4) composability (see Section F.XI.2).

The *architectural view* summarizes the design of the mechanism standardized, semantic service descriptions by proposing a standardized "top-level" ontology for describing any physical cloud logistics capability (see Subsection F.XI.3.1)

Proposition 15. *Physical cloud logistics capabilities will be described by means of a logistics-domain specific, ontology-based description language that uses a standardized "top-level" ontology that consists of five concepts: (1) service functionality, (2) service policies, (3) service metrics, (4) service interface description, and (5) service reachability.*

These top-level concepts are adopted from the SOA-RAF service description model, which is a reference architecture model for SOA and, due to its high level of abstraction, reasonably applicable to the logistics domain. The principle of *standardized service descriptions* is satisfied by only specifying the top-level concepts. *Abstraction* is honored by limiting the basic dimensions of service descriptions to rather few top-level concepts. *Discoverability* is achieved by including service functionality as a top-level concept, which enables logistics consumers to identify capabilities in the repository. *Composability* is supported by positioning the service interface description as a top-level concept.

Service functionality specifies the real-world (non-)logistical transformations that consumers can expect a cloud logistics service to accomplish subject to set of technical assumptions and constraints (see Subsection F.XI.3.2).

Service policies prescribe the conditions and constraints of service use (see Subsection F.XI.3.3). Cloud logistics services descriptions must specify the behavior and performance of cloud logistics services with regard to cloud characteristics: (a) a usage-based pricing policy (pay-per-use), (b) a policy that specifies the lead time for implementing new services (on-demand availability), and (c) a policy that specifies the properties of scaling the capacity of capabilities currently being delivered (rapid elasticity).

Service metrics are the conditions and quantities that can be measured to characterize a service's functions and real-world transformation(s) (see Subsection F.XI.3.4). Cloud logistics service descriptions must include three metrics that determine the cloud characteristics: (a) a metric for measuring actual logistical capacity consumed (pay-per-use), (b) a metric for measuring service lead time for implementing new capabilities (on-demand), and (c) a metric for measuring the speed and precision of scaling the capacity of logistics capabilities currently being provisioned (rapid elasticity).

The *service interface* description contains all information required to interact with a cloud logistics service to achieve the (non-)logistical real-world transformation(s), specified in its service description (see Subsection F.XI.3.5). The service interface for cloud logistics services consists of a virtual interface and a physical interface to enable an expeditious exchange of logistics-relevant information via the cloud-based platform and of physical logistics objects in the real world with and between cloud logistics services.

Cloud logistics *service reachability* refers to the ability of logistics consumers to locate and interact with cloud logistics services and their providers, either by exchanging messages via the cloud-based platform or the logistics object(s) in the real world (see Subsection F.XI.3.6).

2. Viewpoint: An Approach to Developing Service Descriptions of Physical Cloud Logistics Capabilities

The preceding chapter has proposed a mechanism, *servitization*, that identifies a process for coordinating the creation of service-oriented physical cloud logistics capabilities (see Chapter F.X). The last step of that process – description – refers to the actual creation of service descriptions: Defined bundles of virtual logistics resources and capabilities are described using a service description language, yielding service-oriented physical cloud logistics capabilities (see Subsection F.X.3.5).

This viewpoint establishes the convention for creating service descriptions of physical cloud logistics capabilities. Following the proposed mapping of SOC design principles to process steps in the servitization mechanism, this viewpoints establishes the convention that services descriptions are to be created in accordance with the principles of (1) standardized service description, (2) abstraction, (3) discoverability, and (4) composability. By explicitly following these design principles, the creation of service-oriented physical cloud logistics capabilities becomes anchored in existing literature.

3. View: A "Top-level" Ontology for the Descriptions of Physical Cloud Logistics Capabilities

3.1. Proposition

Physical cloud logistics capabilities are described by means of a logistics-domain specific, ontology-based service description language (see Subsection F.IX.3.3). However, such description language has not yet been developed. In fact, this reference architecture also refrains from developing such a language in order to maintain a sufficient level of abstractness to ensure validity for any potential implementation of CLSs. Suppose, for the sake of argument, that this reference architecture proposes a full description language; then, any CLS that uses a different description language would not be covered by this reference architecture, even if it matches all other design propositions and delivers physical logistics capabilities with cloud characteristics.

We therefore constrain the scope of this reference architecture to proposing "top-level" concepts and the most important sub-concepts that any logistics-domain specific, ontology-based description language must have if used to describe physical cloud logistics capabilities. The proposed service description model thus spans the solution space in which any concrete physical cloud logistics capability can be described. In principle, we propose:

Proposition 15. *Physical cloud logistics capabilities will be described by means of a logistics-domain specific, ontology-based description language that uses a standardized "top-level" ontology that consists of five concepts: (1) service functionality, (2) service policies, (3) service metrics, (4) service interface description, and (5) service reachability.*

By specifying this top-level ontology in the reference architecture, the design principle of *standardized service descriptions* is satisfied. The principle of *abstraction* is honored by limiting the fundamental dimensions of service descriptions to rather few top-level concepts. *Discoverability* is achieved by including service functionality as a top-level concept. In this way, logistics consumers can easily identify capabilities that match their concerns when browsing the repository. *Composability* is supported by positioning the service interface description as a top-level concept.

The proposed top-level concepts for describing physical cloud logistics capabilities are adopted from the SOA-RAF service description model (see Subsubsection D.III.4.3.3). By adopting an existing model from SOA, we can anchor the description of physical cloud logistics capabilities in existing literature. Although the SOA-RAF model is rooted in the software domain, we consider the model suitable for describing physical cloud logistics capabilities for two reasons.

First, the SOA-RAF service description model has a high degree of abstractness. This makes it applicable to not only software systems, but also "real" business processes that are performed within a single organization or that stretch across various organizations (business collaborations) (Laskey et al. 2012:69ff.). The SOA-RAF generally identifies

> "the elements and their relationships needed to enable SOA-based systems to be used, realized and owned while avoiding reliance on specific concrete technologies. This positions the work at the more abstract end of the continuum [...] In addition, this Reference Architecture Foundation, as the title illustrates, is intended to provide foundational models on which to build other reference architectures and eventual concrete architectures." (Laskey et al. 2012:10)

Accordingly, the CLS reference architecture can be considered a reference architecture that builds on the SOA-RAF.

Second, the SOA-RAF's suitability can be corroborated by showing that concepts used to structure knowledge about logistics services and the logistics domain can be mapped onto the concepts used in the SOA-RAF model, as shown in Table 31. Analyzing this mapping reveals that concepts included in logistics domain and service ontologies can be well mapped onto the basic concepts of the SOA-RAF service model. The concepts

Authors	Scheuermann & Hoxha (2012)	Augenstein et al. (2010)	Hoxha et al. (2010)	Dong et al. (2008)	Park et al. (2008)	Jung et al. (2008)	Lim et al. (2007)	Lian et al. (2007)
Functionality	logistics service, process, location, logistics object, resource	processes, resources, technology, capabilities	logistics process, service, resource	transport service	logistics process, logistics activity, logistics event, state		logistics context and state	logistics process, action, event, situation, condition, context, object
Policies								
Metrics	logistics KPI	economics	logistics KPI					time duration
Service Interface Description						air freight message		
Service Reachability				transport service area, transport provider address, contact details				location
Concepts not possible to map	logistics company, composite logistics process	description, stakeholder	actor	transport service name and description, transport service provider	process type (atomic vs. composite), processing type (interactive vs. non-interactive)	–	–	–

Table 31.: Mapping Concepts of Selected Logistics Domain Ontologies to SOA-RAF Service Description Model

that cannot be mapped (e.g. stakeholder, actor, atomic/composite services, or the service description itself) are not necessary to describe a service. Rather, they are required to structure knowledge about the entire logistics domain, as intended in these contributions. Hence, this allows to conclude that the SOA-RAF model provides a reasonable starting point for developing service descriptions of cloud logistics services.

The following subsections specify the meaning of each top-level concept and introduce the most important sub-concepts and relationships necessary for describing physical cloud logistics capabilities. As already mentioned, descriptions can be generated by specifying appropriate sub-classes of concepts and assigning values to them. Again, note that due to the abstract nature of this reference architecture, we refrain from developing any more specific service descriptions. FloatBarrier

3.2. Service Functionality

Service functionality expresses the real-world effects that stakeholders can expect a service to accomplish (see Subsubsection D.III.4.3.3). Physical cloud logistics service functionality specifies the real-world logistical transformations that consumers can expect a cloud logistics service to accomplish subject to set of technical assumptions and constraints. Figure 60 depicts the key concepts and relationships that describe the service functionality of cloud logistics services.

Functions of a cloud logistics service are the service's processes that bring about (non-)logistical real-world transformations, which refer to some change in the state of the logistics object(s) (see Scheuermann & Hoxha 2012; Park et al. 2008; Lim et al. 2007:449f.). The functions of cloud logistics services are limited to (non-)logistical processes included in the logistical prototype scenario (see Section F.V.3). These include on the top-level transportation, storage, transshipment, and industry-wide, standardized, value-added, (non-)logistical processes. A concise and hierarchical description of these processes is crucial because they represent the primary criterion for consumers searching for and locating services in the service repository (see Glöckner et al. 2014:193; also see Martin et al. 2005:32f.). In other words, such a hierarchical description is important for enabling *discoverability* of cloud logistics services.

The *(non-)logistical real-world transformations* specify the dimensions along which the state of the logistics object can be changed. Here, "state" is defined as the object's physical situation in terms of its location; at a certain point in time; its transportation, storage, and handling properties (e.g. packaging); its quantity and composition; and the environmental factors to which it is exposed (e.g. temperature, humidity, pressure) (Lim et al. 2007:449; Lian et al. 2007:433; Park et al. 2008; see Section B.III.1). The

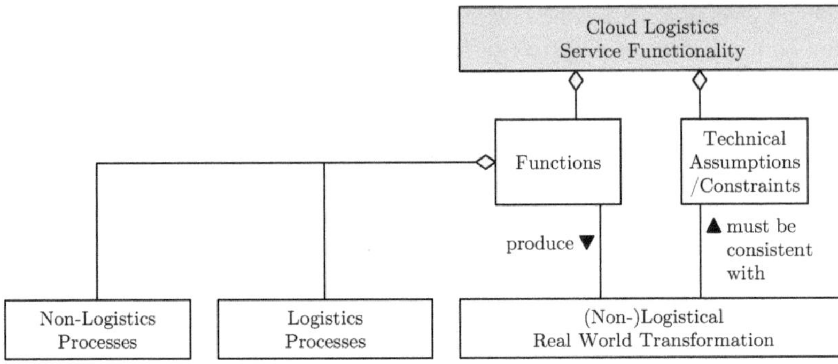

Figure 60.: Cloud Logistics Service Functionality (based on Laskey et al. 2012:50)

scope of transformations that can be achieved depend on the (non-)logistical functions encapsulated by the service.

Technical constraints and assumptions include constraints that (a) arise from the natural laws associated with the physical tangibility of logistics objects and logistics resources, such as aerodynamic drag and gravity that limit the speed and maximum loadable weight, respectively, of trucks and aircrafts, and (b) are imposed on (non-)logistical real-world transformations via service policies. As we argue in the next subsection, this particularly includes policies that constrain transformations by requiring them to be consistent with cloud characteristics.

3.3. Service Policies

Service policies prescribe the conditions and constraints of service use (see Subsubsection D.III.4.3.3). Service policies are asserted by LSPs to specify and constrain the behavior and performance of their offered logistics services (see Glöckner et al. 2014:193). In principle, policies can be established regarding any aspect of a service. Figure 61 depicts a selection of policy classes commonly used for logistics services and a specific set of policies that LSPs must establish for cloud logistics services. Common logistics policies pertain to the general terms and conditions of service provisioning; duties and liabilities of the consumer and LSP; relief from duties in occurrence of force majeure; consequences in case of consumer or LSP non-performance; place of jurisdiction and legislation (see e.g. DSLV Deutscher Speditions- und Logistikverband e.V. 2006); and payment terms (Glöckner et al. 2014:193). In addition, LSPs commonly establish policies regarding the offered QoS, which can include a wide variety of attributes such as order accuracy, order condition, and timeliness (see e.g. Mentzer et al. 2001:84). LSPs can use the same policies as they do for "regular" services. However, LSPs must assert three policies for the

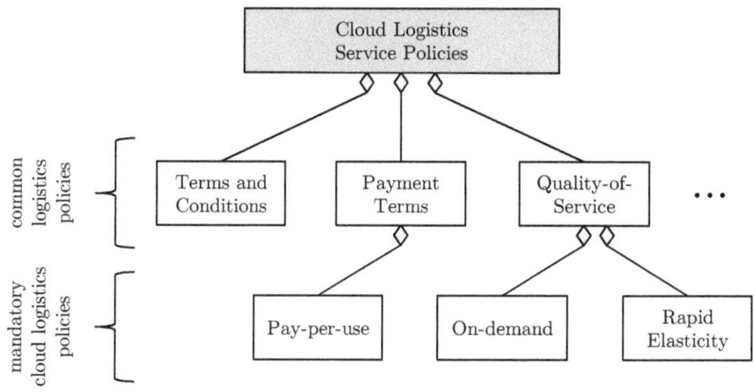

Figure 61.: Cloud Logistics Service Policies

behavior and performance of cloud logistics services with regard to cloud characteristics: (a) a usage-based pricing policy (pay-per-use), (b) a policy that specifies the lead time for implementing new services (on-demand availability), and (c) a policy that specifies the properties of scaling the capacity of capabilities currently being delivered (rapid elasticity). These mandatory policies are denoted as "cloud logistics services policies." Observe that resource pooling and self-service are not service policies. This is because resource pooling applies to all transformations and does not require any further specification, and self-service is independent of real-world transformations.

As policies are asserted by LSPs, inconsistencies may arise between the policies and logistics consumers' preferences. In the case of composite services, there may be further inconsistencies among different policies from different LSPs. In order to resolve these, every attribute in a policy can become the subject of the negotiation between logistics consumers and LSPs. If inconsistencies can be reconciled during negotiations and an agreement can be reached, policies become part of the service contract.

3.4. Service Metrics

Service metrics are the conditions and quantities that can be measured to characterize a service's functions and real-world transformation(s) (see Subsubsection D.III.4.3.3). Logistics service metrics provide measurable quantities for attributes specified in policies and agreed to in service contracts, as depicted in Figure 62. Metrics are also used to measure logistics processes and resulting real-world transformations. Operational measurement data is collected using IoT technologies and stored in a compliance record, which is part of the information repository. The compliance record compares operational data with the conditions and constraints specified in policies and agreed to in the service contract to

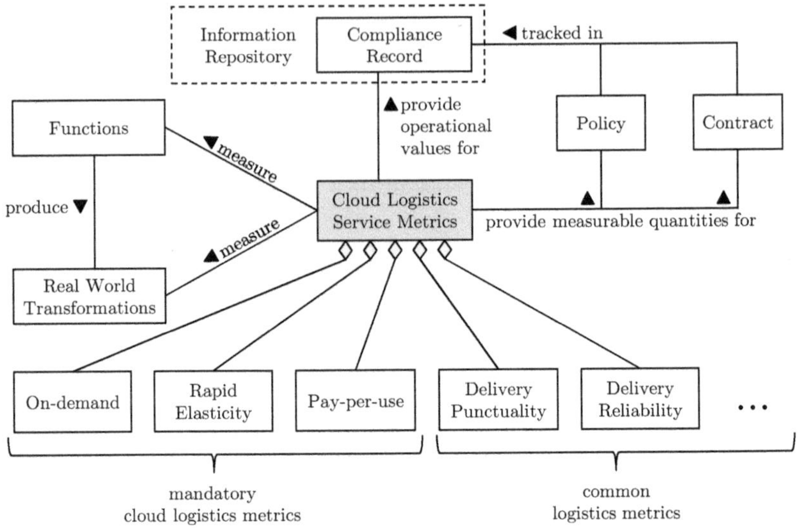

Figure 62.: Cloud Logistics Service Metrics (based on Laskey et al. 2012:52)

determine whether, or the extent to which, the resulting transformations comply with these specifications.

Logistics metrics can take a variety of forms depending on the logistics process, the logistical transformation, or any other measured attribute. For example, transportation metrics include delivery quality, flexibility, and reliability (Hoxha et al. 2010:75). Cloud logistics service descriptions can include any metric used for "regular" logistics services. However, descriptions must include three metrics that determine the cloud characteristics of services: (a) a metric for measuring actual logistical capacity consumed (pay-per-use), (b) a metric for measuring service lead time for implementing new capabilities (on-demand availability), and (c) a metric for measuring the speed and precision of scaling the capacity of logistics capabilities currently being provisioned (rapid elasticity). These metrics are mandatory for every cloud logistics services; they define and measure service compliance pertaining to the essential cloud characteristics. Yet, the types of metrics used for a specific cloud logistics service depend on the specific logistical capabilities offered. Hence, no metric can be proposed at the level of abstraction of this reference architecture. For example, actual resource usage of storage capabilities may be measured based on storage duration and the physical properties of the stored freight. Actual usage of a transportation service may be measured by the distance traveled and the physical properties of the transported freight.

3.5. Service Interface Description

The service interface description contains all the information required to interact with a service in order to achieve the real-world transformation(s), which are specified in its service description (see Subsubsection D.III.4.3.3). Well-defined interfaces of modular logistics services are crucial because complex logistics solutions (composite services) are often delivered by a network of LSPs, and such solutions can be compared to "a relay race, where the baton should be handed over smoothly to the next runner" (Rajahonka 2013:49). The service interface for cloud logistics services consists of a virtual interface and a physical interface. Adopting a twofold interface is motivated by Leukel & Kirn (2011:294) who argue that transformations of service-oriented logistics capabilities are not only limited to logistics-relevant information but also cover the logistics object itself (also see Leukel et al. 2011b:40). The virtual interface specifies how to interact with a service via the cloud-based platform; the physical interface specifies how to interact in the real world. Interactions either take place through messages, which can be exchanged on the cloud-based platform or in the real world, or through the exchange of logistics objects in the real world. In order to model the virtual and physical interfaces, we expand the SOA-RAF service interface model by adding a physical behavior model and logistics object model, as depicted in Figure 63. The proposed interface model thus contains four sub-concepts: (1) virtual behavior model, (2) information model, (3) physical behavior model, and (4) logistics object model. This model aims to enable an expeditious exchange of logistics-relevant information and logistics objects with cloud logistics services. By specifying the critical dimensions of service interfaces in the service description model in this reference architecture, service *composability* can be attained.

The *virtual behavior model* is similar to the behavior model in the SOA-RAF. It specifies the set of permissible actions and their consequences that can be performed against a service via the cloud-based platform (action model) as well as the process that governs the temporal relationships and properties of these actions (process model). The model thus regulates all interactions that can be carried out with the service throughout its life cycle on the cloud-based platform. During the service negotiation phase, the behavior model specifies the set of actions (see "bidding language" for market mechanisms in Blau 2009:90ff.) and the process model specifies their sequence (see "auction process model and architecture" in Blau 2009:102ff.) to facilitate the exchange of services via market mechanisms. The model also specifies the actions, their sequence, and their temporal properties that can be used to trigger service delivery (e.g. request pickup) and request current service status (e.g. shipment tracking). It also identifies actions available after service completion (e.g. filing complaint in the event of service deficiency).

The *information model* pertains to both the virtual and physical interfaces of a cloud

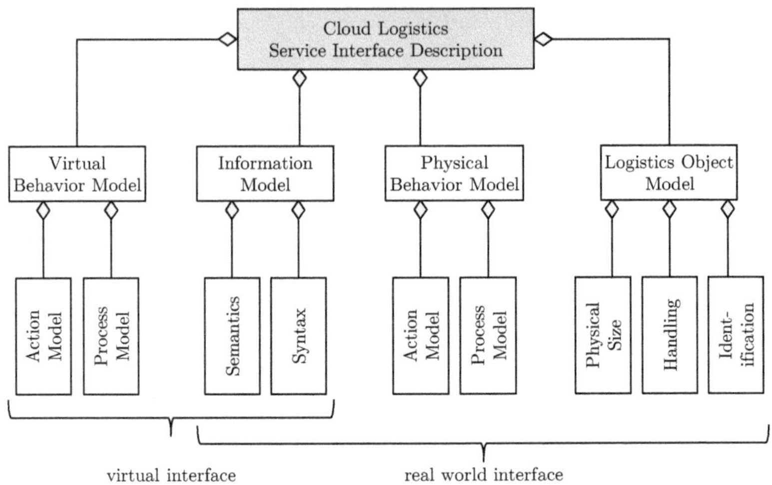

Figure 63.: Logistics Service Interface Description Model

logistics service. It specifies syntax and semantics for and thus standardizes the logistics-relevant information exchanged with a service throughout its life cycle, either via the cloud-based platform (virtual interface) or in the real world (physical interface). During the negotiation and selection phase, the information model specifies the syntax and semantics of proposals exchanged between a consumer and the LSPs of candidate services on the cloud-based platform. During service delivery, the information model pertains to all messages and documents that rush ahead, accompany, or lag behind the physical flow of goods in order to initiate, direct, and monitor this flow (see e.g. Bonfatti et al. 2010; Jung et al. 2008; see "cargo receipts" Zhang & Sun 2013:887). This includes information about the current service status and real-world transformations ("event notifications") (Park et al. 2008:5) as well as physical documents that accompany the logistics object in the real world, such as a CMR transport document, Bill of Lading, Air Way Bill, or International Commercial Invoice.

The *logistics object model* can be conceived of as the "conceptual twin" of the information model. While the information model establishes the requirements for messages and documents to be compatible with a service, the logistics object model establishes the requirements for logistics objects to be compatible with a cloud logistics service. Compatibility of logistics objects can be defined and measured along three dimensions: (a) physical size, as a function of weight and volume (see e.g. Pfohl 2010:59f.; see Corsten & Gössinger 2007:17); (b) storage, transportation, and handling requirements, as a function of sensitivity of goods (Pfohl 2010:59f.), loading units used (e.g. pallets, containers) (Corsten & Gössinger 2007:17), and packaging used; and (c) identification technology as a function of the technology used to tag cargo, such as bar codes and RFID tags. Logistics objects can only be exchanged successfully if the properties of the object align with the physical requirements of a cloud logistics service.

The *physical behavior model* specifies the set of permissible physical actions and their consequences (action model) as well as the temporal relationships and properties of these actions (process model) that can be performed against a logistics service in the real world during the exchange ("handover") of the logistics object at the service interface. The action model specifies activities such as checking freight for completeness and physical integrity, processing freight documents, and signing proof-of-delivery. It also specifies consequences associated with each of these actions, for example the patching of damaged packaging. The process model identifies the temporal sequence and properties of the actions related to exchanging logistics objects, such as checking freight for completeness and damage prior to confirming successful proof-of-delivery.

3.6. Service Reachability

Service reachability is the ability of stakeholders to locate and interact with services (see Subsubsection D.III.4.3.3). Cloud logistics service reachability is the ability of logistics consumers to interact with cloud logistics services and their providers, either by exchanging messages or the logistics object(s) in the real world. Following the SOA-RAF, cloud logistics services reachability can be defined along three basic dimensions: (1) protocols, (2) endpoints, and (3) service presence. However, due to the physical character of cloud logistics services, reachability pertains not only to locating a service and interacting with it via the internet, but also locating and reaching a service in the real world. Hence, reachability gains a geographic component, physical endpoints, as depicted in Figure 64.

Protocols pertain to the interactions between consumers and LSPs on the cloud-based platform. They specify the technical means for establishing a (network) connection to the

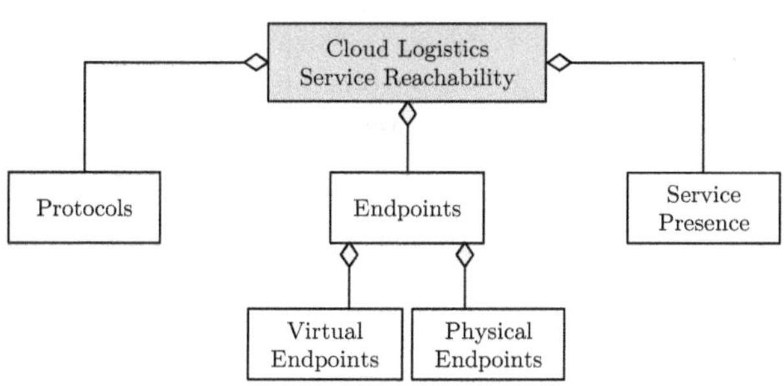

Figure 64.: Cloud Logistics Service Reachability

platform and for exchanging information through this connection. Protocols thus only pertain to services' virtual reachability.

Endpoints can be either virtual or physical. Virtual endpoints specify the addresses to which logistics consumers can send electronic or voice messages. This includes, most importantly, the internet address of the cloud-based platform. It also includes LSPs' contact details, such as email, telephone, and website (see Dong et al. 2008:822). Physical endpoints specify the actual addresses to which logistics consumers can send physical letters or at which logistics objects can be physically exchanged through a human-to-human interaction. This includes the location of a warehouse, an intermodal hub, or a delivery address. Real-world locations may be encoded as regular addresses or as GPS coordinates (Lian et al. 2007:434). Before service negotiation and selection, an LSP may specify physical endpoints only at a higher level of abstraction. For example, in a wide, networked logistical prototype scenario, LSPs may specify in which area of intense economic activity their capabilities are available. Physical endpoints can thus be used to identify and constrain the spatial (or geographical) availability of cloud logistics services (see "transport service area" Dong et al. 2008). During service negotiation and selection, abstract endpoints are specified based on the information provided by the logistics consumer and the LSPs. For example, consumers provide a pick-up address, and LSPs provide a warehouse address.

Service presence refers to the temporal availability of a cloud logistics service. The presence may be different for each action that can be performed at a virtual or physical endpoint. In order to enable on-demand availability and rapid elasticity, virtual endpoints required for negotiation and service selection will need to be available 24 hours a day every day of the week. Physical endpoints required for initiating service delivery need to be available in extended office hours.

Endpoints and service presence are related to the service interface. The service interface description specifies what can be exchanged and how an exchange is carried out, and service reachability specifies where and when an exchange can happen. The service interface and the service reachability are split into two concepts in the service description because the interface is likely to remain unchanged over a given period of time. The reachability, especially the physical endpoints, may differ for each service request, as consumers may request pick-ups at different locations and times.

3.7. Conclusions: Standardized, Semantic Service Descriptions and Market Governance

The preceding section has proposed a five-concept top-level ontology that includes key sub-concepts for the description of physical cloud logistics capabilities. The design of service descriptions is guided by several SOC design principles. By specifying the top-level concepts and key sub-concepts, this reference architecture establishes the foundation for *standardized service descriptions* of physical cloud logistics capabilities. By describing physical cloud logistics capabilities through an ontology containing only few top-level concepts, this reference architecture satisfies the principle of *abstraction*. By specifying the (non-)logistics functionalities in a top-level concept, *discoverability* of physical cloud logistics services is attained. Finally, by specifying the interface in a top-level concept, *composability* of physical cloud logistics services is supported. Recall that designing physical logistics capabilities according to the design principles of service-orientation contributes to delivering them with cloud characteristics from a coordination perspective (see Section F.X.2). We therefore conclude that using the proposed service description model contributes to their delivery with cloud characteristics.

Because service descriptions are the basis for service design and for posting a formal service request, they are also the templates for written contracts between logistics consumers and member LSPs during the negotiation and selection phase. Formalizing the exchange of physical cloud logistics capabilities between these two parties through standardized, semantic service descriptions aligns well with the type of contract law associated with market-based coordination (see Subsection C.III.3.1), which has been proposed as a governance structure between consumers and providers (see Section F.VI.3). Standardized, semantic service descriptions represent contracts that include policies to narrowly prescribe remedies and consequences of non-performance. They also encourage a legalistic interpretation because standardization and well-defined semantics reduce the ambiguity in the meanings of clauses. We thus conclude that in a market-based exchange environment, standardized, semantic service descriptions are a means to exchange physical cloud logistics capabilities in a transaction-cost economizing manner.

XII. Characteristic Mix of Coordination Mechanisms – Part IV: Market Mechanisms

1. Abstract of Chapter

Market mechanisms represent the mode of facilitating economic exchanges on the virtual marketplace on the cloud-based platform.

The *architectural viewpoint* describes the model kinds and establishes the conventions for designing market mechanisms to facilitate such economic exchanges. Specifically, this viewpoint (1) models the exchange and preference environment, (2) assesses the desirability of economic mechanism properties from the perspective of different stakeholders, and (3) assesses the desirability of solution concepts from the perspective of different stakeholders.

The *exchange and preference environment* in CLSs is a (a) multi-attribute (b) combinatorial auction (c) with sequential interdependencies among items that are allocated among (d) rational and intelligent member LSPs with (e) private information about their valuations and production costs for different bundles of services (see Subsection F.XII.2.2).

Four *economic mechanism properties* are commonly deemed desirable for a mechanism that implements a social choice function: (1) interim individual rationality, (2) incentive compatibility (which is a necessary condition for allocative efficiency), (3) allocative efficiency, and (4) (strong or weak) budget balance (see Subsection F.XII.2.3).

Interim individual rationality is desirable for logistics consumers, LSPs, and the LITPAP. It ensures that logistics consumers and LSPs are voluntary willing to participate in exchanges on the virtual market as they will not be worse off by simply participating. This property is also desirable for the LITPAP as it ensures the platform's long-term existence, its independent growth, and the emergence of network effects.

Incentive compatibility is desirable for logistics consumers and the LITPAP. It prevents that LSPs can behave strategically and exploit their information advantage about the true production costs over logistics consumers. This fosters a trustful consumer-provider relationship, thus fostering the long-term voluntary participation of logistics consumers, which makes this property also desirable for the LITPAP. From the perspective of LSPs, the desirability of incentive compatibility is double-edged. On the one hand, it effectively limits their ability to earn excess profits by behaving strategically. On the other hand, LSPs understand that strategic behavior in the short-term alienates consumers in the long-term. Thus, desirability depends on their temporal commitment to the cloud-based platform.

Allocative efficiency is desirable from the LITPAP's and logistics consumers' perspective because it facilitates exchanges such that requested services are allocated to those LSPs that value them most, which means that services are delivered at lowest costs. While this being clearly desirable for consumers, it is also desirable for the LITPAP because it ensures long-term acceptance of the virtual marketplace: exchanges are facilitated in a neutral manner (socially fair), neither favoring consumers nor LSPs. Yet, for LSPs allocative efficiency is double-edged. On the one hand, LSPs recognize the importance of this characteristic for consumers. On the other hand, allocative efficiency may result in allocations in which services are allocated to their competitors, which reduces their income. Thus, desirability from the perspective of LSPs again depends on their temporal commitment to the cloud-based platform.

Strong budget balance is desirable for logistics consumers and LSPs because it implies that no transaction costs are incurred by the mechanism: The logistics consumer's payment is entirely distributed to LSPs. For the LITPAP, strong budget balance is double-edged. On one hand, strong budget balance is desirable because it attracts consumers and LSPs to the virtual marketplace. On the other hand, strong budget balance is not desirable because the LITPAP cannot generate revenue by levying a transaction-based fee.

Weak budget balance is not desirable for logistics consumers and LSPs as such a mechanism incurs transaction costs: a nonnegative share of a logistics consumer's payment is not distributed to the LSPs. For the LITPAP, weak budget balance is desirable because the LITPAP can appropriate some non-negative share of the payment made by the consumer, thus creating a sustainable, scalable business model.

Mechanism design provides two *solution concepts* for the implementation of social choice functions: DSIC and BIC (see Subsection F.XII.2.4). DSIC is preferred by all stakeholders because it is more robust. Robustness translates into reduced information requirements for both the LITPAP when designing the mechanisms and for LSPs when devising their best strategy.

The *architectural view* summarizes the design of the mechanism to facilitate exchanges on the virtual marketplace in CLSs in the following proposition (see Subsection F.XII.3.1):

Proposition 16. *Given multi-attribute combinatorial exchanges, cloud logistics systems will deploy a mechanism that achieves interim individual rationality and weak budget balance at the expense of incentive compatibility and allocative efficiency.*

This proposition may be surprising; after all, it does not suggest an existing mechanism type, but instead proposes a tradeoff between economic mechanism properties. This is because none of the existing mechanism types possess a set of properties for the exchange and preference environment that is desirable for CLSs. More specifically, none achieves (interim) individual rationality and (weak) budget balance simultaneously. However, both are hard constraints given the conditions presumably prevailing in CLSs.

Conditions that determine the effectiveness of the proposed tradeoff of economic mechanism properties relate to the following factors: (1) autonomy of logistics consumers and LSPs, which manifests in their ability to freely decide on whether to participate in CLSs, (2) economic rationality, which manifests in stakeholders' profit-orientation, and (3) environmental hostility, which manifests in low margins and, hence, enforces economic rationality, especially in the case of member LSPs due to low margins in many logistics industry segments (see details in Subsection F.XII.3.2). Given the *autonomy* and *economic rationality* of consumers and LSPs, interim individual rationality is a necessary mechanism property to ensure their voluntary participation. Likewise, budget balance is a necessary property for an economically rational behaving LITPAP to facilitate market-based exchanges. Otherwise the LITPAP may risk to subsidize any exchange. Thus, individual rationality and budget balance are "hard constraints."

If individual rationality and weak budget balance become binding constraints for market mechanisms in CLSs, other desirable economic properties must be sacrificed because of the impossibility results; in other words, tradeoffs between allocative efficiency and incentive compatibility become inevitable. In this context, Blau (2009) proposes a mechanism which we consider as particularly intriguing for the application in CLS for two reasons. For an increasing number of service providers in the system, (a) service providers cannot benefit by deviating from a truth-telling strategy, thus making the mechanism basically strategy-proof (DSIC), and (b) losses in social welfare due to untruthful reporting are reduced (Blau 2009 as cited in van Dinther 2010:13, 15). This is an important and positive result for facilitating mechanism-based exchanges in CLSs because CLSs presumably comprise many LSPs and, due to resource redundancy, multiple LSPs are likely to participate in each exchange. Hence, deviating from a truth-telling strategy is presumably unattractive for LSPs. Consequently, there is a higher chance that logistics consumers participate in CLSs despite losses in efficiency and incentive compatibility.

2. Viewpoint: An Approach to Designing Market Mechanisms

2.1. Preliminary Considerations

Section C.III.5 has introduced the fundamental assumptions, solution concepts, and the mathematical (im-)possibility results of mechanism design. These (im-)possibility results show that, although they are desirable, certain combinations of solution concepts and economic mechanism properties cannot be achieved simultaneously. This viewpoint describes the model kinds and establishes the conventions for designing market mechanisms to facilitate exchanges on the virtual marketplace in CLSs. Specifically, this viewpoint (1) models the exchange and preference environment, (2) assesses the desirability of economic mechanism properties from the perspective of different stakeholders, and (3) assesses the desirability of solution concepts from the perspective of different stakeholders. Modeling the *exchange and preference environment* is crucial because its particularities determine the problem structure for designing a mechanism. Assessing the *desirability of economic mechanism properties* and the *desirability of solution concepts* from different stakeholder perspectives is important because they both determine the extent to which stakeholders can address their financial and logistical concerns. Stakeholders' preferences for these properties and solution concepts ultimately determine which combination is implemented on the virtual marketplace, subject to mathematical possibility.

2.2. Exchange and Preference Environment

Using typical assumptions of mechanism design, logistics consumers, member LSPs, and the LITPAP are assumed to be rational and intelligent. The cost of delivering logistics capabilities is private information held by a given LSP.

In addition, cloud logistics stakeholders are assumed to have quasilinear preferences, which means that their preferences are linear in at least one variable; in this case, financial transfers made or received. Specifically, we assume that LSPs can have non-linear valuations for different bundles of requested services, for example: "I only want A if I also get B" (Parkes 2001:8f.). Such preference structures are convenient in order to account for complementarities and substitution effects that exist between different items (Vries & Vohra 2003:284) and for interdependencies between different items that result from economies of scale and scope, typically inherent in logistics activities. For example, an LSP may only want to transport a consumer's shipment from point A to B if it simultaneously transports another shipment for the same consumer on the same route and if it performs packaging for both shipments before transportation. In the exchange environment, non-linear valuations for service bundles result in a combinatorial resource allocation problem when matching service requests to service offers (candidate services) (Parkes 2001:8f.).

In addition, physical cloud logistics capabilities are assumed to be specified by multiple attributes rather than a single attribute. Attributes can pertain to functional and non-functional service characteristics (see e.g. Schnizler et al. 2008:944). Functional attributes specify the type of logistics process or transformation, non-functional attributes specify the QoS. Every service must include non-functional attributes to specify its characteristics regarding on-demand availability, rapid elasticity, and pay-per-use (see Section F.XI.3). The use of multiple attributes is convenient for capturing the rich preferences of logistics consumers, who want to base their logistics decisions on multiple characteristics relevant to their business context rather than just on price. In the exchange environment, this means that service requests and offers need to be matched in a combinatorial manner subject to multiple attributes simultaneously.

Moreover, in addition to LSPs' non-linear preferences for bundles of services and multi-attribute services, sequential interdependencies between services need to be accounted for because composite services can only generate value if composed in a well-defined sequence, which means that the functional attributes of adjacent services need to match (Blau 2009). In other words, the mechanism needs to account for process modularity (see Section B.IV.3).

In sum, we can conceive of the exchange and preference environment in CLSs as a multi-attribute, combinatorial auction with sequential interdependencies among items that are allocated among rational and intelligent member LSPs with private information about their valuations and production costs for different bundles of services.

Such an exchange environment can be conveniently modeled with a "service value network," which is based on state charts (Blau 2009:85ff.; also see Blau et al. 2009a; Blau et al. 2009b; for a related model see Leukel et al. 2011a). Figure 65 depicts the service value network model. Blau (2009:89f.) defines the service value network model as an acyclic, K-partite, and directed graph that consists of a set of nodes representing service offers and a set of edges denoting technically feasible service compositions. The graph connects a source node with a sink node so that every feasible path between source and sink represents a possible realization of a composite service. Each service offer is specified through multiple attributes, both functional and non-functional. The graph is subdivided into K partitions, each of which represents a candidate pool of service offers that provide the same functionality, but with potentially different non-functional attributes. Each service offer is associated with a service provider through an ownership relationship. Each edge that connects a service offer with its predecessor indicates its provider's variable costs for performing a service given this predecessor. The cost information is private information belonging to the respective service provider. A mechanism deployed to this model must elicit the private information from providers first and then compute the "shortest" path through the network (Blau 2009:105).

2.3. Desirability of Economic Mechanism Properties for Stakeholders

Four economic properties are commonly deemed desirable for a mechanism that implements a social choice function: (1) interim individual rationality, (2) incentive compatibility (which is a necessary condition for allocative efficiency), (3) allocative efficiency, and (4) (strong or weak) budget balance (Subsection C.III.5.4). Different combinations of these properties influence the stakeholders' ability to address their logistical and economic concerns. This is because different combinations result in different service allocations and payments to be made. In order to understand which combination of properties is most in effective coordinating exchanges in CLSs, we need to investigate the desirability of each property from the perspective of each stakeholder. We discuss these properties from the

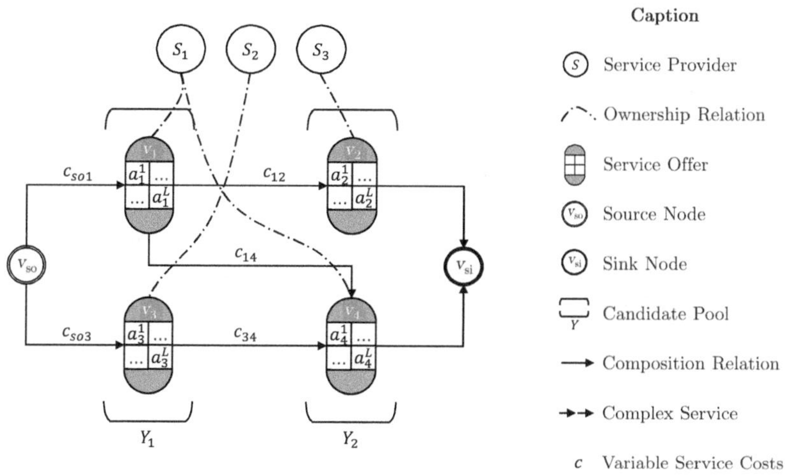

Figure 65.: Service Value Network Model (adopted with minor changes from Blau 2009:90)

perspectives of (1) logistics consumers, who request and buy logistics services; (2) member LSPs, who sell and provide logistics services; and (3) the LITPAP, which mediates the mechanism-based exchange on the virtual marketplace.

Interim individual rationality frames logistics consumers' and LSPs' logistical and financial concerns because it ensures that these parties will not be worse-off if they participate in the mechanism. Therefore, they are willing to join voluntarily. Interim individual rationality also frames the LITPAP's financial concerns. Voluntary participation represents a fundamental prerequisite for the long-term existence of the virtual marketplace, its independent growth, and the emergence of network effects. It thus represents a fundamental prerequisite for making a profit from mediating exchanges on the virtual marketplace. Interim individual rationality is particularly important for the LITPAP since it lacks the authority over logistics consumers and LSPs to coerce them to join the virtual marketplace. Therefore, the only legitimate way to secure their voluntary participation is to offer a mechanism that caters to their intrinsic profit-orientation.

Incentive compatibility frames logistics consumers' and LSPs' logistical and financial concerns because reported preferences determine the allocation of services and the payments to be made. From the perspective of logistics consumers, incentive compatibility is crucial. It creates trust and acceptance of the exchanges facilitated on the virtual marketplace. With incentive compatibility, consumers can be confident that LSPs are not able to systematically exploit their informational advantage regarding the true production costs of services, because LSPs cannot appropriate excessive rents by lying about their preferences. This fosters "a trustful requester-provider-relationship" (Blau et al. 2009a:355). Moreover, this can nurture the perception that this mechanism represents a legitimate way of facilitating exchanges (because it is based on fair business conduct) that leads to legitimate outcomes. We thus conclude that incentive compatibility is an important factor for ensuring the long-term participation of logistics consumers on the marketplace. From the perspective of LSPs, the desirability of incentive compatibility is double-edged. On one hand, incentive compatibility effectively limits their ability to behave strategically and thus systematically exploit their informational advantage over logistics consumers. Strategic behavior is safeguarded by incentives offered for truthful preference revelation. Hence, inventive compatibility limits their ability to address financial concerns by making excess profits. On the other hand, LSPs understand that strategic behavior in the short-term chases logistics consumers away in the long-term. After all, LSPs are aware of that logistics consumers are intelligent and hence aware of the ability LSPs to behave strategically. We thus conclude that the degree to which LSPs perceive incentive compatibility as desirable depends on their temporal commitment to a CLS and their interest in using the CLS as a long-term sales channel. From the LITPAP's perspective, incentive compatibility indirectly frames financial concerns. It is desirable because it fosters logistics consumers' participation and thus also the participation of LSPs. Furthermore, it is a prerequisite for achieving allocative efficiency and thus to securing logistics consumers' long-term market participation.

Allocative efficiency indirectly frames the LITPAP's financial concerns by ensuring long-term acceptance of the marketplace; it frames logistics consumers' and LSPs' financial and logistical concerns through the allocation of service requests to service offers. From the LITPAP's perspective, allocative efficiency is desirable because it refers to a "socially fair" allocation of requested services to LSPs: Each service is allocated to the LSP that values the service most. This translates to services being allocated to the LSPs that offer them with lowest costs. In this way, the LITPAP acts as a "neutral mediator" that focuses on achieving the best system-wide solution, neither biased toward consumers nor LSPs. This is important for securing consumers' and LSPs' long-term participation, and thus long-term success of the marketplace; if the LITPAP was systematically biased, either toward consumers or LSPs, the other stakeholder class would leave the marketplace in the long term. This would reduce the LITPAP's opportunity to generate revenue through membership fees or other services. From the perspective of logistics consumers, allocative efficiency is also desirable because the mechanism allocates requested services to LSPs that report realizing the service request with minimal total costs (Chen et al. 2005:468). From an LSP's perspective, allocative efficiency is double-edged. On one hand, LSPs recognize the importance of allocative efficiency for logistics consumers and its importance for securing a CLSs as a long-term sales channel. On the other hand, allocative efficiency may not necessarily be desirable. While an efficient allocation maximizes the total sum of valuations, it may not necessarily maximize the individual utility of any LSP at the same time. After all, LSPs act strategically, if possible, to influence the allocation of services and thus to increase their profit. Again, as in the case of incentive compatibility, we conclude that the degree to which LSPs consider allocative efficiency desirable depends on their temporal commitment to the virtual marketplace and their intention to use the marketplace as a long-term sales channel.

Strong budget balance frames the financial concerns of logistics consumers, LSPs, and the LITPAP through financial transfers made or received. For logistics consumers and LSPs, strong budget balance is desirable because it implies that no transaction costs are incurred by the mechanism: The logistics consumer's payment is fully distributed to the LSPs. For the LITPAP, strong budget balance is double-edged. On one hand, strong budget balance is desirable because, in combination with allocative efficiency, the service allocation is Pareto optimal, which may be desirable for logistics consumers and LSPs. Furthermore, the mechanism does not run a deficit, which means that the LITPAP is not required to continuously subsidize the mechanism, which is desirable (van Dinther 2010:13). On the other hand, the mechanism does not run a surplus either. Hence, the LITPAP cannot appropriate any share of the payments made by logistics consumers, and transaction based fees cannot be levied. The LITPAP must therefore look for alternative revenue opportunities, such as membership fees or by offering supplementary services.

Weak budget balance also frames the financial concerns of logistics consumers, LSPs, and the LITPAP through financial transfers made and received. For logistics consumers and LSPs, weak budget balance is not desirable; a weakly balanced mechanism can incur transaction costs because a nonnegative share of a logistics consumer's payment is not distributed to the LSPs. By contrast, for the LITPAP, weak budget balance is desirable. This property lets the LITPAP appropriate some fraction of the logistics consumer's payment of each facilitated exchange. In other words, weak budget balance creates a sustainable, volume-based (scalable) business model for the LITPAP. Yet, the amount that the LITPAP can appropriate for each exchange is unknown before the exchange, as it depends on the information reported by the LSPs.

Table 32 summarizes this assessment on the desirability of economic properties from the perspectives of different stakeholders. The table shows that desirability differs between stakeholders, and only interim individual rationality is desirable for all.

Mechanism Property	Logistics Consumer	LSPs	LITPAP
Interim individual rationality	+	+	+
Incentive compatibility	+	+/–	+
Allocative efficiency	+	+/–	+
Strong budget balance	+	+	–
Weak budget balance	–	–	+

Table 32.: Desirability of Economic Properties from Stakeholder Perspectives (+ desirable; +/– double-edged; – less desirable)

2.4. Desirability of Solution Concepts for Stakeholders

Mechanism design provides two solution concepts for the implementation of social choice functions: DSIC and BIC (see Subsection C.III.5.3). This subsection assesses the desirability of these solution concepts from the perspective of the LITPAP, logistics consumers, and member LSPs. Interestingly, the desirability does not differ between the different stakeholders.

To start, let us recall the desirability of both solution concepts for a neutral policy maker. The implementation of a desired social choice function through DSIC is more robust than through BIC and hence preferred by the policy maker: In DSIC, the policy maker can design the mechanism and agents can devise their best strategy independent of their beliefs about the distribution of agent types. As a consequence, the optimal implementation of a social choice function is robust against inconsistent and incorrect beliefs. In addition, in DSIC, the implementation of a social choice function is feasible for any distribution of types of member LSPs. Conversely, BIC is limited to distributions that belong to the exponential family (see Subsection C.III.5.3).

DSIC implementation is more desirable in CLSs due to its higher robustness compared with BIC. In fact, all stakeholders prefer DSIC over BIC because robustness translates into reduced informational requirements and more predictable outcomes: The LITPAP can design the mechanism without any knowledge of or constraints on the member LSP distribution. This is desirable because it reduces design complexity. Moreover, this makes the mechanism applicable to real-world scenarios because the types of member LSPs will most likely follow some ill-behaved, rather than an exponential distribution. Hence, the LITPAP can be more confident that, in DSIC, desired outcomes are actually achieved, thus enabling the LITPAP to better address its financial concerns. Likewise, logistics consumers can be more confident that the requested services are procured at minimum costs, allowing them to better address their logistical and financial concerns. For member LSPs, DSIC "tremendously lowers strategic complexity" (Blau et al. 2009a:355). Unlike in BIC, LSPs can devise their best strategies in DSIC independent of their beliefs about the distribution of member LSPs types and of the strategies played by other member LSPs. This reduction in strategic complexity seems especially desirable in CLSs, as physical cloud logistics capacities are specified by multiple attributes, making the development of an optimal strategy already a complex task (see Subsection F.IX.3.7). Furthermore, DSIC lets LSPs better address their financial concerns because they can be more confident that the strategy they played is indeed their best. We thus conclude that DSIC is more desirable than BIC for all cloud logistics stakeholders.

3. View: A Tradeoff between Desirable Mechanism Properties

3.1. Proposition

The last section has specified the exchange environment, assessed the desirability of economic mechanism properties and solution concepts from different stakeholder perspectives and linked them to stakeholder concerns. This section puts forth a proposition to specify the kind of mechanism via which exchanges are presumably facilitated on the virtual marketplace. Note that, due to the abstract nature of this reference architecture, we do not design an actual mechanism, but focus on identifying a preferable combination of economic properties. Also note that we do not consider issues related to computational tractability, which focuses on whether a mechanism computes the outcome allocation and financial transfers in polynomial time of inputs (on this topic see e.g. Parkes 2001; Blau 2009). Although this is an important issue in mechanism design, it is beyond the scope of this thesis, as we focus on more abstract economic and organizational issues.

Given that the LITPAP establishes, maintains, and evolves the cloud-based platform and the functionality implemented thereon, including the virtual marketplace, we assume that the LITPAP acts as "trustworthy mediator" to facilitate mechanism based exchanges (see "revelation principle" in Subsection C.III.5.3). Member LSPs confidentially and separately report their preferences (either truthfully or untruthfully) to the LITPAP, which aggregates them and determines the allocation of requested services to member LSPs and the payments to be made by the logistics consumer. As the LITPAP is neither a proximate nor distant competitor to member LSPs, LSPs can truthfully reveal their preferences without fearing that their competitive position is threatened, assuming that the LITPAP maintains absolute privacy of the reported preferences as ensured in the cooperation contract.

With regard to the type of mechanism implemented on the virtual marketplace, we propose:

Proposition 16. *Given multi-attribute combinatorial exchanges, cloud logistics systems will deploy a mechanism that achieves interim individual rationality and weak budget balance at the expense of incentive compatibility and allocative efficiency.*

This proposition may be surprising; after all, it does not suggest an existing mechanism type, but instead proposes a tradeoff between economic mechanism properties. This is because none of the existing mechanism types possess the set of properties for the exchange and preference environment that is desirable for CLSs. More specifically, none achieve interim individual rationality and (weak) budget balance simultaneously. Neither the Groves nor dAGVA mechanism can achieve weak budget balance or individual rationality. Clarke mechanisms achieve individual rationality in a fairly general environment, but weak budget balance only if there is no single agent effect (Parkes 2001), which means that any agent can be removed from the mechanism without reducing the total value of the best solution for the remaining agents. However, this assumption does not hold in the CLS exchange environment. In combinatorial exchanges, in which agents have non-linear preferences for bundles of goods, it is not possible to remove any agent without the risk of reducing the total value of the exchange for the remaining agents precisely because of non-linear agent preferences. Hence, Clarke mechanisms are not suitable for facilitating exchanges in CLSs. GVA achieves individual rationality and weak budget balance but only in case of single-attribute combinatorial auctions. However, physical cloud logistics capabilities are specified by multiple attributes, thus making GVA mechanisms unsuitable. In conclusion, existing mechanism types are not suitable for facilitating economic exchanges in CLSs.

The following subsection provides a rationale for the proposed tradeoff of economic properties by investigating the conditions of critical contingency factors.

3.2. Conditions

The factors that determine the effectiveness of the proposed tradeoff of economic mechanism properties are (1) autonomy of logistics consumers and LSPs, which manifests in their ability to freely decide on whether to participate in CLSs, (2) economic rationality, which manifests in stakeholders' profit-orientation (economic concerns), and (3) environmental hostility, which manifests in low margins and, hence, enforces economic rationality, especially in the case of member LSPs due to low margins in many logistics industry segments.

The interaction between autonomy and economic rationality makes interim individual rationality and (weak) budget balance indispensable economic properties for any market mechanism:

> "Budget-balance and individual rationality are hard constraints that the mechanism must satisfy. If these two requirements are not met, the mechanism will not be sustainable; participants will not voluntarily participate in the market if they incur losses and the market operator will not be willing to subsidize the mechanism in the long run" (Stößer 2009:42; see similar Blau 2009:77).

Both properties represent equally "hard constraints" in CLSs. As argued above, *individual rationality* is a desirable property for all stakeholder classes (see Subsection F.XII.2.3). Logistics consumers and member LSPs are autonomous, profit-oriented organizations that only join a CLS if they can expect to be at least as well off as if they had not joined. Furthermore, individual rationality is particularly crucial for LSPs due to the highly competitive business environment. Individual rationality is equally important for the LITPAP, which lacks sufficient authority to coerce either consumers or LSPs to join. Hence, the LITPAP relies on offering a "true" business opportunity to ensure long-term participation and thus the potential to make a profit related to operating the platform.

With regard to *budget balance*, our above assessment has revealed varying degrees of desirability. For the LITPAP, budget balance is desirable because, just like consumers and LSPs, the LITPAP is a profit-oriented organization. Hence, the LITPAP is presumably not willing to continuously subsidize exchanges. More specifically, the LITPAP presumably prefers weak over strong budget balance because this creates a sustainable business model: The LITPAP can appropriate any surplus resulting from exchanges. For the LITPAP, this surplus represents a monetary compensation for developing, maintaining, and evolving the cloud-based platform and for facilitating economic exchanges on the virtual marketplace. Note, however, that the monetary compensation depends on preferences reported by LSPs, thus making it impossible for the LITPAP to know the amount of revenue *ex ante*. In fact, as weak budget balance refers to an inequality condition, the appropriable revenue may be zero for some exchanges. For logistics consumers and LSPs, this surplus represents transaction costs, and is therefore presumably undesirable. Yet these costs may become acceptable if the mechanism based mediation adds "true value" for them, which means that they cannot coordinate a better exchange in another way outside of the CLS. This is particularly the case if exchanges need to be coordinated quickly and involve multiple LSPs with complementary capabilities (composite services). This is likely in logistical prototype scenarios with a wide, networked geographic scope and/or a capability scope consisting of selected compositions (Scenario II, Scenario III, Scenario IV). Logistics consumers can thus conceive of these costs as "coordination costs of on-demand and rapidly elastic provisioning." Member LSPs can conceive of them as "coordination costs for enhancing their problem solving capability" (see Subsection F.V.3.5).

If individual rationality and weak budget balance become binding constraints for market mechanisms in CLSs, other desirable economic properties must be sacrificed because of the impossibility results; in other words, tradeoffs become inevitable. The MS impossibility theorem becomes important in this context. Recall that it states that no mechanism can achieve allocative efficiency, strong budget balance, and individual rationality simultaneously, not in BIC and hence neither in DSIC (see Subsection C.III.5.5). Thus, efficient allocations cannot be reached in CLSs irrespective of incentive compatibility. In this context, Parkes et al. (2001:1161) point to two alternative approaches:

> "(a) impose BB [(weak) budget balance] and IR [individual rationality], and design a fairly efficient but incentive-compatible (or perhaps strategy-proof) scheme. (b) impose BB and IR, and design a fairly efficient and fairly incentive-compatible scheme."

At the level of abstraction used in this reference architecture, no judgment is possible regarding which of these approaches is more reasonable for CLSs. This ultimately depends on the extent of how "fairly efficient" a strategy-proof mechanism might be compared with how "fairly efficient" and "fairly incentive-compatible" an alternative mechanism might be. Yet, with regard to these approaches, Parkes et al. (2001:1161) argue that that the first approach may not be easily extended to combinatorial allocation problems. Whether or not this is theoretically possible remains to be shown; a search of recent literature could not provide any insights into this issue. Hence, given the multi-attribute combinatorial allocation problem encountered in CLSs, we focus on the second approach in the following discussion.

The pivotal question of designing a "suboptimal" or "second-best" mechanism is whether the severity of losses in incentive compatibility and allocative efficiency are acceptable to logistics consumers and member LSPs. Losses in incentive compatibility are particularly critical because they mean that LSPs have an incentive to untruthfully reveal their preferences: They can devise a strategy that manipulates the mechanism's allocation and financial transfers to their advantage by lying to the LITPAP. Of course, this deters logistics consumers from participating in the CLS, as it creates an "institutionalized" opportunity for LSPs to extract rents from consumers. Moreover, losses in incentive compatibility directly translate to losses of allocative efficiency. A loss in allocative efficiency means for logistics consumers that they procure a bundle of services at a higher cost than theoretically necessary. Conversely, LSPs can extract an excessive profit from consumers by untruthfully reporting their preferences (see Subsection F.XII.2.3). Whether such a suboptimal mechanism is acceptable to logistics consumers cannot be determined at the level of abstraction adopted in this reference architecture. On the one hand, it depends on the extent to which LSPs can manipulate the resulting allocation and financial transfers of a mechanism by behaving strategically. On the other hand, it depends on the opportunities logistics consumers have outside the CLS. One can conclude that the higher the losses in incentive compatibility and allocative efficiency, the more likely that logistics consumers can procure comparable services at lower prices outside of the CLS.

Although the preceding paragraphs may have delivered disappointing results, scholars have recently started to develop second-best mechanisms with somewhat promising results. An important contribution in this respect is made by Blau (2009), who develops an interim individual rational and weakly budget balanced mechanism that has "fairly preferable" properties for incentive compatibility and allocative efficiency. The exchange and preference environment of this mechanism matches the conditions encountered in CLSs; in fact, we have used the model of service value networks proposed by Blau (2009) to characterize the exchange and preference environment in CLSs (see Section F.XII.2).

Blau's (2009) proposed mechanism is particularly intriguing for CLS application for two reasons. For an increasing number of service providers in the system, (a) service providers cannot benefit by deviating from a truth-telling strategy, thus making the mechanism basically strategy-proof (DSIC), and (b) losses in social welfare due to untruthful reporting are reduced (Blau 2009 as cited in van Dinther 2010:13, 15). This is an important and positive result for facilitating mechanism-based exchanges in CLSs because CLSs presumably comprise many LSPs and, due to resource redundancy, multiple LSPs are likely to participate in each exchange. Hence, deviating from a truth-telling strategy is presumably unattractive for LSPs. Consequently, there is a higher chance that logistics consumers participate in CLSs despite losses in efficiency and incentive compatibility. This is also a positive result for the LITPAP because it can partially resolve the issue of losses in incentive compatibility and allocative efficiency by setting incentives for platform growth (van Dinther 2010:13, 15; also see Conte et al. 2010:7). However, this represents a double-edged result for member LSPs. On one hand, it reduces their ability to extract excessive rents from logistics consumers. On the other hand, it increases the likelihood of logistics consumers joining the marketplace, and it reduces informational requirements and strategic complexity. LSPs can largely devise their best strategy without the need of making assumptions about the types and actions of other member LSPs, reducing the complexity for participating in a mechanism based exchange. Moreover, reducing strategic complexity can increase the speed of bidding and thus contribute to on-demand delivery.

We thus conclude that Blau's (2009) proposed mechanism is a promising candidate for application in CLS and a good starting point for further research. In fact, research in this field would benefit not only CLSs, but also the wider field of mechanism-based exchanges of computing services.

3.3. Conclusions: Acceptable Intensity of Price Competition

The preceding subsections have proposed a tradeoff between economic mechanism properties of market mechanisms. Market mechanisms assume a central role in CLSs, as they govern the primary value exchanges between logistics consumers and member LSPs. Formalizing the process of how economic decisions are reached enables the rapid reconciliation of multiple stakeholder's diverse preferences. This is critical for logistics consumers to quickly sign service contracts and thus to attain on-demand delivery from a coordination perspective. This is also important for member LSPs, as they can rapidly integrate their capabilities with capabilities of other member LSPs in order to collectively deliver complex logistics solutions with low lead time. Thus, market mechanisms are crucial instruments for increasing member LSPs' problem solving abilities (see Subsection F.V.3.5), which may be a source of competitive advantage for them over non-member LSPs. Through the lens of the RBV for interconnected firms, market mechanisms represent a device for accessing other LSPs' resources and distributing rents generated through cooperation.

However, market-based coordination generally exposes sellers to intense price competition, especially if price is the only attribute to differentiate them from competitors, as in the case of commodities. While this general market argument also applies to CLSs, the intensity of price competition is, at least to some extent, reduced by describing physical cloud logistics capabilities using multiple attributes, all visible in the service descriptions. Thus, member LSPs have some options to differentiate themselves from competing member LSPs by offering capabilities with different attributes. Price becomes one of several selection criteria. In fact, the relative importance of price as a selection criterion decreases with an increasing number of attributes that are relevant for logistics consumers when selecting capabilities. Specifically, member LSPs may be able to achieve competitive advantage by offering capabilities with lower lead times (on-demand availability) and quicker capacity scaling (rapid elasticity). Still, price competition remains in effect for each combination of attributes. Hence, if multiple member LSPs offer capabilities with very similar attributes, they may only be able to achieve competitive advantage through lower production costs.

The relative importance of price differs among the logistical prototype scenarios (see Subsection F.V.3.3). Importance specifically depends on each scenario's capability scope. Prices become more important the higher the homogeneity of physical cloud logistics capabilities offered by a prototype scenario.

Accordingly, price competition is intense in Scenario Ia and Ib, as, in both scenarios, capabilities are highly homogeneous. Due to the localized geographic scope of these scenarios, any other member LSP will be a direct competitor.

In Scenario II, price competition is also intense because transportation, transshipment, pick-up, and delivery capabilities provide little room for differentiation. Due to the wide, networked geographic scope of this scenario, member LSPs are only in direct competition with member LSPs that offer capabilities for the same transportation link or within the same area of intense economic activity, but not with the remaining member LSPs.

In Scenario III, price competition is intense for transshipment and (last-mile) distribution capabilities, due to these capabilities' homogeneity. Price competition is less intense for storage and SVACs, as providers have more possibilities to differentiate themselves from their competitors by tailoring their offerings to several quality attributes, such as lower picking-and-packing times. Due to the localized geographic scope, member LSPs are in direct competition with any other member LSP that offers similar capabilities.

In Scenario IV, price competition is intense among LSPs that offer transportation and transshipment capabilities, due to their homogeneity. By contrast, price competition is less intense among LSPs that offer storage capabilities and SVACs because these providers have the opportunity to differentiate their offerings through different quality attributes. Due to the wide, networked geographic scope of this scenario, member LSPs only compete directly with those that offer similar capabilities in the same geographic region, thus creating several "pockets of competition."

We thus conclude that CLSs are likely to expose member LSPs to a higher degree of price competition than other sales channels. However, prices are not everything. The use of multiple attributes to specify physical cloud logistics services provides some remedies. Moreover, the ability of market mechanisms to enable member LSPs to integrate their capabilities quickly with those of other LSPs, and thus increase their problem solving ability, provides some compensation for increased price competition. Whether these compensations are enough to make LSPs voluntarily join a CLS cannot be determined on the level of abstraction in this reference architecture. Still, this presents a crucial research question to be answered before successfully introducing CLSs.

XIII. Summary of Propositions

The preceding nine chapters have described a CLS reference architecture. Specifically, they have described both the design process (in the form of viewpoints) and the design outcome (in the form of views). The cloud logistics reference architecture has been designed on different levels of granularity. While the value creation logic focuses on CLSs from an aggregate perspective, the physical logistics infrastructure, structural governance, centralization, formalization, and the characteristic mix of coordination mechanisms each focus on a specific design dimensions. Servitization, standardized, semantic service descriptions, and market mechanisms each focus on a specific mechanism in the characteristic mix of coordination mechanisms. Together these nine viewpoints and views provide a comprehensive understanding of the reference architecture design process and its outcome. Table 33 summarizes the design propositions of the reference architecture of CLSs. These propositions can be understood as recommendations to practitioners who aim to implement a CLS. For scholars, they represent direct starting points to further investigate these design choices with the objective to provide further theoretical support or empirical evidence and, if required, suggest adjustments.

View	Proposition
Value Creation Logic	Cloud logistics systems are a special type of meso-logistics systems that comprise logistics consumers, LSPs, and a LITPAP and that offer Logistics-as-a-Service: Logistics consumers are provided with access to a shared, open pool of service-oriented physical and virtual logistics capabilities that can be made available in an on-demand, rapidly elastic, and pay-per-use manner with minimal provider interaction by means of formalized mechanisms via a universally accessible cloud computing technology based platform.
Physical Logistics Infrastructure	Cloud logistics systems will operate in an environment characterized by a particularly high degree of temporal specificity of logistics resources and capabilities.
	Cloud logistics systems will make use of warehouses and transshipment points that are located within or in the vicinity of areas of high economic intensity.
	Cloud logistics systems will make use of air and road transport.
	Cloud logistics systems will provide transportation links in, in the vicinity of, and between areas of intense economic activity.
	Cloud logistics systems will make use of logistical resources that are compatible with either a standardized unit load or an industry-wide standardized handling interface.
	Cloud logistics systems will offer transportation (spatial), storage (temporal), transshipment (quantity and composition), and industry-wide standardized value-added (non-)logistical transformations of logistics objects.
	Cloud logistics systems will use personnel with the capacity to deploy the resources required to achieve core logistical and industry-wide standardized, value-added, (non-)logistical transformations of logistics objects that use standardized unit-loads or an industry-wide standardized handling interface.

	The physical infrastructure of any cloud logistics system will match one of five logistical prototype scenarios, with each scenario's capability scope being either narrow or comprising selected compositions and its geographic scope either localized or wide, networked.
Structural Governance	Cloud logistics systems represent a unilateral contract-based alliance consisting of a LITPAP and a network of horizontally cooperating LSPs; value exchanges between logistics consumers and LSPs participating in a CLS are facilitated through a virtual marketplace via the price mechanism.
Centralization	Lead LITPAP governance will be an effective distribution of authority for network-level decisions in cloud logistics systems.
Formalization	Cloud logistics systems will formalize (a) the relationship between the LITPAP and LSPs that intend to offer logistics capabilities on the virtual marketplace by means of a written "framework" cooperation contract and (b) the characteristic mix of coordination mechanisms in the applications implemented on the cloud-based platform (computer-formalized mechanisms).
Characteristic Mix of Coordination Mechanisms	Cloud logistics systems will use a characteristic mix of formalized coordination mechanisms implemented on a cloud-based platform and consisting of programmed routines, system-supported skills, system-supported supervision, implicit coordination, price, and service-orientation all structured along the life cycle of cloud logistics services, where service-orientation is an underlying mechanism that supports all life cycle phases.
Servitization	Servitization refers to a programmed routine that consists of four steps: (1) identification, (2) resource and capability modeling, (3) encapsulation, and (4) description. These steps incorporate the IoT and are guided by the design principles of service-orientation.
Standardized, Semantic Service Descriptions	Physical cloud logistics capabilities will be described by means of a logistics-domain specific, ontology-based description language that uses a standardized "top-level" ontology that consists of five concepts: (1) service functionality, (2) service policies, (3) service metrics, (4) service interface description, and (5) service reachability.

Market Mechanisms	Given multi-attribute combinatorial exchanges, cloud logistics systems will deploy a mechanism that achieves interim individual rationality and weak budget balance at the expense of incentive compatibility and allocative efficiency.

Table 33.: Cloud Logistics Reference Architecture – Summary of Design Propositions

Part G. Conclusions and Outlook

The term "cloud logistics" has emerged recently in academia and practice. This term combines the terms "cloud computing" and "logistics," and inherits and blends their meanings, which obscures the resulting meaning(s). In this context, this thesis' contribution is twofold.

This thesis' first contribution is the systematic identification and synthesis of existing (non-)academic cloud logistics knowledge. This review has identified four distinct but related meanings of "cloud logistics." Where appropriate, the knowledge associated with each meaning has been structured in an ontological manner by specifying key concepts and their interrelations. The identified meanings are distinct, as they refer to different phenomena; they are related, as they share a cloud-based platform as common element. Several research opportunities have been proposed for each meaning. Due to the novelty of these fields, many more are likely to exist and worthwhile to pursue.

Based on this systematic knowledge review, this thesis' primary contribution is the advancement of knowledge about the fourth meaning of "cloud logistics," which can be summarized by the term Logistics-as-a-Service. Specifically, this thesis has synthesized cloud logistics knowledge with existing logistics, organizational, and cloud computing literature in order to design a reference architecture of a CLS, a special type of meso-logistics system that delivers physical logistics capabilities with cloud characteristics. This reference architecture identifies nine perspectives at different levels of granularity from which CLSs can be considered and designed. These perspectives anchor the design of CLSs in the literature by using existing methods, models, and constructs. Moreover, these perspectives collectively provide a comprehensive understanding of how the design principles and concepts of cloud computing can be transferred to logistics system design and what the outcome of this design process "looks like." The reference architecture thus shows which physical logistics capabilities (service models) can be offered in an on-demand, rapidly elastic, and pay-per-use manner based on a shared resource pool and can be made available through a universally accessible cloud-based platform in a self-service manner (essential cloud characteristics) and how such systems need to be governed and orga-

F. Jaekel, *Cloud Logistics*, Produktion und Logistik,
https://doi.org/10.1007/978-3-658-22837-8_7

nized (deployment model). However, the goal of delivering physical logistics capabilities with cloud characteristics significantly limits the scope of capabilities and the geographies within which they can be made available. Still, the capabilities that can be delivered with cloud characteristics reconcile the objectives of logistical efficiency and effectiveness, enabling logistics consumers to realize agile logistical responses, which may help them gain competitive advantage in dynamic environments.

Although we identified several logistics systems that share some elements with CLSs, the proposed reference architecture can largely be conceived of as a *gedankenexperiment* of an innovative logistics system that, so far, exists in our minds only. The reference architecture hence becomes both an "object of design" and "object of inquiry."

As an object of design the reference architecture expresses both the design process and the design outcome in a standardized model for architecture descriptions; the reference architecture thus becomes well-structured and easily accessible to both scholars and practitioners in the fields of logistics and logistics-related computer science. This is crucial given the novelty and interdisciplinary character of cloud logistics. The reference architecture therefore has the potential to establish a common language among scholars and practitioners of these fields. Moreover, the reference architecture is of critical relevance for practitioners because it provides an abstract solution template which practitioners could immediately follow when implementing a (prototype) CLS.

As an object of inquiry, it provides an intermediate point of reference or, in other words, a foundational structure for scholars' research efforts. In fact, immediate future efforts from scholars should focus on supporting practitioners to implement a prototype system in order to collect evidence that allows the propositions of this reference architecture to be corroborated, refined, and/or refuted. The first prototype implementation would also play an important role in investigating, and potentially validating, the economic viability of this new logistics concept. In addition, researchers should focus on further transferring the principles of virtualization and service-oriented design to the logistics domain. Specifically, researchers should focus on developing a suitable ontology-based description language that takes into account the particularities of the logistics domain, enabling coordination through standardized, semantic service descriptions. This research would benefit not only the field of cloud logistics, but also the general field of service-oriented logistics and potentially the field of cloud manufacturing.

Future research should also more deeply investigate the impact CLS participation has on stakeholders. It should also focus on how the delivery of modular, service-oriented logistics capabilities may impact LSPs' the organizational structure and how LSPs could gain competitive advantage over other LSPs within and outside of a CLS. For logistics consumers, research should focus on identifying specific business contexts in which the use of physical cloud logistics capabilities can be a potential source of competitive advantage. This should also include identifying industries that may particularly benefit from cloud logistics capabilities. For the LITPAP, research should focus on how IT vendors can develop into central players in the logistics industry, especially by focusing on how they overcome the initial "chicken-and-egg" problem of connecting a critical number of logistics consumers and LSPs to the cloud-based platform. In sum, the advantageousness for stakeholders to participate in CLSs will ultimately determine the diffusion and success of such special type of logistics system.

Future research should also extend beyond the immediate field of cloud logistics. After all, logistics systems connect the fundamental economic sectors of production and consumption. In addition to cloud logistics, scholars adopt the concepts of virtualization and service-orientation for production systems, thus creating cloud manufacturing systems (see e.g. He & Xu 2014; Xu 2012) or the semantic (integration of) supply chains (see e.g. Ameri & Patil 2012; Leukel et al. 2011a; Leukel et al. 2011b; Leukel & Kirn 2008). Based on this context, investigating potential commonalities, differences, and opportunities among these concepts with an eye toward cross-fertilization is a worthwhile research avenue. In fact, the question of whether integrating cloud logistics with cloud manufacturing can form a "cloud supply chain" is a particular intriguing field for future research. Furthermore, scholars should also investigate potential similarities, differences, and opportunities in relation to the concept of "Industry 4.0". Like in cloud logistics, this concept is based on connecting distributed production resources and to collecting information via the IoT.

In conclusion, this reference architecture has established an intermediate point for reference of both practitioners and scholars. As such, it aspires to be among the first contributions to cloud logistics, rather than the last. Moreover, this architecture hopes to create additional momentum among practitioners and scholars to test and investigate this new logistical phenomenon about which we so far know very little.

Bibliography

ABAWAJY, J. (2009), Determining Service Trustworthiness in Intercloud Computing Environments, in *10th International Symposium on Pervasive Systems, Algorithms, and Networks (ISPAN)*, pp. 784–788.

AERTSEN, F. (1993), Contracting out the Physical Distribution Function: A Trade-off between Asset Specificity and Performance Measurement, *International Journal of Physical Distribution & Logistics Management* 23(1), pp. 23–29.

AGARWAL, A., SHANKAR, R., & TIWARI, M. K. (2007), Modeling agility of supply chain, *Industrial Marketing Management* 36(4), pp. 443–457.

AGRAWAL, D., ABBADI, A., DAS, S., & ELMORE, A. J. (2011), Database Scalability, Elasticity, and Autonomy in the Cloud, in Jeffrey Xu Yu, Myoung Ho Kim, & Rainer Unland, eds., *16th international conference on Database systems for advanced applications (DASFAA)*, volume 6587 of *Lecture Notes in Computer Science*, pp. 2–15 (Springer, Berlin, Heidelberg).

AKERMAN, A., & TYREE, J. (2006), Using ontology to support development of software architectures, *IBM Systems Journal* 45(4), pp. 813–825.

AKKIRAJU, R., FARRELL, J., MILLER, J., NAGARAJAN, M., AMIT SHETH, & VERMA, K. (2005), Web Service Semantics - WSDL-S, `http://knoesis.org/library/download/WSDL-S-V1.pdf` (Accessed: 20.12.2015).

AKW/JÖ. (2011), Fakten versus Emotionen, *LOGISTIK HEUTE* 33(11), pp. 42–45.

ALBERS, S., SCHWEIGER, B., & GIBB, J. (2013), A process model of strategic network member acquisition and retention, *Academy of Management Proceedings* 2013(1), pp. 11492.

ALBERS, S. (2005), *The design of alliance governance systems* (Kölner Wissenschaftsverlag, Kölner Wissenschaftsverlag).

ALBERS, S. (2010), Configurations of Alliance Governance Systems, *Schmalenbach Business Review* 62(3), pp. 204–233.

ALBERS, S., & KLAAS-WISSING, T. (2012), Organisation of multilateral LTL alliances, *International Journal of Logistics Research and Applications: A Leading Journal of Supply Chain Management* 15(3), pp. 181–198.

F. Jaekel, *Cloud Logistics*, Produktion und Logistik,
https://doi.org/10.1007/978-3-658-22837-8

ALBERS, S., WOHLGEZOGEN, F., & ZAJAC, E. J. (2016), Strategic Alliance Structures: An Organization Design Perspective, *Journal of Management* 42(3), pp. 582–614.

ALCHIAN, A. A., & DEMSETZ, H. (1972), Production, Information Costs, and Economic Organization, *American Economic Review* 62(5), pp. 777–795.

ALTER, C., & HAGE, J. (1993), *Organizations working together*, volume 191 of *Sage library of social research* (Sage Publications, Newbury Park, CA).

AMERI, F., & PATIL, L. (2012), Digital manufacturing market: A semantic web-based framework for agile supply chain deployment, *Journal of Intelligent Manufacturing* 23(5), pp. 1817–1832.

AMIT, R., & SCHOEMAKER, P. J. H. (1993), Strategic Assets and Organizational Rent, *Strategic Management Journal* 14(1), pp. 33–46.

ANDERSÉN, J. (2011), Strategic resources and firm performance, *Management Decision* 49(1), pp. 87–98.

ANDERSSON, D., & NORRMAN, A. (2002), Procurement of logistics services—a minutes work or a multi-year project?, *European Journal of Purchasing & Supply Management* 8(1), pp. 3–14.

ANGELOV, S., GREFEN, P., & GREEFHORST, D. (2012), A framework for analysis and design of software reference architectures, *Information and Software Technology* 54(4), pp. 417–431.

ARGYRES, N. S. (1999), The Impact of Information Technology on Coordination: Evidence from the B-2 "Stealth" Bomber, *Organization Science* 10(2), pp. 162–180.

ARMBRUST, M., STOICA, I., ZAHARIA, M., FOX, A., GRIFFITH, R., JOSEPH, A. D., KATZ, R., KONWINSKI, A., LEE, G., PATTERSON, D., & RABKIN, A. (2010), A view of cloud computing, *Communications of the ACM* 53(4), pp. 50–58.

ARNOLD, U. (2000), New dimensions of outsourcing: a combination of transaction cost economics and the core competencies concept, *European Journal of Purchasing & Supply Management* 6(1), pp. 23–29.

ARNOLD, U. (2011), Project LOGICAL, `http://www.project-logical.eu/?lang=en` (Accessed: 12.08.2015).

ARNOLD, U. (2014)a, Cloud Logistics - Anwenderperspektive, Anforderungen aus der Logistik und Lösungsansatz des Projekts LOGICAL, *Wirtschaftsinformatik & Management* 6(1), pp. 16–27.

ARNOLD, U. (2014)b, Cloud Logistics - ein neues Buzzword oder wie bewertet die Anwenderseite diese Entwicklung?, *Wirtschaftsinformatik & Management* 6(1), pp. 10–15.

ARNOLD, U., OBERLÄNDER, J., & SCHWARZBACH, B. (2012), LOGICAL—Development of Cloud Computing Platforms and Tools for Logistics Hubs and Communities, in Maria Ganzha, Leszek A. Maciaszek, & Marcin Paprzycki, eds., *Proceedings of the Federated Conference on Computer Science and Information Systems*, pp. 1083–1090 (Polskie Towarzystwo Informatyczne, Warsaw).

ARNOLD, U., OBERLÄNDER, J., & SCHWARZBACH, B. (2013), Advancements in Cloud Computing for Logistics, in M. Ganzha, Leszek Maciaszek, & M. Paprzycki, eds., *Proceedings of the Federated Conference on Computer Science and Information Systems*, pp. 1055–1062 (IEEE).

ARROW, K. J. (1969), The Organization of Economic Activity: Issues Pertinent to the Choice of Market versus Nonmarket Allocation, in U.S. Joint Economic Committee, 91st Congress, 1st Session, ed., *The Analysis And Evaluation Of Public Expenditures: The PBB System*, pp. 47–64 (U.S. Government Printing Office, Washington, DC).

ARROW, K. J. (1979), The Property Rights Doctrine and Demand Revelation Under Incomplete Information, in Tibor Scitovsky, & Michael J. Boskin, eds., *Economics and human welfare*, Economic theory, econometrics, and mathematical economics (Academic Press, New York).

ARSANJANI, A., LIANG-JIE ZHANG, ELLIS, M., ALLAM, A., & CHANNABASAVAIAH, K. (2007), S3: A Service-Oriented Reference Architecture, *IT Professional* 9(3), pp. 10–17.

ATZORI, L., IERA, A., & MORABITO, G. (2010), The Internet of Things: A survey, *Computer Networks* 54(15), pp. 2787–2805.

AUGENSTEIN, C., MÜLLER, H., & FRANCZYK, B. (2010), Developing a unified service model for collaborative modeling of logistics services, in *Advanced Information Technologies for Management (AITM)*.

AUMANN, R. J. (1976), Agreeing to Disagree, *The Annals of Statistics* 4(6), pp. 1236–1239.

AXIT AG. (2012), Expert knowledge and best practices for your successful cloud logistics project, `http://www.thesupplychaincloud.com/wp-content/uploads/2014/07/AXIT-Xperience-Booklet.pdf` (Accessed: 12.08.2015).

AZEEZ, A., PERERA, S., GAMAGE, D., LINTON, R., SIRIWARDANA, P., LEELARATNE, D., WEERAWARANA, S., & FREMANTLE, P. (2010), Multi-tenant SOA Middleware for Cloud Computing, in *IEEE 3rd International Conference on Cloud Computing (CLOUD)*, pp. 458–465.

BADGER, L., GRANCE, T., PATT-COMER, R., & VOAS, J. (2012), Cloud Computing Synopsis and Recommendations, *Recommendations of the National Institute of Standards and Technology* .

BAKER, G., GIBBONS, R., & MURPHY, K. J. (1999), Informal authority in organizations, *Journal of Law, Economics, and Organization* 15(1), pp. 56–73.

BAKOS, J. Y. (1991), A Strategic Analysis of Electronic Marketplaces, *MIS Quarterly* 15(3), pp. 295–310.

BAKOS, J. Y. (1997), Reducing Buyer Search Costs: Implications for Electronic Marketplaces, *Management Science* 43(12), pp. 1676–1692.

BAKOS, Y. (1998), The Emerging Role of Electronic Marketplaces on the Internet, *Communications of the ACM* 41(8), pp. 35–42.

BALDWIN, C. Y., & CLARK, K. B. (1997), Managing in an Age of Modularity, *Harvard Business Review* 75(5), pp. 84–93.

BALLARDT, A. (2014), Whitepaper - Eurasian Cloud Logistics, http://issuu.com/andrejballardt/docs/eurasian_cloud_logistics_whitepaper (Accessed: 12.08.2015).

BALLOU, R. H. (1999), *Business logistics management*, fourth edition (Prentice-Hall, Upper Saddle River, New Jersey).

BANYAI, A. (2014), Cloud Logistics, *Advanced Logistic System* 8(1), pp. 11–16.

BARNEY, J., WRIGHT, M., & KETCHEN, D. J. JR. (2001), The resource-based view of the firm: Ten years after 1991, *Journal of Management* 27(6), pp. 625–641.

BARNEY, J. (1991), Firm Resources and Sustained Competitive Advantage, *Journal of Management* 17(1), pp. 99–120.

BASK, A., LIPPONEN, M., RAJAHONKA, M., & TINNILÄ, M. (2011), Modularity in logistics services: A business model and process view, *International Journal of Services and Operations Management* 10(4), pp. 379–399.

BASK, A., LIPPONEN, M., RAJAHONKA, M., & TINNILÄ, M. (2010), The concept of modularity: diffusion from manufacturing to service production, *Journal of Manufacturing Technology Management* 21(3), pp. 355–375.

BAUMGARTEN, H., DARKOW, I.-L., & ZADEK, H. (2004), *Supply Chain Steuerung und Services: Logistik-Dienstleister managen globale Netzwerke — Best Practices* (Springer, Berlin, Heidelberg, New York).

BEEKUN, R. I., & GLICK, W. H. (2001), Organization Structure from a Loose Coupling Perspective: A Multidimensional Approach, *Decision Sciences* 32(2), pp. 227–250.

BENGTSSON, M., & KOCK, S. (2000), "Coopetition" in Business Networks—to Cooperate and Compete Simultaneously, *Industrial Marketing Management* 29(5), pp. 411–426.

BERDICA, K. (2002), An introduction to road vulnerability: what has been done, is done and should be done, *Transport Policy* 9(2), pp. 117–127.

BERGLUND, M., VAN LAARHOVEN, P., SHARMAN, G., & WANDEL, S. (1999), Third-Party Logistics: Is There a Future?, *International Journal of Logistics Management* 10(1), pp. 59–70.

BERNERS-LEE, T. (1998), What the Semantic Web can represent, http://www.w3.org/DesignIssues/RDFnot.html (Accessed: 25.05.2015).

BERNERS-LEE, T., & FISCHETTI, M. (1999), *Weaving the Web: The original design and ultimate destiny of the World Wide Web*, first edition (HarperCollins, San Francisco).

BERNERS-LEE, T., HENDLER, J., & LASSILA, O. (2001), The Semantic Web, *Scientific American* 284(5), pp. 34–43.

BERNSTEIN, D., LUDVIGSON, E., SANKAR, K., DIAMOND, S., & MORROW, M. (2009), Blueprint for the Intercloud - Protocols and Formats for Cloud Computing Interoperability, in *2009 Fourth International Conference on Internet and Web Applications and Services*, pp. 328–336.

BERNSTEIN, D., & VIJ, D. (2010), Intercloud Security Considerations, in *IEEE Second International Conference on Cloud Computing Technology and Science*, pp. 537–544.

BERNSTEIN, D., VIJ, D., & DIAMOND, S. (2011), An Intercloud Cloud Computing Economy - Technology, Governance, and Market Blueprints, in *2011 Annual SRII Global Conference*, pp. 293–299.

BEVERE, L., SEILER, T., ZIMMERLI, P., & FEYEN, H. (2013), Natural catastrophes and man-made disasters in 2012: A year of extreme weather events in the US, http://media.swissre.com/documents/sigma2_2013_EN.pdf (Accessed: 20.10.2015).

BEZEMER, C.-P., & ZAIDMAN, A. (2010), Multi-tenant SaaS applications: maintenance dream or nightmare?, in Andrea Capiluppi, Anthony Cleve, & Naouel Moha, eds., *Proceedings of the Joint ERCIM Workshop on Software Evolution (EVOL) and International Workshop on Principles of Software Evolution (IWPSE)*, pp. 88–92 (ACM Press, New York City, NY).

BHASKAR, R. (1986), *Scientific Realism and Human Emancipation* (Verso, London).

BHATIA, G., LANE, C., & WAIN, A. (2013), Building Resilience in Supply Chains: An Initiative of the Risk Response Network In collaboration with Accenture, http://www3.weforum.org/docs/WEF_RRN_MO_BuildingResilienceSupplyChains_Report_2013.pdf (Accessed: 20.12.2016).

BIEBERSTEIN, N., BOSE, S., WALKER, L., & LYNCH, A. (2005), Impact of service-oriented architecture on enterprise systems, organizational structures, and individuals, *IBM Systems Journal* 44(4), pp. 691–708.

BIERWIRTH, C., SCHNEIDER, S., & KOPFER, H. (2002), Elektronische Transportmärkte Aufgaben, Entwicklungsstand und Gestaltungsoptionen, *Wirtschaftsinformatik* 44(4), pp. 335–344.

BILOTKACH, V. (2007), Airline Partnerships and Schedule Coordination, *Journal of Transport Economics and Policy* 41(3), pp. 413–425.

BLAU, B., CONTE, T., & VAN DINTHER, C. (2010), A multidimensional procurement auction for trading composite services, *Electronic Commerce Research and Applications* 9(5), pp. 460–472.

BLAU, B., DINTHER, C., CONTE, T., XU, Y., & WEINHARDT, C. (2009)a, How to Coordinate Value Generation in Service Networks - A Mechanism Design Approach, *Business & Information Systems Engineering* 1(5), pp. 343–356.

BLAU, B., KRÄMER, J., CONTE, T., & VAN DINTHER, C. (2009)b, Service value networks, in *IEEE Conference on Commerce and Enterprise Computing (CEC)*, pp. 194–201.

BLAU, B. S. (2009), *Coordination in Service Value Networks: A Mechanism Design Approach*, Doctoral Thesis.

BLEEKE, J., & ERNST, D. (1995), Is Your Strategic Alliance Really a Sale?, *Harvard Business Review* 73(1), pp. 97–105.

BLOOS, M., & KOPFER, H. (2011), On the Formation of Operational Transport Collaboration Systems, in H.-J Kreowski, B. Scholz-Reiter, & K.-D Thoben, eds., *Dynamics in logistics*, pp. 191–201 (Springer, Berlin and Heidelberg).

BOGATIN, D. (2006), Google CEO's new paradigm: 'cloud computing and advertising go hand-in-hand', http://www.zdnet.com/article/google-ceos-new-paradigm-cloud-computing-and-advertising-go-hand-in-hand/ (Accessed: 10.05.2015).

BÖHMER, M. (2012), User Group - Cloud Computing für Logistik, http://www.cloudzone-karlsruhe.de/messe-karlsruhe-cloudzone/2012/media/data/Logistics_Mall_-_User_Group.pdf (Accessed: 12.08.2015).

BÖHMER, M. (2013), Neues vom Marktplatz, in Michael ten Hompel, ed., *Software in der Logistik - Prozesse steuern mit Apps*, pp. 40–42 (Huss-Verlag, München).

BONDI, A. B. (2000), Characteristics of scalability and their impact on performance, in Murray Woodside, Hassan Gomaa, & Danny Menasce, eds., *Proceedings of the second international workshop on software and performance (WOSP 2000)*, pp. 195–203 (ACM, New York City, NY).

BONFATTI, F., MONARI, P. D., & MARTINELLI, L. (2010), Proposal for a Logistic Domain Ontology, in *IST-Africa 2010 Conference Proceedings*, pp. 1–8 (IEEE).

BORYS, B., & JEMISON, D. B. (1989), Hybrid Arrangements as Strategic Alliances: Theoretical Issues in Organizational Combinations, *The Academy of Management Review* 14(2), pp. 234–249.

BÖTTCHER, M., & KLINGNER, S. (2011), Providing a method for composing modular B2B services, *Journal of Business & Industrial Marketing* 26(5), pp. 320–331.

BOURLAKIS, C., & BOURLAKIS, M. (2005), Information technology safeguards, logistics asset specificity and fourth-party logistics network creation in the food retail chain, *Journal of Business & Industrial Marketing* 20(2), pp. 88–98.

BOWERSOX, D. J., & CLOSS, D. J. (1996), *Logistical management: The integrated supply chain process*, McGraw-Hill series in marketing (McGraw-Hill Companies, New York).

BOWERSOX, D. J., CLOSS, D. J., & HELFERICH, O. K. (1986), *Logistical management: A systems integration of physical distribution, manufacturing support, and materials procurement*, third edition (Macmillan and Collier Macmillan, New York and London).

BRETZKE, W.-R. (2012), Logistikdienstleistungen, in Peter Klaus, Winfried Krieger, & Michael Krupp, eds., *Gabler Lexikon Logistik*, pp. 372–378 (Gabler, Wiesbaden).

BROWN, T. L., & POTOSKI, M. (2005), Transaction Costs and Contracting: The Practitioner Perspective, *Public Performance & Management Review* 28(3), pp. 326–351.

BUIJS, P., & WORTMANN, J. C. (2014), Joint operational decision-making in collaborative transportation networks: The role of IT, *Supply Chain Management: An International Journal* 19(2), pp. 200–210.

BUNGE, M. (1996), *Finding philosophy in social science* (Yale University Press, New Haven, CT).

BUNGE, M. (1997), Mechanism and Explanation, *Philosophy of the Social Sciences* 27(4), pp. 410–465.

BUNGE, M. (2000), Systemism: the alternative to individualism and holism, *The Journal of Socio-Economics* 29(2), pp. 147–157.

BURSTEIN, M., BUSSLER, C., ZAREMBA, M., FININ, T., HUHNS, M. N., PAOLUCCI, M., SHETH, A. P., & WILLIAMS, S. (2005), A Semantic Web Services Architecture, *IEEE Internet Computing* 9(5), pp. 72–81.

BUSINESSCLOUD.DE. (2012), Self-Service, `http://www.business-cloud.de/cloud-self-service-hilf-dir-selbst-dann-hilft-dir-die-cloud/` (Accessed: 10.03.2015).

BUYYA, R., YEO, C. S., & VENUGOPAL, S., eds. (2008), *Market-Oriented Cloud Computing: Vision, Hype, and Reality for Delivering IT Services as Computing Utilities: 2008 10th IEEE International Conference on High Performance Computing and Communications*.

BUYYA, R., RANJAN, R., & CALHEIROS, R. N. (2010), InterCloud: Utility-Oriented Federation of Cloud Computing Environments for Scaling of Application Services, in Ching-Hsien Hsu, Laurence Yang, Yong Hyuk Park, & Sang-Soo Yeo, eds., *Algorithms and Architectures for Parallel Processing*, volume 6081, pp. 13–31 (Springer, Berlin, Heidelberg).

BUYYA, R., YEO, C. S., VENUGOPAL, S., BROBERG, J., & BRANDIC, I. (2009), Cloud computing and emerging IT platforms: Vision, hype, and reality for delivering computing as the 5th utility, *Future Generation Computer Systems* 25(6), pp. 599–616.

CALHEIROS, R. N., TOOSI, A. N., VECCHIOLA, C., & BUYYA, R. (2012), A coordinator for scaling elastic applications across multiple clouds, *Future Generation Computer Systems* 28(8), pp. 1350–1362.

CAMBRIDGE UNIVERSITY PRESS. (2016), Cambridge Dictionaries Online, `http://dictionary.cambridge.org/dictionary/english/` (Accessed: 12.08.2016).

CAMPAGNOLO, D., & CAMUFFO, A. (2010), The Concept of Modularity in Management Studies: A Literature Review, *International Journal of Management Reviews* 12(3), pp. 259–283.

CAMPBELL, D. J. (1988), Task Complexity: A Review and Analysis, *The Academy of Management Review* 13(1), pp. 40–52.

CARBONE, V., & STONE, M. A. (2005), Growth and relational strategies used by the European logistics service providers: Rationale and outcomes, *Transportation Research Part E: Logistics and Transportation Review* 41(6), pp. 495–510.

CERF, V. G. (2013), Loose Couplings, *IEEE Internet Computing* 17(2), pp. 95–96.

CHANDRASEKARAN, B., JOSEPHSON, J. R., & BENJAMINS, V. R. (1999), What are ontologies, and why do we need them?, *IEEE Intelligent Systems* 14(1), pp. 20–26.

CHANDRASHEKAR, A., & SCHARY, P. B. (1999), Toward the Virtual Supply chain: The Convergence of IT and Organization, *The International Journal of Logistics Management* 10(2), pp. 27–40.

CHEN, F. C.-Y., & CHEN, C. (2003)a, The effects of strategic alliances and risk pooling on the load factors of international airline operations, *Transportation Research Part E: Logistics and Transportation Review* 39(1), pp. 19–34.

CHEN, H., & CHEN, T.-J. (2003)b, Governance structures in strategic alliances: transaction cost versus resource-based perspective, *Journal of World Business* 38(1), pp. 1–14.

CHEN, R. R., ROUNDY, R. O., ZHANG, R. Q., & JANAKIRAMAN, G. (2005), Efficient Auction Mechanisms for Supply Chain Procurement, *Management Science* 51(3), pp. 467–482.

CHERBAKOV, L., GALAMBOS, G., HARISHANKAR, R., KALYANA, S., & RACKHAM, G. (2005), Impact of service orientation at the business level, *IBM Systems Journal* 44(4), pp. 653–668.

CHI, T. (1994), Trading in strategic resources: Necessary conditions, transaction cost problems, and choice of exchange structure, *Strategic Management Journal* 15(4), pp. 271–290.

CHILD, J. (1972), Organizational Structure, Environment and Performance: The Role of Strategic Choice, *Sociology* 6(1), pp. 1–22.

CHILD, J. (1997), Strategic Choice in the Analysis of Action, Structure, Organizations and Environment: Retrospect and Prospect, *Organization Studies* 18(1), pp. 43–76.

CHOW, G., HEAVER, T. D., & HENRIKSSON, L. E. (1994), Logistics Performance: Definition and Measurement, *International Journal of Physical Distribution & Logistics Management* 24(1), pp. 17–28.

CHOY, K. L., LEE, W. B., & LO, V. (2003), Design of an intelligent supplier relationship management system: a hybrid case based neural network approach, *Expert Systems with Applications* 24(2), pp. 225–237.

CHRISTOPHER, M., & PECK, H. (2004), Building the Resilient Supply Chain, *The International Journal of Logistics Management* 15(2), pp. 1–14.

CISCO SYSTEMS INC. (2011), The Need for Service Catalog Design in Cloud Services Development, `http://www.cisco.com/en/US/services/ps2961/ps10364/ps10370/ps11104/need-for-cloud-services-catalog_whitepaper.pdf` (Accessed: 14.08.2013).

CLARKE, E. H. (1971), Multipart pricing of public goods, *Public Choice* 11(1), pp. 17–33.

CLEMENTS, P., & BASS, L. (2010), The Business Goals Viewpoint, *IEEE Software* 27(6), pp. 38–45.

CLEMENTS, P., BACHMANN, F., BASS, L., GARLAN, D., IVERS, J., LITTLE, R., MERSON, P., NORD, R., & STAFFORD, J. (2010), *Documenting software architectures: Views and beyond*, second edition (Addison-Wesley, Upper Saddle River, NJ).

CLEMONS, E. K., & ROW, M. C. (1993), Limits to Interfirm Coordination through Information Technology: Results of a Field Study in Consumer Packaged Goods Distribution, *Journal of Management Information Systems* 10(1), pp. 73–95.

CLOUD LOGISTICS. (2013), Company website, `http://www.gocloudlogistics.com/` (Accessed: 03.12.2013).

CLOUD PALLET STORAGE. (2015), Company website, `http://www.cloud-ls.com/` (Accessed: 12.08.2015).

CLOUTIER, R., MULLER, G., VERMA, D., NILCHIANI, R., HOLE, E., & BONE, M. (2010), The Concept of Reference Architectures, *Systems Engineering* 13(1), pp. 14–27.

CLSCM. (2002), Supply Chain Vulnerability: Executive Report: On Behalf of: Department of Transport, Local Government and the Regions, Home Office, Department of Trade and Industry, `http://www.som.cranfield.ac.uk/som/dinamic-content/research/lscm/downloads/Vulnerability_report.pdf` (Accessed: 22.09.2012).

COASE, R. H. (1937), The Nature of the Firm, *Economica* 4(16), pp. 386–405.

COLAJANNI, M. (2012), Opportunity of cloud computing for logistics, http://www.project-logical.eu/download/2012-10-30_Opportunity_of_cloud_computing_for_logistics.pdf (Accessed: 12.08.2015).

CONTE, T., BLAU, B., & KNAPPPER, R. (2010), Networked Mechanism Design Incentive Engineering in Service Value Networks as Exemplified by the Co-Opetition Mechanism, in *AMCIS 2010 Proceedings*, p. Paper 396.

CONTRACTOR, F. J., LORANGE, P., MANAGEMENT, R. U. G. S. O., & SCHOOL, W. (1988), *Cooperative strategies in international business*, Lexington Books (Lexington Books).

COOPER, J. C. (1993), Logistics Strategies for Global Businesses, *International Journal of Physical Distribution & Logistics Management* 23(4), pp. 12–23.

CORSTEN, H., & GÖSSINGER, R. (2007), *Modularisierung von Logistikdienstleistungen*, volume Nr 84 of *Schriften zum Produktionsmanagement* (Lehrstuhl für Produktionswirtschaft, Universität Kaiserslautern).

COYLE, J. J., BARDI, E. J., & LANGLEY, C. J. (1988), *The management of business logistics*, fourth edition (West Publishing Company, St. Paul).

COYLE, J. J., BARDI, E. J., & LANGLEY, C. J. (2003), *The management of business logistics: A supply chain perspective*, 7th edition (South-Western/Thomson Learning, Mason, Ohio).

CRUIJSSEN, F., BRÄYSY, O., DULLAERT, W., FLEUREN, H., & SALOMON, M. (2007)a, Joint route planning under varying market conditions, *International Journal of Physical Distribution & Logistics Management* 37(4), pp. 287–304.

CRUIJSSEN, F., COOLS, M., & DULLAERT, W. (2007)b, Horizontal cooperation in logistics: Opportunities and impediments, *Transportation Research Part E: Logistics and Transportation Review* 43(2), pp. 129–142.

CRUIJSSEN, F., DULLAERT, W., & FLEUREN, H. (2007)c, Horizontal Cooperation in Transport and Logistics: A Literature Review, *Transportation Journal* 46(3), pp. 22–39.

CUI, L., & HERTZ, S. (2011), Networks and capabilities as characteristics of logistics firms, *Industrial Marketing Management* 40(6), pp. 1004–1011.

DA SILVEIRA, G., BORENSTEIN, D., & FOGLIATTO, F. S. (2001), Mass customization: Literature review and research directions, *International Journal of Production Economics* 72(1), pp. 1–13.

DAFT, R. L. (1983), *Organization Theory and Design* (West Publishing Company).

DAFT, R. L. (2001), *Essentials of Organization Theory and Design*, second edition (South-Western College Publishing).

DAHLMAN, C. J. (1979), The Problem of Externality, *Journal of Law and Economics* 22(1), pp. 141–162.

DANILUK, D., RAHN, J., & WOLF, M.-B. (2014), Logistics Mall – Logistiksoftware aus der Cloud, *Wirtschaftsinformatik & Management* 6(1), pp. 28–37.

DARESTANI, S. A., GHADERI, H., LEMAN, Z., & ISMAIL, M. Y. (2012), Horizontal collaboration in logistics: A feasible task for group purchasing, *International Journal of Procurement Management* 5(1), pp. 43–54.

DAS, T. K., & TENG, B.-S. (2000), A Resource-Based Theory of Strategic Alliances, *Journal of Management* 26(1), pp. 31–61.

DAS, T., & TENG, B.-S. (2001), A risk perception model of alliance structuring, *Journal of International Management* 7(1), pp. 1–29.

DAS, T., & TENG, B.-S. (2003), Partner analysis and alliance performance, *Scandinavian Journal of Management* 19(3), pp. 279–308.

DASGUPTA, P., HAMMOND, P., & MASKIN, E. (1979), The Implementation of Social Choice Rules: Some General Results on Incentive Compatibility, *The Review of Economic Studies* 46(2), pp. 185–216.

DASH, R. K., JENNINGS, N. R., & PARKES, D. C. (2003), Computational-mechanism design: A call to arms, *IEEE Intelligent Systems* 18(6), pp. 40–47.

D'ASPREMONT, C., & GÉRARD-VARET, L.-A. (1979), Incentives and incomplete information, *Journal of Public Economics* 11(1), pp. 25–45.

DAUGHERTY, P. J. (2011), Review of logistics and supply chain relationship literature and suggested research agenda, *International Journal of Physical Distribution & Logistics Management* 41(1), pp. 16–31.

DAY, G. S. (1994), The Capabilities of Market-Driven Organizations, *Journal of Marketing* 58(4), pp. 37–52.

DE VITA, G., TEKAYA, A., & WANG, C. L. (2010), Asset specificity's impact on outsourcing relationship performance: A disaggregated analysis by buyer–supplier asset specificity dimensions, *Journal of Business Research* 63(7), pp. 657–666.

DE VITA, G., TEKAYA, A., & WANG, C. L. (2011), The Many Faces of Asset Specificity: A Critical Review of Key Theoretical Perspectives, *International Journal of Management Reviews* 13(4), pp. 329–348.

DELFMANN, W. (1995), Logistische Segmentierung: Ein modellanalytischer Ansatz zur Gestaltung logistischer Auftragszyklen, in Horst Albach, & Helmut Koch, eds., *Dynamik und Risikofreude in der Unternehmensführung*, pp. 171–202 (Gabler, Wiesbaden).

DELFMANN, W. (2012), Kernelemente der Logistikkonzeption, in Peter Klaus, Winfried Krieger, & Michael Krupp, eds., *Gabler Lexikon Logistik*, pp. 267–271 (Gabler, Wiesbaden).

DELFMANN, W., ALBERS, S., & GEHRING, M. (2002), The impact of electronic commerce on logistics service providers, *International Journal of Physical Distribution & Logistics Management* 32(3), pp. 203–222.

DELFMANN, W., ALBERS, S., KERSTEN, W., & SCHOLZ-REITER, B. (2011), Call for Pappers - 6th ISSL - Coordinated Autonomous Systems, `http://www.bvl.de/misc/filePush.php?id=15746&name=ISSL12+Call+for+Papers` (Accessed: 12.08.2015).

DELFMANN, W., DANGELMAIER, W., GÜNTHNER, W., KLAUS, P., OVERMEYER, L., ROTHENGATTER, W., WEBER, J., & ZENTES, J. (2010), Towards a science of logistics: cornerstones of a framework of understanding of logistics as an academic discipline, *Logistics Research* 2(2), pp. 57–63.

DELFMANN, W., & JAEKEL, F. (2012), The Cloud - Logistics for the Future?, in Werner Delfmann, & Thomas Wimmer, eds., *Coordinated Autonomous Systems*, Schriftenreihe Wirtschaft & Logistik, pp. 5–26 (DVV Media Group, Hamburg).

DENG, H. S. Y., & FANG, C. (2014), Research on the Situation and Prospect of ICT Application in China Modern Logistics, *Applied Mechanics and Materials* 568-570, pp. 1639–1642.

DEWETT, T. (2001), The role of information technology in the organization: A review, model, and assessment, *Journal of Management* 27(3), pp. 313–346.

DHL. (2014), Logistics Trend Radar: Cloud Logistics (T02), `http://www.dhl.com/content/dam/downloads/g0/about_us/logistics_insights/DHL_Logistics-TrendRadar_2014.pdf` (Accessed: 12.08.2015).

DIAO, P.-P. (2014), Research on the B2C E-business Logistics and Distribution Model Based on "Cloud Logistics", *Logistics Sci-Tech* 38(5).

DIERICKX, I., & COOL, K. (1989), Asset Stock Accumulation and Sustainability of Competitive Advantage, *Management Science* 35(12), pp. 1504–1511.

DILLON, T., WU, C., & CHANG, E. (2010), Cloud Computing: Issues and Challenges, in *2010 24th IEEE International Conference on Advanced Information Networking and Applications*, pp. 27–33.

DING, L.-F. (2014), The Integration of Agriculture, Logistics and Commerce: A Logistic Pattern of Agricultural Products in the Context of Cloud Logistics, *China Business and Market* 28(6).

DITTMANN, L. (2006), *Der angemessene Grad an Visibilität in Logistik-Netzwerken: die Auswirkungen von RFID*, Doctoral Thesis, Universität St.Gallen.

DIWAN, M. A. (2015), Cloud Community in E-Clusters: towards Sustainable Logistics Clusters, in *International Martime Transport & Logistics Conference (MARLOG 4)*.

DONG, H., HUSSAIN, F. K., & CHANG, E. (2008), Transport service ontology and its application in the field of semantic search, in *2008 IEEE International Conference on Service Operations and Logistics, and Informatics (SOLI)*, pp. 820–824.

DRAZIN, R., & VAN DE VEN, A. H. (1985), Alternative Forms of Fit in Contingency Theory, *Administrative Science Quarterly* 30(4), pp. 514–539.

DSLV DEUTSCHER SPEDITIONS- UND LOGISTIKVERBAND E.V. (2006), General terms and conditions of logistics-services providers, `http://www.dslv.org/dslv/web.nsf/gfx/3D34AFDEC915859AC1257B47004AFF5E/$file/DSLV%20Logistik%20AGBs%202006%20englisch.pdf` (Accessed: 30.07.2015).

DU, X.-J. (2014), The Application of Cloud Logistics Based on Cloud Computing in the Express Industry, *Journal of Zhejiang Wanli University* 27(2).

DUNCAN, R. B. (1972), Characteristics of Organizational Environments and Perceived Environmental Uncertainty, *Administrative Science Quarterly* 17(3), pp. 313–327.

DUSCHEK, S. (2004), Inter-Firm Resources and Sustained Competitive Advantage, *Management Revue* 15(1), pp. 53–73.

DYER, J. H., & SINGH, H. (1998), The Relational View: Cooperative Strategy and Sources of Interorganizational Competitive Advantage, *The Academy of Management Review* 23(4), pp. 660–679.

EBERS, M. (1999), Explaining Inter-Organizational Network Formation, in Mark Ebers, ed., *The Formation of Inter-Organizational Networks*, pp. 3–40 (Oxford University Press, Oxford, NY).

EBERS, M., & OERLEMANS, L. (2016), The Variety of Governance Structures Beyond Market and Hierarchy, *Journal of Management* 42(6), pp. 1491–1529.

EHRENBERG, D., & LUDWIG, A. (2014), Cloud Logistics - Innovationspotenziale durch Virtualisierung von Logistikeinheiten nutzen, *Wirtschaftsinformatik & Management* 6(1), pp. 6–9.

EISENHARDT, K. M., & SCHOONHOVEN, C. B. (1996), Resource-Based View of Strategic Alliance Formation: Strategic and Social Effects in Entrepreneurial Firms, *Organization Science* 7(2), pp. 136–150.

EMERY, D., & HILLIARD, R. (2009), Every architecture description needs a framework: Expressing architecture frameworks using ISO/IEC 42010, in *Joint Working IEEE/IFIP Conference on Software Architecture & European Conference on Software Architecture (WICSA/ECSA 2009)*, pp. 31–40.

ENDREI, M. (2004), *Patterns: Service-oriented architecture and web services*, IBM Redbooks, first edition (IBM, International Technical Support, Poughkeepsie).

ERL, T. (2008), *SOA: Principles of service design* (Prentice Hall, Upper Saddle River, NJ).

ERRAMILLI, M. K., & RAO, C. P. (1993), Service Firms' International Entry-Mode Choice: A Modified Transaction-Cost Analysis Approach, *Journal of Marketing* 57(3), pp. 19–38.

ETZIONI, A. (1964), *Modern organizations* (Prentice-Hall, Englewood Cliffs, NJ).

FAGUI, L., KUN, L., & YANG, Z. (2008), Semantic Web Services and Its Application in Third-Party Logistics, in *2008 International Workshop on Education Technology and Training & 2008 International Workshop on Geoscience and Remote Sensing*, pp. 626–630.

FISHER, M. L. (1997), What is the right supply chain for your product?, *Harvard Business Review* 75, pp. 105–117.

FITZSIMMONS, J. A., NOH, J., & THIES, E. (1998), Purchasing business services, *Journal of Business & Industrial Marketing* 13(4/5), pp. 370–380.

FJELDSTAD, Ø. D., SNOW, C. C., MILES, R. E., & LETTL, C. (2012), The Architecture of Collaboration, *Strategic Management Journal* 33(6), pp. 734–750.

FLEISCH, E. (2010), What is the internet of things? An economic perspective, *Economics, Management, and Financial Markets* 5(2), pp. 125–157.

FORD, J. D., & SLOCUM, J. W. (1977), Size, Technology, Environment and the Structure of Organizations, *Academy of Management Review* 2(4), pp. 561–575.

FOSS, N. J. (1999), Networks, capabilities, and competitive advantage, *Scandinavian Journal of Management* 15(1), pp. 1–15.

FOSTER, I., ZHAO, Y., RAICU, I., & LU, S. (2008), Cloud Computing and Grid Computing 360-Degree Compared, in *Grid Computing Environments Workshop (GCE)*, pp. 1–10.

FOWLER, G. A., & WORTHEN, B. (2009), The Internet Industry Is on a Cloud – Whatever That May Mean – Forget ASP and Web 2.0: Tech Companies Push Cirrus, Stratus, Other Cumulo-Nebulous Lingo, *Wall Street Journal, Eastern edition* 26.03.2009, pp. A.1.

FOWLER, M. (2001), Reducing coupling, *IEEE Software* 18(4), pp. 102–104.

FREICHEL, S. L. K. (1992), *Organisation von Logistikservice-Netzwerken: Theoretische Konzeption und empirische Fallstudien*, volume Bd. 4 of *Unternehmensführung und Logistik* (E. Schmidt, Berlin).

GALANTE, G., & BONA, L. C. E. D. (2012), A Survey on Cloud Computing Elasticity, in *2012 IEEE/ACM Fifth International Conference on Utility and Cloud Computing*, pp. 263–270.

GANTZIA, D., & SKLATINIOTI, M. E. (2014), *Cloud computing in the 3PL industry: A profound insight into the benefits & challenges of cloud-based services: A two fold approach*, Master Thesis, Jönköping University.

GARETTE, B., & DUSSAUGE, P. (2000), Alliances versus acquisitions: Choosing the right option, *European Management Journal* 18(1), pp. 63–69.

GARG, S. K., VECCHIOLA, C., & BUYYA, R. (2013), Mandi: a market exchange for trading utility and cloud computing services, *The Journal of Supercomputing* 64(3), pp. 1153–1174.

GAWER, A. (2014), Bridging differing perspectives on technological platforms: Toward an integrative framework, *Research Policy* 43(7), pp. 1239–1249.

GAXSYS GMBH. (2013), Cloud Logistics: Wir denken Logistik vernetzt, http://gaxsys.com/wp-content/uploads/2013/04/Cloud_Logistics.pdf (Accessed: 20.01.2014).

GAXSYS GMBH. (2015), Company website, http://gaxsys.com/ (Accessed: 03.08.2015).

GENG, Y., & LI, J. (2012), An research of electronic commerce logistics model based on cloud logistics and warehousing, in *2012 International Symposium on Management of Technology (ISMOT)*, pp. 631–635.

GERO, J. S. (1990), Design Prototypes: A Knowledge Representation Schema for Design, *AI Magazine* 11(4), pp. 26–36.

GERO, J. S., & KANNENGIESSER, U. (2004), The situated function–behaviour–structure framework, *Design Studies* 25(4), pp. 373–391.

GIBBARD, A. (1973), Manipulation of Voting Schemes: A General Result, *Econometrica* 41(4), pp. 587–601.

GLÖCKNER, M., AUGENSTEIN, C., & LUDWIG, A. (2014), Metamodel of a Logistics Service Map, in Witold Abramowicz, & Angelika Kokkinaki, eds., *Business information systems*, volume 176 of *Lecture Notes in Business Information Processing*, pp. 185–196 (Springer International Publishing).

GOLDBERG, R. P. (1974), Survey of virtual machine research, *Computer* 7(6), pp. 34–45.

GOMEZ-PEREZ, A., & CORCHO, O. (2002), Ontology languages for the Semantic Web, *IEEE Intelligent Systems* 17(1), pp. 54–60.

GONG, X.-L., & YANG, R. (2012), Explore the Application of "Cloud Computing" and "Cloud Logistics" to Logistics, *China Business and Market* 26(10).

GOTTSCHALK, K. (2000), Web Services architecture overview: The next stage of evolution for e-business, http://www.ibm.com/developerworks/library/w-ovr/ (Accessed: 24.04.2015).

GOTTSCHALK, K., GRAHAM, S., KREGER, H., & SNELL, J. (2002), Introduction to Web services architecture, *IBM Systems Journal* 41(2), pp. 170–177.

GRANDORI, A. (1997)a, An Organizational Assessment of Interfirm Coordination Modes, *Organization Studies* 18(6), pp. 897–925.

GRANDORI, A., & SODA, G. (1995), Inter-firm Networks: Antecedents, Mechanisms and Forms, *Organization Studies* 16(2), pp. 183–214.

GRANDORI, A. (1997)b, Governance Structures, Coordination Mechanisms and Cognitive Models, *Journal of Management & Governance* 1(1), pp. 29–47.

GRANDORI, A. (2000), Part II: Coordination Mechanisms: A Definition of Coordination, *Organization & Economic Behaviour* pp. 89–98.

GRANOVETTER, M. (1985), Economic Action and Social Structure: The Problem of Embeddedness, *American Journal of Sociology* 91(3), pp. 481–510.

GRANT, R. M. (1991), The Resource-Based Theory of Competitive Advantage: Implications for Strategy Formulation, *California Management Review* 33(3), pp. 114–135.

GRANT, R. M. (2010), *Contemporary strategy analysis*, 7th edition (John Wiley & Sons, Hoboken, NJ).

GREEN, J., & LAFFONT, J.-J. (1977), Characterization of Satisfactory Mechanisms for the Revelation of Preferences for Public Goods, *Econometrica* 45(2), pp. 427–438.

GRIEGER, M. (2003), Electronic marketplaces: A literature review and a call for supply chain management research, *European Journal of Operational Research* 144(2), pp. 280–294.

GRINTER, R. E., HERBSLEB, J. D., & PERRY, D. E. (1999), The geography of coordination: Dealing with Distance in R&D Work, in Stephen Hayne, ed., *Proceedings of the International ACM SIGGROUP Conference on Supporting Group Work*, pp. 306–315 (ACM, New York).

GROTH, L. (1999), *Future organizational design: The scope for the IT-based enterprise* (John Wiley & Sons, Chichester).

GROVES, T. (1973), Incentives in Teams, *Econometrica* 41(4), pp. 617–631.

GRUBER, T. R. (1993), A translation approach to portable ontology specifications, *Knowledge Acquisition* 5(2), pp. 199–220.

GT NEXUS. (2014), Company website, `http://www.gtnexus.com/` (Accessed: 20.04.2014).

GUBBI, J., BUYYA, R., MARUSIC, S., & PALANISWAMI, M. (2013), Internet of Things (IoT): A vision, architectural elements, and future directions, *Future Generation Computer Systems* 29(7), pp. 1645–1660.

GUDEHUS, T. (2012), *Logistik 2: Netzwerke, Systeme und Lieferketten* (Springer Berlin Heidelberg, Berlin, Heidelberg).

GUINARD, D., TRIFA, V., KARNOUSKOS, S., SPIESS, P., & SAVIO, D. (2010), Interacting with the SOA-Based Internet of Things: Discovery, Query, Selection, and On-Demand Provisioning of Web Services, *IEEE Transactions on Services Computing* 3(3), pp. 223–235.

GULATI, R. (1998), Alliances and Networks, *Strategic Management Journal* 19(4), pp. 293–317.

GULATI, R. (1999), Network location and learning: The influence of network resources and firm capabilities on alliance formation, *Strategic Management Journal* 20(5), pp. 397–420.

GULATI, R., & GARGIULO, M. (1999), Where Do Interorganizational Networks Come From?, *American Journal of Sociology* 104(5), pp. 1439–1493.

GULATI, R., NOHRIA, N., & ZAHEER, A. (2000), Strategic networks, *Strategic Management Journal* 21(3), pp. 203–215.

GULATI, R., PURANAM, P., & TUSHMAN, M. (2012), Meta-organization design: Rethinking design in interorganizational and community contexts, *Strategic Management Journal* 33(6), pp. 571–586.

GULATI, R., & SINGH, H. (1998), The Architecture of Cooperation: Managing Coordination Costs and Appropriation Concerns in Strategic Alliances, *Administrative Science Quarterly* 43(4), pp. 781–814.

GULATI, R., SYTCH, M., & MEHROTRA, P. (2008), Breaking Up is Never Easy: Planning for Exit in a Strategic Alliance, *California Management Review* 50(4), pp. 147–163.

HAAS, H. (2004), Web Services Glossary, `http://www.w3.org/TR/ws-gloss/` (Accessed: 24.05.2015).

HABERMAS, J. (1970), Ein Literaturbericht: zur Logik der Sozialwissenschaften, in Jürgen Habermas, ed., *Zur Logik der Sozialwissenschaften*, pp. 71–310 (Suhrkamp, Frankfurt am Main).

HAGIU, A. (2009), Two-Sided Platforms: Product Variety and Pricing Structures, *Journal of Economics & Management Strategy* 18(4), pp. 1011–1043.

HAGIU, A. (2014), Strategic Decisions for Multisided Platforms, *MIT Sloan Management Review* 55(2), pp. 71–80.

HAGIU, A., & SPULBER, D. (2013), First-Party Content and Coordination in Two-Sided Markets, *Management Science* 59(4), pp. 933–949.

HÅKANSSON, H. (1982), *International marketing and purchasing of industrial goods: An Interaction Approach* (John Wiley & Sons, Chichester).

HALL, R. H. (2002), *Organizations: Structures, processes, and outcomes*, 8th edition (Prentice Hall, Upper Saddle River, NJ).

HALL, R. H., JOHNSON, N. J., & HAAS, J. E. (1967), Organizational Size, Complexity, and Formalization, *American Sociological Review* 32(6), pp. 903–912.

HANNIG, U. (2012), Survey on the Benefits of Cloud Computing for Logistics, https://axit.de/images/Whitepaper_Download/AXIT_Survey_on_the_Benefits_of_Cloud_Computing_for_Logistics.pdf (Accessed: 10.01.2013).

HAPAG-LLOYD. (2010), Container Specification, http://www.hapag-lloyd.com/downloads/press_and_media/publications/Brochure_Container_Specification_en.pdf (Accessed: 03.03.2014).

HARSANYI, J. C. (1967), Games with Incomplete Information Played by "Bayesian" Players, I–III Part I. The Basic Model, *Management Science* 14(3), pp. 159–182.

HARSANYI, J. C. (1968)a, Games with Incomplete Information Played by "Bayesian" Players Part II. Bayesian Equilibrium Points, *Management Science* 14(5), pp. 320–334.

HARSANYI, J. C. (1968)b, Games with Incomplete Information Played by 'Bayesian' Players, Part III. The Basic Probability Distribution of the Game, *Management Science* 14(7), pp. 486–502.

HASSAN, A. E., & HOLT, R. C. (2000), A reference architecture for Web servers, in *Proceedings of the 7th Working Conference on Reverse Engineering*, pp. 150–159.

HE, W., & XU, L. (2014), A state-of-the-art survey of cloud manufacturing, *International Journal of Computer Integrated Manufacturing* 28(3), pp. 239–250.

HE, X., JIANG, J., & WEI, G. (2015), The cloud logistics modeling and validation based on Pi calculus, in *International Conference on Information Sciences, Machinery, Materials and Energy (ICISMME 2015)*, pp. 1721–1727.

HEANEY, B. (2010), Cloud Logistics: Solution For Networks, http://blogs.aberdeen.com/supply-chain-management/cloud-logistics-solution-for-enabling-multi-enterprise-cross-channel-logistics-networks/ (Accessed: 04.09.2015).

HELLSTRÖM, D., & SAGHIR, M. (2007), Packaging and logistics interactions in retail supply chains, *Packaging Technology and Science* 20(3), pp. 197–216.

HENNART, J.-F. (1988), A transaction costs theory of equity joint ventures, *Strategic Management Journal* 9(4), pp. 361–374.

HENNART, J.-F. (1991), The Transaction Costs Theory of Joint Ventures: An Empirical Study of Japanese Subsidiaries in the United States, *Management Science* 37(4), pp. 483–497.

HENNART, J.-F. (1993), Explaining The Swollen Middle: Why Most Transactions Are a Mix of "Market" And "Hierarchy", *Organization Science* 4(4), pp. 529–547.

HERBST, N. R., KOUNEV, S., & REUSSNER, R. (2013), Elasticity in Cloud Computing: What It Is, and What It Is Not, in *Proceedings of the 10th International Conference on Autonomic Computing (ICAC 13)*, pp. 23–27, San Jose, CA.

HERNÁNDEZ, S., PEETA, S., & KALAFATAS, G. (2011), A less-than-truckload carrier collaboration planning problem under dynamic capacities, *Transportation Research Part E: Logistics and Transportation Review* 47(6), pp. 933–946.

HERR, S., & KLAUS, P. (2012), Leistungstiefenoptimierung in der Logistik, in Peter Klaus, Winfried Krieger, & Michael Krupp, eds., *Gabler Lexikon Logistik*, pp. 314–318 (Gabler, Wiesbaden).

HERTZ, S. (1993), *The Internationalization Processes of Freight Transport Companies: Towards a Dynamic Network Model of Internationalization*, Doctoral thesis, Stockholm School of Economics.

HERTZ, S., & ALFREDSSON, M. (2003), Strategic development of third party logistics providers, *Industrial Marketing Management* 32(2), pp. 139–149.

HESSE, M., & RODRIGUE, J.-P. (2004), The transport geography of logistics and freight distribution, *Journal of Transport Geography* 12(3), pp. 171–184.

HILLIARD, R. (1999), Views and Viewpoints in Software Systems Architecture, in *First Working IFIP Conference on Software Architecture*.

HITT, M. A., DACIN, M. T., LEVITAS, E., ARREGLE, J.-L., & BORZA, A. (2000), Partner Selection in Emerging and Developed Market Contexts: Resource-Based and Organizational Learning Perspectives, *Academy of Management Journal* 43(3), pp. 449–467.

HOFMEISTER, C., KRUCHTEN, P., NORD, R. L., OBBINK, H., RAN, A., & AMERICA, P. (2007), A general model of software architecture design derived from five industrial approaches, *Journal of Systems and Software* 80(1), pp. 106–126.

HOLTKAMP, B., STEINBUSS, S., GSELL, H., LOEFFELER, T., & SPRINGER, U. (2010), Towards a Logistics Cloud, in *2010 Sixth International Conference on Semantics, Knowledge and Grids*, pp. 305–308.

HOXHA, J., SCHEUERMANN, A., & BLOEHDORN, S. (2010), An Approach to Formal and Semantic Representation of Logistics Services, in Kerstin Schill, Bernd Scholz-Reiter, & Lutz Frommberger, eds., *Proceedings of the Workshop on Artificial Intelligence and Logistics (AILog) at the 19th European Conference on Artificial Intelligence (ECAI)*, pp. 73–78.

HU, X., & ZHANG, S. (2015), Game Analysis on Logistics Cloud Service Combination, in Zhenji Zhang, Zuojun Max Shen, Juliang Zhang, & Runtong Zhang, eds., *LISS 2014*, pp. 127–132 (Springer, Berlin, Heidelberg).

HURWICZ, L. (1972), On informationally decentralized systems, in Jacob Marschak, C. B. McGuire, Roy Radner, & Kenneth Joseph Arrow, eds., *Decision and organization*, Studies in mathematical and managerial economics, pp. 297–336 (North-Holland Pub. Co., Amsterdam).

HURWICZ, L. (1975), On the existence of allocation systems whose manipulative Nash equilibria are pareto-optimal, in *3rd World Congress of the Econometric Society*.

HURWICZ, L., & WALKER, M. (1990), On the Generic Nonoptimality of Dominant-Strategy Allocation Mechanisms: A General Theorem That Includes Pure Exchange Economies, *Econometrica* 58(3), pp. 683–704.

IATA. (2014), IATA ULD Regulations (ULDR): The Foundation for Improved Aircraft Operations and Safety, https://www.iata.org/whatwedo/workgroups/Pages/uldp.aspx (Accessed: 04.04.2013).

IHDE, G. B. (2001), *Transport, Verkehr, Logistik: Gesamtwirtschaftliche Aspekte und einzelwirtschaftliche Handhabung*, Vahlens Handbücher der Wirtschafts- und Sozialwissenschaften, third edition (Vahlen, München).

IMA RWTH AACHEN UNIVERSITY. (2011), Cloud Logistic, www.cloud-logistic.de (Accessed: 03.12.2013).

IRELAND, R. D., HITT, M. A., & VAIDYANATH, D. (2002), Alliance Management as a Source of Competitive Advantage, *Journal of Management* 28(3), pp. 413–446.

ISERMANN, H. (1998), *Logistik: Gestaltung von Logistiksystemen*, second edition (mi, Verl. Moderne Industrie, Landsberg/Lech).

ISO/IEC 42010. (2007), Systems and software engineering – Recommended practice for architectural description of software-intensive systems, http://www.iso.org/iso/catalogue_detail.htm?csnumber=45991 (Accessed: 20.01.2014).

ISO/IEC/IEEE 42010. (2011)a, Getting Started with ISO/IEC/IEEE 42010, http://www.iso-architecture.org/42010/getting-started.html (Accessed: 20.01.2016).

ISO/IEC/IEEE 42010. (2011)b, International Standard for Systems and software engineering – Architecture description: ISO/IEC/IEEE 42010:2011(E) (Revision of ISO/IEC 42010:2007 and IEEE Std 1471-2000) pp. 1–46.

ISO/IEC/IEEE 42010. (2011)c, Templates for using the Standard, http://www.iso-architecture.org/42010/templates/ (Accessed: 20.01.2016).

JAEGER, P., & LINDENLAUB, R. (2013), Cloud Logistics (United States Patent Application Publication), https://www.google.com/patents/US9378255 (Accessed: 20.11.2015).

JARILLO, J. C. (1988), On strategic networks, *Strategic Management Journal* 9(1), pp. 31–41.

JASPERNEITE, J. (2012), Was hinter Begriffen wie Industrie 4.0 steckt, http://www.computer-automation.de/steuerungsebene/steuern-regeln/artikel/93559/ (Accessed: 10.04.2014).

JHARKHARIA, S., & SHANKAR, R. (2007), Selection of logistics service provider: An analytic network process (ANP) approach, *Omega* 35(3), pp. 274–289.

JIANG, Y.-S., LI, X.-X., & TAN, H. (2013), Logistics Network System Construction Based on Cloud Computing Processing, *Journal of Guangdong Education Institute* 33(3).

JIAO, Z. (2014), Operations Mode and Development Trend of Online Shopping Logistics in China, in Bing-lian Liu, Shao-ju Lee, Ling Wang, Ya Xu, & Xiang Li, eds., *Contemporary Logistics in China*, Current Chinese Economic Report Series, pp. 213–233 (Springer, Berlin, Heidelberg).

JONES, C., HESTERLY, W. S., & BORGATTI, S. P. (1997), A General Theory of Network Governance: Exchange Conditions and Social Mechanisms, *The Academy of Management Review* 22(4), pp. 911–945.

JONES, E. C., & CHUNG, C. A. (2008), *RFID in logistics: A practical introduction / Erick C. Jones, Christopher A. Chung* (CRC Press and Taylor & Francis, Boca Raton, London, New York).

JONES, S. (2005), Toward an Acceptable Definition of Service, *IEEE Software* 22(3), pp. 87–93.

JÖRGL, T. (2012), Mensch im Mittelpunkt, *LOGISTIK HEUTE* 34(4), pp. 52–53.

JUNG, J.-G., PARK, S.-B., CHA, S.-J., & GEUN-SIK JO, G.-S. (2008), Semantic Web Service for Freight Management System, in *IEEE International Workshop on Semantic Computing and Applications (IWSCA)*, pp. 132–138.

JURKOVICH, R. (1974), A Core Typology of Organizational Environments, *Administrative Science Quarterly* 19(3), pp. 380–394.

JÜTTNER, U., & MAKLAN, S. (2011), Supply chain resilience in the global financial crisis: an empirical study, *Supply Chain Management: An International Journal* 16(4), pp. 246–259.

KALE, P., DYER, J., & SINGH, H. (2001), Value creation and success in strategic alliances, *European Management Journal* 19(5), pp. 463–471.

KERSTEN, W., SINGER, C., & LOPEZ CASTELLANOS, M. A. (2012), Cloud Logistics as a Risk Management Strategy - An Agent Based Simulation Scenario, in Werner Delfmann, & Thomas Wimmer, eds., *Coordinated Autonomous Systems*, Schriftenreihe Wirtschaft & Logistik, pp. 256–275 (DVV Media Group, Hamburg).

KIESER, A., & WALGENBACH, P. (2007), *Organisation*, fifth edition (Schäffer-Poeschel, Stuttgart).

KIEWITT, A. (2013), Aufstieg in die Wolke, *LOGISTIK HEUTE* 35(6), pp. 26–31.

KING, J., & WENGER, K. (2013), Vier Erfolgsfaktoren, in Michael ten Hompel, ed., *Software in der Logistik - Prozesse steuern mit Apps*, pp. 44–46 (Huss-Verlag, München).

KLAAS, T. (2002), *Logistik-Organisation: Ein konfigurationstheoretischer Ansatz zur logistikorientierten Organisationsgestaltung*, first edition (Deutscher Universitäts-Verlag, Wiesbaden).

KLAAS-WISSING, T., & ALBERS, S. (2010), Cooperative versus corporate governance of LTL networks, *International Journal of Logistics Research and Applications: A Leading Journal of Supply Chain Management* 13(6), pp. 493–506.

KLAUS, P., KRIEGER, W., & KRUPP, M., eds. (2012), *Gabler Lexikon Logistik: Management logistischer Netzwerke und Flüsse*, fifth edition (Gabler, Wiesbaden).

KLAUS, P., & MÜLLER, S., eds. (2012), *The Roots of Logistics: A Reader of Classical Contributions to the History and Conceptual Foundations of the Science of Logistics* (Springer, Berlin and Heidelberg).

KNGINE. (2015), Query: "What day is today?", `http://www.kngine.com/#Know!q=what%20day%20is%20today` (Accessed: 25.04.2015).

KOCH, H. (2010), Developing dynamic capabilities in electronic marketplaces: A cross-case study, *The Journal of Strategic Information Systems* 19(1), pp. 28–38.

KOGUT, B. (1988), Joint ventures: Theoretical and empirical perspectives, *Strategic Management Journal* 9(4), pp. 319–332.

KOVÁCS, G. (2009), The structure, modules, services, and operational process of modern electronic freight and warehouse exchanges, *Periodica Polytechnica Transportation Engineering* 37(1-2), pp. 33–38.

KOVÁCS, G. (2010), Possible methods of application of electronic freight and warehouse exchanges in solving the city logistics problems, *Periodica Polytechnica Transportation Engineering* 38(1), pp. 25–28.

KRAFZIG, D., BANKE, K., & SLAMA, D. (2004), *Enterprise SOA: Service-oriented architecture best practices* (Prentice Hall PTR, Indianapolis, IN).

KRAJEWSKA, M. A., KOPFER, H., LAPORTE, G., ROPKE, S., & ZACCOUR, G. (2008), Horizontal Cooperation among Freight Carriers: Request Allocation and Profit Sharing, *The Journal of the Operational Research Society* 59(11), pp. 1483–1491.

KRISHNA, V. (2010), *Auction theory*, second edition (Academic Press, Burlington, MA).

KRUCHTEN, P. B. (1995), The 4+1 View Model of architecture, *IEEE Software* 12(6), pp. 42–50.

KRUCHTEN, P. (2004), An Ontology of Architectural Design Decisions in Software Intensive Systems, in Jan Bosch, ed., *Proceedings of the 2nd Groningen Workshop on Software Variability Management*, pp. 54–61.

KRULIS-RANDA, J. S. (1977), *Marketing-Logistik: Eine systemtheoretische Konzeption der betrieblichen Warenverteilung und Warenbeschaffung*, volume 21 of *Schriftenreihe des Instituts für Betriebswirtschaftliche Forschung an der Universität Zürich* (Paul Haupt, Bern and Stuttgart).

KRUPP, M., & KLAUS, P. (2012), Beziehungsnetzwerke, in Peter Klaus, Winfried Krieger, & Michael Krupp, eds., *Gabler Lexikon Logistik*, pp. 72–77 (Gabler, Wiesbaden).

LAGO, P., AVGERIOU, P., & HILLIARD, R. (2010), Software Architecture: Framing Stakeholders' Concerns, *IEEE Software* 27(6), pp. 20–24.

LAI, K.-H. (2004), Service capability and performance of logistics service providers, *Transportation Research Part E: Logistics and Transportation Review* 40(5), pp. 385–399.

LAMBERT, D. M., EMMELHAINZ, M. A., & GARDNER, J. T. (1999), Building Successful Logistics Partnerships, *Journal of Business Logistics* 20(1), pp. 165–181.

LANGLOIS, R. N. (1992), External Economies and Economic Progress: The Case of the Microcomputer Industry, *Business History Review* 66(1), pp. 1–50.

LASKEY, K., BROWN, P., ESTEFAN, J. A., MCCABE, F., & THORNTON, D., eds. (2012), *Reference Architecture Foundation for Service Oriented Architecture 1.0*, OASIS Committee Specification.

LAVIE, D. (2006), The Competitive Advantage of Interconnected Firms: An Extension of the Resource-Based View, *The Academy of Management Review* 31(3), pp. 638–658.

LAWRENCE, P. R., & LORSCH, J. W. (1967), *Organization and Environment: Managing differentiation and integration* (Harvard University Press, Boston).

LEE, C., & CHAN, T. M. (2009), Development of RFID-based Reverse Logistics System, *Expert Systems with Applications* 36(5), pp. 9299–9307.

LEE, H. L. (2004), The Triple-A Supply Chain, *Harvard Business Review* 82(10), pp. 102–112.

LEIBLEIN, M. J. (2003), The Choice of Organizational Governance Form and Performance: Predictions from Transaction Cost, Resource-based, and Real Options Theories, *Journal of Management* 29(6), pp. 937–961.

LEITNER, R., MEIZER, F., PROCHAZKA, M., & SIHN, W. (2011), Structural concepts for horizontal cooperation to increase efficiency in logistics, *CIRP Journal of Manufacturing Science and Technology* 4(3), pp. 332–337.

LEMOINE, W., & DAGNÆS, L. (2003), Globalisation strategies and business organisation of a network of logistics service providers, *International Journal of Physical Distribution & Logistics Management* 33(3), pp. 209–228.

LEUKEL, J., & KIRN, S. (2008), A Supply Chain Management Approach to Logistics Ontologies in Information Systems, in Witold Abramowicz, & Dieter Fensel, eds., *Business Information Systems*, volume 7 of *Lecture Notes in Business Information Processing*, pp. 95–105 (Springer, Berlin).

LEUKEL, J., KIRN, S., & SCHLEGEL, T. (2011)a, Supply Chain as a Service: A Cloud Perspective on Supply Chain Systems, *IEEE Systems Journal* 5(1), pp. 16–27.

LEUKEL, J., JACOB, A., KARAENKE, P., KIRN, S., & KLEIN, A. (2011)b, Individualization of Goods and Services: Towards a Logistics Knowledge Infrastructure for Agile Supply Chains, in *Association for the Advancement of Artificial Intelligence Spring Symposium: AI for Business Agility*, pp. 39–49.

LEUKEL, J., & KIRN, S. (2011), A Service-Oriented Approach to Freight Routing in Intermodal Transport Systems, in *10th International Conference on Wirtschaftsinformatik*, pp. 292–301.

LEUKEL, J., & SCHEUERMANN, A. (2014), Cloud Logistics ist mehr als Logistiksoftware aus der Cloud, *Wirtschaftsinformatik & Management* 6(1), pp. 38–45.

LEVY, D. L. (1997), Lean production in an international supply chain, *Sloan Management Review* 38, pp. 94–102.

LI, C., ZHANG, X., & LI, L. (2014)a, Construction and Operation of Regional Logistics Public Information Platform Based on Cloud Computing, *The Open Cybernetics & Systemics Journal* 8(1), pp. 22–28.

LI, C., ZHANG, X., & LI, L. (2014)b, Research on Comparative Analysis of Regional Logistics Information Platform Operation Mode Based on Cloud Computing, *International Journal of Future Generation Communication and Networking* 7(2), pp. 73–80.

LI, H. Z., & HUANG, X. (2013), Research on Communication Carriers' Development in Cloud Logistics Age, *Advanced Materials Research* 753-755, pp. 2825–2828.

LI, W., ZHONG, Y., WANG, X., & CAO, Y. (2013), Resource virtualization and service selection in cloud logistics, *Journal of Network and Computer Applications* 36(6), pp. 1696–1704.

LI, X.-G., & DING, H. (2013), New Terminal Distribution Model Under Cloud Logistics: Logistics Supermarket, *Logistics Sci-Tech* 37(5).

LIAN, P., PARK, D.-W., & KWON, H.-C. (2007), Design of Logistics Ontology for Semantic Representing of Situation in Logistics, in *Second Workshop on Digital Media and its Application in Museum & Heritages*, pp. 432–437.

LIAO, H., & FU, T. (2014), Study on the characteristic books logistics system, in Y. Yang, & M. Ma, eds., *3rd International Conference on Green Communications and Networks*, WIT Transactions on Information and Communication Technologies, pp. 99–104 (WIT PressSouthampton, UK).

LIEB, R., & BENTZ, B. A. (2005), The Use of Third-Party Logistics Services by Large American Manufacturers: The 2004 Survey, *Transportation Journal* 44(2), pp. 5–15.

LIM, S.-S., PARK, D.-W., & KWON, H.-C. (2007), Ontology-Based Semantic Representation of Context in Port Supply Chain, in *Sixth International Conference on Advanced Language Processing and Web Information Technology*, pp. 446–451.

LIMTHANMAPHON, B., & ZHANG, Y. (2003), Web service composition with case-based reasoning, in Klaus-Dieter Schewe, & Xiaofang Zhou, eds., *Proceedings of the 14th Australasian Database Conference*, pp. 201–208.

LIN, Y., LUO, J., & ZHOU, L. (2010), Modular logistics service platform, in *2010 IEEE International Conference on Service Operations and Logistics and Informatics (SOLI)*, pp. 200–204.

LIN, Y., & PEKKARINEN, S. (2011), QFD-based modular logistics service design, *Journal of Business & Industrial Marketing* 26(5), pp. 344–356.

LINDBERG, N., & NORDIN, F. (2008), From products to services and back again: Towards a new service procurement logic, *Industrial Marketing Management* 37(3), pp. 292–300.

LIU, C.-L., & LYONS, A. C. (2011), An analysis of third-party logistics performance and service provision, *Transportation Research Part E: Logistics and Transportation Review* 47(4), pp. 547–570.

LIU, F., TONG, J., MAO, J., BOHN, R. B., MESSINA, J. V., BADGER, M. L., & LEAF, D. M. (2011)a, NIST Cloud Computing Reference Architecture, `https://www.nist.gov/node/597546` (Accessed: 20.01.2014).

LIU, N., & LI, X. (2012), A Resource Virtualization Mechanism for Cloud Manufacturing Systems, in Marten J. van Sinderen, ed., *Enterprise interoperability*, volume 122 of *Lecture Notes in Business Information Processing*, pp. 46–59 (Springer, Berlin and New York).

LIU, N., LI, X., & SHEN, W. (2014), Multi-granularity resource virtualization and sharing strategies in cloud manufacturing, *Journal of Network and Computer Applications* 46, pp. 72–82.

LIU, N., LI, X., & WANG, Q. (2011)b, A resource & capability virtualization method for cloud manufacturing systems, in *IEEE International Conference on Systems, Man, and Cybernetics (SMC)*, pp. 1003–1008.

LOGATA GMBH. (2012), Noch nie war Logistik so einfach. Das Mall-Prinzip der Logistik, `https://mmp.logistics-mall.com/web/mmp/purchase` (Accessed: 23.09.2012).

LOGATA GMBH. (2013), Website: Logistics Mall, `http://mmp.logistics-mall.com/` (Accessed: 17.12.2013).

LOHTIA, R., BROOKS, C. M., & KRAPFEL, R. E. (1994), What constitutes a transaction-specific asset?: An examination of the dimensions and types, *Journal of Business Research* 30(3), pp. 261–270.

LONG, Y.-Q. (2014), Cloud Logistics Sharing Mechanism Based on Cloud Federation, in *2014 Seventh International Joint Conference on Computational Sciences and Optimization (CSO)*, pp. 147–150.

LORENZEN, P. (1969), *Normative Logic and Ethics*, volume 2 (Zürich, Bibliographisches Institut).

LOVELOCK, C., & GUMMESSON, E. (2004), Whither Services Marketing?: In Search of a New Paradigm and Fresh Perspectives, *Journal of Service Research* 7(1), pp. 20–41.

LUDWIG, A. (2014), Engineering und Management kundenindividueller Logistikdienste nach dem Cloud-Prinzip, *Wirtschaftsinformatik & Management* 6(1), pp. 46–55.

LUSCH, R., VARGO, S., & TANNIRU, M. (2010), Service, value networks and learning, *Journal of the Academy of Marketing Science* 38(1), pp. 19–31.

LYONS, B. R. (1995), Specific investment, economies of scale, and the make-or-buy decision: A test of transaction cost theory, *Journal of Economic Behavior and Organization* 26(3), pp. 431–443.

MACHAMER, P., DARDEN, L., & CRAVER, C. F. (2000), Thinking about Mechanisms, *Philosophy of Science* 67(1), pp. 1–25.

MACKENZIE, C. M., LASKEY, K., MCCABE, F., BROWN, P. F., & METZ, R., eds. (2006), *Reference model for service oriented architecture 1.0*, OASIS Committee Specification.

MACKIE, J. L. (1974), *The Cement of the Universe: A Study of Causation* (Oxford University Press).

MACNEIL, I. R. (1974), Restatement (Second) of Contracts and Presentiation, *Virginia Law Review* 60(4), pp. 589–610.

MACNEIL, I. R. (1977), Contracts: adjustment of long-term economic relations under classical, neoclassical, and relational contract law, *Nw. UL Rev.* 72, pp. 854–906.

MAIER, M., EMERY, D., & HILLIARD, R. (2001), Software architecture: introducing IEEE Standard 1471, *Computer* 34(4), pp. 107–109.

MALONE, T. W., YATES, J., & BENJAMIN, R. I. (1987), Electronic markets and electronic hierarchies, *Communications of the ACM* 30(6), pp. 484–497.

MARCH, J. G., & SIMON, H. A. (1958), *Organizations* (John Wiley & Sons, New York, NY).

MARTIN, D., BURSTEIN, M., HOBBS, J., LASSILA, O., MCDERMOTT, D., MCILRAITH, S., NARAYANAN, S., PAOLUCCI, M., PARSIA, B., PAYNE, T., SIRIN, E., SRINIVASAN, N., & SYCARA, K. (2004), OWL-S: Semantic Markup for Web Services, `http://www.w3.org/Submission/OWL-S/` (Accessed: 26.06.2014).

MARTIN, D., PAOLUCCI, M., MCILRAITH, S., BURSTEIN, M., MCDERMOTT, D., MCGUINNESS, D., PARSIA, B., PAYNE, T., SABOU, M., SOLANKI, M., SRINIVASAN, N., & SYCARA, K. (2005), Bringing Semantics to Web Services: The OWL-S Approach, in David Hutchison, Takeo Kanade, Josef Kittler, Jon M. Kleinberg, Friedemann Mattern, John C. Mitchell, Moni Naor, Oscar Nierstrasz, C. Pandu Rangan,

Bernhard Steffen, Madhu Sudan, Demetri Terzopoulos, Dough Tygar, Moshe Y. Vardi, Gerhard Weikum, Jorge Cardoso, & Amit Sheth, eds., *Semantic Web Services and Web Process Composition (SWSWPC)*, pp. 26–42 (Springer, Berlin, Heidelberg).

MAS-COLELL, A., WHINSTON, M. D., & GREEN, J. R. (1995), *Microeconomic theory* (Oxford University Press, New York).

MATKOVIC, P., TUMBAS, P., & PAVLICEVIC, V. (2014), Decision Making in Logistics Processes Supported by Cloud Computing, *International Scientific Journal of Management Information Systems* 9(1), pp. 11–20.

MATROS, R. (2012), *Der Einfluss von Cloud Computing auf IT-Dienstleister: Eine fallstudienbasierte Untersuchung kritischer Einflussgrössen* (Springer Gabler, Wiesbaden).

MAYER, A. (2007), *Modularisierung der Logistik: Ein Gestaltungsmodell zum Management von Komplexität in der industriellen Logistik*, first edition (Universitätsverlag der Technischen Universität Berlin, Berlin).

MCAFEE, A., & BRYNJOLFSSON, E. (2012), Big data: The Management Revolution, *Harvard Business Review* 90(10), pp. 61–67.

MCILRAITH, S. A., & MARTIN, D. L. (2003), Bringing semantics to Web services, *IEEE Intelligent Systems* 18(1), pp. 90–93.

MCILRAITH, S. A., SON, T. C., & HONGLEI ZENG. (2001), Semantic Web services, *IEEE Intelligent Systems* 16(2), pp. 46–53.

MELL, P. M., & GRANCE, T. (2011), The NIST Definition of Cloud Computing, *Recommendations of the National Institute of Standards and Technology* .

MENTZER, J. T. (1991), An Efficiency/Effectiveness Approach to Logistics Performance Analysis, *Journal of Business Logistics* 12(1), pp. 33–62.

MENTZER, J. T., FLINT, D. J., & HULT, G. T. M. (2001), Logistics Service Quality as a Segment-Customized Process, *Journal of Marketing* 65(4), pp. 82–104.

MENTZER, J. T., FLINT, D. J., & KENT, J. L. (1999), Developing a Logistics Service Quality Scale, *Journal of Business Logistics* 20(1), pp. 9–32.

MENTZER, J. T., MIN, S., & MICHELLE BOBBITT, L. (2004), Toward a unified theory of logistics, *International Journal of Physical Distribution & Logistics Management* 34(8), pp. 606–627.

MERRIAM-WEBSTER. (2016), `http://www.merriam-webster.com/` (Accessed: 20.10.2016).

MEUTER, M. L., OSTROM, A. L., ROUNDTREE, R. I., & BITNER, M. J. (2000), Self-Service Technologies: Understanding Customer Satisfaction with Technology-Based Service Encounters, *Journal of Marketing* 64(3), pp. 50–64.

MEYER, J., STAHLBOCK, R., & VOSS, S. (2012), Slow Steaming in Container Shipping, in *2012 45th Hawaii International Conference on System Sciences (HICSS)*, pp. 1306–1314.

MIDORO, R., & PITTO, A. (2000), A critical evaluation of strategic alliances in liner shipping, *Maritime Policy & Management* 27(1), pp. 31–40.

MIETZNER, R., LEYMANN, F., & PAPAZOGLOU, M. P. (2008), Defining Composite Configurable SaaS Application Packages Using SCA, Variability Descriptors and Multi-tenancy Patterns, in *The Third International Conference on Internet and Web Applications and Services*, pp. 156–161.

MIHAILESCU, M., & TEO, Y. M. (2010), Dynamic Resource Pricing on Federated Clouds, in *10th IEEE/ACM International Conference on Cluster, Cloud and Grid Computing (CCGrid)*, pp. 513–517.

MIHAILESCU, M., & TEO, Y. M. (2012), The Impact of User Rationality in Federated Clouds, in *12th IEEE/ACM International Symposium*, pp. 620–627.

MILANOVIC, N., & MALEK, M. (2004), Current solutions for Web service composition, *IEEE Internet Computing* 8(6), pp. 51–59.

MILES, R. E., & SNOW, C. C. (1986), Organizations: New Concepts for New Forms, *California Management Review* 28(3), pp. 62–73.

MILES, R. E., & SNOW, C. C. (1992), Causes of Failure in Network Organizations, *California Management Review* 34(4), pp. 53–72.

MINTZBERG, H. (1979), *The structuring of organizations* (Prentice-Hall, Englewood Cliffs, NJ).

MINTZBERG, H. (1980), Structure in 5's: A Synthesis of the Research on Organization Design, *Management Science* 26(3), pp. 322–341.

MORENO-VOZMEDIANO, R., MONTERO, & LLORENTE, I. M. (2012), IaaS Cloud Architecture: From Virtualized Datacenters to Federated Cloud Infrastructures, *Computer* 45(12), pp. 65–72.

MORRIS, M. H., KOÇAK, A., & ÖZER, A. (2007), Coopetition as a Small Business Strategy: Implications for Performance, *Journal of Small Business Strategy* 18(1), pp. 35–55.

MOSAVI, M. (2012), Keine Transaktion, keine Kosten, *LOGISTIK HEUTE* 34(5), pp. 16–17.

MURPHY, P. R., & POIST, R. F. (2000), Third-Party Logistics: Some User versus Provider Perspectives, *Journal of Business Logistics* 21(1), pp. 121–133.

MYERSON, R. B. (1979), Incentive Compatibility and the Bargaining Problem, *Econometrica* 47(1), pp. 61–74.

MYERSON, R. B. (1983), Mechanism Design by an Informed Principal, *Econometrica* 51(6), pp. 1767–1797.

MYERSON, R. B. (1988), Mechanism Design, `https://www.financialtrustindex.org/research/math/papers/796.pdf` (Accessed: 26.01.2015).

MYERSON, R. B. (1991), *Game theory: Analysis of conflict* (Harvard University Press, Cambridge, MA).

NAKAGAWA, E. Y., OQUENDO, F., & BECKER, M. (2012), RAModel: A Reference Model for Reference Architectures, in *2012 Joint Working IEEE/IFIP Conference on Software Architecture (WICSA) and European Conference on Software Architecture (ECSA)*, pp. 297–301.

NANDIRAJU, S., & REGAN, A. (2003), Freight Transportation Electronic Marketplaces: A Survey of the Industry and Exploration of Important Research Issues, `https://escholarship.org/uc/item/9fj2c4jw` (Accessed: 10.06.2015).

NARAHARI, Y., GARG, D., NARAYANAM, R., & PRAKASH, H. (2009), *Game theoretic problems in network economics and mechanism design solutions*, Advanced information and knowledge processing (Springer, London).

NASSIMBENI, G. (1998), Network structures and co-ordination mechanisms, *International Journal of Operations & Production Management* 18(6), pp. 538–554.

NEHM, A., & VERES-HOMM, U. (2012), Logistikimmobilien, in Peter Klaus, Winfried Krieger, & Michael Krupp, eds., *Gabler Lexikon Logistik*, pp. 380–387 (Gabler, Wiesbaden).

NEXUS DISTRIBUTION. (2016), Company Website, `http://www.nexusdistribution.com/value-added-logistics` (Accessed: 20.10.2016).

NIE, K., & FAN, X. F. (2015), Research on the Architecture and Application Schema of Cloud Warehousing Mode, in Kennis Chan, & J. Yeh, eds., *International conference on computer information systems and industrial applications (CISIA 2015)*, pp. 367–370.

NIEBUER, A. (1996), *Qualitätsmanagement für Logistikunternehmen* (Deutscher Universitäts-Verlag and Gabler, Wiesbaden).

NOHRIA, N. (1992), Is a Network Perspective a Useful Way of Studying Organizations?, in Nitin Nohria, & Robert G. Eccles, eds., *Networks and organizations*, pp. 1–22 (Harvard Business School Press, Boston, MA).

NORRMAN, A., & JANSSON, U. (2004), Ericsson's proactive supply chain risk management approach after a serious sub-supplier accident, *International Journal of Physical Distribution & Logistics Management* 34(5), pp. 434–456.

NOWICKA, K. (2014), Smart City Logistics on Cloud Computing Model, *Procedia - Social and Behavioral Sciences* 151, pp. 266–281.

O'Leary, M. B., & Cummings, J. N. (2007), The spatial, temporal, and configurational characteristics of geographic dispersion in teams, *MIS Quarterly* 31(3), pp. 433–452.

Oliver, A. L., & Ebers, M. (1998), Networking Network Studies: An Analysis of Conceptual Configurations in the Study of Inter-organizational Relationships, *Organization Studies* 19(4), pp. 549–583.

Oliver, C. (1990), Determinants of Interorganizational Relationships: Integration and Future Directions, *The Academy of Management Review* 15(2), pp. 241–265.

Opresnik, M. (2014), Cloud Logistics eröffnen Einsparpotenziale, *Controlling & Management Review* 58(3), pp. 73.

Oracle. (2002), Web Logic Server 7.0: Programming WebLogic Web Services, http://docs.oracle.com/cd/E13222_01/wls/docs70/webserv/client.html (Accessed: 24.05.2015).

Orlikowski, W. J., & Robey, D. (1991), Information Technology and the Structuring of Organizations, *Information Systems Research* 2(2), pp. 143–169.

Ouchi, W. G. (1979), A Conceptual Framework for the Design of Organizational Control Mechanisms, *Management Science* 25(9), pp. 833–848.

Owens, D. (2010), Securing Elasticity in the Cloud, *Communications of the ACM* 53(6), pp. 46–51.

Pagel, P. (2014), Überall Wolken, *Wirtschaftsinformatik & Management* 6(1), pp. 1.

Palay, T. M. (1984), Comparative Institutional Economics: The Governance of Rail Freight Contracting, *The Journal of Legal Studies* 13(2), pp. 265–287.

Pan, Y.-m., & Tong, Y.-l. (2014), The Loan Pricing Study of Supply Chain Finance Based on Cloud Muster House, *Value Engineering* 33(6).

Panayides, P. M., & So, M. (2005), Logistics service provider–client relationships, *Transportation Research Part E: Logistics and Transportation Review* 41(3), pp. 179–200.

Pankratz, G. (2003), Zweiseitige kombinatorische Auktionen in elektronischen Transportmärkten – Potenziale und Probleme, https://core.ac.uk/download/pdf/11559115.pdf (Accessed: 16.05.2015).

Papazoglou, M. P., Traverso, P., Dustdar, S., & Leymann, F. (2007), Service-Oriented Computing: State of the Art and Research Challenges, *Computer* 40(11), pp. 64–71.

Papazoglou, M. P., Traverso, P., Dustdar, S., & Leymann, F. (2008), Service-oriented Computing: a Research Roadmap, *International Journal of Cooperative Information Systems* 17(2), pp. 223–255.

PAPAZOGLOU, M. P. (2003), Service-Oriented Computing: Concepts, Characteristics and Directions, in *Proceedings of the Fourth International Conference on Web Information Systems Engineering (WISE)*, pp. 3–12.

PARK, D.-W., LEE, G.-T., & KWON, H.-C. (2008), Ontological Approach to Integration of Event-Centric Logistics Information into EPC Network, in *IEEE International Workshop on Semantic Computing and Applications (IWSCA)*, pp. 1–8.

PARK, S. H. (1996), Managing an Interorganizational Network: A Framework of the Institutional Mechanism for Network Control, *Organization Studies* 17(5), pp. 795–824.

PARK, S. H., & RUSSO, M. V. (1996), When Competition Eclipses Cooperation: An Event History Analysis of Joint Venture Failure, *Management Science* 42(6), pp. 875–890.

PARKES, D. C., KALAGNANAM, J. R., & ESO, M. (2001), Achieving budget-balance with Vickrey-based payment schemes in exchanges, in *Proceedings of the 17th international joint conference on Artificial Intelligence (IJCAI)*, volume 2, pp. 1161–1168.

PARKES, D. C. (2001), *Iterative Combinatorial Auctions: Achieving Economic And Computational Efficiency*, Doctoral Thesis.

PARKHE, A. (1993), Strategic Alliance Structuring: A Game Theoretic and Transaction Cost Examination of Interfirm Cooperation, *The Academy of Management Journal* 36(4), pp. 794–829.

PATIL, A. A., OUNDHAKAR, S. A., SHETH, A. P., & VERMA, K. (2004), Meteor-s web service annotation framework, in Stuart Feldman, Mike Uretsky, Marc Najork, & Craig Wills, eds., *Proceedings of the 13th International Conference on the World Wide Web*, pp. 553–562.

PAUROBALLY, S., TAMMA, V., & WOOLDRDIGE, M. (2007), A Framework for Web service negotiation, *ACM Transactions on Autonomous and Adaptive Systems* 2(4), pp. Article 14.

PAUTASSO, C., & WILDE, E. (2009), Why is the web loosely coupled?, in Juan Quemada, Gonzalo León, Yoelle Maarek, & Wolfgang Nejdl, eds., *Proceedings of the 18th international conference on World wide web*, pp. 911–920.

PAYNE, T., & LASSILA, O. (2004), Semantic Web Services, *IEEE Intelligent Systems* 19(4), pp. 14–15.

PEKKARINEN, S., & ULKUNIEMI, P. (2008), Modularity in developing business services by platform approach, *The International Journal of Logistics Management* 19(1), pp. 84–103.

PENROSE, E. (1959), *The Theory of the Growth of the Firm* (Blackwell, Oxford).

PERSSON, G., & VIRUM, H. (2001), Growth Strategies for Logistics Service Providers: A Case Study, *The International Journal of Logistics Management* 12(1), pp. 53–64.

PETERAF, M. A. (1993), The Cornerstones of Competitive Advantage: A Resource-Based View, *Strategic Management Journal* 14(3), pp. 179–191.

PETTIT, T. J., FIKSEL, J., & CROXTON, K. L. (2010), Ensuring Supply Chain Resilience: Development of a Conceptual Framework, *Journal of Business Logistics* 31(1), pp. 1–21.

PFOHL, H.-C. (2010), *Logistiksysteme: Betriebswirtschaftliche Grundlagen*, 8th edition (Springer, Berlin and Heidelberg).

PFOHL, H.-C. (2016), *Logistikmanagement: Konzeption und Funktionen*, third edition (Springer, Berlin).

PFOHL, H.-C., & ZÖLLNER, W. (1997), Organization for logistics: The contingency approach, *International Journal of Physical Distribution & Logistics Management* 27(5/6), pp. 306–320.

PHILLIPS, A. (1960), A Theory of Interfirm Organization, *The Quarterly Journal of Economics* 74(4), pp. 602–613.

PIERINGER, M. (2012), Logistik aus der Wolke, *LOGISTIK HEUTE* 34(7-8), pp. 44–45.

POHLMANN, M., JAEGER, P., BOSE, A., DE VRIES, R., LINDENLAUB, R., HIRTLE, M., KIROV, K., & BUSSIEK, T. (2013), Preassembled, Raplidly Deployable Systems and Components, `https://www.google.com/patents/US20130305244` (Accessed: 20.11.2015).

PONOMAROV, S. Y., & HOLCOMB, M. C. (2009), Understanding the concept of supply chain resilience, *The International Journal of Logistics Management* 20(1), pp. 124–143.

POOLEY, J., & STENGER, A. J. (1992), Modeling and Evaluating Shipment Consolidation in a Logistics System, *Journal of Business Logistics* 13(2), pp. 153–174.

POON, T. C., CHOY, K. L., CHOW, H. K., LAU, H. C., CHAN, F. T., & HO, K. C. (2009), A RFID case-based logistics resource management system for managing order-picking operations in warehouses, *Expert Systems with Applications* 36(4), pp. 8277–8301.

PORTER, M. E., & FULLER, M. B. (1986), Coalitions and Global Strategy, in M. E. Porter, ed., *Competition in Global Industries*, pp. 315–344 (Harvard Business School Press).

PORTER, M. E. (1980), *Competitive strategy: Techniques for analyzing industries and competitors* (Free Press, New York).

POWELL, W. W. (1987), Hybrid Organizational Arrangements: New Form or Transitional Development?, *California Management Review* 30(1), pp. 67–87.

PRAHALAD, C. K., & HAMEL, G. (1990), The Core Competence of the Corporation, *Harvard Business Review* 68(3), pp. 79–91.

PRATER, E., BIEHL, M., & SMITH, M. A. (2001), International supply chain agility - Tradeoffs between flexibility and uncertainty, *International Journal of Operations & Production Management* 21(5/6), pp. 823–839.

PREIST, C., ESPLUGAS-CUADRADO, J., BATTLE, S. A., GRIMM, S., & WILLIAMS, S. K. (2005), Automated Business-to-Business Integration of a Logistics Supply Chain Using Semantic Web Services Technology, in David Hutchison, Takeo Kanade, Josef Kittler, Jon M. Kleinberg, Friedemann Mattern, John C. Mitchell, Moni Naor, Oscar Nierstrasz, C. Pandu Rangan, Bernhard Steffen, Madhu Sudan, Demetri Terzopoulos, Dough Tygar, Moshe Y. Vardi, Gerhard Weikum, Yolanda Gil, Enrico Motta, V. Richard Benjamins, & Mark A. Musen, eds., *The Semantic Web – ISWC 2005*, volume 3729, pp. 987–1001 (Springer, Berlin, Heidelberg).

PRIDE, W. M., & FERRELL, O. C. (2003), *Marketing: Concepts and strategies*, 12th edition (Houghton Mifflin Co, Boston).

PROVAN, K. G., & KENIS, P. (2008), Modes of Network Governance: Structure, Management, and Effectiveness, *Journal of Public Administration Research and Theory* 18(2), pp. 229–252.

PROVAN, K. G., FISH, A., & SYDOW, J. (2007), Interorganizational Networks at the Network Level: A Review of the Empirical Literature on Whole Networks, *Journal of Management* 33(3), pp. 479–516.

PUGH, D. S., HICKSON, D. J., HININGS, C. R., MACDONALD, K. M., TURNER, C., & LUPTON, T. (1963), A Conceptual Scheme for Organizational Analysis, *Administrative Science Quarterly* 8(3), pp. 289–315.

PUGH, D. S., HICKSON, D. J., HININGS, C. R., & TURNER, C. (1968), Dimensions of Organization Structure, *Administrative Science Quarterly* 13(1), pp. 65–105.

PUGH, D. S., HICKSON, D. J., HININGS, C. R., & TURNER, C. (1969), The Context of Organization Structures, *Administrative Science Quarterly* 14(1), pp. 91–114.

QUANTUMID TECHNOLOGIES INC. (2015), Company Website, `http://www.qidtech.com/` (Accessed: 12.08.2015).

QUINN, J. B., & HILMER, F. G. (1994), Strategic Outsourcing, *Sloan Management Review* 35(4), pp. 43–55.

RAJAHONKA, M. (2013), Views of logistics service providers on modularity in logistics services, *International Journal of Logistics Research and Applications: A Leading Journal of Supply Chain Management* 16(1), pp. 34–50.

RAJAHONKA, M., BASK, A., & LIPPONEN, M. (2013), Modularity and customisation in LSPs' service strategies, *International Journal of Services and Operations Management* 16(2), pp. 174–204.

RASMUSEN, E. (2007), *Games and information: An introduction to game theory*, fourth edition (Blackwell Pub., Malden, MA and Oxford).

RAUE, J. S., & WALLENBURG, C. (2013), Alike or not? Partner similarity and its outcome in horizontal cooperations between logistics service providers, *Logistics Research* 6(4), pp. 217–230.

RAUE, J. S., & WIELAND, A. (2015), The interplay of different types of governance in horizontal cooperations, *The International Journal of Logistics Management* 26(2), pp. 401–423.

RAZZAQUE, M. A., & SHENG, C. C. (1998), Outsourcing of logistics functions: a literature survey, *International Journal of Physical Distribution & Logistics Management* 28(2), pp. 89–107.

REITER, S. (1977), Information and Performance in the (New) Welfare Economics, *The American Economic Review* 67(1), pp. 226–234.

RENY, P. J. (2001), Arrow's theorem and the Gibbard-Satterthwaite theorem: a unified approach, *Economics Letters* 70(1), pp. 99–105.

RESCHER, N. (1997), *Objectivity: the obligations of impersonal reason* (University of Notre Dame Press).

REUER, J. J., & ARIÑO, A. (2007), Strategic Alliance Contracts: Dimensions and Determinants of Contractual Complexity, *Strategic Management Journal* 28(3), pp. 313–330.

RHENUS. (2017), Best High-Tech-Handling, `http://www.rhenus.com/en/services/value-added-logistics/high-tech-handling/` (Accessed: 08.01.2017).

RINDFLEISCH, A. (2000), Organizational Trust and Interfirm Cooperation: An Examination of Horizontal Versus Vertical Alliances, *Marketing Letters* 11(1), pp. 81–95.

ROCHWERGER, B., BREITGAND, D., LEVY, E., GALIS, A., NAGIN, K., LLORENTE, I. M., MONTERO, R., WOLFSTHAL, Y., ELMROTH, E., CACERES, J., BEN-YEHUDA, M., EMMERICH, W., & GALAN, F. (2009), The Reservoir model and architecture for open federated cloud computing, *IBM Journal of Research and Development* 53(4), pp. 4:1–4:11.

ROCHWERGER, B., BREITGAND, D., EPSTEIN, A., HADAS, D., LOY, I., NAGIN, K., TORDSSON, J., RAGUSA, C., VILLARI, M., CLAYMAN, S., LEVY, E., MARASCHINI, A., MASSONET, P., MU, H., & TOFETTI, G. (2011), Reservoir - When One Cloud Is Not Enough, *Computer* 44(3), pp. 44–51.

ROGERS, D. S., & DAUGHERTY, P. J. (1995), Warehousing firms: The impact of alliance involvement, *Journal of Business Logistics* 16(2), pp. 249–269.

ROSENBERG, N. (1972), *Technology and American economic growth* (Harper & Row).

ROSENBLUM, M., & GARFINKEL, T. (2005), Virtual machine monitors: current technology and future trends, *Computer* 38(5), pp. 34–43.

ROSENBLUM, M. (2004), The Reincarnation of Virtual Machines, *Queue* 2(5), pp. 34–40.

ROUSE, M. (2007), Definition: encapsulation, http://searchnetworking.techtarget.com/definition/encapsulation (Accessed: 22.07.2015).

ROUWENHORST, B., REUTER, B., STOCKRAHM, V., VAN HOUTUM, G. J., MANTEL, R. J., & ZIJM, W. (2000), Warehouse design and control: Framework and literature review, *European Journal of Operational Research* 122(3), pp. 515–533.

ROWLEY, J. (2000), *The principles of warehouse design*, volume 4, second edition.

ROZANSKI, N., & WOODS, E. (2011), *Software systems architecture: Working with stakeholders using viewpoints and perspectives*, second edition (Addison-Wesley, Upper Saddle River, NJ).

RUMELT, R. P. (1984), Towards a Strategic Theory of the Firm, in Robert Lamb, ed., *Competitive strategic management*, pp. 556–570 (Prentice-Hall, Englewood Cliffs, NJ).

RÜMENAPP, T. (2002), *Strategische Konfigurationen von Logistikunternehmen: Ansätze zur konsistenten Ausrichtung in den Dimensionen Strategie, Struktur und Umwelt*, first edition (Deutscher Universitäts-Verlag, Wiesbaden).

SALIN, V., & JR, R. M. N. (2003), A cold chain network for food exports to developing countries, *International Journal of Physical Distribution & Logistics Management* 33(10), pp. 918–933.

SANCHEZ, R. (1995), Strategic flexibility in product competition, *Strategic Management Journal* 16, pp. 135–159.

SÄNGER, F. (2004), *Elektronische Transportmärkte: Untersuchungen zur Optimierung des Straßengütertransports*, first edition (Deutscher Universitäts-Verlag, Wiesbaden).

SANTUCCI, G., & LANGE, S. (2008), Internet of Things in 2020: Roadmap for the Future, http://www.smart-systems-integration.org/public/documents/publications/Internet-of-Things_in_2020_EC-EPoSS_Workshop_Report_2008_v3.pdf (Accessed: 12.01.2016).

SATTERTHWAITE, M. A. (1975), Strategy-proofness and Arrow's conditions: Existence and correspondence theorems for voting procedures and social welfare functions, *Journal of Economic Theory* 10(2), pp. 187–217.

SCHERER, A. G. (2006), Kritik der Organisation oder Organisation der Kritik? - Wissenschaftstheoretische Bemerkungen zum kritischen Umgang mit Organisationstheorien, in Alfred Kieser, & Mark Ebers, eds., *Organisationstheorien*, pp. 19–62 (Kohlhammer, Stuttgart).

SCHEUERMANN, A., & HOXHA, J. (2012), Ontologies for Intelligent Provision of Logistics Services, in *7th International Conference on Internet and Web Applications and Services (ICIW)*, pp. 106–111.

SCHILLING, M. A. (2000), Toward a general modular systems theory and its application to interfirm product modularity, *Academy of Management Review* 25(2), pp. 312–334.

SCHILLING, M. A., & STEENSMA, H. K. (2001), The use of modular organizational forms: an industry-level analysis, *Academy of Management Journal* 44(6), pp. 1149–1168.

SCHMIDT, H. (2013), Volle Flexibilität voraus, in Michael ten Hompel, ed., *Software in der Logistik - Prozesse steuern mit Apps*, pp. 32–35 (Huss-Verlag, München).

SCHMOLTZI, C., & WALLENBURG, C. M. (2011), Horizontal cooperations between logistics service providers: motives, structure, performance, *International Journal of Physical Distribution & Logistics Management* 41(6), pp. 552–576.

SCHMOLTZI, C., & WALLENBURG, C. M. (2012), Operational Governance in Horizontal Cooperations of Logistics Service Providers: Performance Effects and the Moderating Role of Cooperation Complexity, *Journal of Supply Chain Management* 48(2), pp. 53–74.

SCHNIZLER, B., NEUMANN, D., VEIT, D., & WEINHARDT, C. (2008), Trading grid services – a multi-attribute combinatorial approach, *European Journal of Operational Research* 187(3), pp. 943–961.

SCHOLZ-REITER, B., WINDT, K., & FREITAG, M. (2004), Autonomous logistic processes: New demands and first approaches, in L. Monostori, ed., *Proceedings of the 37th CIRP international seminar on manufacturing systems*, pp. 357–362.

SCHREYÖGG, G. (1980), Contingency and Choice in Organization Theory, *Organization Studies* 1(4), pp. 305–326.

SCHULTE, C. (2005), *Logistik: Wege zur Optimierung der supply chain*, fourth edition (Vahlen, München).

SELVIARIDIS, K., & SPRING, M. (2007), Third party logistics: a literature review and research agenda, *The International Journal of Logistics Management* 18(1), pp. 125–150.

SHADBOLT, N., BERNERS-LEE, T., & HALL, W. (2006), The Semantic Web Revisited, *IEEE Intelligent Systems* 21(3), pp. 96–101.

SHAN, W., & HAMILTON, W. (1991), Country-specific advantage and international cooperation, *Strategic Management Journal* 12(6), pp. 419–432.

SHEFFI, Y. (2012), *Logistics clusters: Delivering value and driving growth* (MIT Press, Cambridge, MA).

SHELANSKI, H. A., & KLEIN, P. G. (1995), Empirical Research in Transaction Cost Economics: A Review and Assessment, *Journal of Law, Economics, & Organization* 11(2), pp. 335–361.

SHEN, Y. Y., & QIAN, Y. (2014)a, A Study of New E-Commerce Logistics Mode Based on Cloud Computing Technology, *Advanced Materials Research* 1049-1050, pp. 1848–1852.

SHEN, Y., & QIAN, Y. (2014)b, Research on Innovative Mode of Logistics Network in E-commerce Era, in *International Conference on Mechatronics, Control and Electronic Engineering (MCE)*, pp. 495–499.

SHI, S. Y., MO, R., YANG, H.-C., CHANG, Z.-Y., & CHEN, Z.-F. (2007), An implementation of modelling resource in a manufacturing grid for resource sharing, *International Journal of Computer Integrated Manufacturing* 20(2-3), pp. 169–177.

SIMATUPANG, T. M., WRIGHT, A. C., & SRIDHARAN, R. (2002), The knowledge of coordination for supply chain integration, *Business Process Management Journal* 8(3), pp. 289–308.

SIMON, H. A. (1957), *Administrative behavior*, second edition (Macmillan, New York).

SINGH, K. (1997), The Impact of Technological Complexity and Interfirm Cooperation on Business Survival, *The Academy of Management Journal* 40(2), pp. 339–367.

SITKIN, S. B., & BIES, R. J. (1993), The Legalistic Organization: Definitions, Dimensions, and Dilemmas, *Organization Science* 4(3), pp. 345–351.

SKJOETT-LARSEN, T. (2000), Third party logistics – from an interorganizational point of view, *International Journal of Physical Distribution & Logistics Management* 30(2), pp. 112–127.

SMITH, J. E., & NAIR, R. (2005), The architecture of virtual machines, *Computer* 38(5), pp. 32–38.

SPECTER, S. P. (2012), Reduce supply chain costs with modular bulk containers, *Modern Materials Handling* 67(9), pp. 14.

SRINIVASAN, K., KEKRE, S., & MUKHOPADHYAY, T. (1994), Impact of Electronic Data Interchange Technology on JIT Shipments, *Management Science* 40(10), pp. 1291–1304.

STANK, T. P., & DAUGHERTY, P. J. (1997), The impact of operating environment on the formation of cooperative logistics relationships, *Transportation Research Part E: Logistics and Transportation Review* 33(1), pp. 53–65.

STANK, T. P., GOLDSBY, T. J., VICKERY, S. K., & SAVITSKIE, K. (2003), Logistics Service Performance: Estimating its Influence on Market Share, *Journal of Business Logistics* 24(1), pp. 27–55.

STEIN, A. (2012), Umschlagsprozesse in der Logistik, in Peter Klaus, Winfried Krieger, & Michael Krupp, eds., *Gabler Lexikon Logistik*, pp. 600–606 (Gabler, Wiesbaden).

STEINBUSS, S., & WEISSENBERG, N. (2013), Service Design and Process Design for the Logistics Mall Cloud, in Xiaoyu Yang, & Lu Liu, eds., *Principles, methodologies, and service-oriented approaches for cloud computing*, pp. 186–206 (Business Science Reference, Hershey, PA).

STINNES, A. (2012), Ist Cloud Logistics nur Wasserdampf?, *LOGISTIK HEUTE* 34(9), pp. 8.

STINNES, A. (2013), Schritt für Schritt in die Wolke, in Michael ten Hompel, ed., *Software in der Logistik - Prozesse steuern mit Apps*, pp. 36–39 (Huss-Verlag, München).

STOCK, G. N., GREIS, N. P., & KASARDA, J. D. (1999), Logistics, strategy and structure: A conceptual framework, *International Journal of Physical Distribution & Logistics Management* 29(4), pp. 37–52.

STÖSSER, J. (2009), *Market-Based Scheduling in Distributed Computing Systems*, Doctoral Thesis, Universität Karlsruhe.

STRASSER, S., EVELAND, J. D., CUMMINS, G., DENISTON, O. L., & ROMANI, J. H. (1981), Conceptualizing the Goal And System Models of Organizational Effectiveness —Implications for Comparative Evaluation Research, *Journal of Management Studies* 18(3), pp. 321–340.

STRAUBE, F. (2004), *E-Logistik: Ganzheitliches Logistikmanagement* (Springer, Berlin, Heidelberg, New York).

STUDER, R., BENJAMINS, V., & FENSEL, D. (1998), Knowledge engineering: Principles and methods, *Data & Knowledge Engineering* 25(1-2), pp. 161–197.

SUGERMAN, J., VENKITACHALAM, G., & LIM, B.-H. (2001), Virtualizing I/O Devices on VMware Workstation's Hosted Virtual Machine Monitor, in *Proceedings of the 2001 USENIX Annual Technical Conference*, pp. 1–14.

SUM, C.-C., & TEO, C.-B. (1999), Strategic posture of logistics service providers in Singapore, *International Journal of Physical Distribution & Logistics Management* 29(9), pp. 588–605.

SUN, G., WANG, X.-Y., WANG, H., & ZHAO, J. (2015)a, Construction of Regional Logistics Information Platform Based on Cloud Computing, in *International Conference on Computational Science and Engineering (ICCSE)*, pp. 108–113.

SUN, H.-L., & LIU, Y.-X. (2013), Fresh Agricultural Products Supply Chain's Optimization and Upgrading with the Support of Cloud Logistics, *Prices Monthly* 34(7).

SUN, W., & TAO, L. (2013), Cloud Logistics Mode-Based Location of Regional Logistics Distribution Center, in Qiyuan Peng, & Kelvin C. P. Wang, eds., *Fourth International Conference on Transportation Engineering*, pp. 2477–2483.

SUN, X., CHU, X., & WU, Z. (2015)b, Logistics Hybrid Cloud Services Based on the Theory of Collaborative Management, in Zhenji Zhang, Zuojun Max Shen, Juliang Zhang, & Runtong Zhang, eds., *LISS 2014*, pp. 357–367 (Springer, Berlin, Heidelberg).

SWISSRE. (2016), Sigma Explorer, `http://www.sigma-explorer.com/` (Accessed: 25.04.2016).

SYDOW, J. (1992), *Strategische Netzwerke: Evolution und Organisation*, first edition (Gabler, Wiesbaden).

SYDOW, J. (2010), Management von Netzwerkorganisationen – Zum Stand der Forschung, in Jörg Sydow, ed., *Management von Netzwerkorganisationen*, pp. 373–470 (Gabler Verlag, Wiesbaden).

TADELIS, S. (2013), *Game theory: An introduction* (Princeton University Press, Princeton and Oxford).

TANG, C. (2006), Robust strategies for mitigating supply chain disruptions, *International Journal of Logistics Research and Applications: A Leading Journal of Supply Chain Management* 9(1), pp. 33–45.

TANG, C., & TOMLIN, B. (2008), The power of flexibility for mitigating supply chain risks, *International Journal of Production Economics* 116(1), pp. 12–27.

TATE, K. (1996), The elements of a successful logistics partnership, *International Journal of Physical Distribution & Logistics Management* 26(3), pp. 7–13.

TEICHMANN, G. (2014), Cloud-basierte Kollaborationen in der Logistik, *Wirtschaftsinformatik & Management* 6(1), pp. 56–65.

THE ECONOMIST. (2008), A survey of corporate IT: Let it rise, http://www.economist.com/node/12411882 (Accessed: 15.06.2014).

THOMAS, F. (2011), Cloud Logistics: Radikal anders, http://www.beschaffung-aktuell.de/home/-/article/16537505/30563604/Radikal-anders/art_co_INSTANCE_0000/maximized/ (Accessed: 18.12.2013).

THOMAS, F., & UNRUH, V. (2010), Cloud Logistics – Neue Trends in der Informationstechnologie, http://www.mm-logistik.vogel.de/management-it/articles/278012/ (Accessed: 17.12.2013).

THOMPSON, J. D. (1967), *Organizations in action; Social science bases of administrative theory* (McGraw-Hill, New York).

THORELLI, H. B. (1986), Networks: Between markets and hierarchies, *Strategic Management Journal* 7(1), pp. 37–51.

TOLBERT, P. S., & HALL, R. H. (2009), *Organizations: Structures, processes, and outcomes*, 10th edition (Pearson/Prentice Hall, Upper Saddle River, NJ).

TORDSSON, J., MONTERO, R. S., MORENO-VOZMEDIANO, R., & LLORENTE, I. M. (2012), Cloud brokering mechanisms for optimized placement of virtual machines across multiple providers, *Future Generation Computer Systems* 28(2), pp. 358–367.

TRANFIELD, D., DENYER, D., & SMART, P. (2003), Towards a Methodology for Developing Evidence-Informed Management Knowledge by Means of Systematic Review, *British Journal of Management* 14(3), pp. 207–222.

TSALGATIDOU, A., & PILIOURA, T. (2002), An Overview of Standards and Related Technology in Web Services, *Distributed and Parallel Databases* 12(2-3), pp. 135–162.

TSANG, E. W. K. (1998), Motives for strategic alliance: A resource-based perspective, *Scandinavian Journal of Management* 14(3), pp. 207–221.

TSUI, A. S. (1990), A Multiple-Constituency Model of Effectiveness: An Empirical Examination at the Human Resource Subunit Level, *Administrative Science Quarterly* 35(3), pp. 458–483.

TUMMEL, C., HAUCK, E., & JESCHKE, S. (2014), CloudLogistic – A Formal Description of Exclusive Freight Routes Based on a Multi-Zone Model with Elliptic Isochrones, in Sabina Jeschke, Ingrid Isenhardt, Frank Hees, & Klaus Henning, eds., *Automation, Communication and Cybernetics in Science and Engineering 2013/2014*, pp. 671–680 (Springer International Publishing, Cham).

TUSHMAN, M. L. (1979), Work Characteristics and Subunit Communication Structure: A Contingency Analysis, *Administrative Science Quarterly* 24(1), pp. 82–98.

TUSHMAN, M. L., & ANDERSON, P. (1986), Technological Discontinuities and Organizational Environments, *Administrative Science Quarterly* 31(3), pp. 439–465.

TYAN, J. C., WANG, F.-K., & DU, T. C. (2003), An evaluation of freight consolidation policies in global third party logistics, *Omega* 31(1), pp. 55–62.

URBAN, J. (2015), The Cloud Technology Opportunity for Logistics Service Providers, `http://talkinglogistics.com/2015/04/14/the-cloud-technology-opportunity-for-logistics-service-providers/` (Accessed: 12.08.2015).

VAN DE VEN, A. H. (1976), On the Nature, Formation, and Maintenance of Relations among Organizations, *The Academy of Management Review* 1(4), pp. 24–36.

VAN DINTHER, C. (2010), *Designing service Value Networks: Coordination, Competition and Cooperation*, Kumulative Habilitationsschrift, Karlsruher Institut für Technologie.

VAN HOEK, R. I. (2000), The Role of Third–Party Logistics Providers in Mass Customization, *The International Journal of Logistics Management* 11(1), pp. 37–46.

VAN LAARHOVEN, P., BERGLUND, M., & PETERS, M. (2000), Third–party logistics in Europe – five years later, *International Journal of Physical Distribution & Logistics Management* 30(5), pp. 425–442.

VANNESTE, B. S., & PURANAM, P. (2010), Repeated Interactions and Contractual Detail: Identifying the Learning Effect, *Organization Science* 21(1), pp. 186–201.

VAQUERO, L. M., RODERO-MERINO, L., CACERES, J., & LINDNER, M. (2009), A break in the clouds, *ACM SIGCOMM Computer Communication Review* 39(1), pp. 50–55.

VARGO, S. L., & LUSCH, R. F. (2008), Why "Service"?, *Journal of the Academy of Marketing Science* 36(1), pp. 25–38.

VERMA, A., KUMAR, G., & KOLLER, R. (2010), The cost of reconfiguration in a cloud, in Lucy Cherkasova, & Rajeev Rastogi, eds., *11th International Middleware Conference Industrial Track*, pp. 11–16.

VERNIER, D., AMIES, A., SANCHEZ, J., & DONG ZHENG, X. (2011), Monitor Services in the Cloud: Discover the secrrets of monitoring cloud services through tips, tricks, and tools, http://www.ibm.com/developerworks/cloud/library/cl-monitorcloudservices/cl-monitorcloudservices-pdf.pdf (Accessed: 14.08.2013).

VERSTREPEN, S., COOLS, M., CRUIJSSEN, F., & DULLAERT, W. (2009), A dynamic framework for managing horizontal cooperation in logistics, *International Journal of Logistics Systems and Management* 5(3/4), pp. 228–248.

VICKREY, W. (1961), Counterspeculation, Auctions, and Competitive Sealed Tenders, *The Journal of Finance* 16(1), pp. 8–37.

VLAAR, P. W. L., VAN DEN BOSCH, F. A. J., & VOLBERDA, H. W. (2007), Towards a Dialectic Perspective on Formalization in Interorganizational Relationships: How Alliance Managers Capitalize on the Duality Inherent in Contracts, Rules and Procedures, *Organization Studies* 28(4), pp. 437–466.

VOORSLUYS, W., BROBERG, J., & BUYYA, R. (2011), Introduction to Cloud Computing, in Rajkumar Buyya, James Broberg, & Andrzej Gościński, eds., *Cloud computing*, pp. 1–41 (Wiley, Hoboken, N.J).

VOSS, C. A., & HSUAN, J. (2009), Service Architecture and Modularity, *Decision Sciences* 40(3), pp. 541–569.

VRIES, S. D., & VOHRA, R. V. (2003), Combinatorial Auctions: A Survey, *INFORMS Journal on Computing* 15(3), pp. 284–309.

V&V DABELSTEIN LOGISTIK GMBH. (2015), Company website: Cloud Logistics - Was bedeutet Cloud Logistics? Welche Vorteile bietet Cloud Logistics?, http://www.dabelstein.com/cms/cloud-logistics.html (Accessed: 12.08.2015).

WALLENBURG, C. M., & RAUE, J. S. (2011), Conflict and its governance in horizontal cooperations of logistics service providers, *International Journal of Physical Distribution & Logistics Management* 41(4), pp. 385–400.

WANG, J., ZHANG, X., HU, X., & ZHAO, J. (2015)a, Cloud Logistics Service Mode and its Several Key Issues, *Journal of System and Management Sciences* 5(1), pp. 67–83.

WANG, J., ZHANG, X., HU, X., & ZHAO, J. (2015)b, Survey on Logistics Service Mode Based on Cloud Computing, in Runtong Zhang, Zhenji Zhang, Kecheng Liu, & Juliang Zhang, eds., *LISS 2013*, pp. 321–327 (Springer, Berlin, Heidelberg).

WANG, J. (2014), Exploring the Development of Agricultural Products Cloud Logistics Based on E-commerce, *China Business and Market* 28(11).

WANG, Q., ZANTOW, K., LAI, F., & WANG, X. (2006), Strategic postures of third-party logistics providers in mainland China, *International Journal of Physical Distribution & Logistics Management* 36(10), pp. 793–819.

WANG, Q., LV, H., & JIANG, Y. (2012)a, Study on Cloud Logistics System Framework and Application Mode, *Telecommunications Science* 57(3).

WANG, X., LI, W., ZHONG, Y., & ZHAO, W. (2012)b, Research on cloud logistics-based one-stop service platform for logistics center, in *IEEE 16th International Conference on Computer Supported Cooperative Work in Design (CSCWD)*, pp. 558–563.

WANG, Y., POTTER, A., NAIM, M., & BEEVOR, D. (2011), A case study exploring drivers and implications of collaborative electronic logistics marketplaces, *Industrial Marketing Management* 40(4), pp. 612–623.

WANG, Y., XIA, H., HAYASHI, M., & YOSHIDA, T. (2005), Adding semantic support to Web services in distribution logistics, in *Proceedings of the International Conference on Services Systems and Services Management (ICSSSM)*, volume 1, pp. 566–572.

WANG, Z., DONG, S., LI, Z., LI, Y., LI, J., & CHENG, H. (2015)c, Traffic Patterns in the Silk Road Economic Belt and Construction Modes for a Traffic Economic Belt across Continental Plates, *Journal of Resources and Ecology* 6(2), pp. 79–86.

WEERAWARANA, S. (2005), *Web services platform architecture: SOAP, WSDL, WS-Policy, WS-Addressing, WS-BPEL, WS-Reliable Messaging, and more / Sanjiva Weerawarana ... [et al.]* (Prentice Hall PTR, Upper Saddle River, N.J.).

WEICK, K. E. (1976), Educational Organizations as Loosely Coupled Systems, *Administrative Science Quarterly* 21(1), pp. 1–19.

WEISSENBERG, N., & SPRINGER, U. (2014), Cloud Process Modeling for the Logistics Mall-Object-Aware BPM for Domain Experts, *Open Journal of Mobile Computing and Cloud Computing* 1(2), pp. 31–49.

WERNERFELT, B. (1984), A resource-based view of the firm, *Strategic Management Journal* 5(2), pp. 171–180.

WHITTINGTON, R. (1988), Environmental Structure and Theories of Strategic Choice, *Journal of Management Studies* 25(6), pp. 521–536.

WIELAND, A. (2012), Dealing with supply chain risks: Linking risk management practices and strategies to performance, *International Journal of Physical Distribution & Logistics Management* 42(10), pp. 887–905.

WILDER, B. (2012), *Cloud architecture patterns: Using microsoft azure* (O'Reilly Media).

WILLIAMS, B. (2012), *The Economics of Cloud Computing : An Overview For Decision Makers* (Cisco Press, Indianapolis, IN).

WILLIAMS, S. R. (1999), A Characterization of Efficient, Bayesian Incentive Compatible Mechanisms, *Economic Theory* 14(1), pp. 155–180.

WILLIAMSON, O. E. (1979), Transaction-Cost Economics: The Governance of Contractual Relations, *Journal of Law and Economics* 22(2), pp. 233–261.

WILLIAMSON, O. E. (1981), The Economics of Organization: The Transaction Cost Approach, *American Journal of Sociology* 87(3), pp. 548–577.

WILLIAMSON, O. E. (1983), Credible Commitments: Using Hostages to Support Exchange, *American Economic Review* 73(4), pp. 519–540.

WILLIAMSON, O. E. (1985), *The economic institutions of capitalism: Firms, markets, relational contracting* (Free Press, New York and London).

WILLIAMSON, O. E. (1988), The Logic of Economic Organization, *Journal of Law, Economics, & Organization* 4(1), pp. 65–93.

WILLIAMSON, O. E. (1991), Comparative Economic Organization: The Analysis of Discrete Structural Alternatives, *Administrative Science Quarterly* 36(2), pp. 269–296.

WILLIAMSON, O. E. (2002), The Theory of the Firm as Governance Structure: From Choice to Contract, *The Journal of Economic Perspectives* 16(3), pp. 171–195.

WILLIAMSON, O. E., & OUCHI, W. G. (1981), The markets and hierarchies program of research: origins, implications, prospects, in Andrew H. van de Ven, & William F. Joyce, eds., *Perspectives on organization design and behavior*, pp. 347–370 (Wiley, New York).

WINDT, K., BÖSE, F., & PHILIPP, T. (2008), Autonomy in production logistics: Identification, characterisation and application, *Robotics and Computer-Integrated Manufacturing* 24(4), pp. 572–578.

WINDT, K., & HÜLSMANN, M. (2007), Changing Paradigms in Logistics - Understanding the Shift from Conventional Control to Autonomous Cooperation and Control, in Michael Hülsmann, & Katja Windt, eds., *Understanding autonomous cooperation and control in logistics* (Springer, Berlin Heidelberg).

XU, X., PAN, S., & BALLET, E. (2012), Game theoretic contribution to horizontal cooperation in logistics, in *4th international conference on information systems, logistics and supply chain (ILS)*, pp. 10–19.

XU, X. (2012), From cloud computing to cloud manufacturing, *Robotics and Computer-Integrated Manufacturing* 28(1), pp. 75–86.

YAN, S., & CHEN, C.-H. (2007), Coordinated scheduling models for allied airlines, *Transportation Research Part C: Emerging Technologies* 15(4), pp. 246–264.

YAO, Y., YANG, F., & SU, S. (2006), Flexible Decision Making in Web Services Negotiation, in J. Euzenat, & J. Domingue, eds., *Artificial Intelligence: Methodology, Systems, and Applications (AIMSA)*, volume 4183, pp. 108–117 (Springer, Berlin, Heidelberg).

YOUSEFF, L., BUTRICO, M., & DA SILVA, D. (2008), Toward a Unified Ontology of Cloud Computing, in *Grid Computing Environments Workshop (GCE)*, pp. 1–10.

ZACHARIA, Z. G., SANDERS, N. R., & NIX, N. W. (2011), The Emerging Role of the Third-Party Logistics Provider (3PL) as an Orchestrator, *Journal of Business Logistics* 32(1), pp. 40–54.

ZAHEER, A., & VENKATRAMAN, N. (1994), Determinants of Electronic Integration in the Insurance Industry: An Empirical Test, *Management Science* 40(5), pp. 549–566.

ZAHEER, A., & VENKATRAMAN, N. (1995), Relational governance as an interorganizational strategy: An empirical test of the role of trust in economic exchange, *Strategic Management Journal* 16(5), pp. 373–392.

ZEITHAML, V. A., PARASURAMAN, A., & BERRY, L. L. (1985), Problems and Strategies in Services Marketing, *Journal of Marketing* 49(2), pp. 33–46.

ZHANG, H., HONG, Q., ZHAO, G., & ZHOU, L.-Y. (2014), Research of logistics park service resource sharing and configuration mode based on cloud logistics services, *Application Research of Computers (Chinese)* 31(2).

ZHANG, Q., CHENG, L., & BOUTABA, R. (2010), Cloud computing: state-of-the-art and research challenges, *Journal of Internet Services and Applications* 1(1), pp. 7–18.

ZHANG, Q., VONDEREMBSE, M. A., & LIM, J.-S. (2005), Logistics flexibility and its impact on customer satisfaction, *The International Journal of Logistics Management* 16(1), pp. 71–95.

ZHANG, X., & YUAN, Z. (2013), The Platform and Collaborative Operating Model of "Wisdom Cloud Logistics" in Online Shopping Era, *Forum on Science and Technology in China* 29(7).

ZHANG, Y., & SUN, Q. (2013), Design of Standardized Data Process in 4PL Freight Management, in *International Conference on Education Technology and Information System (ICETIS)*, pp. 886–889.

ZHAO, M., & WU, J. (2013), Security and Strategies in Air Logistics Based on Cloud Computing, in *International Workshop on Cloud Computing and Information Security (CCIS)*, pp. 449–452.

ZHOU, G., VAN HUI, Y., & LIANG, L. (2011), Strategic alliance in freight consolidation, *Transportation Research Part E: Logistics and Transportation Review* 47(1), pp. 18–29.

FSC
www.fsc.org
MIX
Papier | Fördert
gute Waldnutzung
FSC® C083411

Kolibri 360 GmbH
Ferdinand-Jühlke-Straße 7
99095 Erfurt, Deutschland
produktsicherheit@kolibri360.de